DATE DUE

STRATIGRAPHIC MICROPALEONTOLOGY OF
ATLANTIC BASIN AND BORDERLANDS

FURTHER TITLES IN THIS SERIES

This book is produced by a photographic offset process directly from the manuscript.
Thus, the publisher is not responsible for any errors appearing in the book.

Developments in Palaeontology and Stratigraphy, 6

STRATIGRAPHIC MICROPALEONTOLOGY OF ATLANTIC BASIN AND BORDERLANDS

Edited by

F.M. SWAIN

Department of Geology, University of Delaware, Newark, Del. (U.S.A.)

and

Department of Geology and Mineralogy, University of Minnesota, Minneapolis, Minn. (U.S.A.)

ELSEVIER SCIENTIFIC PUBLISHING COMPANY
Amsterdam — Oxford — New York   1977

ELSEVIER SCIENTIFIC PUBLISHING COMPANY
335 Jan van Galenstraat
P.O. Box 211, Amsterdam, The Netherlands

Distributors for the United States and Canada:

ELSEVIER NORTH-HOLLAND INC.
52, Vanderbilt Avenue
New York, N.Y. 10017

Library of Congress Cataloging in Publication Data

Main entry under title:

Stratigraphic micropaleontology of Atlantic basin
    and border lands.

    (Developments in palaeontology and stratigraphy ;
6)
    Papers presented at a symposium supported by and
convened at the University of Delaware, June 14-16,
1976.
    Bibliography: p.
    Includes indexes.
    1. Micropaleontology--North Atlantic Ocean--
Congresses. 2. Micropaleontology--North America--
Congresses. 3. Geology, Stratigraphic--Congresses.
4. Geology--North Atlantic Ocean--Congresses.
5. Geology--North America--Congresses. I. Swain,
Frederick Morrill, 1916-   II. Delaware.
University, Newark. III. Series.
QE719.S83   560'.921   77-915
ISBN 0-444-41554-8

Printed in The Netherlands

PREFACE

A symposium on the stratigraphic micropaleontology of the Atlantic basin and margins was convened at the University of Delaware, June 14—16, 1976. The symposium was attended by 56 people, and 27 papers were presented.

The volume represents the published proceedings of the symposium and consists of 26 papers by 28 authors. Although a serious attempt was made to provide at least minimum coverage of all major micropaleontological groups in and around the Atlantic, these efforts were not successful in the case of diatoms, Paleozoic calcareous Foraminifera, coccoliths and Paleozoic and Cenozoic palynology. The volume includes summaries and discussions of the stratigraphic and geographic distribution of Paleozoic, Mesozoic and Cenozoic Foraminifera, Radiolaria and Ostracoda; Paleozoic and Mesozoic conodonts; Paleozoic Chitinozoa and acritarchs; Mesozoic dinoflagellates, and palynomorphs; and Cenozoic silicoflagellates. The discussions of individual papers represent transcripts of tapes made at the meeting, written questions submitted at the meeting, and subsequent modifications of questions and answers by questioners and authors, through correspondence.

Several participants in the symposium were unable to submit final manuscripts or plan to publish their papers elsewhere; G.L. Williams and J.P. Bujak, Geological Survey of Canada (Cenozoic palynology), Stefan Gartner, Texas A. and M. University (Cenozoic coccoliths), F.M. Gradstein, Geological Survey of Canada (Jurassic Foraminifera), and Piero Ascoli, Geological Survey of Canada (Biostratigraphy of Scotian Shelf).

The papers on South Atlantic Cretaceous Ostracoda and Paleogene and Neogene Foraminifera contain more details on localities than most of the other papers. It was decided to retain this information because some of it is not easily accessible to workers in the Northern Hemisphere.

It was originally planned that a summary chapter be prepared on Atlantic stratigraphic micropaleontology, with the collaboration of all the authors. Limitations on the length of the book have prevented this being done.

ACKNOWLEDGEMENTS

I am greatly indebted to the Graduate School and the College of Arts and Sciences of the University of Delaware for support of the symposium and to Dr. John C. Kraft, Chairman of the Geology Department who made the arrangements for support. Michelle Mayrath prepared illustrations for my ostracode and acritarch papers, and also provided invaluable assistance in editorial matters. Teresa Grilli typed or re-typed many of the manuscripts. Patricia Barlow, Barbara Frank, and Sarah Cornell also assisted in secretarial and editorial work.

F.M. SWAIN

CONTENTS

EARLY PALEOZOIC OSTRACODA OF EASTERN CANADA

M. J. Copeland

Geological Survey of Canada, Ottawa, Ontario, K1A 0E8

Abstract

Pre-Middle Devonian Ostracoda occur in three northeast-trending structural provinces of eastern Canada. They comprise biogeographically and stratigraphically distinctive leperditicopid, palaeocopid and podocopid assemblages dominated by Ordovician hollinids in the northern (Anticosti Basin) province and by Siluro-Devonian beyrichiids in all three provinces. Correlation is possible only between endemic North American Early Silurian beyrichiid assemblages of the northern and central (Appalachian Belt) provinces; the southern (Fundy Belt) province displays the typical Beyrichienkalk fauna of northern Europe.

Introduction

No continuous sequence of pre-Middle Devonian palaeocopid or podocopid ostracode faunas has been described from eastern Canada. This is due in part to facies variation and discontinuities in the geological record but it is anticipated that future study will reveal a more complete zonation than is now possible. Distinct depositional, temporal, geographic and tectonic control is evident, however, in the ostracode assemblages presently known. These faunas are contained in three generally parallel structural provinces comprising an older platformal sequence in Anticosti Basin and two younger, shallow water sequences, those of the "Appalachian Belt" in Gaspé Peninsula and southern Quebec and the "Fundy Belt" of Maritime Canada and coastal New England (Figure 1).

Assemblages of North American aspect occur on either side of the northern limit of Acadian orogenic deformation. The northern, undeformed St. Lawrence Platform – Anticosti Basin faunal sequence reflects an initially provincial Middle Ordovician ostracode assemblage, modified somewhat by Late Ordovician boreal elements, and a subsequently endemic North American Early Silurian ostracode fauna of Appalachian aspect. This endemic fauna, bearing the only palaeocopid ostracodes presently known to provide correlation across the Acadian Front (Figure 2), also occurs in basal strata of the deformed Silurian sequence in Gaspé Peninsula and marks initiation of the typical northeastern North American Siluro-Devonian Appalachian Belt ostracode succession. It differs completely from the more stratigraphically restricted European beyrichiid fauna of the Fundy Belt to the south. These faunas are separated in central New Brunswick by a

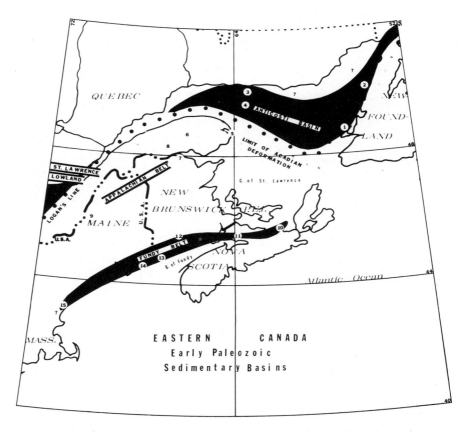

1. Port-au-Port Peninsula 2. Table Point 3. Mingan Islands
4. Anticosti Island 5. Forillon Peninsula 6. Mt. Albert area
7. Dalhousie 8. Lake Matapedia 9. Moose River Synclinorium
10. Arisaig 11. Portapique River 12. Jones Creek
13. Eastport 14. Coast of Maine (general) 15. Newburyport

Figure 1. Early Paleozoic structural provinces of eastern Canada and northeastern
United States, with localities mentioned.

wide zone of Silurian turbidites that may represent a marine trough that effectively
separated contemporaneous ostracode faunas of North American and European
aspect. The time-restricted Late Silurian-Early Devonian Fundy Belt assemblage,
of distinctive north European Beyrichienkalk affinity, is contained in a segment
of the European continental plate, left behind after collision with the North American
plate had resulted in closure of the proto-Atlantic Ocean and its subsequent
reopening during the Mesozoic.

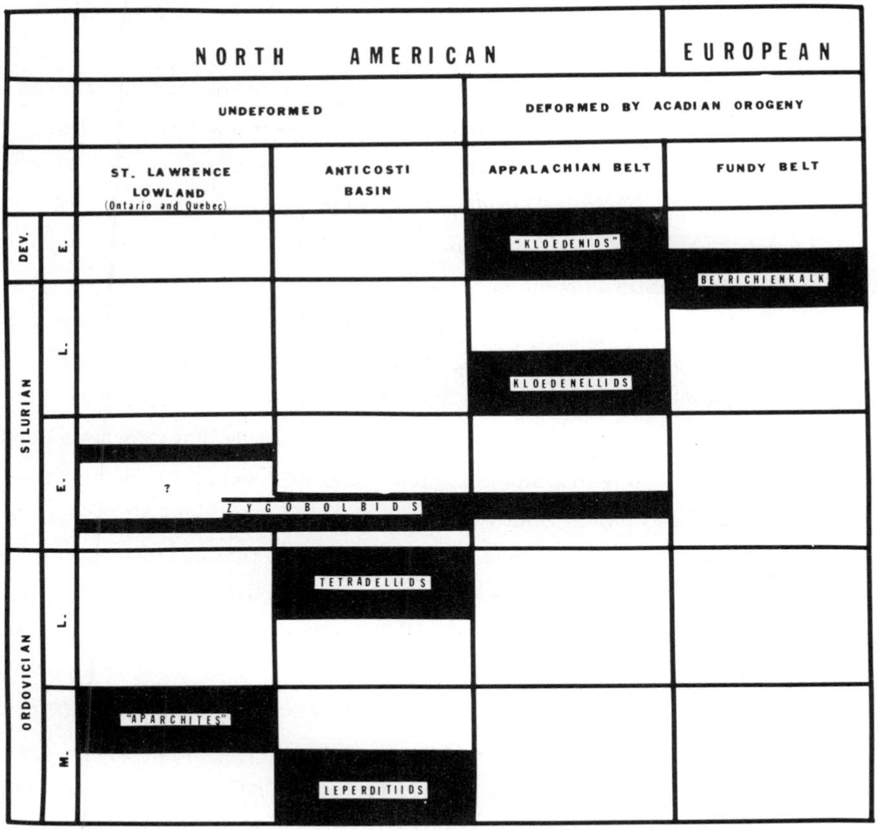

Figure 2. Faunal and tectonic relationships of ostracode assemblages recognized in eastern Canada

History of Investigations

Ostracoda have been recorded from relatively few localities in Anticosti Basin (Fig. 1, locs. 1-4). Chief among the early investigators were Billings (1865, 1866) and Jones (1858, 1890a, b, 1891). During the present century, Ulrich and Bassler (1923), Bassler (*in* Twenhofel, 1928) and Copeland (1970a, b, 1973, 1974a) presented a Late Ordovician and Early Silurian ostracode zonation for Anticosti Island, and Berdan (*in* Whittington and Kindle, 1963) reported on a Middle Ordovician fauna from western Newfoundland (Fig. 1, loc. 2). Additional information on younger Early Paleozoic ostracode faunas may be forthcoming when samples of strata from beneath the Gulf of St. Lawrence between Anticosti

Island, Quebec (Fig. 1, loc. 4) and Port-au-Port Peninsula, Newfoundland (Fig. 1, loc. 1) are studied.

Middle Ordovician ostracode faunas from St. Lawrence Lowland and related areas surrounding the Precambrian shield of eastern Canada are recorded in Kay (1934), Carter (1957) and Copeland (1965, 1970a, 1976, in press). This is probably the most extensively distributed Early Paleozoic ostracode fauna in northern North America as it is also reported from Yukon Territory and Districts of Mackenzie and Franklin (Copeland, 1974b, in press).

Within the Appalachian Belt the earliest report of Devonian ostracodes was by Jones (1889) from collections made along the southern shore of Chaleur Bay in northern New Brunswick (Fig. 1, loc. 7). This fauna was redescribed and numerous additional species were recorded by Clarke (1909) and Copeland (1962) from the vicinity of Campbellton and Dalhousie, New Brunswick. In northern Maine, south of Quebec City, Berdan (in Boucot, 1961) reported a small beyrichiid fauna from the Moose River Synclinorium (Fig. 1, loc. 9) that has proved slightly older than an Early Devonian fauna from Forillon Peninsula (Fig. 1, loc. 5) that was identified independently by Berdan and Copeland and reported in Burk (1964) and Boucot (1965).

Because of its European aspect, ostracode faunas from the Fundy Belt (Fig. 1, locs. 10-15) attracted the attention of several early workers during the mid-nineteenth century. Honeyman (1859, 1864), Dawson (1860, et seq.), Hall (1860) and Jones (1870, 1881a, b) each recorded part of this Beyrichienkalk fauna from Arisaig, Nova Scotia (Fig. 1, loc. 10). In this century, McLearn (1924) and Copeland (1960, 1964) defined the position of this fauna within the thick stratigraphic sequence present in that area. Only recently have additional localities been reported in Nova Scotia and New Brunswick (Fig. 1, locs. 11, 12), and Berdan (1966, 1971, in Brookins et al., 1973) recorded a longer ranging ostracode fauna from coastal Maine (Fig. 1, locs. 13, 14) extending southwestward to Massachusetts (Fig. 1, loc. 15).

The Ostracode Fauna

The earliest ostracode fauna of St. Lawrence Lowland-Anticosti Basin is Whiterockian (Llanvernian) of the Table Head Formation of western Newfoundland. 'Eoleperditia' of the general 'Eoleperditia' bivia (White) group is most significant in that it represents a circum-cratonal North American platform fauna that occurs also in Nevada and southwestern District of ⠄⠄ackenzie (Copeland, 1974b). On Mingan Islands (Twenhofel, 1938) and between Quebec and Montreal, a small, relatively unknown ostracode fauna has been reported. There appears to be little similarity between this and other Chazyan ostracode faunas with which it has been correlated.

Middle Ordovician ostracodes of Wildernessian-Barneveldian (Caradocian) age are widespread within the platformal facies north and west of St. Lawrence River (Plate 1). These may be generally designated as the mid-continental Decorah fauna, after the area in Minnesota and Iowa from which they were described by Kay (1934) and later workers. Ostracodes of this assemblage occur in two temporal subassemblages, with older Wildernessian elements in Minnesota, Michigan and southern Ontario, which could be termed 'southern', typified by the genera 'Aparchites', Dicranella, Eurychilina and Ceratopsis; and younger, Barneveldian elements in northern Ontario, Foxe Basin and Baffin Island, of 'boreal' aspect, typified by the genera Oepikella, Oepikium, Distobolbina, Steusloffina and Levisulculus. This younger fauna shows distinct north European affinities, but may be slightly older than its first appearance in Scandinavia, and marks initiation of a modified North American-European fauna.

Late Ordovician faunas are best known from Anticosti Island (Fig. 3, Plate 1) in strata of the Vauréal and Ellis Bay formations. There, also, a mixing of North American and Baltic faunas is apparent. Typical North American hollinaceans such as Tetradella and Anticostiella occur with European Carinobolbina and Foramenella. The upper limit of this Maysvillian-Richmondian (Ashgillian) fauna is established by the last occurrence of tetradellid ostracodes, which marks the close of the Ordovician in northeastern North America. On Anticosti Island a conformable sequence of Late Ordovician to Early Silurian strata is present. We have been unable to establish an ostracode succession across this interval as about fifty to sixty-five feet (15-20 m) of strata, barren of palaeocopid ostracodes separate the youngest Ordovician tetradellids and the oldest Silurian beyrichiids. This is the presumed glacial or Cherokee Discontinuity (Dennison and Head, 1975) postulated by Berry and Boucot (1973), among others, to explain a faunal hiatus during which time glacio-eustatic changes in sea-level caused near complete destruction of all faunas throughout northern North America.

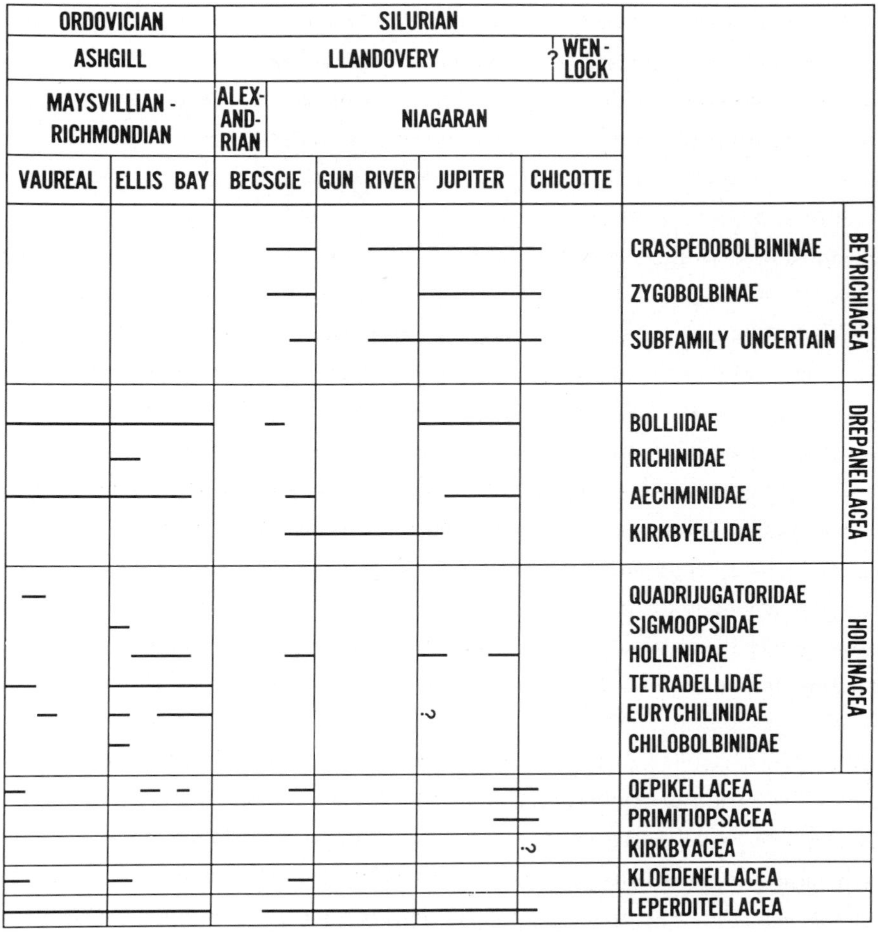

ORDOVICIAN		SILURIAN					
ASHGILL		LLANDOVERY			? WEN-LOCK		
MAYSVILLIAN - RICHMONDIAN		ALEX-AND-RIAN	NIAGARAN				
VAUREAL	ELLIS BAY	BECSCIE	GUN RIVER	JUPITER	CHICOTTE		
						CRASPEDOBOLBININAE	BEYRICHIACEA
						ZYGOBOLBINAE	
						SUBFAMILY UNCERTAIN	
						BOLLIIDAE	DREPANELLACEA
						RICHINIDAE	
						AECHMINIDAE	
						KIRKBYELLIDAE	
						QUADRIJUGATORIDAE	HOLLINACEA
						SIGMOOPSIDAE	
						HOLLINIDAE	
						TETRADELLIDAE	
						EURYCHILINIDAE	
						CHILOBOLBINIDAE	
						OEPIKELLACEA	
						PRIMITIOPSACEA	
						KIRKBYACEA	
						KLOEDENELLACEA	
						LEPERDITELLACEA	

Figure 3. Stratigraphic distribution of palaeocopid ostracodes, Anticosti Island, Quebec.

The transition between Ordovician and Silurian marks one of the most distinct lines of extinction and reestablishment of Early Paleozoic ostracode faunas in eastern North America. The sudden appearance of beyrichiids with the early Niagaran brachiopod *Virgiana* in the upper member of the Becscie Formation is startling. It is unlikely that this beyrichiid fauna arose from a loculate tetradellid; it is far more plausible to assume a eurychilinid or piretellid ancestry. However, the beyrichiids *Zygobursa* from Anticosti Island and *Zygocosta* from southern

Ontario mark the earliest occurrence of this fauna in eastern North America (Plate 2).

The Ulrich and Bassler zygobolbine succession (1923) that was established with the appearance of *Zygobursa* (Copeland, 1970a) continued to well within the Llandoverian on Anticosti Island and, in the Appalachian Belt and platform facies from Maryland to southern Ontario, to the top of the Wenlockian with *Drepanellina clarki.* This fauna is known from nowhere else in the world; indeed only one zygobolbid genus is known to occur elsewhere than in eastern North America. Other beyrichiids occur with the zygobolbids in the Gun River and Jupiter formations but the top of the Anticosti Island succession is marked by *Zygobolba decora,* zone fossil to the top of the Lower Clinton Group in the Appalachian Belt (Fig. 2).

Relatively little has been published on the ostracode micropaleontology of the Appalachian Belt in Gaspé Peninsula. The earliest assemblage (Fig. 1, loc. 6), *Zygobolba decora* (Billings) in the Awantjish Formation, is similar to that of the upper Jupiter Formation of Anticosti Island. Deeper water, graptolite-bearing clastic strata of Wenlockian age occur there and it is only in the Ludlovian, calcareous Sayabec Formation (Fig. 1, loc. 8) that an ostracode fauna, here termed kloedenellid, initiated, in eastern Canada, what may be called the Appalachian 'false *Kloedenia*' assemblage (Plate 3). The older or kloedenellid fauna is typified by primitiopsids, kloedenellids and some thlipsurids as well as a few smaller beyrichiids. This fauna is also present in the Hardwood Mountain Formation, late Ludlovian, of northwestern Maine (Fig. 1, loc. 9). The younger or 'false *Kloedenia*' fauna occurs throughout shallow water Late Cayugan to Helderbergian (Pridolian-Emsian) strata of the Appalachian Belt (Berdan *et al.*, 1969). This fauna contains such distinctive genera as *Kloedeniopsis, Pintopsis, Cornikloedenia, Welleria* and *Zygobeyrichia* and thlipsurids of many types. It is not yet as fully defined in eastern Canada as that demonstrated by Berdan for the Cobleskill, Coeymans-Manlius, Becraft-Port Ewen-Glenerie and Schoharie formations of New York State but it does occur in the St. Albans Formation of Forillon Peninsula (Fig. 1, loc. 5), Dalhousie Beds of northern New Brunswick (Fig. 1, loc. 7) and in strata of the Grand Grève Formation of Gaspé.

Although faunas of Llandoverian to Ludlovian ages have been reported from the Fundy Belt, the earliest datable ostracodes from this province are beyrichiids typical of, and in most cases conspecific with those of the shallow water benthonic Beyrichienkalk of the Baltic Province and Downtonian of Great Britain (Plate 3). True *Kloedenia* and *Frostiella* occur in the Stonehouse Formation of northeastern Nova Scotia (Fig. 1, loc. 10) with *Londinia* only in the lower part of the formation and *Nodibeyrichia* only in the upper part. *Londinia* and *Frostiella* occur in the Jones Creek Formation of southern New Brunswick (Fig. 1, loc. 12) and the Leighton Shale Member of the Pembroke Formation near Eastport, Maine (Fig. 1, loc. 13). Also near Eastport, *Nodibeyrichia* has been found in the overlying Hersey Shale Member of the Pembroke Formation, thus equating with the upper Stonehouse Formation. These faunas are considered as Pridolian and no younger marine ostracodes are as yet identified from that part of the Fundy Belt in Canada. In the Eastport Quadrangle of Maine, however, Berdan has discovered a most interesting later ostracode fauna that is presently being studied. Equivalence with the Gedinnian of Podolia and Germany is quite possible for this fauna.

Provincialism of Ostracode Faunas

Questions arise as to the distinct faunal separation of the North American (St. Lawrence Lowland-Anticosti Basin-Appalachian) and European (Fundy) provinces. We can speculate very little on the Ordovician faunal history of the Fundy Belt but do know that the provincial ostracode faunas of northeastern North American and northern European aspect were in contact from at least the late stages of the Middle Ordovician. As the continents continued to approach, the Anticosti Basin, sheltered to the north and east by the Canadian Shield, remained undeformed by the Acadian Orogeny. This stable, shallow, platformal area may have received marine sediments and supported a rich ostracode fauna almost as long as the deformed Appalachian Belt. This will only be known if post mid-Llandoverian calcareous marine strata are found in the Gulf of St. Lawrence area between Anticosti Island and Newfoundland.

It would seem logical that an even more complete physical connection may have existed between platformal North American and European faunal elements during the latest Ordovician and earliest Silurian due to eustatic fall of sea level as a result of glacial activity. The ancestral, benthonic beyrichiid stock could have been able to migrate through shallow waters to inhabit shores of both approaching continents. In this case it might appear that a west-to-east migration along the Appalachian-Caledonian belt took place as North American beyrichiids seem to have appeared earlier than their European counterparts. Return to normal sea level during the later Llandoverian isolated these rapidly evolving faunas, which were separated, possibly by a remnant subsiding trough of deep water, relict of the proto-Atlantic Ocean, across which the large, cruminate, benthonic beyrichiids could not migrate. This trough today is represented by the thick sequence of Middle and Late Silurian turbidites extending through the central parts of New Brunswick and Maine that is covered in eastern New Brunswick by continental strata of Devonian and younger Paleozoic age.

References

Bassler, R. S., 1928. Ostracoda: in Twenhofel, W. H., Geology of Anticosti Island: Geol. Surv. Can., Mem. 154, p. 340-350.

Berdan, J. M., 1966. Baltic Ostracodes from Maine: U. S. Geol. Surv., Prof. Paper 550-A, p. 111.

------, 1971. Silurian to Early Devonian ostracodes of European aspect from the Eastport Quadrangle, Maine: Geol. Soc. Amer., (abs.) Northeastern Section, p. 18.

------, Berry, W. B. N., Boucot, A. J., Cooper, G. A., Jackson, D. E.,

Johnson, J. G., Klapper, G., Lenz, A. C., Martinsson, A., Oliver, W. A., Jr.,

Rickard, L. V. and Thorsteinsson, R., 1969. Siluro-Devonian boundary in North America: Geol. Soc. Amer. Bull., v. 84, p. 275-284.

Berry, W. B. N. and Boucot, A. J., 1973. Glacio-eustatic control of Late Ordovician-Early Silurian platform sedimentation and faunal changes: Geol. Soc. Amer. Bull., v. 84, p. 275-284.

Billings, E., 1865. Palaeozoic Fossils. Vol. 1, Containing descriptions and figures of new or little known species of organic remains from the Silurian rocks: Geol. Surv. Can., p. 299, 300.

------, 1866. Catalogues of Silurian fossils of the Island of Anticosti, with descriptions of some new genera and species: Geol. Surv. Can.

Boucot, A. J., 1961. Stratigraphy of the Moose River Synclinorium, Maine: U. S. Geol. Surv., Bull. 1111-E.

------, 1965. Silurian stratigraphy of Gaspé Peninsula, Québec: Bull. Am. Assoc. Petrol. Geols., v. 49, n. 12, p. 2295-2316.

Brookins, D. G., Berdan, J. M. and Stewart, D. B., 1973. Isotopic and paleontologic evidence for correlating three volcanic sequences in the Maine coastal volcanic belt: Geol. Soc. Amer. Bull., v. 84, p. 1619-1628.

Burk, C. F., Jr., 1964. Silurian stratigraphy of Gaspé Peninsula, Québec: Bull. Am. Assoc. Petrol. Geols., v. 48, n. 4, p. 437-464.

Carter, G. F. E., 1957. Ordovician Ostracoda from the St. Lawrence Lowlands of Québec: unpubl. Ph. D. thesis, McGill Univ., Montreal.

Clarke, J. M., 1909. Early Devonic history of New York and eastern North America: N. Y. State Mus., Mem. 9, pt. 2.

Copeland, M. J., 1960. Ostracoda from the Upper Silurian Stonehouse Formation, Arisaig Nova Scotia, Canada: Palaeont., v. 3, pt. 1, p. 93-103.

------, 1962. Ostracoda from the Lower Devonian Dalhousie beds, northern New Brunswick: Geol. Surv. Can., Bull. 91, p. 18-51, pls. V-X.

10

Copeland, M.J., 1964. Stratigraphic distribution of Upper Silurian Ostracoda, Stonehouse Formation, Nova Scotia: Geol. Surv. Can., Bull. 117, p. 1-13, pl. I.

------, 1965. Ordovician Ostracoda from Lake Timiskaming, Ontario: Geol. Surv. Can., Bull. 127.

------, 1970a. Two new genera of beyrichiid Ostracoda from the Niagaran (Middle Silurian) of Eastern Canada: Geol. Surv. Can., Bull. 187, p. 1-7, pl. I, figs. 1, 2.

------, 1970b. Ostracoda from the Vauréal Formation (Upper Ordovician) of Anticosti Island, Québec: *ibid.*, p. 15-29, pls. IV-V.

------, 1973. Ostracoda from the Ellis Bay Formation (Ordovician), Anticosti Island, Québec: Geol. Surv. Can., Paper 72-43.

------, 1974a. Silurian Ostracoda from Anticosti Island, Québec: Geol. Surv. Can., Bull. 241.

------, 1974b. Middle Ordovician Ostracoda from southwestern District of Mackenzie: Geol. Surv. Can., Bull. 244.

------, 1976. Leperditicopid ostracodes as Silurian biostratigraphic indices: Geol. Surv. Can., Paper 76-IB, p. 83-88.

------, in press. Early Paleozoic Ostracoda from southwestern District of Mackenzie and Yukon Territory: Geol. Surv. Can., Bull.

------, in press. Ordovician Ostracoda from southeastern District of Franklin: *in*, Bolton, T.E., Sanford, B.V., Copeland, M.J. Barnes, C.R. and Rigby, J.K., Geology of Ordovician rocks, Melville Peninsula and region, southeastern District of Franklin: Geol. Surv. Can., Bull. 269.

Dawson, J.W., 1860. (1868, 1878, 1891). The geology of Nova Scotia, New Brunswick and Prince Edward Island or Acadian Geology: London.

Dennison, J.M. and Head, J.W., 1975. Sea level variations interpreted from the Appalachian Basin Silurian and Devonian: Am. J. Sci., v. 275, p. 1089-1120.

Hall, J., 1860. Description of new species of fossils from Silurian rocks of Nova Scotia; Can. Nat. Geol., v. 5, p. 144-159.

Honeyman, D., 1859. Abstract of a paper on the fossiliferous rocks of Arisaig: Trans. Nova Scotia Lit. and Sci. Soc., p. 19-29.

------, 1864. On the geology of Arisaig, Nova Scotia: Quart. J. Geol. Soc. London, v. 20, p. 333-345.

Jones, T.R., 1858. On the Palaeozoic bivalve Entomostraca of Canada: Geol. Surv. Can., Figures and descriptions of Canadian organic remains, Dec. III, p. 91-102, pl. XI.

------, 1870. Notes on some Entomostraca from Arisaig: *in*, Honeyman, D., Notes on the geology of Arisaig, Nova Scotia: Quart. J. Geol. Soc. London, v. 26, p. 492.

------, 1881a. Notes on some Palaeozoic Entomostraca: Nova Scotian Inst. Nat. Sci., Proc. and Trans., v. 5, n. 3, p. 313, 314.

------, 1881b. Notes on some Palaeozoic bivalved Entomostraca: Geol. Mag., dec. 2, v. 8, p. 337-347.

------, 1889. Notes on the Palaeozoic bivalved Entomostraca. No. XXVII. On some North-American (Canadian) species: Ann. Mag. Nat. Hist., ser. 6, n. 3, p. 373-387.

------, 1890a. On some Palaeozoic Ostracoda from North America, Wales and Ireland; Quart. J. Geol. Soc. London, v. 46, p. 1-31.

------, 1890b. On some Devonian and Silurian Ostracoda from North America, France and the Bosphorus: Quart. J. Geol. Soc. London, v. 46, p. 534-536.

Jones, T.R., 1891. On some Ostracoda from the Cambro-Silurian, Silurian, and Devonian rocks: Contrib. Can. Micro-Pal. III, Geol. Nat. Hist. Surv. Canada, p. 59-99.

Kay, G.M., 1934. Mohawkian Ostracoda: species common to Trenton faunules from the Hull and Decorah formations: J. Paleontol., v. 8, n. 3, p. 328-343.

McLearn, F.H., 1924. Palaeontology of the Silurian rocks of Arisaig, Nova Scotia: Geol. Surv. Can., Mem. 137.

Twenhofel, W.H., 1938. Geology and paleontology of the Mingan Islands, Québec: Geol. Soc. Amer. Sp. Papers, n. 11, p. 65-67.

Ulrich, E.O. and Bassler, R.S., 1923. Ostracoda: in Maryland Geol. Surv., Silurian volume, p. 500-704.

Whittington, H.B. and Kindle, C.H., 1963. Middle Ordovician Table Head Formation, western Newfoundland: Geol. Soc. Amer., Bull., v. 74, p. 745-758.

Explanation of Plates

(GSC-Geological Survey of Canada; USNM-United States National Museum)

PLATE 1: (*1-19* Middle Ordovician, *20-27* Late Ordovician)
1. Norochilina nora Copeland, x20, GSC 17056, *2. Oepikella labrosa* Copeland, x10, GSC 17048d, *3. Tetradella ulrichi* Kay, x60, USNM 216123, *4. Ceratopsis quadrifida* (Jones), x20, USNM 216115, *5. Scofieldia bilateralis* (Ulrich), x10, GSC 18654, *6. Bassleratia typa* Kay, x28, GSC 18657, *7. Balticella* sp., x28, GSC 18655, *8. Euprimitia labiosa* (Ulrich), x10, GSC 18652, *9. Dicranella marginata* Ulrich, x16, GSC 17078, *10. Distobolbina teicherti* Copeland, x50, GSC 41919, *11. Steusloffina ulrichi* Teichert, x20, USNM 216128, *12. Thomasatia falcicosta* Kay, x28, GSC 48245, *13. Krausella rawsoni* Roy, x20, USNM 216129, *14, 17, Tetradella buckensis* Guber, x40, x50, GSC 41924, 41926, *15. Levisulculus michiganensis* Kesling, x60, USNM 216120, *16. Glymmatobolbina? spinosa* Copeland, x20, USNM 216131, *18. Monoceratella decorata* Copeland, x37, ventral view, USNM 216094, *19. Eurychilina subradiata* Ulrich, x20, GSC 17035c, *20. Jonesites semilunatus* (Jones), x40, GSC 31568, *21. Euprimitia gamachei* Copeland, x43, GSC 31556, *22. Foramenella phippsi* Copeland, x43, GSC 31526, *23. Tetradella anticostiensis* Copeland, x50, GSC 31554, *24. Platybolbina shaleri* Copeland, x43, GSC 31538, *25. Anticostiella ellisensis* Copeland, x43, GSC 31529, *26. Foramenella phippsi* Copeland, x30, GSC 31411, *27. Tetradella thomasi* Copeland, x43, GSC 31539.

PLATE 2: (*1–4, 7–29* Silurian, Anticosti Island, *5, 6* Devonian)
1. Conbathella inornata Copeland, x30, GSC 32470, *2. Conbathella biporata*
Copeland, x32, GSC 32518, *3. Conbathella equilateralis* Copeland, x32, GSC
32504, *4. Ulrichia (Ulrichia) verticalis* Copeland, x33, GSC 32922, *5. Venzavella*
sp., x37, GSC 48246, *6. Kirkbyella (Berdanella) obliqua* Coryell and Cuskley,
x50, GSC 48247, *7, 8. Anticostiella pustulosa* Copeland, x33, GSC 32564, 32565,
9. Eustephanella? jupiterensis Copeland, x33, GSC 66763, *10. Punctobeecherella*
punctata Copeland, x33, GSC 33025, *11. Zygobolba twenhofeli* Ulrich and Bassler,
x17, GSC 33087, *12, 23. Zygobolba decora* Ulrich and Bassler, x15 and x20, GSC
34698, 33126, *13, 19. Bolbineossia (Brevibolbineossia) berdanae* Copeland, x16.5
and x15, GSC 32944, 32951, *14. Bolbibollia papillosa* Copeland, x33, GSC 33057,
15. Bolbibollia labrosa Ulrich and Bassler, x24, GSC 33043, *16. Zygobolba robusta*
Ulrich and Bassler, x17, GSC 33117, *17, 18. Zygobursa praecursor* Copeland, x25
and x26, GSC 24389, 24390, *20, 24, 29. Apatobolbina whiteavesi* Copeland, x33,
x16 and x33, GSC 32952, 32954, 32960, *21. Craspedobolbina (Mitrobeyrichia)*
boltoni Copeland, x22, GSC 33012, *22. Zygobolba anticostiensis* Ulrich and Bassler,
x15, GSC 33123, *25, 27. Bolbineossia (Bolbineossia) pineaulti* Copeland, x30, GSC
32398, 32392, *26, 28. Anticostibolbina jupiterensis* Copeland, x30, GSC 32415,
32416.

PLATE 3: (*1–4, 6–9* Silurian, Arisaig, *5* Silurian, Pinto, Maryland, *11, 15* Silurian,
Lake Matapedia, *19–23* Devonian, Dalhousie, *10, 12–14, 16–18* Devonian,
Forillon Peninsula)
1, 2. Macrypsilon salterianum (Jones), x16, GSC 14566, a, *3. Kloedenia wilkensiana*
(Jones), x13, GSC 14513, *4. Beyrichia (Nodibeyrichia) pustulosa* Hall, x9, GSC
14503, *5. Pintopsis tricornis* (Ulrich and Bassler), x16, USNM 142251, *6. Hemsiella*
maccoyiana mclearni Copeland, x15, GSC 14561, *7, 8. Londinia arisaigensis*
Copeland, x10, GSC 14562, 14563, *9. Hemsiella maccoyiana sulcata* (Reuter),
x16, GSC 14511, *10. Kloedeniopsis* sp., x20, GSC 48248, *11. Garniella concentrica*
Berdan, x40, GSC 48249, *12. Dizygopleura* sp., x40, GSC 48250, *13. Thlipsuropsis*
inaequalis (Ulrich and Bassler), x50, GSC 48251, *14. Kloedenella* sp., x30, GSC
48252, *15. Leiocyamus* sp., x40, GSC 48253, *16. Janusella biceratina* Roth, x20,
GSC 48254, *17. Neothlipsura* sp., x52, GSC 48255, *18. Eucraterellina* sp. cf. *E.*
oblonga (Ulrich and Bassler), x55, GSC 48256, *19. Mesomphalus magnificus*
Copeland, x15, GSC 14537j, *20. Kloedeniopsis retifera* (Ulrich and Bassler), x15,
GSC 14540d, *21. Arikloedenia newbrunswickensis* (Copeland), x13.5, GSC
14541d, *22. Thlipsura* cf. *T. v-scripta* (Jones and Holl), x30, GSC 14518, *23.*
Strepulites dalhousiensis Copeland, x30, GSC 14523.

PLATE 1

13

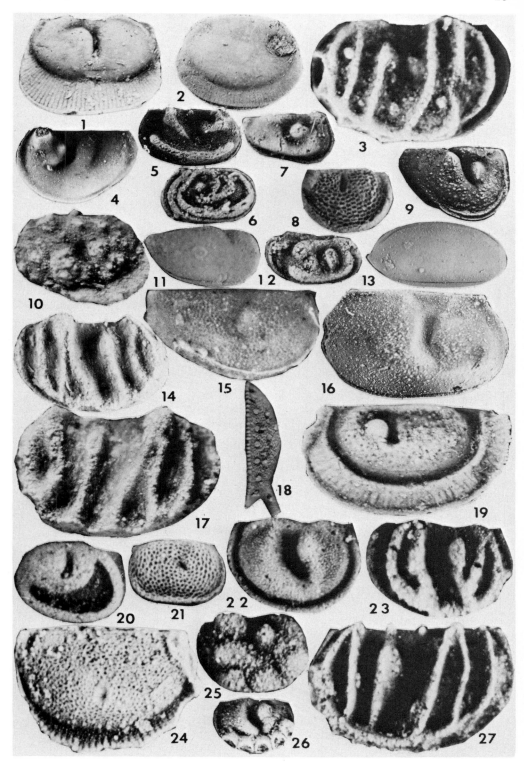

PLATE 2

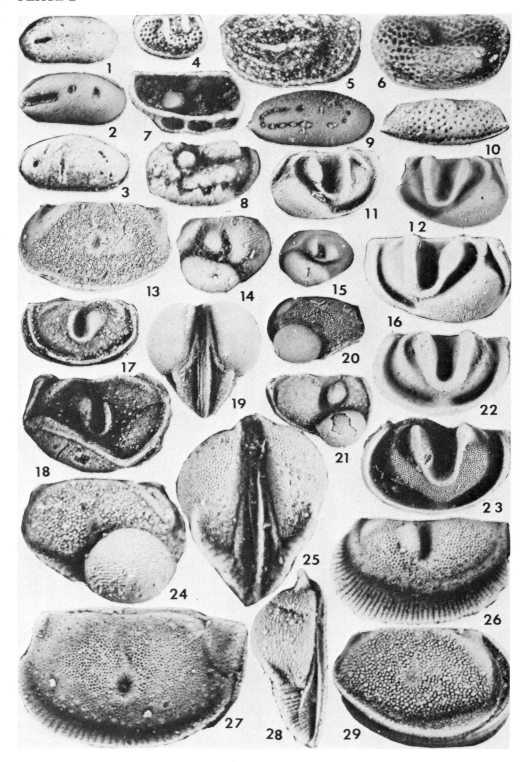

PLATE 3

15

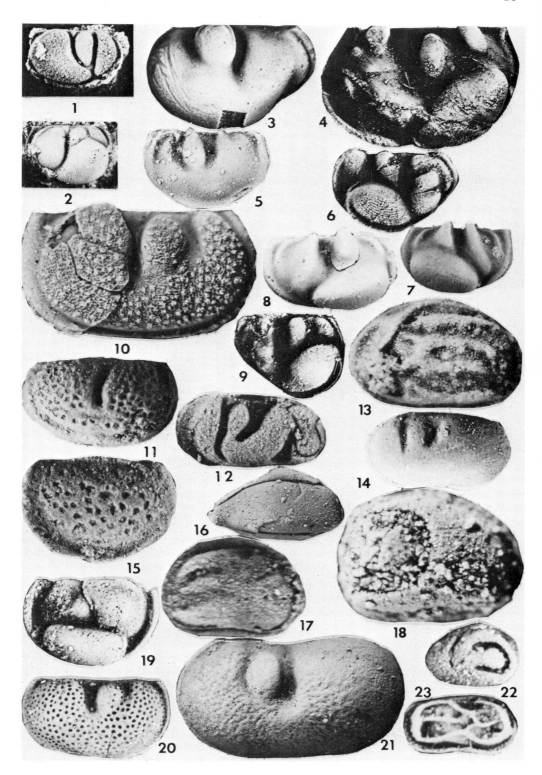

Discussion

Dr. B. K. Holdsworth: To what extent is the apparent provincialism of Ordovician ostracodes perhaps a function of environment and facies rather than
one of geographic separation?

Dr. M. J. Copeland: North American Middle Ordovician ostracodes are quite
different from their European counterparts. This is less obvious in Late
Ordovician faunas, at least those I have studied in eastern and Arctic
Canada. Undoubtedly environment and facies have a greater affect on some
faunal groups than on others. This is certainly evident in those groups
that depend on a pelagic stage of their distribution. The large, heavy
Siluro-Devonian beyrichiaceans of eastern North America on which Jean
Berdan and I base most of our concept of provincialism by geographic separation were, however, benthonic, shallow water facies dwellers. Their
cruminal type of dimorphism could have been a response to this high energy type of environment and provided protection for both the eggs and juve-
niles, precluding a pelagic stage in their development. Distribution of
such forms were therefore limited to migration through shallow water environments. The appearance of beyrichiids at about the same time in both
the North American and European assemblages cannot have been coincidental.
There must have been an initial, shallow water connection between the continents in the Late Ordovician or Early Silurian across which the primitive beyrichiacean stock, whatever it was, migrated and, we think, the
subsequent development of a marine barrier such as a trough that effectively divided the fauna and permitted independent development of North
American and European assemblages in shallow waters on either side.

Dr. Jean Berdan: The beyrichiacean ostracodes of the Hardwood Mountain Formation of western Maine are of Appalachian affinities and represent, according to A. J. Boucot, an environment of shallow water deposition possibly around island arcs. The beyrichiacean ostracodes of the Eastport
area in eastern Maine are entirely different; they are related to those of
the European Beyrichienkalk but also represent a very shallow water environment associated with volcanics. Both types of ostracodes also occur in
platform environments and the differences between them are therefore considered provincial.

Dr. J. E. Conkin: What is the age of the Pridolian beds and their presumed
equivalent, the Downtonian, on the basis of ostracodes?

Copeland: I think most ostracode workers would consider them to be Late
Silurian.

Dr. I. G. Sohn: Murray, you have Kirkbyacea? in one of your slides. What
genus was that?

Copeland: A doubtful Roundyella? sp. from near the base of the Chicotte
Formation of Anticosti Island. This is at best a questionable determination based on relatively few specimens.

  One comment concerning ostracodes from the Lower Mississippian Banff
Formation. Green (1963, Research Council of Alberta, Bull. 11) found that
many of the silicified ostracodes obtained by etching in dilute hydrochlo-
ric acid differed from those of the non-silicified fauna. If you're working in Nevada on the silicified fauna, I strongly suggest you don't etch
everything.

Sohn: We have known that for years.(Sohn, 1950, Geol. Soc. America Bull.,
v. 61, p. 1504). This is something we discussed at the Pau ostracode con-
vention. The silicified fauna that Jordan has in Germany, Jean Berdan has
from the Devonian and I have from the Mississippian seem to have a great
deal in common, namely, all the ornaments and spines are preserved, which
you do not obtain when you beat them out of the other rocks.

Copeland: I wasn't thinking of the morphological aspect, I was thinking of
the taxonomic. In limestone of the Banff Formation, and I expect in many
other similar rock units, species of ostracodes are preserved either as

calcareous or siliceous remains, but, usually not both. Also, numerous specimens of other fossil groups such as trilobites and brachiopods may survive acid disintegration but the ostracodes do not, even though they were visible in hand specimens. Why there is this preferential silicification I don't know. It must be due to the original composition and structure of the shell combined with the type of acid that is used. In some studies, to arrive at a complete list of the contained ostracodes, you have to combine both methods - chemical and physical.

EARLY PALEOZOIC OSTRACODA OF THE ATLANTIC MARGIN, OTHER THAN EASTERN CANADA

by
F. M. Swain
University of Delaware, Newark, Delaware 19711; and
University of Minnesota, Minneapolis, Minnesota 55455

Abstract

Cambrian Archaeocopida and Ordovician and Silurian ostracodes are better developed in the marginal areas of the North Atlantic than those of the South Atlantic where rocks of those ages are only sparsely represented.

The most useful assemblages biostratigraphically are those of the middle Ordovician and middle Silurian of both the eastern and western North Atlantic.

Résumé

Les Archaeocopida cambriens et les ostracodes ordoviciens et siluriens sont mieux évolués dans les régions de bord de l'Atlantique du Nord que dans l'Atlantique du Sud, où les rochers de ces ères ne sont que peu représentes.

Les assemblages les plus utiles biostratigraphiquement sont ceux de mi-ordovicien et mi-silurien des régions est et ouest de l'Atlantique du Nord.

Introduction

Important early contributions to knowledge of early Paleozoic Ostracoda were made in Europe by Jones and Holl (1869), Barrande (1872), Krause (1891), Richter (1869) and Gurich (1896) and in the eastern United States by Ulrich (1890, 1891), and Whitfield (1890) to mention a few. During the early decades of the twentieth century major European works included those by Bonnema (1909) and Kummerow (1924). In the United States significant works included those of Ulrich and Bassler (1908, 1913a, 1913b, 1923), Swartz (1932, 1933, 1936), and Kay (1934, 1940).

In Europe major papers on Ordovician and Silurian ostracodes in Scandinavia and the Baltic region were published by Hessland (1949), Henningsmoen (1953a, 1953b, 1954a, 1954b, 1954c) Jaanusson (1957) and Sarv (1959). In the United States papers dealing with Ordovician and

Silurian ostracodes include: Swartz and Whitmore (1956); Swain (1953, 1957, 1962). In subsequent years many papers have appeared on European Ordovician and Silurian ostracodes as discussed below, but only a few in the United States. For the literature on Canadian faunas see Copeland (this volume).

This article will briefly review the major early Paleozoic ostracode faunas of the Atlantic margins other than those of Canada which are dealt with by Dr. Murray Copeland, herein.

Cambrian

Archaeocopids occur throughout the Cambrian System in the United Kingdom. In the early Cambrian, Aluta, Beyrichiona, Bradoria, Dielymella, and Indiana are represented (Cobbold and Pocock, 1934; Cobbold, 1936). The middle Cambrian contains, in addition, Entomidiella?, Hipponicharion, and Svealuta (Cobbold and Pocock, 1934; Jones, 1856; Taylor and Rushton, 1971). The late Cambrian has yielded further to the preceding: Cyclotron, Falites, Hesslandona, and Vestrogothia (A. W. A. Rushton, written communication, 6 February 1976).

Elsewhere in the Atlantic borderlands, archaeocopids are poorly known. A few occur in the early and middle Cambrian of maritime Canada (Copeland this volume) and New York (Ulrich and Bassler, 1931).

Ordovician

The distribution of Ordovician Ostracoda around the "Proto-Atlantic" Ocean is shown in Figure 1. The continental reconstruction is from Smith, Briden, and Drewery (1973).

Early Ordovician

Early Ordovician Canadian (Tremadocian? + Arenigian) ostracodes are poorly known from eastern North America, but isochilinids · such as I. gregaria (Whitfield) (Table 1, fig. 2-11) Isochilina seelyi (Whitfield) (Table 1, fig. 2-12) and I. cristata (Whitfield), (Table 1, fig. 2-10) and eoleperditiids are represented in carbonate facies (Whitfield, 1890, Bassler and Kellett, 1934). Undescribed leperditellids and other smaller species occur in the calcitic facies (Swain, 1957), but widespread early Ordovician dolomites are barren of ostracodes. Isochilinids also occur in the Arenigian of northern Greenland (Poulsen 1929, 1934).

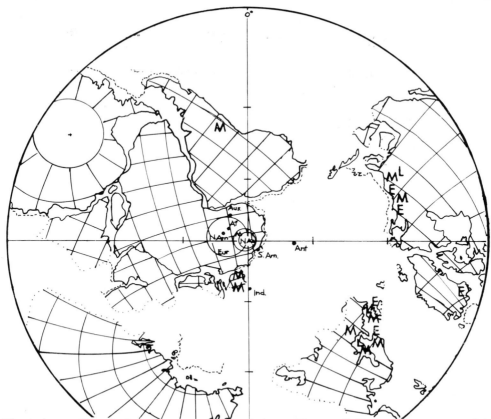

Fig. 1.   Cambrian-Lower Ordovician Map, S-Pole Stereographic
Projection after Smith, Briden and Drewery (1973), showing Ordovician
ostracode distribution about the "Proto-Atlantic " Ocean.   E, Early
Ordovician; M, Middle Ordovician; L, Late Ordovician.   Circles and
dots show interpretations of South polar positions as determined from
various places, with size of circle proportional to degree of uncertainty.

In Sweden (Hessland, 1949), a diverse early Arenigian ostracode
assemblage includes Conchoides minuta Hessland (Table 1, fig. 2-5) and
related species, Glossomorphites tenuilimbata (Hessland) (Table 1), and
related forms, Primitiella brevisulcata Hessland (Table 1, fig. 2-17),
Aulacopsis spp. (Table 1, figs. 2-1, 2-5) Protallinella grewingki (Bock)
(Table 1, fig. 2-18), and others.   The Arenigian of Norway (Asaphus
Series) has yielded a variety of beyrichiacean ostracodes in a mainly
shaly facies (Henningsmoen, 1954a).   In Wales and central England
entomidellids, beyrichiids, and tetradellids have been recorded from
Arenigian beds (Table 1, fig.2).   Additional early Ordovician ostracodes
not mentioned above are listed in Table 1 and illustrated in figure 1
(Henningsmoen, 1954a; Ulrich and Bassler, 1908; Jones, 1884).   An
important early Ordovician ostracode fauna has been described from
the Estonian S.S.R. (Sarv, 1959)

Figure 2

1. <u>Aulacopsis bifissurata</u> Hessland. Right valve, x23, early Ordovician, Leskusanget, Sweden (Hessland, 1949).
2. <u>Aulacopsis monofissurata</u> Hessland. Right valve, x23, early Ordovician, Stenberg, Sweden (Hessland, 1949).
3. <u>Beyrichia barrandiana</u> Jones. Left valve, x3, early Ordovician, Mynyddgarw, Wales (Jones, 1885).
4. <u>Ceratocypris longispina</u> Hessland. Right valve, x23, early Ordovician, Leskusanget, Sweden (Hessland, 1949).
5. <u>Conchoides minuta</u> Hessland. Right valve, x23, early Ordovician, Stenberg, Sweden (Hessland, 1949).
6. <u>Ctenentoma plana</u> Hessland. Right valve, x23, early Ordovician, Leskusanget, Sweden (Hessland, 1949).
7. <u>Entomidella marrii</u> Jones. Right valve, xl.5, early Ordovician (distorted), Nantlle, Wales (Jones, 1884).
8. <u>Eurychilina dorsotuberculata</u> Hessland. Right valve, x23, early Ordovician, Silberberg, Sweden. (Hessland, 1949).
9. <u>Glossopsis depressolimbata</u> Hessland. Right valve, x23, early Ordovician, Leskusanget, Sweden (Hessland, 1949).
10. <u>Isochilina cristata</u> (Whitfield). Right valve x8, early Ordovician, Lake Champlain, Vermont (Swain, 1957).
11. <u>Isochilina gregaria</u> (Whitfield). Right valve, x6, early Ordovician, Lake Champlain, Vermont (Swain, 1957).
12. <u>Isochilina seelyi</u> (Whitfield), Left valve, x6, early Ordovician, Lake Champlain, Vermont (Swain, 1957).
13. <u>Laccochilina dorsoplicata</u> Hessland. Right valve, x23, early Ordovician, Silberberg, Sweden.
14. <u>Nanopsis nanella</u> (Moberg and Segerberg). Right valve, x38, early Ordovician, Slemmenstad, Norway (Henningsmoen, 1954).
15. <u>Ogmoopsis nodulifera</u> Hessland. Right valve, x23, early Ordovician, Silverberg, Sweden (Hessland, 1949).
16. <u>Pinnatulites procera</u> (Kummerow). Right valve, x23, early Ordovician, Leskusanget, Sweden (Hessland, 1949).
17. <u>Primitiella brevisulcata</u> Hessland. Right valve, x23, early Ordovician, Leskusanget, Sweden, (Hessland, 1949).
18. <u>Protallinella grewingki</u> (Bock). Right valve, x23, early Ordovician, Stenberg, Sweden (Hessland, 1949).
19. <u>Steusloffia acuta</u> (Krause). Left valve, xl5, early Ordovician? drift boulders, Mark Brandenberg, Germany (Ulrich and Bassler, 1908).
20. <u>Steusloffia polynodulifera</u> Hessland. Left valve, x23, early Ordovician, Born-Dadran, Sweden (Hessland, 1949).

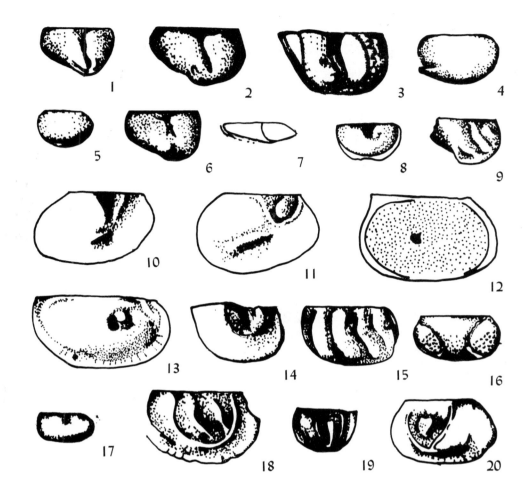

Middle Ordovician

The ostracodes of the Chazyan (Llanvirnian and Llandeilan) of the eastern United States are represented by a lower <u>Bullatella kauffmanensis</u> Zone and an upper <u>Monoceratella</u> teres Zone (Table 2, figs. 3-6, 21,) Swain 1957, 1962). Many other leperditiids, aparchitids, opikellids, leperditellids, eurychilinids, budnianellids and primitive cyprinids are also represented (Swain 1957, 1962) some of which are listed in Table 2 and illustrated in figure 3. An extensive silicified Chazyan fauna is represented in Virginia (Kraft, 1962) and in New York (Swain, 1962).

Black Riveran (early Caradocian) ostracodes are poorly known in the eastern United States although well represented in the central United States (Kay 1934, 1940; Swain, Cornell, and Hansen, 1961). Trentonian (middle Caradocian) shaly limestones of New York are

Table 1. Early Ordovician Ostracoda, Atlantic Margins
E, Eastern Atlantic; W, Western Atlantic

Species	Loc*	Tremad	Canadian Arenig
Beyrichiona triceps	GB	- -E- -	
Nanopsis nanella	N,S	- -E- -	
Primitia? sp.	S	- -E- -	
Aulacopsis spp.	N,S		- -E- - - -?- - - -?- -
Conchoides minuta	S		- -E- -
Glossomorphites tenuilimbata	S		- -E- -
Primitiella brevisulcata	S		- -E- -
Protallinella grewingki	N		- -E- - - -?- - - -?- -
Ceratocypris longispina	S		- -E- -
Eurychilina dorsotuberculata	S		- -E- -
Glossopsis depressolimbata	S		- -E- -
Laccochilina dorsoplicata	S		- -E- -
Ogmoopsis nodulifera	S		- -E- -
Pinnatulites procera	S		- -E- -
Steusloffia cf. polynodulifera	S		- -E- -
Beyrichia barrandiana	GB		- - - - - - - E - - - - - - - - -
Conchoprimitia eos	N		- - - - - - - E - - - - - - - - -
Ctentoma plana	S		- - - - - - - E - - - - - - - - -
Glossmorphites sp.	N		- - - - - - - E - - - - - - - - -
Isochilina cristata	US		- -?- - - - - -W- - - - - - - -
Isochilina gregaria	US		- -?- - - - - -W- - - - - - - -
Isochilina seelyi	US		- -?- - - - - -W- - - - - - - -
Rigidella erratica	N		- - - - - - - E- - - - - - - - -
Steusloffia acuta	N		- - - - - - - E- - - - - - - - -
Tetradella? sp.	GB		- - - - - - - E - - - - - - - - -
Isochilina arctica	Gr	- -?- -	- W - - - - -?- - - - - - - -

* GB, Great Britian; N, Norway, S, Sweden; US, United States;
  Gr, Greenland

exemplified by Bassleratia typica Kay (Table 2, fig.2-3), Bollia subaequata Ulrich (Table 2, fig. 3-4), Primitiella constricta Ulrich (Table 2, fig. 3-27), Thomasatia falcicosta (Table 2, fig. 3-3) and other forms (Kay 1934, 1940; Swain et al., 1961).

In limestone facies of central and southern Sweden, the Llanvirnian (Aseri Stage) contains among others, Piretia geniculata Jaanusson (Table 2, fig. 3-23) and Laccochilina (L.) bulbata Jaanusson (Table 2, fig. 3-19). The overlying early Llandeilan (Lasnamae Stage) is typified by Euprimites effusus Jaanusson (Table 2, fig. 2-12) and Actinochilina spp. (Table 2, fig. 3-1) (Jaanusson,1957). The succeeding Uhaku Stage

Table 2. Middle Ordovician Ostracoda of Atlantic Margins
E, Eastern Atlantic; W, Western Atlantic

Species	Loc.	Mohawkian Chazyan Llanv.	Lland.	B.R. Early	Tr. Caradoc.
Bullatella kauffmanensis	US	--W--			
Laccochilina bulbata	S	--E--			
Piretia geniculata	S	--E--			
Budnianella shenandoense	US		-------W		
Elliptocyprites parallela	US		-------W		
Eographiodactylus eos	US		-------W		
Eurychilina strasburgensis	US		-------W		
Krausella variata	US		-------W--?		
Primitiella champlainensis	US		-------W		
Actinochilina spp.	S		--E--		
Balticella spp.	US,S		-E,W-		
Conchoprimites? spp.	S		--E--		
Euprimites effusus	S		--E--		
Euprimites bursellus	S		--E--		
Euprimites suecicus	S		--E--		
Hesperidella spp.	S		--E--		
Parapyxion spp.	S		--E--		
Primitiella spp.	S		--E--		
Pyxion sp.	S		--E--		
Sigmobolbina sigmoidea	S		--E--		
Steusloffia linnarssoni	S		--E--		
Tallinella dimorpha	S		--E--		
Tvarenella spp.	S		--E--		
Ulrichia? bipunctata	GB		--E--		
Monoceratella teres	US		---------W--		
Euprimites locknensis	S,G		--E--------?		
Platybolbina spp.	S,E		--E--------?		
Steusloffia costata	S,E		--E--------?		
Tetradella complicata	GB		--E--------?		
Tallinella trident	N,E		--E--------?		
Ullerella triplicata	N,E		--E--------?		
Ctenobolbina spp.	E,F,P		?-------E,W------?		
Bolbina spp.	G			--------E--------	
Laccochilina paucigranosa	S			--------E--------	
Primitia simplex	GB			--------E--------	
Primitia? bicornis	GB			--------E--------	
Uhakiella spp.	G			--------E--------	
Bassleratia typa	US				---W---
Bollia subaequata	US				---W----
Primitiella constricta	US				---W----
Thomasatia falcicosta	US				---W----

contains, among others Laccochilina (L.) paucigranosa Jaanusson (Table 2, fig. 3-20), Euprimites bursellus Jaanusson (Table 2, fig. 3-13), Tallinella dimorpha Jaanusson (Table 2, fig. 3-32), Steusloffia linnarssoni Jaanusson (Table 2, fig. 3-31), and Sigmobolbina sigmoidea Jaanusson (Table 2, fig. 3-29). In the late Llandeilan (Kukruse Stage) two facies assemblages occur: a carbonate facies, with species of Platybolbina (Table 2), Tvarenella (Table 2, fig. 3-36), Euprimites (Table 2, fig. 3-15), Hesperidella (Table 2, fig. 3-17) and Balticella (Table 2, fig. 3-2), (Jaanusson 1957); and a mudstone facies with Conchoprimites? (Table 2, fig. 3-7), Actinochilina (Table 2, fig. 3-1) and Parapyxion (Table 2); both facies are represented by Euprimites locknensis Jaanusson (Table 2, fig. 3-14), and Steusloffia costata Jaanusson (Table 2, fig. 3-20) (Jaanusson 1957). A large middle Ordovician fauna is present in the Kuckruse and other formations of Estonia (Bonnema, 1909; Sarv, 1959).

Figure 3

1. Actinochilina suecica (Thorslund). Right valve of holotype, x19, middle Ordovician, Ludibundus beds, Kinnekulle, Sweden (Jaanusson, 1957).

2. Balticella deckeri (Harris). Right valve, x13, middle Ordovician, Strasburg Junction, Virginia (Kraft, 1962).

3. Bassleratia typa Kay. Left valve, holotype, x23, middle Ordovician, Northumberland County, Ontario (Kay, 1934).

4. Bollia subaequata Ulrich. Left valve, x23, middle Ordovician, Church, Iowa (Kay, 1940).

5. Budnianella shenadoense Kraft. Right valve, holotype, x20, middle Ordovician, Tumbling Run, Virginia (Kraft, 1940).

6. Bullatella kauffmanensis (Swain). Right valve, x16, middle Ordovician, Marion, Pennsylvania (Swain, 1957).

7. Conchoprimitia leperditioidea Thorslund. Right valve, x15, middle Ordovician, Siljan, Sweden (Jaanusson, 1957).

8. Ctenobolbina minor kuckersiana (Bonnema). Left valve, x11, middle Ordovician, early Caradocian, Kukruse, Estonia (Bonnema, 1909).

9. Ctenbolbina ornata latimarginata (Bonnema). Left valve, x11, middle Ordovician, early Caradocian, Kukruse, Estonia (Bonnema, 1909).

10. Elliptocyprites longula Swain. Right valve, holotype, x36, middle Ordovician, Lake Champlain, Vermont (Swain, 1962).

11. Eographiodactylus eos Kraft. Left valve, x26, middle Ordovician, Tumbling Run, Virginia (Kraft, 1962).

12. Euprimites effusus Jaanusson. a, Left valve, of heteromorph, x19, b, left valve of heteromorph, middle Ordovician, Oland Seby, Sweden (Jaanusson, 1957.)

13.   Euprimites bursellus Jaanusson. Left valve of heteromorph, x19, middle Ordovician, Ostergotland, Sweden (Jaanusson, 1957.)

14.   Euprimites locknensis (Thorslund). Right valve of heteromorph, x19, middle Ordovician, Jamtland, Sweden (Jaanusson, 1957).

15.   Euprimites suecicus (Thorslund). Left valve, of heteromorph, x19, middle Ordovician, Jamtland, Sweden (Jaanusson, 1957).

16.   Eurychilina strasburgensis Kraft. Left valve of female, x13, middle Ordovician, Strasburg Junction, Virginia (Kraft, 1962).

17.   Hesperidella esthonica (Bonnema). Right valve of heteromorph, x28, middle Ordovician, Ludibundus Limestone, Tvaren area, Sweden (Jaanusson, 1957).

18.   Krausella variata Kraft. Right valve, x13, middle Ordovician, Strasburg Junction, Virginia (Kraft, 1962).

19.   Laccochilina (Laccochilina) bulbata Jaanusson. Left valve of tecnomorph, x19, middle Ordovician, Ostergotland, Sweden (Jaanusson, 1957).

20.   Laccochilina (Laccochilina) paucigranosa Jaanusson. Left valve of heteromorph, x19, middle Ordovician, erratic boulder Erken No. 10, South Bothnia, Sweden (Jaanusson, 1957.)

21.   Monoceratella teres Teichert. Left valve, x13, middle Ordovician, Strasburg Junction, Virginia (Kraft, 1962).

22.   Parapyxion subovatum (Thorslund). Left valve, x23, middle Ordovician Ludibundus beds, Vasterogotland, Sweden (Jaanusson, 1957).

23.   Piretia geniculata Jaanusson. Left valve of heteromorph, x26, middle Ordovician, Ostergotland, Sweden (Jaanusson, 1957).

24.   Primitia simplex (Jones). Right valve, x6, middle Ordovician, near Coimbra, Portugal (Jones, 1855).

25.   Primitia? bicornis (Jones). Left valve, x6, middle Ordovician, Shropshire, England (Jones, 1855).

26.   Primitiella champlainensis Swain. Right valve, x16, middle Ordovician, Lake Champlain, Vermont (Swain, 1962).

27.   Primitiella constricta Ulrich. Left valve, x23, middle Ordovician, Cannon Falls, Minnesota (Kay, 1940).

28.   Pyxion carinatum (Hadding). Right valve, x19, middle Ordovician, Scania, Sweden (Jaanusson, 1957).

29.   Sigmobolbina sigmoidea Jaanusson. a, Left valve of tecnomorph, x19; b, diagrammatic drawing of heteromorph of species, x30, middle Ordovician, Oland, Sweden (Jaanusson, 1957).

30.   Steusloffia costata (Linnarsson). Right valve, x11, middle Ordovician, Ludibundus limestone, Siljan district, Sweden (Jaanusson, 1957).

31.   Steusloffia linnarsoni (Krause). Left valve, x11, middle Ordovician, Oland, Sweden (Jaanusson, 1957).

32.   Tallinella dimorpha Opik. Right valve, x11, middle Ordovician, Siljan district, Sweden (Jaanusson, 1957).

Figure 3

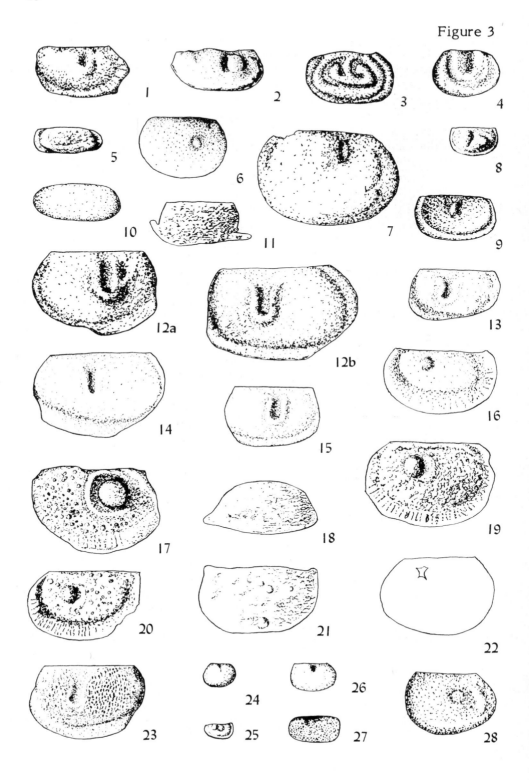

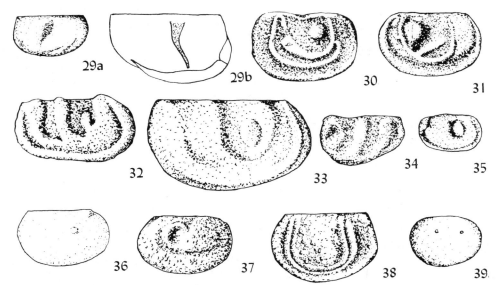

A few middle Ordovician ostracodes of primitiid and tetradellid types are known from France, Spain, and Portugal (Bassler and Kellett 1934). A great many species have also been described from drift boulders presumed to be middle Ordovician in northern Germany (Kummerow, 1921). Ostracoda are scarce in the United Kingdom prior to the Caradocian. Primitiids and tetradellids are common in the Caradocian and Ashgillian of Wales (Jones, 1855; Harper, 1947; A.W.A. Rushton, written communication, 6 February 1976). Late Llandeilan and early Caradocian beds in San Juan Province, Argentina contain Ctenobolbina?, Parenthatia, Balticella, and Anisochilina, all of which are reminiscent of Scandinavian and middle Appalachian Assemblages (Baldis and Rossi de Garcia 1972).

Figure 3. (Continued)
33. Tallinella trident Henningsmoen. Left valve, x16, middle Ordovician, Sandsvaer, Norway (Henningsmoen, 1953).
34. Tetradella complicata Salter. Right valve, x9, middle Ordovician, early Llandeilan, Pembrokeshire, Wales(Harper, 1947).
35. Thomasatia falcicosta Kay. Right valve, x23, middle Ordovician, Cannon Falls, Minnesota (Kay, 1940).
36. Tvarenella carinata (Thorslund). Right valve of tecnomorph, x19, middle Ordovician, Sodermanland, Sweden (Jaanusson, 1957).
37. Uhakiella coelodesma Opik. Right valve of male, x15, middle Ordovician, early Caradocian, Purtsejogi River, Estonia.
38. Ullerella triplicata Henningsmoen. Left valve, x13, middle Ordovician, Ringerike, Norway (Henningsmoen, 1953).
39. Ulrichia bipunctata (Jones and Holl). Right valve, x15, middle Ordovician, Llandeilan, Brecknockshire, Wales (Jones and Holl, 1869).

Table 3. Late Ordovician Ostracoda of Atlantic Margins
E, Eastern Atlantic; W, Western Atlantic

Species	Loc.	Eden-Mays Late Caradoc.	Rich-Gam Ashgill
Ceratopsis chambersi	US	- - - - W - - ?	
Ctenobolbina ciliata	US	- - - - W - - ?	
Tetradella aff. quadrilirata	US	- - - - W - - ?	
Aparachites maccoyi	GB		- - E - -
Macronotella spp.	N		- - E - -
Platybolbina cf. plana	N		- - E - -
Primitia sanctipatricii	GB		- - E - -

Late Ordovician

There are sparse late Ordovician Edinian, Maysvillian, and Richmondian (Late Caradocian and Ashgillian) ostracode faunas in clastic lithofacies of the eastern United States (Butts, 1940). A few species are listed in Table 3 and illustrated in figure 4, but in general these faunas are poorly known.

Figure 4

1.   Aparchites maccoyii Jones and Holl. Right valve, x8?. late Ordovician, Caradocian (Bala), Chair of Kildare, Ireland (Jones and Holl 1868).

2.   Ceratopsis chambersi (Miller). Left valve, internal mold, x9, late? Ordovician, Martinsburg Shale, Roanoke County, Virginia. (Butts, 1941).

3.   Ctenbolbina ciliata (Emmons). Left valve, internal mold, x9, late? (Ordovician, Martinsburg Shale, Roanoke County, Virginia. (Butts, 1941)

4.   Macronotella cf. M. praelonga (Steusloff). Right valve, x8, lage Ordovician, Asker, Norway (Henningsmoen, 1954).

5.   Platybolbina cf. P. elongata (Krause). Left valve, x15, late Ordovician, Oslo, Norway (Henningsmoen, 1954).

6.   Platybolbina cf. P. plana (Krause). Left valve, x15, late Ordovician, Oslo, Norway (Henningsmoen, 1954).

7.   Primitia sanctipatricii Jones and Holl. Right valve, x15, late Ordovician, Caradocian (Bala), Chair of Kildare, Ireland.

8.   Tetradella sp. aff. T. quadrilirata (Hall and Whitfield). Right valve, x9, late? Ordovician, Martinsburg Shale, Roanoke County, Virginia. (Butts, 1941).

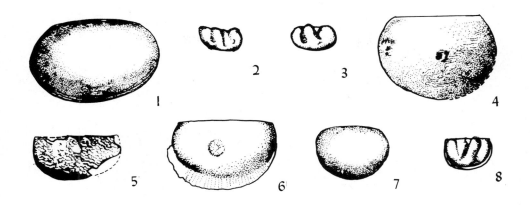

A hiatus occurs between middle Ordovician and late Ordovician ostracode faunas in Norway, the Tretaspis beds being nearly barren (Henningsmoen, 1954b). The overlying Dalmanitina beds (Ashgillian) contain a varied fauna including Platybolbina cf. P. plana Krause (Table 3, fig. 4-6), P. cf. P. elongata (Krause) (Table 3, fig. 4-5), Macronotella spp. (Table 3, fig. 4-4) and others.

Ashgillian deposits of the United Kingdom have yielded a dozen or more species including Aparchites maccoyii (Salter) (Table 3, fig. 4-1) and Primitia sanctipatricii Salter (Table 3, fig. 4-7). The late Ordovician of Estonian S.S.R. has a major ostracode fauna (Sarv, 1959). Other regions bordering the Atlantic have few or no recorded late Ordovician ostracodes.

Silurian

The distribution of Silurian Ostracoda around the "ProtoAtlantic" Ocean plotted on the map for the Lower Devonian of Smith, Briden, and Drewery (1973) is shown in Figure 5.

Early Llandoverian (Medinan) ostracodes are poorly developed in eastern United States owing to unfavorable sandy facies but a few leperditiids occur, i.e. Leperditia cylindrica Hall (Bassler and Kellett, 1934). Late Llandoverian C_3-C_6 beds (Clinton Group) of the Applachian region United States are abundantly supplied with ostracodes and have been divided into the following zygobolbid zones in ascending order (Ulrich and Bassler, 1923): Zygobolba erecta Zone (Table 4, fig 6-35),Zygobolba anticostiensis Zone (Table 4, fig. 6-33), Zygobolba decora Zone (Table 4, fig. 6-34), Zygobolbina emaciata Zone (Table 4, fig. 6-36), Mastigobolbina lata Zone (Table 4, fig. 6-19), Zygosella postica Zone (Table 4, fig. 6-37), Bonnemaia rudis Zone (Table 4, fig. 6-6), Mastigobolbina typus Zone (Table 4, fig. 6-20), and Paraechmina spinosa Zone (Table 4, fig. 6-25). All of these occur in

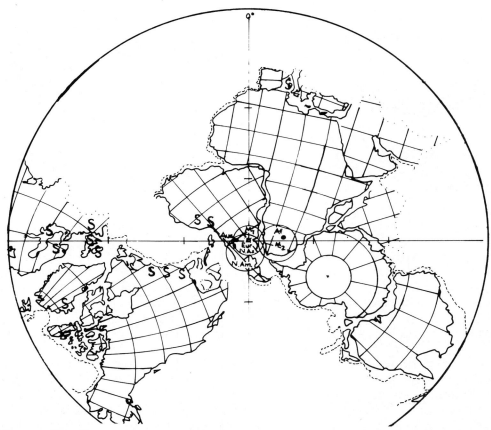

Figure 5. Map of Lower Devonian, S-pole stereographic projection, after Smith, Briden, and Drewery (1973), showing Silurian ostracode distribution about the "Protoatlantic" Ocean. Circles and dots show interpretations of south polar positions as determined from various places, with size of circle proportional to degree of uncertainty.

the Rose Hill Shale and other Clinton equivalents. The succeeding Wenlockian beds (Rochester Shale) contain another zone, the Drepanellina clarki (Table 4, fig. 6-11). Ludlovian beds (McKenzie and Wills Creek Formations) have abundant small kloedenellids such as: Dizygopleura acuminata Ulrich and Bassler (Table 4, fig. 6-10), Eukloedenella umbonata Ulrich and Bassler (Table 4, fig. 6-12), and Kloedenella scapha Ulrich and Bassler (Table 4, fig 6-15) Pridolian shaly limestones have large Kloedenia (Table 4, fig. 6-16), Pintopsis (Table 4), Welleria (Table 4, fig. 6-3), Welleriopsis (Table 4, fig 6-32), and Lophokloedenia (Table 4, fig. 6-18), which are indigenous to the American Appalachians at the species level but have a number of European affinities (Berdan 1962, 1970, 1971a; Martinsson 1970; Swartz and Whitmore, 1956).

Table 4a. Early Silurian Ostracoda of Atlantic Margins
E, Eastern Atlantic; W, Western Atlantic

Species	Loc.	Medinian Llandoverian Early	Early Niagaran Late
Leperditia cylindrica	US	- - W - -	
Zygobolba erecta	US	? - W - ?	
Zygobolba anticostiensis	US	? - W - ?	
Zygobolba decora	US	? - W - ?	
Zygobolbina emaciata	US	? - W - ?	
Mastigobolbina lata	US	? - W - ?	
Zygosella postica	US	? - W - ?	
Bonnemaia rudis	US	? - W -?	
Mastigobolbina typus	US	? - W?	
Paraechmina spinosa	US	? - W	
Arcuaria sineclivula	B	? - - - - - - - - - E - - - - - - -?	
Beyrichia paucituberculata	N?	? - - - - - - - - - E - - - - - - -?	
Craspedobolbina armata	N?	? - - - - - - - - - E - - - - - - -?	
Neobeyrichia regnans	B?	? - - - - - - - - - E - - - - - - -?	
Nodibeyrichia scissa	GB,B	? - - - - - - - - - E - - - - - - -?	
Platybolbina? bulbosa	N?	? - - - - - - - - - E - - - - - - -?	
Primitiella? parallela	N?	? - - - - - - - - - E - - - - - - -?	

Table 4b. Middle Silurian Ostracoda of Atlantic Margins
E, Eastern Atlantic; W, Western Atlantic

Species	Loc.	Wenlockian M. Niagaran	Ludlovian Late Niag.- Early Cay.
Apatobolbina platygaster	N	- - - E - - -	
Beyrichia kloedeni	GB	- - - E - - -	
Craspedobolbina variolata	GB,N	- - - E - - -	
Drepanellina clarki	US		- - - W - - -
Microcheilinella variolaris	B	- - - E - - -	
Beyrichia ringerikensis	N		- - - E - - -
Craspedobolbina percurrens	B		- - - E - - -
Dizygopleura acuminata	US		- - - W- - -
Entomozoe? marstoniana	GB		- - - E - - -
Eukloedenella umbonata	US		- - - W- - -
Hemsiella anterovelata	B		- - - E - - -
Hemsiella maccoyana	GB		- - - E - - -
Kloedenella scapha	US		- - - W- - -
Neobeyrichia nutans	B		- - - E - - -
Primitiopsis planifrons	N		- - - E - - -
Signetopsis semicircularis	N		- - - E - - -
Macrypsilon salterianum	GB		- - - E - - -
Welleria obliqua	US		- -W- -

Table 4c. Late Silurian Ostracode of Atlantic Margins
E, Eastern Atlantic; W, Western Atlantic

Species	Loc.	Pridolian (Downtonian) Middle and Late Cayugan
Amphitoxis curvata	B	? - - - - E - - - -?
Frostiella groenvalliana	US,GB,N	? - - - - -E,W- - - ?
Hemsiella aff. maccoyana	US,B	? - - - - -E,W- - - ?
Kloedenia crassipunctata	US	? - - - - W - - - ?
Leperditia scalaris	US	? - - - - W - - - ?
Londinia arisaigensis	US,GB	? - - - - -E,W- - - ?
Londinia aff. arisaigensis	B,US	? - - - - -E,W- - - ?
Londinia kiesowi	B	? - - - - E - - - -?
Lophokloedenia manliensis	US	? - - - - W - - - ?
Macrypsilon aff. salterianum	US	? - - - - W - - - ?
Nodibeyrichia tuberculata	B	? - - - - E - - - -?
Pintopsis spp.	US,B	? - - - - -W,E- - - ?
Richteria? sp.	US	? - - - - W - - - ?
Sleia equestris	GB,B,US	? - - - - -E,W- - - ?
Welleriopsis jerseyensis	US	? - - - - E - - - -?

In Norway (Oslo region) the Llandoverian is represented by
Beyrichia (Eobeyrichia) spp. (Table 4, figs. 6-4), Platybolbina spp.
(Table 4, figs. 6-26), Primitiella? spp. (Table 4, fig. 6-7) (Henningsmoen, 1954c). The Wenlockian contains few ostracodes, among them
Apatobolbina platygaster Kummerow (Table 4, fig. 6-2) (Henningsmoen, 1954c). The Ludlovian contains abundant Primitiopsis spp.
(Table 4, fig. 6-28), Signetopsis spp. (Table 4, fig. 6-29), and
Neobeyrichia ringerikensis (Henningsmoen) (Table 4, fig. 6-5) (Henningsmoen, 1954c). The Ludlovian of northwestern Europe is generally
characterized by "amphitoxidine" ostracodes: Londinia sp. (Table 4,
fig. 6-21), Lophoctenella spp. (Table 4), and Sleia sp. (Table 4, fig. 6-30)
(Martinsson 1962). fauna occurs in late Wenlockian? and Ludlovian
beds (Berdan, 1967, 1970, 1971; Martinsson, 1970). The later Ludlovian
(Leintwardinian and Whitcliffian Stages) in northern England comprises
zones based on Nodibeyrichia (Table 4, fig. 6-24), Neobeyrichia (Table
4, fig. 6-23), Cyptopholobus, Juviella, and Hemsiella (Table 4, fig. 6-14)
(Shaw, 1971). The succeeding Downtownian (Pridolian) is characterized
by the kloedenid Frostiella groenvalliana Martinsson (Table 4, fig. 6-13)
(Shaw, 1971). A similar fauna occurs in New York, Maine and Nova
Scotia (Martinsson, 1970; Berdan, 1970, 1972).

An interesting undifferentiated Silurian ostracode assemblage
characterized by entomids was long ago recorded from Sardinia
(Canavari, 1899, 1900). The entomids are generally believed to be
pelagic ostracodes (Rozhdestvenskaja, 1971) and although the family is
known from the Ordovician, the Sardinian Silurian entomids seem to
represent one ot the earliest pelagic assemblages.

Elsewhere in the Atlantic basin, except in eastern Canada which is discussed herein by Dr. Copeland, Silurian ostracodes are only slightly known; a few forms have been described from France (Tromelin and Lebesconte 1876) and from Brazil (Clarke, 1869) (Table 4).

Figure 6

1. Amphitoxis curvata Martinsson. Left valve, female, x45, Pridolian, Snoder Gotland (Martinsson, 1962).
2. Apatobolbina platygaster (Kummerow). Left valve, xll, Wenlockian, Ringerike, Norway (Henningsmoen, 1954).
3. Beyrichia (Beyrichia) cf. B. kloedeni M'Coy). Left valve, x23, Wenlockian, Shropshire, England (Martinsson, 1962).
4. Beyrichia (Eobeyrichia) paucituberculata Henningsmoen. Left valve, xl4, Llandoverian, Gunneklev, Norway (Henningsmoen, 1954).
5. Beyrichia (Neobeyrichia) ringerikensis Henningsmoen. Left valve, x7, Ludlovian, Oslo, Norway (Henningsmoen, 1954).
6. Bonnemaia rudis Ulrich and Bassler. Right valve, x6, Niagaran, Powell Mountain, Tennessee (Ulrich and Bassler, 1923).
7. Craspedobolbina armata Heningsmoen. Left valve, xll, Llandoverian, Oslo, Norway (Henningsmoen, 1950).
8. Craspedobolbina (Mitrobeyrichia) percurrens Martinsson. Left valve of female, x30, Ludlovian, Follingbo, Sweden (Martinsson, 1962).
9. Craspedobolbina (Mitrobeyrichia) variolata Martinsson. Left valve of female, x30, Wenlockian, Djupvik, Sweden, (Martinsson, 1962).
10. Dizygopleura acuminata Ulrich and Bassler. Left valve, xl2, late Niagaran, Cumberland, Maryland. (Ulrich and Bassler, 1923).
11. Drepanellina clarkei Ulrich and Bassler. Left valve of male, x9, Niagaran, Cumberland, Maryland. (Ulrich and Bassler, 1923).
12. Eukleodenella umbonata Ulrich and Bassler. Left valve, exfoliated posteriorly, xl2, late Niagaran, Flintstone, Maryland, (Ulrich and Bassler, 1923).
13. Frostiella groenvilliana Martinsson. Left valve of female, x23, Downtonian-Pribolian, Scania, Sweden (Martinsson, 1962).
14. Hemsiella maccoyana (Jones). Left valve of heteromorph, x30, Ludlovian, Kirkby Moor Flags, England (Shaw, 1971).
15. Kloedenella scapha Ulrich and Bassler. Left valve, xl5, Late Niagaran, Great Cacapon, Maryland (Ulrich and Bassler, 1923).
16. Kloedenia crassipunctata Swartz and Whitmore. Left valve of female, xl5, Cayugan, Austin's Glen, New York (Swartz and Whitmore, 1956).
17. Leperditia scalaris praecedens Ulrich and Bassler. Left valve,

Figure 6 (continued)

x5, Cayugan, Pinto, Maryland (Ulrich and Bassler, 1923).

18. Lophokloedenia manliensis (Weller). Left valve of male, x15, Cayugan, Tristate, New York (Swartz and Whitmore, 1956).

19. Mastigobolbina lata (Hall). Right valve of male, x6, Niagaran, New Hartford, New York (Ulrich and Bassler, 1923).

20. Mastigobolbina typus Ulrich and Bassler. Right valve of female, x5, Niagaran, Six Mile House, Maryland (Ulrich and Bassler, 1923).

21. Londinia kiesowi (Krause). Right valve of female, x23, Pridolian, erratic boulder, central Baltic area (Martinsson, 1962).

22. Macrypsilon salterianum (Jones). Right valve, x30, Ludlovian, Kirkby Moor Flags, England (Shaw, 1971).

23. Neobeyrichia (Neobeyrichia) nutans (Kiesow). Left valve of female, x23, Ludlovian, Hammarudden, Sweden (Martinsson, 1962).

24. Nodibeyrichia scissa Martinsson. Left valve of female, x23, Llandoverian, Hulte, Sweden (Martinsson, 1962).

25. Paraechmina spinosa (Hall). Left valve, x15, Niagaran, Rochester, New York (Ulrich and Bassler, 1923).

26. Platybolbina? bulbosa Henningsmoen. Left valve, x11, Llandoverian, Oslo, Norway (Henningsmoen, 1954).

27. Primitiella? cf. P. parallela Kummerow. Left valve, x11, Llandoverian, Gunneklev, Norway (Henningsmoen, 1954).

28. Primitiopsis cf. P. planifrons Jones. Left valve, x11, Ludlovian, Oslo region, Norway (Henningsmoen, 1954).

29. Signetopsis semicircularis (Krause). Right valve, x17, Ringerike, Norway, (Henningsmoen, 1954).

30. Sleia equestris Martinsson. Left valve of female, x34, Pridolian, Sles, Gotland (Martinsson, 1962).

31. Welleria obliqua Ulrich and Bassler. Right valve of female, x9, Cayugan, Hancock, Maryland (Ulrich and Bassler, 1923).

32. Welleriopsis jerseyensis (Weller). Holotype left valve of male, x14, Cayugan, Nearpass Quarries, New Jersey (Swartz and Whitmore 1956).

33. Zygobolba anticostiensis Ulrich and Bassler. Right valve of male, x6, Niagaran, Anticosti Island, Quebec (Ulrich and Bassler, 1923).

34. Zygobolba decora (Billings). Right valve of female, x23, Niagaran, Anticosti Island, Quebec (Martinsson, 1962).

35. Zygobolba erecta Ulrich and Bassler. Left valve of male, x6, Niagaran, Cherrytown, Pennsylvania (Ulrich and Bassler, 1923).

36. Zygobolbina emaciata Ulrich and Bassler. Right valve of female, x6, Niagaran, Cove Gap, Pennsylvania (Ulrich and Bassler, 1923).

37. Zygosella postica Ulrich and Bassler. Right valve of male, x6, Niagaran, Cove Gap, Pennsylvania (Ulrich and Bassler, 1923).

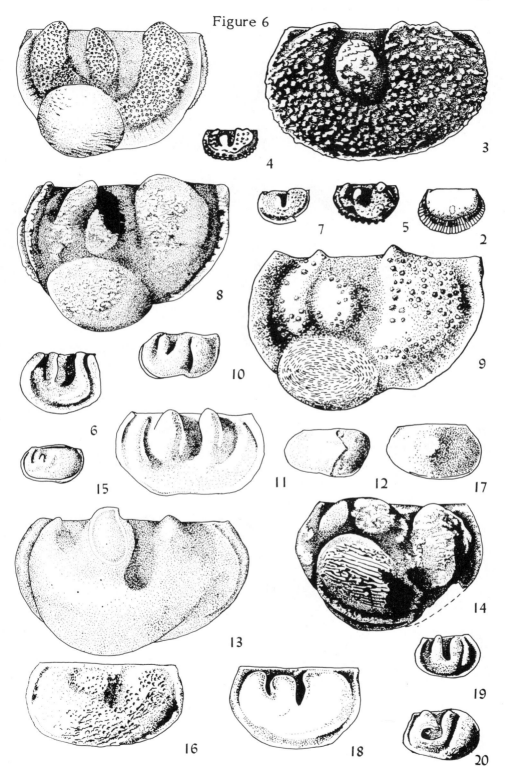

Figure 6

Figure 6 (continued)

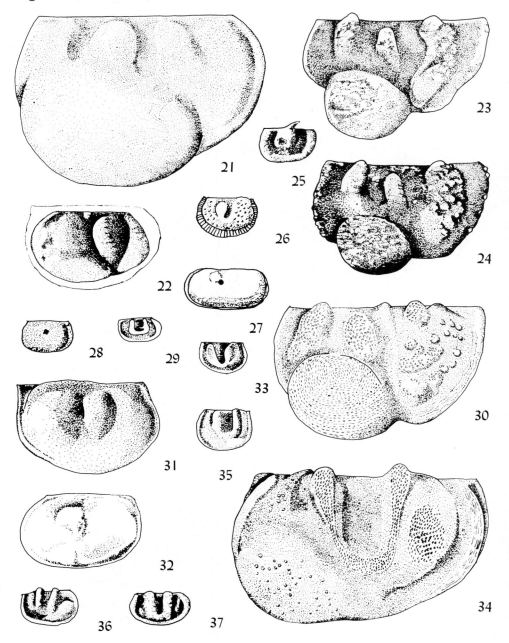

Summary and Conclusions

Early Paleozoic ostracodes, are plentifully represented in the marginal northern Atlantic region, but are poorly known or generally sparse in occu.rrence in the southern Atlantic, where early Paleozoic rocks are restricted in distribution. Although ostracodes occur in small numbers in most of the marine facies of the north Atlantic margins, they are most abundant and biostratigraphically useful in the time-absorbent, slowly-deposited rock units and are sparse in thick clastics or massive carbonate units.

Cambrian archaeocopids, although poorly known, appear to have limited biostratigraphic possibilities in the Atlantic region. Early Ordovician ostracodes are presently or potentially useful as assemblage zone faunas in the North Atlantic region, but differ in important respects on either side of the Atlantic probably owing to minimal transoceanic connections during that Epoch.

The middle Ordovician radiation of northwestern European ostracodes that took place from Europe to the present western North Atlantic and in some instances to the southwestern Atlantic was a striking event and probably represents increasing proximity of not only trans-protoatlantic plates, but connections with Gondwanaland as well. In late Ordovician time unfavorable epicontinental facies obscured ostracode faunal interrelationships but they apparently remained similar in the North Atlantic area.

The hindrance posed by unfavorable facies to knowledge of ostracode faunas continued through the early Silurian, but in the late Llandoverian and Wenlockian well developed ostracode zones appeared in both the eastern United States and part of western Europe. The Ludlovian faunas on the other hand are not well matched in the United States and western Europe, perhaps owing to arid-facies controls. The Pridolian-Downtonian Age marks a return to trans-protoatlantic similarities of ostracodes in the United States and western Europe.

References Cited

Baldis, Bruno and Rossi de Garcia, Elsa, 1972, Algunos ostracodes der Llandeiliense - Caradociense de la Republica Argentina. Revista Espanola de Micropaleontologia, v. 4, no. 1, p. 19-22.

Barrande, Joachim, 1872, Système Silurien du centre de la Bohème. Ire partie. Recherches paléontologiques. Supplement au vol. 1, Trilobites. Crustaces divers et Poissons. Plates in separate atlas. Prague and Paris, (fide Bassler and Kellett, 1934).

Bassler, R. S., and Kellett, Betty, 1934, Bibliographic Index of Paleozoic Ostracoda. Geol. Soc. Amer. Spec. Paper 1, 500 p.

Berdan, J. M., 1967, Baltic ostracodes from Maine. (in) Geol. Survey Research, 1966. U.S. Geol. Survey Prof. Paper 550-A, p. A111.

_____, 1970, American ostracode zonation: in Correlation of the North American Silurian rocks. Geol. Soc. Amer., Spec. Paper, No. 102, p. 39-40, 1970.

_____, 1971, Some ostracodes from the Schoharie formation (lower Devonian) of New York. in Paleozoic perspectives; a paleontological tribute to G. Arthur Cooper, Smithson. Contrib. Paleobiol., No. 3, p. 161-174, illus., 1971.

_____, 1972, Brachiopoda and Ostracoda of the Cobleskill Limestone (Upper Silurian) of Central New York. U.S. Geol. Surv., Prof. Paper, No. 730, 44 p., illus. (incl. sketch map), 1972.

Bonnema, J. H., 1909, Beitrag zur Kenntnis der Ostrakoden der Kuckersschen Schicht (C_2). Mitt. Min. Geol. Inst. Univ. Groningen, vol. 2, pt. 1, p. 1-84, pls. 1-8. Leipzig and Groningen.

Butts, C., 1940, Geology of the Appalachian Valley in Virginia. Geol. Survey Bull. 52, p. 333-335.

Canavari, M., 1899, Ostracodi Siluriani di Sardegna. Soc. Toscana Sci. Nat. Pisa., Pr. Verb., v. 11, 1899, art. 5, p. 150-153.

_____, 1900, Fauna dei calcari nerastri con Cardiola ed Orthoceras di Xea San Antonia in Sardegna I. Palaeont. ital., 1899, v. 5, p. 187-210, 2 tav. Pisa.

Clarke, J. M., 1899, The Paleozoic faunas of Para, Brazil, I. The Silurian fauna of the Rio Trombetas. II. The Devonian Mollusca of the state of Para. Mus. Nac. Rio de Janeiro, Arch. v. 10 (author's English edition), p. 1-100, pls. 1-8, 1900, Albany.

Cobbold, E. S., 1936, The Conchostracea of the Cambrian area of Comley, Shropshire with a note on a new variety of Atops reticulatus (Walcott). Geol. Soc. London, Q. J. no. 367, v. 92, pt. 3, p. 221-235, 2 pls.

_____, and Pocock, R. W., 1934, The Cambrian area of Rushton (Shropshire). Roy. Soc. London. Philos. Tr. B501, v. 223, p. 305-409.

Gurich, George, 1896, Das Palaeozoicum im Pölnischen Mittel gebirges. Russ. Kais. Min. Ges., St. Petersburg, Verh., ser. 2, v. 32, p. 374-392, pl. 10, fig. 15, pl. 14, pl. 15 (part).

Harper, J. C., 1940, The upper Valentian ostracode fauna of Shropshire. Ann. Mag. Nat. Hist., v. 11, p. 385-400.

_____, 1947, Tetradella complicata (Salter) and some Caradoc species of the genus. Geol. Mag., v. 84, p. 345-353, pl. 10.

Henningsmoen, Gunnar, 1953a, Classification of Paleozoic straight-hinged ostracods. Norsk Geol. Tidsskrift, v. 31, p. 185-288, 12 text figs.

_____, 1953b, The Middle Ordovician of the Oslo region, Norway, 4. Ostracoda. Norsk Geol. Tidsskrift, v. 32, p. 35-56, 5 pls., 1 text fig.

_____, 1954a, Lower Ordovician Ostracoda from the Oslo region, Norway. Norsk. Geol. Tidsskrift, v. 33, nos. 1, 2, p. 41-68, 2 pls.

_____, 1954b, Upper Ordovician ostracods from the Oslo region, Norway. Norsk Geol. Tidsskrift, v. 33, nos. 1, 2, p. 69-108, pls. 1-6.

_____, 1954c, Silurian ostracods from the Oslo region, Norway. 1. Beyrichiacea with a revision of the Beyrichiidae. Norsk. Geol. Tidsskrift, v. 34, no. 1, p. 15-71, 8 pls., 5 text figs.

Hessland, Ivar, 1949, Investigations of the Lower Ordovician of the Siljan District, Sweden. Bull. Geol. Inst. Uppsala, v. 33, 408 p., 26 pls.

Jaanusson, Valdar, 1957, Middle Ordovician ostracodes of central and southern Sweden. Bull. Geol. Inst. Uppsala University, v. 37, p. 176-442, 15 pls.

Jones, T. R., 1855, Notes on the Paleozoic bivalved Entomostraca, No. 1, Some species of Beyrichia of the upper Silurian limestones of Scandinavia. Ann. Mag. Nat. Hist. Ser. 2, v. 16, p. 31-92, pl. 5.

_____, 1856, Notes on the Paleozoic bivalved Entomostraca, No. 3, Some species of Leperditia. Ann. Mag. Nat. Hist., ser. 2, v. 17, p. 81-101, pls. 6, 7.

_____, 1884, Notes on the Palaeozoic bivalved Entomostraca, No. 17, Some North American Leperditiae and allied forms. Ann. Mag. Nat. Hist. Ser. 5, v. 14, p. 339-347.

_____, and Holl, H. B., 1869, Notes on Palaeozoic bivalved Entomostraca, No. 9, Some Silurian species. Ann. Mag. Nat. Hist., ser. 4, vol. 3, p. 211-227, pls. 14, 15.

Kay, G. M., 1934, Mohawkian Ostracoda: species common to Trenton faunules from the Hull and Decorah Formations. Jour. Paleontology, v. 8, p. 328-343, pls. 44-46.

_____, 1940, Ordovician Mohawkian Ostracoda: lower Trenton Decorah fauna. Jour. Paleontology, v. 14, p. 234-269, pls. 29-34.

Kraft, J. C., 1962, Morphologic and systematic relationships of some middle Ordovician Ostracoda. Geol. Soc. Amer. Mem. 86, 104 p., 19 pl., 15 text figs.

Krause, Aurel, 1891, Beitrag zur Kenntniss der Ostrakoden-Fauna in Silurian Diluvialgeschieben. Deutsch. Geol. Ges., Zeitschr., v. 43, p. 488-521, pls. 29-33.

Kummerow, E., 1924, Beitrage zur Kenntnis der Ostracoden und Phyllocariden aus nordischen Diluvialgeschieben. Preuss. Geol. Landes., Jahrb., 1923, vol. 44, p. 405-433, pls. 21,22.

Lundin, Robert F., 1971, Possible paleoecological significance of Silurian and early Devonian ostracode faunas from midcontinental and northeastern North America (with discussion): (in) Oertli, H. J., ed., Paleoecology of ostracodes, Cent. Rech. Pau, Bull., vol. 5, (suppl.), p. 853-868.

Martinsson, Anders, 1962, Ostracods of the Family Beyrichiidae from the Silurian of Gotland. Bull. Geol. Inst. Univ. Uppsala, v. 41, p. 1-369, 203 text-figs.

_____, 1970, Correlation with Europe: (in) Correlation of the North American Silurian rocks, Geol. Soc. Amer., Spec. Paper, No. 102, p. 41-44, 1970.

Poulsen, C., 1929, The Cambrian, Ozarkian, and Canadian faunas of northwest Greenland, Medd. om Gronland, v. 70 (Jubilaeumsekpeditionen Nord om Gronland 1920-23, no. 2), p. 308-316, pl. 21.

_____, 1934, The Silurian faunas of North Greenland. 1. The fauna of the Cape Schuchert Formation. Medd. om Gronland, v.72, (Jubilaeumsekpeditionen Nord om Gronland, 1920-23) p. 1-46, pl. 1-3.

Rozhdestvenskaja, A. A., 1971, Ostracods and paleogeographic conditions of their distribution in a late Devonian basin in the east of the Russian Platform. (in) Oertli, H. J., ed., Paleoecologie Ostracodes, Bull. Centre Rech. Pau. - SNPA, v. 5, suppl., p. 763-768.

Sarv, L., 1959, Ordovician ostracods in the Estonia SSR. Eesti Nsvteaduste Akademmia Geoloogia Instidundi Unrimused, v. 4, 206 p., 32 pls.

Shaw, R. W. L., 1971, Ostracoda from the Underbarrow, Kirkby Moor and Scout Hill Flags (Silurian) near Kendal, Westmorland. Palaeontology, vol. 14, part 4, p. 595-611, illus., 1971.

Smith, A. G., Briden, J. C., and Drewery, G. E., 1963, Phanerozoic world maps. Palaeontol. Assoc. Spec. Paper. 12, p. 1-42.

Swain, F. M., 1953, Ostracoda from the Camden Chert, western Tennessee. Jour. Paleontology, v. 27, p. 237-284, pls. 37-39, 21 text figs.

_____, 1957, Early Middle Ordovician Ostracoda of the eastern United States, Part I. Stratigraphic data and description of Leperditiidae, Aparchitidae, and Leperditellidae. Jour. Paleontology, v. 31, p. 528, 590, pls. 59-62.

_____, 1962, Early Middle Ordovician Ostracoda of the eastern United States, Part II.

_____, Cornell, J. and Hansen, D., 1961, Ostracoda of the Decorah Shale of Minnesota. Jour. Paleontology, v. 45, p. 519-545.

Swartz, F. M., 1933, Dimorphism and orientation in ostracods of the Family Kloedenellidae from the Silurian of Pennsylvania. Jour. Paleontology, v. 7, p. 231-260, pls. 28-30.

_____, 1939, Keyser Limestone and Helderberg Group, (in) B. Willard et al., The Devonian of Pennsylvania, Penn. Geol. Survey 4th Ser. Bull. G 19, p. 29-72.

_____, and Whitmore, F. C., Jr., 1956, Ostracoda of the Silurian Decker and Manlius limestones in New Jersey, and eastern New York. Jour. Paleontology, v. 30, p. 1029-1091, pls. 103-110, 3 text figs.

Tromelin, G. and Lebesconte, P., 1876, Essai d'un catalogue raisonne des fossiles Siluriens des Departments de Maine-et-Loire, de la Loire inférieure et du Morbihan. Assoc. Francaise l'Avanc. Sci., C.R., 4th Session, 1875, p. 623, Paris.

Tschigova (Chizhova), V. A., 1971, Geographical distribution of ostracods in the European sea basin at Famennian time: (in) Oertli, H. J., ed., Paleoecology of ostracodes, Cent. Rech. Pau, Bull., vol. 5, (suppl), p. 755-761.

Ulrich, E.O.,1890, 1891, New and little known American Paleozoic Ostracoda. Cincinnati Soc. Nat. Hist., Jour., v. 13, p. 104-137, (1890), p. 173-211, (1891), 8 pls.

_____, and Bassler, R. S., 1908, New American Paleozoic Ostracoda. Preliminary revision of the Beyrichiidae, with descriptions of new genera. U.S. Nat. Mus., Pr., v. 35, p. 277-340, figs. 1-61, pls. 37-44.

_____, 1923, Maryland Geological Survey, Silurian volume, 794 p., 27 figs., 67 pls. Paleozoic Ostracoda: their morphology, classification and occurrence, p. 271-391. Systematic paleontology of Silurian deposits (Ostracoda) p. 500-704, pls. 36-65, Baltimore.

_____, 1931, Cambrian bivalved Crustacea of the Order Conchostraca. U.S. Nat. Mus. Proc., v. 78, Part 4, p. 1-130, pl. 1-10.

Additional References Not Cited

Abushik, A. F., 1967, The Importance of ostracods in drawing the Siluro-Devonian boundary in the European part of the USSR. (in) Oswald, D. H., ed., Internat. Sympos. on Devonian System, Calgary, Alberta, Alta. Soc. Pet. Geol., v. 2, p. 875-884.

Brookins, Douglas G; Berdan, Jean M.; and Stewart, David B., 1973, Isotopic and Paleontologic Evidence for correlating three volcanic sequences in the Maine coastal volcanic belt. Geol. Soc. Am., Bull., v. 84, no. 5, p. 1619-1628.

Guber, A. L. and Jaanusson, Valdar, 1964, Ordovician ostracodes with posterior domiciliar dimorphism. Bull. Geol. Inst. Univ. Uppsala, v.42, no. 53, 43 p., 6 pls.

Gurevich, K. Ya., 1971, (Silurian-Devonian boundary and Stage Subdivision of the Lower Devonian in the Lvov Paleozoic Depression): Meyhdunar. Simp. Granitsa Silura Devona, Biostratigr. Silura, Nizhega Srednego Devona, Tr., no. 3, v. 1, p. 79-85.

Krandievsky, V. S., 1963, Fauna ostracod siluriiskikh vidkladiv Podillya, Akad. Nauk. Ukrain. RSR, Inst. Geol. Nauk, Kiev, 148 pp.

_____, 1971, (The Silurian-Devonian Boundary of Volyn-Podolia based on Ostracods and Graptolites): Mezhdunar. Simp. Granitsa Silura Devona, Biostratigr. Silura, Nizhego Srednego Devona, Tr., no. 3, v. 1, p. 108-113, (inc. Engl. sum.).

Le Fevre, J., 1967, (Succession of Ostracod and Conodont Associations in the Silurian Lower Devonian, and Eifelian from Several Sections in France and the Sahara). (in) Colloque sur le devonien inferieur et ses limites (Rennes, 16-24 Septembre 1964), Fr., Bur. Rech. Geol. Minieres, mem., no. 33, p. 373-389.

Martinsson, A., 1963, Kloedenia and related Ostracode genera in the Silurian and Devonian of the Baltic area and Britain. Geol. Inst. Univ. Uppsala, Bull. 42, p. 1-63.

_____, 1965a, The Siluro-Devonian ostracode genus Nodibeyrichia and faunally associated kloedeniinen. Geol Foren. Forhandl., v. 87, p. 109-138.

_____, 1965b, Remarks on the Silurian ostracode genus Craspedobolbina from the Baltic area and Britain. Geol. Foren. Stockholm, Forh. v. 87, pt. 3, no. 522, p. 314-325, illus., 1965.

_____, 1967, The succession and correlation of ostracode faunas in the Silurian of Gotland. Geol. Foren. Forhandl., v. 89, p. 350-386.

_____, 1968a, An Appalachian species of the Silurian Ostracode Genus Craspedobolbina. Geol. Foren. Stockholm, Forh., v. 90, pt. 2, no. 633, p. 302-208, illus., 1968.

_____, 1968c, The Appalachian species of the Silurian Ostracode Genus Craspedobolbina. Geol. Foren. Forhand, Stockholm, v. 90, p. 302-308, 6 text-figs.

Petersen, L. E.; and Lundin, R. F., 1974, Thlipsura martinssoni, a new ostracode species from the Silurian of England. J. Paleontol., v. 48, no. 2, p. 357-360.

Pranskevichus, A. A., 1970, Silurian ostracod assemblages of the south Baltic Region and their correlation value: Acad. Sci. USSR, Dokl., Earth Sci. Sect., vol. 192, p. 83-85.

_____, 1971, (New ostracodes of Wenlockian age in the southern Baltic region): (in) Paleontologiya i stratigrafiya Pribaltiki i Belorussii, sbornik III, p. 51-60 (incl. Engl. sum.), illus., Lit. Nauchno-Issled. Geologorazc. Inst. Vilnius., 1971.

_____, 1971, (Paleontological characteristics of lower Silurian in the southern Baltic region based on studies of Ostracods): (in) Paleontologiya i stratigrafiya Pribaltiki i Belorussii, shomik III, p. 61-70 Inst., Vilnius, 1971.

_____, 1972, (Silurian ostracods of the southern Baltic sea): Liet. Geol. Mokslinio Tyrimo Inst., Tr., no. 15, 180 p. (incl. Lith., Engl. sum.), illus. (Incl. sketch map), 1972.

_____, 1972, (New Silurian rishonids from the South Baltic area): (in) Novyye vidy drevnikh rasteniy i bespozvonochnykh SSSR, p. 272-275, illus., Akad. Nau, SSSR, Nauch. Soviet Probl., Moscow, 1972.

_____, 1972, (Silurian Ostracoda of the south Baltic region): Vilnius, Geol., Inst., Darb., no. 15, 280 p. (incl. Lith., Engl. sum.), illus. (incl. sketch map), 1972.

_____, 1972, Ostracods from the Upper Silurian of the Southern Baltic region: Geol. Foeren. Stockh., Foerh., v. 94, part 3, no. 550, p. 439-447, ilus., 1972.

_____, 1972, New ostracods of the Llandoverian of Lithuania. Geol. Foren. Stockh., Foerh., v. 94, part 3, no. 550, p. 435-438, illus., 1972.

_____, 1973, New late Silurian ostracodes from the South of the Baltic region: Paleontol. J., v. 7, no. 1, p. 32-42, illus., 1973.

_____, 1973, (New late Silurian ostracods from the South of the Baltic region): Paleontol. Zh., no. 1, p. 39-47, illus., 1973.

_____, 1973, New late Silurian ostracodes from the south of the Baltic region): Paleontological Journal, v. 7, no. 1, p. 32-41, pls. 3, 4.

Robardet, M.; Henry, J. L.; Nion, J.; et al., 1972, (Pont de Daer (Caradocian) Formation in the Domfront and Sees Synclines, Normandy): Soc. Geol. Nord., Ann., v. 92, no. 3, p. 117-137 (incl. Engl. sum), illus, 1972.

Sarv, L., 1973, (Zonation of the Silurian in the section of the Kalvariya Well based on ostracods): Eesti NSVstead. Akad., Toim., Keemia Geol., v. 22, no. 2, p. 88-91, illus., 1973.

Schallreuter, R., 1967, (New ostracods from Ordovician boulders). Geologie (Berlin), v. 16, no. 5, p. 615-631 (incl. Engl. Russ. sum.) illus., 1967.

46

_____, 1968, (Ordovician Podocopida; Beecherellidae). Neues Jahrb. Geol. Palaontol., Abh., v. 131, no. 1, p. 82-96 (incl. Engl. sum.), illus., 1968.

_____, 1969, Neue Ostracoden aus ordovizischen Geschieben, II. Sonderdruck aus Geol. Jg. 18, H.2, Akad-Verlag Berlin, p. 203-215, 3 pls.

_____, 1971, Ostrakoden aus Ojlemyrgeschrieben (Ordoviz) N. jb. Geol.Paläont, Mh, Jg. 1971, H. 7, p. 423-431, 1 pl.

_____, 1971 (Asymmetric Ordovician Ostracods): Neues Jahrb. Geol. Palaeontol., Monatsh, no. 4, p. 249-260 (incl. Engl. sum.), illus., 1971.

_____,1971, (Ostracods from the Ojlemyr flint, Ordovician): Neues Jahrb. Geol. Palaeontol., Monatsh, no. 4, p. 423-431 (incl. Engl. sum.), illus., 1971.

_____, 1972, (Drepanellacea (Ostracoda, Beyrichicopida) from the middle Ordovician Backstein Limestone boulders: 4, Laterophores hystrix n. sp., Pedomphalella germanica n. sp. and Easchmidtella fragosa): Dtsch. Ges. Geol. Wiss., Ber., Reihe A. Geol. Palaontol., v. 17, no. 1, p. 139-145, illus., 1972.

Shaw, R. W. L., 1971, The faunal stratigraphy of the Kirky Moor flags of the type area near Kendal, Westmorland: Geol. J., v. 7, part 2, p. 359-380, illus. (incl. geol. map 1:63, 360), 1971.

Sidaravichene, N.V., 1971, (New Ostracods from the middle and upper Ordovician of Lithuania), (in) Paleontologiya i stratigrafiya Pribaltiki i Belorusii, sbornik III, p. 23-36 (incl. Engl. sum.), illus. (incl. sketch map), Lit. Naerchno-Issled. Geologorazv. Inst., Vilnius, 1971.

Whitfield, R. P., 1890, Observations on some imperfectly known fossils from the calciferous sandrock of Lake Champlain and descriptions of several new forms. Am. Mus. Nat Hist. Bull. 2, 56-80., pl. 13, figs. 1-6.

Zbikowska, B., 1973, (Upper Silurian Ostracods from the Leba Elevation, northern Poland): (in) Prace zwiazane y problematylca struktur wg lebnych Polski, Acta Geol. Pol., v. 23, no. 4, p. 607-644, (incl. Engl. sum.), illus., 1973.

Zenger, D. H., 1971, Uppermost Clinton (Middle Silurian) Stratigraphy and Petrology, East-Central New York: N.Y. State Mus. Sci. Serv., Bull., no. 417, 58 p. illus. (incl. sketch map), 1971.

Discussion

Dr. S. Bergström: I was most interested to learn about the Baltic affinities of the ostracodes of the San Juan Formation of Argentina, and reference to it as being of Llanvirnian and Llandeilan age. The conodont fauna of this unit also shows Baltic affinities but, as noted by Serpagli (1974), it suggests a late Arenigian rather than younger age of the formation. Can the San Juan ostracodes be used to date the formation, and if so, do they show correlation with Llanvirnian and Llandeilan strata in northwestern Europe rather than with Arenigian units?

Swain: The few species of ostracodes from the San Juan Formation are not definitive in age, but are more suggestive of middle than of early Ordovician. Reconstructions of continental position in the Ordovician by Smith et al. (1963) show that shallow marine migrations between the Baltic area and western South America could reasonably have occurred.

Dr. M. C. Keen: In your abstract you mentioned brackish water ostracods in the Ludlovian. When did brackish water ostracods first appear? How do you recognise them and are there any fresh-water ones?

Swain: Jean Berdan and I have discussed this several times. Certain elongate, sulcate (eukloedenellid ostracodes of the Silurian may represent either brackish or high-salinity forms; they occur in fossiliferous red beds (Bloomsburg) as described by Hoskins from central Pennsylvania.

Dr. J. Berdan: I would agree with that, and also I think that the difference in this context, between the Appalachian fauna, dominated by kloedenellids, and that to the west, such as the Henryhouse Formation where kloedenellids are not abundant, indicates that there was something peculiar about the environment in the Appalachian area at that time.

Swain: One of the problems is that this was a time during which evaporite conditions occurred in the Appalachian region, so it is difficult to know whether these are high or low salinity forms. As far as freshwater forms are concerned, we have indications of freshwater species in the Carboniferous and perhaps in the Devonian.

Berdan: I think they are earlier and appear in the
   middle or late Silurian.   Murray Copeland has
   more information.
Dr. J. E. Conkin: Does the Pridolian represent a
   pulsation (transgression) of the seas onto the
   continents?   If so what is the bearing of this
   on the Silurian-Devonian boundary?
Swain: The result of a late Silurian transgression in the Appalachians, as well as a return to
   a more favorable benthic environment seems to
   have been the production of a transitional Siluro-Devonian population of ostracodes and other
   invertebrates in such units as the Keyser Limestone.

PALEOZOIC SMALLER FORAMINIFERA
OF THE NORTH AMERICAN ATLANTIC BORDERLANDS

by

James E. Conkin, University of Louisville, Lou. Ky. 40208

and

Barbara M. Conkin, Jefferson Community College, Lou. Ky. 40202

Abstract

Species of Paleozoic smaller Foraminifera, combined with physical evidence of diastrophism (paracontinuity), may be used in precise placement of time-stratigraphic boundaries in the eastern United States as with the Devonian-Mississippian and Kinderhookian-Osagean boundaries. Evolutionary lineages are recognized in the Devonian and Mississippian and certain stratigraphic intervals within the Lower and Middle Paleozoic are delimited by foraminiferal zones; nevertheless, even though the stratigraphic value of Paleozoic smaller Foraminifera has been demonstrated, their full potential remains to be realized.

Agglutinated Foraminifera frequently occur in shale and limestone containing silt or fine sand; however, their distribution varies in diverse lithologies within single genetic units. Sparry bioclastic calcarenites rarely contain agglutinated Foraminifera, but often bear calcareous forms.

Earliest textulariinid Foraminifera, consisting of simple, free or attached, tubular, globular, spherical, or hemispherical chambers, arose by mid-Ordovician. Tubular to hemitubular attached forms appeared in late Ordovician. Aperturred planoconvex, tubular, irregularly coiled and planispirally coiled forms and globular forms with various projections, arose in early Silurian. Multichambered forms with enlarging chambers began in late Devonian. Highly organized, attached hemitubular, and well-organized multichambered forms, appeared in early Mississippian and evolved into more complex multichambered forms (including attached forms) in Pennsylvanian and Permian. Microgranular calcareous (fusulininid) foraminiferans arose in Middle Devonian and increased in number and complexity in Mississippian, Pennsylvanian, and Permian. Calcareous porcelaneous (miliolinid) tests appeared in early Pennsylvanian and calcareous hyaline (rotaliinid) tests evolved in early Permian.

Résumé

Les espèces de petits foraminifères paléozoïques, unies à la preuve physique du diastrophism (paracontinuité), peuvent être utilisées dans l'emplacement précis des frontières de temps-stratigraphiques dans la région est des Etats-Unis comme avec les frontières dévonien-mississippiens et kinderhookien-osagéens. Les lignages évolutionnistes se sont reconnus dans le dévonien et le mississippien. Certains intervalles stratigraphiques dans le paléozoïque inférieur et le mi-paléozoïque sont delimités par des zones foraminiférales; néanmoins, malgré la valeur stratigraphique démontrée des petits foraminifères paléozoïques, leur potentialité reste à être réalisée.

Des foraminifères agglutinés souvent se trouvent dans des schistes argileux et des calcaires contenant des sables fins ou des vases. Cependant, leur distribution se varie dans de diverses lithologies à l'intérieur de simples unités génétiques. Il est rare que les sparry calcarenites bioclastiques contiennent des foraminifères agglutinés mais souvent donnent des formes calcaires.

Par le mi-ordovicien, les plus anciens foraminifères textulariinid apparaissaient. Ils contenaient des chambres, soit simples, soit libres ou liées, soit tubulaires, globulaires, sphériques ou hémisphériques. Des formes liés tubulaires à l'hémi-tubulaires apparaissaient pendant l'ordovicien récent. Des formes planoconvexes à apertures, ainsi que tubulaires, lovés irrégulièrement et lovés planispiralement et des formes globulaires à projections diverses apparaissaint pendant le début du silurien. Des formes multichambrés avec des chambres aggrandissantes commençaient pendant le dévonien récent. Des formes multichambrés liés, hémitubulaires, assez complexes et organisés apparaissaient pendant le début du mississippien et se sont évolués en des formes multichambrés plus complexes (y compris les formes liés) pendant le pennsylvanien et le permien. Des foraminifères microgranulaires calcaires (fusulininid) apparaissaient dans le midevonien et se sont augmentés en nombre et complexité pendant le mississippien, le pennsylvanien et le permien. Des tests calcaires porcelanaires (miliolinid) apparaissaient pendant le début du pennsylvanien. Des tests calcaires hyalines (rotaliinid) se sont évolués pendant le début du permien.

Introduction

Agglutinated and microgranular, porcelaneous, and hyaline calcareous (excluding fusulinid, endothyrid, and tournayellid) foraminiferans may be considered "smaller" Foraminifera; however, even these may require sectioning for identification.

Paleozoic smaller Foraminifera from the North American Atlantic borderlands (Texas to New York and Ontario), have been studied for nearly 50 years, but the quality of the work is uneven.

Reviews of North American late Devonian-early Mississippian agglutinated Foraminifera (Conkin and Conkin, 1967 and 1970) and smaller Foraminifera of the pre-Pennsylvanian (Conkin and Conkin, 1973a) stressed their stratigraphic significance.

Agglutinated Foraminifera occur abundantly in shale (Conkin, 1961, p. 234) which contains silt and/or fine-grained sand, as well as in limestone (even sublithographic) if fine-grained detrital material is present, but they may be unevenly distributed in diverse lithologies within a single genetic unit; limestone lenses and impure ironstone concretions may contain many specimens, while surrounding shales, silty shales or shaly siltstones bear essentially none. Sparry bioclastic calcarenites rarely contain agglutinated forms, but calcareous foraminiferans are frequently found therein. The dominance of agglutinated (textulariinid) Foraminifera in early Paleozoic resulted from their early evolution and pioneer occupation of shallow marine benthonic niches; with evolution in the Devonian of calcareous Foraminifera, competition ensued and increased significantly from Carboniferous onward, with calcareous groups progressively replacing agglutinated forms in "normal epicontinental" marine environments.

Paleozoic Genera Of Smaller Foraminifera

In preparing generic keys and ascertaining ranges of genera (Text-figs. 1-3), we have exercised our best judgment in determining the validity of records of stratigraphic occurrence and taxonomic assignment. The keys serve to clarify generic identities and relationships. We consider several genera recorded from the Paleozoic to be junior synonyms (in parentheses) of the following: Calcitornella (Apterrinella, Calcivertella and Plummerinella), Hemisphaerammina (Fairliella, Metamorphina, and "Webbinella"), Ammovertella (Ammodiscella), Oryctoderma (Crithionina), Ordovicina (Croneisella, Gastroammina, and Shidelerella), Nodosinella (Monogenerina), Palaeotextularia (Paratextularia), Climacammina (Cribrostomum) Saccammina (Proteonina, Raibosammina and Thekammina), Lunucammina (Geinitzina), and Tholosina (Amphicervicis). Other genera are so ill understood (Agathamminoides, Kerionammina, Tuberitina, Stacheia, and Stacheoides) that no useful statement can be made. Some genera have been recorded erroneously from the Paleozoic (Astrammina, Astrorhiza, Bullapora, and

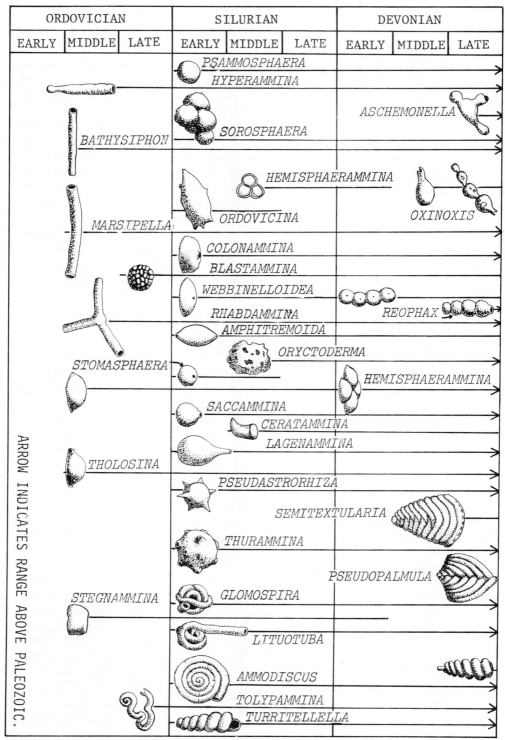

TEXT-FIGURE 1. GENERIC RANGES OF PALEOZOIC SMALLER

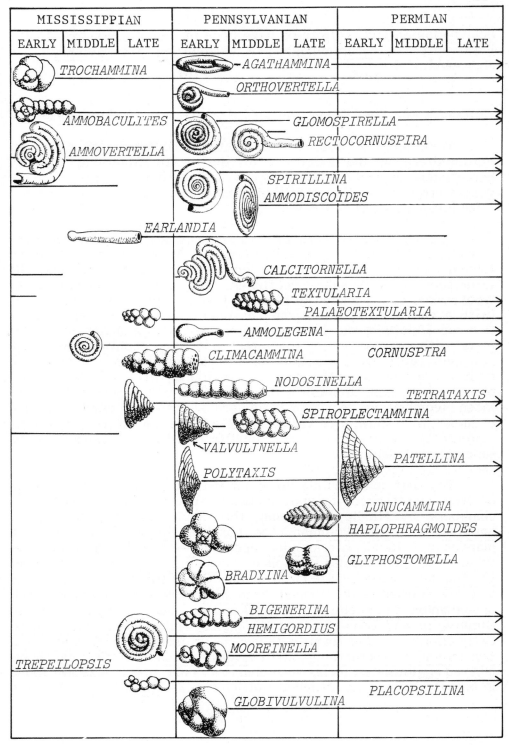

MISSISSIPPIAN			PENNSYLVANIAN			PERMIAN		
EARLY	MIDDLE	LATE	EARLY	MIDDLE	LATE	EARLY	MIDDLE	LATE

FORAMINIFERA IN THE NORTH AMERICAN ATLANTIC BORDERLANDS.

Schizammina); and still others have been incorrectly assigned to the Foraminifera (Thuramminoides, a radiolarian, and Weikkoella, a charophyte).

Ranges And First Occurrences Of Some Genera Of Foraminifera

Text-figure 2 presents the stratigraphic ranges of genera of Foraminifera in the Paleozoic of the North American Atlantic borderlands arranged systematically. Text-figure 3 presents the genera arranged according to their first occurrences (see Text-figure 1 for illustrations of genera).

The agglutinated Foraminifera are represented by the suborder Textulariina, with two superfamilies, the Ammodiscacea (globular, tubular, or irregular tests with one or more apertures) and the Lituolacea (multilocular, rectilinear, enrolled or uncoiled, or trochospiral coil modified to biserial, tests). The suborder Fusulinina (with a nonlamellar, microgranular, calcite wall) is represented by two superfamilies, the Parathuramminacea (with globular or tubular tests with one or more apertures) and the Endothyracea (with multiple, rectilinear, enrolled, or biserial chambers and single to multiple apertures). The suborder Miliolina (with an imperforate, porcelaneous, calcite wall) is represented by the superfamily Miliolacea (comprising forms with a proloculus and tubular second chamber, planispirally or variously coiled) and the suborder Rotaliina (calcareous, hyaline, perforate, lamellar tests) is represented by the superfamily Spirillinacea.

The first textulariinid Foraminifera had simple tests consisting of free or attached, tubular, globular, spherical, or hemispherical chambers; by Middle Ordovician, the tubular Bathysiphon, Hyperammina, and Rhabdammina, the box-like Stegnammina, and the planoconvex Hemisphaerammina and Tholosina had appeared. The tubular to hemitubular attached Tolypammina and the globular Blastammina arose in late Ordovocian. Other simple forms appeared in early Silurian: the globular Psammosphaera and Sorosphaera and planoconvex Webbinelloidea, as well as the tubular, irregularly coiled Glomospira and Lituotuba and the planispirally coiled Ammodiscus. Early Siluian globular forms with various projections are Saccammina, Lagenammina, Thurammina, and Pseudastrorhiza. The first multichambered agglutinated forms with chambers enlarging in size are found in late Devonian: the attached multichambered Oxinoxis and the free

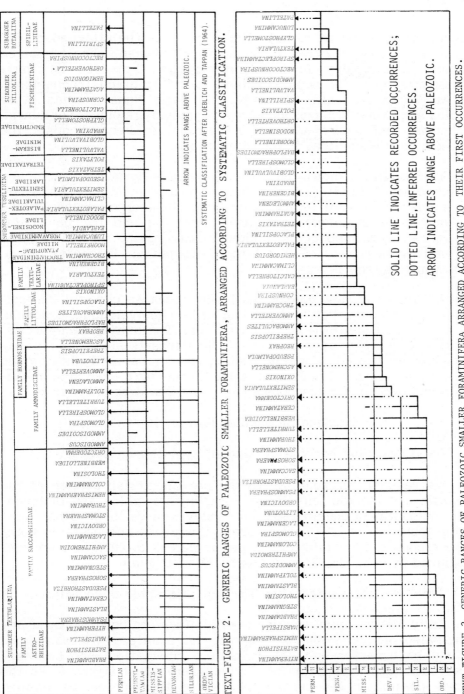

TEXT—FIGURE 2. GENERIC RANGES OF PALEOZOIC SMALLER FORAMINIFERA, ARRANGED ACCORDING TO SYSTEMATIC CLASSIFICATION.

SYSTEMATIC CLASSIFICATION AFTER LOEBLICH AND TAPPAN (1964).

ARROW INDICATES RANGE ABOVE PALEOZOIC.

SOLID LINE INDICATES RECORDED OCCURRENCES;
DOTTED LINE, INFERRED OCCURRENCES.
ARROW INDICATES RANGE ABOVE PALEOZOIC.

TEXT—FIGURE 3. GENERIC RANGES OF PALEOZOIC SMALLER FORAMINIFERA ARRANGED ACCORDING TO THEIR FIRST OCCURRENCES.

Reophax (first abundant in early Mississippian). The attached, spiral, tubular Trepeilopsis is first found in late Devonian. The organized attached hemitubular form, Ammovertella, appeared in early Mississippian when the well-organized multichambered Trochammina and Ammobaculites also appeared. In early Pennsylvanian other complex agglutinated forms appeared: Haplophragmoides, Textularia, Bigenerina, and Mooreinella. Finally, in late Pennsylvanian, the attached, multichambered Placopsilina arose.

The earliest known calcareous foraminiferan, the complex, multichambered Semitextularia, first appeared in Middle Devonian. Two simpler calcareous forms are the two-chambered (proloculus and tubular second chamber) rectilinear Earlandia and the planispiral Cornuspira which appear in middle Mississippian. More complex calcareous genera are found in late Mississippian: Palaeotextularia, Climacammina, and Tetrataxis.

A number of new calcareous forms appeared in Early Pennsylvanian: Nodosinella, Polytaxis, Valvulinella, Globivalvulina, Bradyina, Agathammina, Calcitornella, and Spirillina. Textularia, Hemigordius, and Rectocornuspira appeared in middle Pennsylvanian, and, in late Pennsylvanian two additional multichambered genera, Glyphostomella and Lunucammina. Finally, in early Permian, a complex, multichambered rotaliinid calcareous genus, Patellina, was introduced.

Zonation By Smaller Foraminifera

And Devonian Early Mississippian Evolutionary Lineages

We restrict our zonation to the pre-Chesterian portion of the Paleozoic and include only those Foraminifera which are taxonomically valid, widely distributed, and occupy well-defined, rather short stratigraphic intervals. Text-figure 4 illustrates and presents notes on the recognized foraminiferal zones as well as evolutionary lineages recognized by us in the Devonian-early Mississippian.

Smaller Foraminifera And Time-Stratigraphic Boundaries

Early Mississippian smaller Foraminifera are best known of any in the Paleozoic and a number of species have quite restricted ranges, and can be used for age determination, correlation, and boundary placement. Using agglutinated Foraminifera (Text-figure 5) as the biologic parameter and paracontinuities as the physical parameter, we have succeeded in precise placement of the long-sought Devonian-Mississippian boundary, as well as the Kinderhookian-Osagean boundary, in east-central United States (Conkin and Conkin, 1973b and 1975).

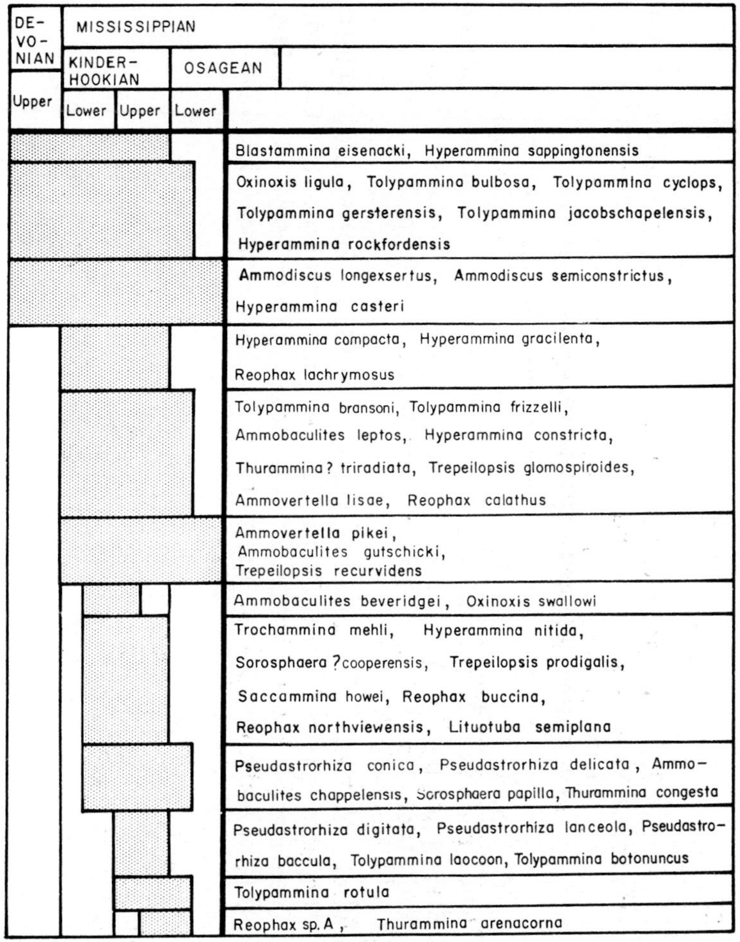
DE-VO-NIAN
MISSISSIPPIAN
KINDER-HOOKIAN
OSAGEAN
Upper
Lower
Upper
Lower
Blastammina eisenacki, Hyperammina sappingtonensis
Oxinoxis ligula, Tolypammina bulbosa, Tolypammina cyclops,
Tolypammina gersterensis, Tolypammina jacobschapelensis,
Hyperammina rockfordensis
Ammodiscus longexsertus, Ammodiscus semiconstrictus,
Hyperammina casteri
Hyperammina compacta, Hyperammina gracilenta,
Reophax lachrymosus
Tolypammina bransoni, Tolypammina frizzelli,
Ammobaculites leptos, Hyperammina constricta,
Thurammina? triradiata, Trepeilopsis glomospiroides,
Ammovertella lisae, Reophax calathus
Ammovertella pikei,
Ammobaculites gutschicki,
Trepeilopsis recurvidens
Ammobaculites beveridgei, Oxinoxis swallowi
Trochammina mehli, Hyperammina nitida,
Sorosphaera ?cooperensis, Trepeilopsis prodigalis,
Saccammina howei, Reophax buccina,
Reophax northviewensis, Lituotuba semiplana
Pseudastrorhiza conica, Pseudastrorhiza delicata, Ammo-
baculites chappelensis, Sorosphaera papilla, Thurammina congesta
Pseudastrorhiza digitata, Pseudastrorhiza lanceola, Pseudastro-
rhiza baccula, Tolypammina laocoon, Tolypammina botonuncus
Tolypammina rotula
Reophax sp. A, Thurammina arenacorna

TEXT-FIGURE 5. RANGES OF KINDERHOOKIAN AGGLUTINATED FORAMINIFERA IN THE UPPER DEVONIAN, KINDERHOOKIAN, AND LOWER OSAGEAN OF NORTH AMERICA.

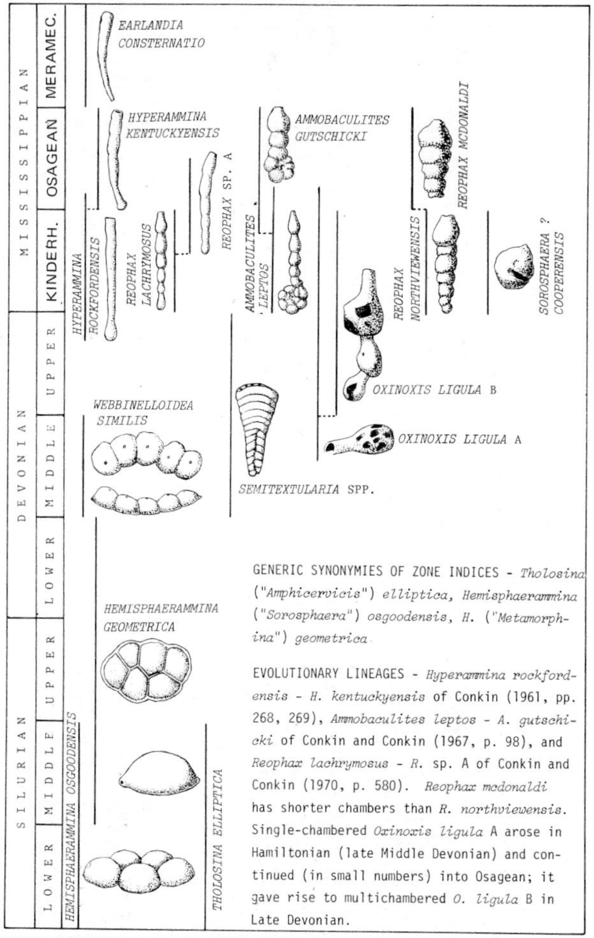
MISSISSIPPIAN
MERAMEC.
OSAGEAN
KINDERH.
DEVONIAN
UPPER
MIDDLE
LOWER
SILURIAN
UPPER
MIDDLE
LOWER
EARLANDIA CONSTERNATIO
HYPERAMMINA KENTUCKYENSIS
HYPERAMMINA ROCKFORDENSIS
REOPHAX LACHRYMOSUS
REOPHAX SP. A
AMMOBACULITES LEPTOS
AMMOBACULITES GUTSCHICKI
REOPHAX NORTHVIEWENSIS
REOPHAX MCDONALDI
SOROSPHAERA ? COOPERENSIS
OXINOXIS LIGULA B
OXINOXIS LIGULA A
SEMITEXTULARIA SPP.
WEBBINELLOIDEA SIMILIS
HEMISPHAERAMMINA GEOMETRICA
HEMISPHAERAMMINA OSGOODENSIS
THOLOSINA ELLIPTICA
GENERIC SYNONYMIES OF ZONE INDICES - *Tholosina* ("*Amphicervicis*") *elliptica*, *Hemisphaerammina* ("*Sorosphaera*") *osgoodensis*, *H.* ("*Metamorphina*") *geometrica*
EVOLUTIONARY LINEAGES - *Hyperammina rockfordensis* - *H. kentuckyensis* of Conkin (1961, pp. 268, 269), *Ammobaculites leptos* - *A. gutschicki* of Conkin and Conkin (1967, p. 98), and *Reophax lachrymosus* - *R.* sp. A of Conkin and Conkin (1970, p. 580). *Reophax mcdonaldi* has shorter chambers than *R. northviewensis*. Single-chambered *Oxinoxis ligula* A arose in Hamiltonian (late Middle Devonian) and continued (in small numbers) into Osagean; it gave rise to multichambered *O. ligula* B in Late Devonian.

TEXT-FIGURE 4. FORAMINIFERAL ZONATION IN THE LOWER AND MIDDLE PALEOZOIC AND EVOLUTIONARY LINEAGES IN THE DEVONIAN-LOWER MISSISSIPPIAN.

References Cited

Conkin, J. E., 1961. Mississippian smaller Foraminifera of Kentucky, southern Indiana, northern Tennessee, and south-central Ohio: Bull. Amer. Paleont., v. 43, no. 196, p. 129-368.

_____, and Conkin, B. M., 1967. Arenaceous foraminifera as a key to Upper Devonian and Lower Mississippian relationships in the type Mississippian area: Essays in Paleontology and Stratigraphy. R. C. Moore Commemorative Volume, Spec. Pap. Univ. Kansas, Dept. of Geol., no. 2, p. 85-101.

_____, and _____, 1970. North American Kinderhookian (Lower Mississippian) Arenaceous Foraminifera: Compte Rendu 6e Congress Intern. Strat. Geol. Carbonif., Sheffield, 1967, v. 2, p. 575-584.

_____, and _____, 1973a. Pre-Pennsylvanian Foraminifera of North America: XXII Int'l. Geol. Congress, India, 1964, p. 319-335.

_____, 1973b. The paracontinuity and the determination of the Devonian-Mississippian boundary in the type Lower Mississippian area of North America: Univ. Louisville Studies in Paleont. and Strat. No. 1, Univ. Lou. Reprod. Serv., 36 pp.

_____, and _____, 1975. The Dovonian-Mississippian and Kinderhookian-Osagean boundaries in the east-central United States are paracontinuities: Univ. of Louisville Studies in Paleont. and Strat. No. 4, Univ. of Lou. Reprod. Serv., 54 pp.

Loeblich, A. R., Jr., and Tappan, H., 1964. Sarcodina, chiefly "Thecamoebians" and Foraminifera. In: Moore, R. C., Ed., Treatise on Invertebrate Paleontology. New York: Geol. Soc. Amer., and Univ. Kans. Press, pt. C, v. 1, p. i-xxxi, C1-C510a.

Discussion

Dr. R. A. Olsson: Have you done any work on the facies relationships of these fauna?

Dr. J. E. Conkin: Well, we didn't have space in the paper to
go into facies. Agglutinate foraminiferans can, and do, occur in great abundance in some Paleozoic limestones, such as
the Upper Devonian Louisiana Limestone of northeastern Missouri and western Illinois (sublithographic limestone bearing much fine-grained silt) which carries a magnificent agglutinate foraminiferal fauna. Here's a paper [Conkin and
Conkin, 1964, Devonian Foraminifera, Part 1 - The Louisiana
of Missouri and Illinois: Bull. Amer. Pal., v. 47, no. 213,
pp. 53-105, pls. 12-15, text-figs. 1-5, charts 1-3] on it.
I don't say there's no ecological control for there is always some. For example, in sparry bioclastic limestone there
are few agglutinate foraminiferans for often not enough detrital grains are available for building their tests; the
Lower Mississippian (Osagean) Burlington Limestone of Missouri and Illinois bears no agglutinate foraminiferans for this
reason. In the essential equivalent of the Burlington Limestone, the New Providence Shale of northwestern Kentucky and
southern Indiana, agglutinate foraminiferans are found. In
some shales and siltstones units in the Osagean Brodhead Formation of Kentucky and Indiana (in which very few and/or
poorly preserved arenaceous foraminiferans are found), there
are occasional intercalated limestone or ironstone nodules
and lenses which produce, upon acidization, well preserved
agglutinate foraminiferans. Some agglutinate species have
quite restricted stratigraphic ranges such as Hyperammina
kentuckyensis which is a definitive Osagean marker occurring
in both limestones and shales. In general, we may say that
in Mississippian limestone facies, the tournayellids and en-
dothyrids are dominant while in the shale facies the agglutinate foraminiferans dominate.

LATE PALEOZOIC OSTRACODES OF WESTERN EUROPE AND NORTH AMERICA: A REVIEW

Luis C. Sanchez De Posada
Universidad de Oviedo, Oviedo, Espana

ABSTRACT

It is perhaps due to the fact that ostracodes have proved to be successful in so many different environments adapting themselves to every variation in temperature, salinity, water energy, substratum, depth, that they often seem less useful for biostratigraphical applications, at least as far as long-distance correlations are concerned. On the one hand we observe several long-ranging species, which persist in diachronous facies and on the other hand there exist many groups which are able to survive small facies changes with small changes in their morphology.

The biostratigraphic use of Paleozoic ostracodes is further hampered by the fact that up to now we have only the rather isolated results of work on ostracodes from different areas and environments at our disposal. This situation may improve within the forthcoming years as has been shown by the discovery of so-called silicified Devonian faunas and Devonian-Carboniferous entomozoacean faunas. However, at the present time these are only known from northwestern Europe, Spain, Portugal, and North America. At the same time attempts are being made to come to long-distance correlations of Upper Devonian and Dinantian assemblages between the USSR and western Europe by Russian ostracodologists.

In Europe, the attention of Devonian to Permian ostracode workers has up to now mainly been concentrated on descriptive palaeontology and in the last decade on palaeoecology.

RÉSUMÉ

C'est un fait que les ostracodes ont réussi à s'adapter à beaucoup d'environnements différents et à beaucoup de variations de la température, de la salinité, de l'énergie d'eau, de la soustrate et de la profondeur. C'est peut-être à cause de ceci que les ostracodes semblent être moins utiles quant aux applications biostratigraphiques, du moins en ce qui touche les corrélations de longue-portée. D'un côté, on remarque plusieurs espèces à grande distribution, qui persistent dans des facies diachroneux. De l'autre côté, il y a beaucoup de groupes qui arrivent à surmonter les changements de petits facies avec des changements minimes dans leur morphologie.

De plus, l'utilization biostratigraphique des ostracodes paléozoïques est entravée par le fait que jusqu'ici, on n'avait eu que des résultats isolés sur des ostracodes provenant des régions et des environnements différents. Cette situation pourrait s'améliorer dans l'avenir avec le découvert des soi-disant faunes dévoniennes et des faunes entomozoacéennes du Dévonien-Carbonifère. Cependent, pour le moment, elles ne sont connues que pour la partie nordouest de l'Europe, l'Espagne, le Portugal et l'Amérique du Nord. En même temps, des ostracodologistes russes font des efforts pour arriver aux corrélations de longue-portée entre l'U.R.S.S. et l'Europe de l'Ouest pour les assemblages du Dévonien Supérieur et du Dinantien.

En Europe, jusqu'ici l'intérêt des recherches sur les ostracodes du Dévonien jusqu'au Permien a été placé sur la paléontologie descriptive et dans les dix dernières années, sur la paléoécologie.

INTRODUCTION

There exists an extensive literature on late Paleozoic ostracodes from both North America and western Europe. Amongst the many papers we may quote the work by JONES AND KIRBY on European ostracodes in the past century, and that of ULRICH and BASSLER at the end of the 19th and beginning of the 20th century, who investigated the North American faunas. Their fundamental studies have been continued by a large number of ostracode workers since the First World War.

In Europe, late Paleozoic ostracodologists have been especially active in Germany and surrounding areas. It was through their investigations, that a first attempt was made to use entomozoan and silicified ostracode faunas for biostratigraphical purposes. The past ten years have moreover shown a growing interest in the palaeoecology of these ostracodes.

In North America, work on Devonian to Permian ostracodes has been mainly concentrated on descriptive palaeontology rather than on biostratigraphy, although in the last decade there is a tendency to review their value for infra- and interbasinal correlations.

In comparison with post-Paleozoic ostracodes the study of Paleozoic ostracodes has been less intensive on the whole as shown by the papers read during the past five ostracode symposia at Naples, Hull, Pau, Newark, and Hamburg. It should be kept in mind of course, that 12 Paleozoic ostracode workers met separately after the Hamburg symposium at Gotland, where eight non-published progress reports have been discussed. It is also curious to notice how several younger disciplines within the field of Paleozoic micropalaeontology, such as conodont, acritarchs, chitinozoans, palynolomorphs and foraminifers have yielded world-wide applicable correlations, whereas the biostratigraphic value of ostracodes is still in discussion.

In this context it appears to be impossible to present at this moment uniform biostratigraphic ostracode range charts which can be used on both sides of the Atlantic.

LOWER DEVONIAN

Canada and Alaska

Lower Devonian to Mississippian ostracodes in Canada were reviewed by COPELAND (1972).

Gedinnian assemblages, containing several thlipsurid and beyrichiid genera and Poloniella species have been described from Quebec (BERDAN & COPELAND in BURK, 1965) and New Brunswick (COPELAND, 1962). Upper Emsian ostracodes, including many beyrichiids and the first diversified hollinomorphs have been discovered in Alaska and Yukon Territory (BERDAN & COPELAND, 1973). The latter fauna also contains several genera which are known from the Lower Devonian of Europe, such as Acanthoscapha,

Tricornina, Beecherelia, Berounella, Praepilatina and Poloniella.

U.S.A.

Gedinnian faunas have been described by among others
LUNDIN (1968) from Oklahoma and Tennessee (cf. also WILSON,
1935) with many thlipsurid species and Poloniella. Emsian
thlipsurid faunas have been described from Tennessee (BASSLER,
1941) and Siegenian thlipsurid faunas have been described from
West Virginia (ULRICH and BASSLER, 1913) and Pennsylvania
(SWARTZ, 1932).

Germany and Czechoslovakia

When JORDAN (1970) reviewed the Paleozoic ostracodes of
Central Europe, he listed several species of Tricornina,
Acanthoscapha and hollinomorphs from the Siegenian-Emsian of
the German Democratic Republic, including an unidentified species of Polyzygia from the Emsian. BECKER & BLESS (1974) considered several species of Zygobeyrichia, Kozlowskiella,
Poloniella, Carinokloedenia and Bassleratia to have some biostratigraphical value for a regional subdivision of Emsian
strata in the Rhenisch Mountains. Kozlowskiella also occurs
in the Upper Emsian of Czechoslovakia (PRIBYL, 1962).

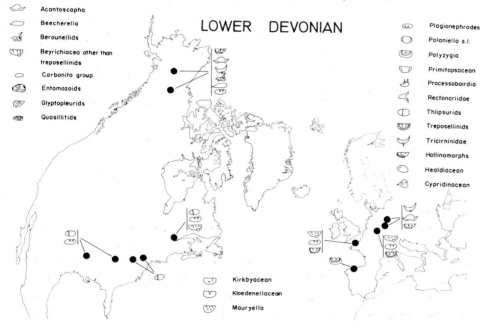

Fig. 1. Schematic distribution of some ostracod groups
in Lower Devonian strata

France

The beyrichiid ostracodes which have been described from
the Siegenian of Normandy (WEYANT, 1965, 1966) together with a
thlipsurid species and a diversified assemblage of Polyzygia
show little in common with either North America (except for a
local species of Octonaria) or German faunas.

Carnic Alps

BANDEL & BECKER, 1975, studied silicified ostracodes from
the Upper Silurian to Lower Carboniferous of the Central Carnic
Alps. The faunas are characterized by the predominance of
distinct, relatively few and long-ranging systematic units.

Spain

MICHEL (1972) described five species of Polyzygia from
Siegenian and Emsian of northwestern Spain.

Conclusions

Beyrichiid ostracods s. l. and Poloniella s. l. are known
from both sides of the Atlantic, although no clear relationships even at generic level can be observed. Thlipsurid ostracodes except Polyzygia appear to be widespread in the Lower
Devonian of North America but they tend to become rare in the
Emsian. This group of ostracodes is practically absent in
western Europe, where only one species has been described from
Normandy. In western Europe the thlipsurid genus Polyzygia
is abundant in Normandy and Spain with one species occurring in
German Democratic Republic. Genera like Acanthoscapha, Beecherella, Tricornina, and Praepilatina, which seem to be restricted to rather quiet, low energy facies (cf. JORDAN, 1970) occur
at several places in Lower Devonian strata where the environmental conditions appear to have been favorable for such assemblages. They contain few marker species in general and several
of the genera range into Upper Devonian and Dinantian.

MIDDLE DEVONIAN

Canada

The Eifelian of Ellesmere Island has yielded several species of Poloniella (WEYANT, 1968). BRAUN, 1966 described
Givetian assemblages from the Northwest Territories with primitsiopsid and quasillitid forms and Poloniella. McGILL (1967)
described poorly diagnostic ostracod faunas from the Givetian
of the Northwest Territories, which he compared with faunas of

similar age in Germany. McGILL (1966) discovered primitiopsid ostracodes from the Givetian of Alberta.

U.S.A.

Rich and diversified hollinomorph assemblages have been described from the Lower Eifelian of Ohio (KESLING & PETERSON, 1958), the Givetian of Michigan (KESLING & TABOR, 1953; KESLING, 1952; KESLING & McMILLAN, 1952) and the Givetian of Ontario (KESLING, 1952). The Givetian of Ontario has also yielded quasillitid ostracodes and several species of Poloniella (cf. STUMM & WRIGHT, 1958). Treposellinid (Hibbardia), quasillitid and thlipsurid ostracodes have been found in the Givetian of New York (SWARTZ & ORIEL, 1948). Thlipsurid and hollinomorph ostracodes from the Eifelian of Pennsylvania have been described by SWARTZ & SWAIN (1941). KESLING & KILGORE (1952) and KESLING & WEIS (1953), discovered thlipsurid and quasillitids from the Givetian of Michigan. KESLING (1954) described Poloniella from the Upper Eifelian of Ohio. ULRICH (1891) and KESLING (1955) described treposellinid species from the Eifelian of Kentucky and the Givetian of New York, respectively.

Germany & Poland

Endemic black shale faunas of Upper Eifelian age occur in Thuringia with high percentages of Tricornina, Hlubocepina and healdiid ostracodes (K. ZAGORA, 1967). Significant markers are the genera Jenningsina and Poloniella. Silicified ostracode assemblages of presumably low-energy environments with Acanthoscapha, Beecherella, Berounella, Tricornina, Jenningsina and several hollinomorphs occur in the Eifelian of the Harz (BLUMENSTENGEL, 1969) and Vogtland (JORDAN, 1965). Treposellinid (Kozlowskiella, Parakozlowskiella), hollinomorphs, quasillitid, primitiopsid and Polyzygia appear to be important regional guides in the Eifelian of Poland (ADAMCZAK, 1968) and the Eifelian-Givetian boundary of the Rhenish Mountains (BECKER, 1964, 1965, 1969, 1970, GROOS, 1969, KROMMELBEIN, 1950, 1953, 1955, BECKER & BLESS, 1974).

Czechoslovakia

POKORNY (1951) described some ostracodes occurring in Middle Devonian of Czechoslovakia, among others fuesillitids, hollinomorphs and primitiopsids.

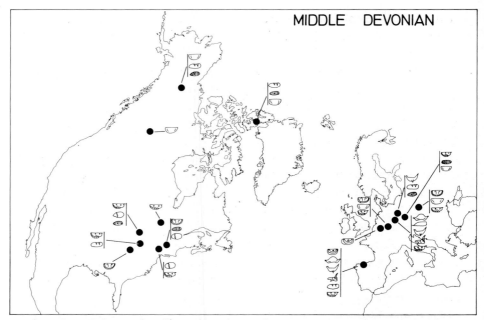

Fig. 2. Schematic distribution of some ostracode groups in Middle Devonian strata.

Belgium

BULTIYNCK (1967) cited _Polyzygia_ from the Eifelian of the Dinant Basin.

Spain

The genus _Polyzygia_ is widely distributed in Eifelian and Givetian strata of NW Spain (MICHEL, 1972). Primitsiopsid, hollinomorphs and quasillitid ostracodes together with the genera _Acanthoscapha_, _Tricornina_, _Berounella_, _Poloniella_ and _Praepilatina_ occur in the Eifelian of NW Spain (BECKER & SANCHEZ DE POSADA, and BECKER et al., in Press).

Conclusions

Treposellinid ostracodes are distinctive elements from the Middle Devonian in both North America and Europe. As in the case of the Lower Devonian thlipsurid genera, treposellinid genera appear to exclude each other on both sides of the Atlantic. _Hibbardia_ and _Treposella_ being restricted to North America, whereas _Kozlowskiella_ and _Parakozlowskiella_ occur in Europe (Germany and Poland) and Australia (PRIBYL, 1962). It should be noted, however, that it is not clear to me, whether _Hibbardia_ appears to possess a more distinct dolonoid scar than

Parakozlowskiella. Whether this phenomenom should be considered to be a specific or generic character remains open for discussion.

Poloniella s.l. (including Dizyglopeura and Framella as subgenera) and quasillitids are characteristic of many Middle Devonian strata. However species are usually restricted to a single basin and do not permit long distance correlations. The same is true of hollinomorphs, which according to BECKER & BLESS (1971) appear to have migrated in time and space from North America to Europe during the Devonian and Carboniferous, the different genera showing a clear diachronism.

The low energy environments containing Tricornina, Acanthoscapha, Berounella, Beecherella, have not been described in North America. The similarity at the specific level between certain German and Spanish Lower Fifelian faunas may be due to similar facies conditions hence their stratigraphical use remains uncertain.

UPPER DEVONIAN

Canada

Large Frasnian ostracode faunas from Alberta and Northwest Territories containing amongst others several species of Plagionephrodes, and quasillitids (Eriella, Quasillites) seem to have some similarities with Russian Platform faunas of similar age. LORANGER, 1954, 1963, 1965; McGILL, 1963; BRAUN, 1963, 1968; LORANGER, 1971, describes a Frasnian fauna containing amongst others several entomozoan ostracods which may permit comparison with entomozoan faunas from NW Europe. The apparent absence of hollinomorphs is remarkable.

U.S.A.

GIBSON (1955) described a fauna from the Cerro Gordo Formation of Iowa containing primitiopsid ostracodes. STEWART & HENDRIX (1945) described rather heterogeneous ostracode assemblages containing some entomozoan species from the Frasnian of Ohio. KESLING & PLOCH, (1960) studied a probably endemic cypridinacean species from the Upper Devonian Black Limestone (Frasnian/Famennian). The absence of hollinomorphs is again remarkable.

Germany and Poland

BLUMENSTENGEL (1965) described a rich Frasnian/Famennian

silicified fauna from Thuringia, with hollinomorphs, Acanthoscapha, Tricornina, Berounella, Processobairdia among others.
This fauna has shown to be useful for regional zonation. The
Frasnian fauna from the Harz described by BLUMENSTENGEL (1969)
also contains Acanthoscapha, Berounella, Tricornina and hollinomorphs but there is very little similarity between the two
faunas at specific level. However GRUNDEL (1962) BLUMENSTENGEL
(1954) described a rich entomozoan fauna from Thuringia which
can be directly compared with the Frasnian/Famennian entomozoan
faunas from the Rhenisch Mountains (RABIEN, 1954).

KRÖMMELBEIN (1954); BECKER (1967, 1968, 1970) GROOS, (1969),
have described rich and diversified ostracod assemblages from
the Frasnian of the Rhenisch Mountains which contained several
species of Polyzygia, Plagionephrodes, Primitiopsacea,
Poloniella, and hollinomorphs.

Entomozoan ostracodes also appear in the Upper Devonian
of Poland (cf. JORDAN, 1970).

Papers by RABIEN, (especially RABIEN, 1954) represent an
important attempt to obtain a zonation with entomozoan ostracods and this zonation with minor modifications seems to be
mainly accepted in Europe.

Fig. 7 shows the Upper Devonian and lowermost Carboniferous entomozoan zonation after BECKER & BLESS (1974). The
Richterina (R.) aff. latior Zone is proposed by BECKER & BLESS
(op. cit.) and characterized by the presence of ostracodes described as R. (R.) aff. latior by GRUNDEL (1961, 1963) JORDAN
& BLESS (1970) and SANCHEZ DE POSADA & BLESS (1974). This
zone is supposed to be equivalent to the Middle-Upper Tournaisian and part of the Visean.

France and Belgium

LETHIERS (1970, 1972, 1974) described ostracode faunas
ranging from Frasnian to Tournaisian age containing, amongst
others, Polyzygia, Quasillitidae, and several entomozoan species. BECKER & BLESS (1974) noticed that the stratigraphical
range of some of the species described by LETHIERS does not coincide with that observed by themselves suggesting that Upper
Famennian benthonic ostracodes of the Dinant Basin had been
clearly facies controlled in their lateral and vertical distribution.

LETHIERS (1974b) proposed a preliminary zonation for ostracodes from Frasnian to lowermost Tournaisian in the Dinant Basin. However, as yet this zonation has not been tested outside
of the area in which it was established.

Spain

Little is known about Upper Devonian ostracodes in Spain.
BLESS & MICHEL (1967) studied a silicified ostracod fauna from
northwestern Spain containing among others Acratia, Bohemina,
Ceratacratia, Processobairdia, Rectoplacera and Tricornina.
This fauna has been compared with the fauna described by
BLUMENSTENGEL (1965) from the Upper Devonian in Thuringia,
the age established from the ostracodes is in agreement with
that obtained from conodonts and cephalopods. In addition
MICHEL (1972) has described 2 species of Polyzygia in Frasnian
strata of northwestern Spain.

Conclusions

Upper Devonian entomozoan ostracode faunas seems to be
very useful for stratigraphical purposes, at least in Europe,
and entomozoan zonation may be correlated with the orthochrono-
logical scale. Benthonic ostracodes are occasionally useful in
regional correlations.

LOWER CARBONIFEROUS EUROPEAN FAUNAS AND
MISSISSIPPIAN AMERICAN FAUNAS

Recent progress in correlation techniques in Carboniferous
strata of Europe and North America coupled with the difficulty
of not knowing the exact stratigraphical level from which some
ostracode faunas (described many years ago) have been obtained,
has necessitated that Chester ostracodes from North America
should be treated together with Lower Carboniferous European
faunas.

Canada

GREEN (1963) described rich and diversified ostracodes
assemblages from the Banff Formation of Alberta containing
among others hollinomorphs, Kirkbyacea, Beyrichiopsidae, Kloedenellacea, Healdiidae, Paraparchitacea, rich quasillitids,
some Berounellidae, Beecherellidae, Bythocytheridae and Entomozoidae. LORANGER (1958) divided the Rundle Formation into a

70

Lower <u>Criboconcha</u> zone and an Upper <u>Paraparchites</u> <u>carbonarius</u>
zone. Other Mississippian ostracodes in Canada were described
by COPELAND (1957), BELL (1960) and BLESS & JORDAN (1971).

U.S.A.

Carboniferous and Permian ostracodes described in U.S.A.
were compiled by ECHOLS & CREATH, 1959. Most of the known
Carboniferous occurrences are in the interior region outside
the area of this paper.

Great Britain

JONES (1884) and JONES & KIRKBY (1874) described in many
publications a great number of ostracodes in the British Isles
covering the whole Carboniferous at the end of last century.
Most of these species need revision now. As far as we know
only a few of the lower Carboniferous species have been restudied. ANDERSON (1970) dealt with species of the genus <u>Carbonita</u>.

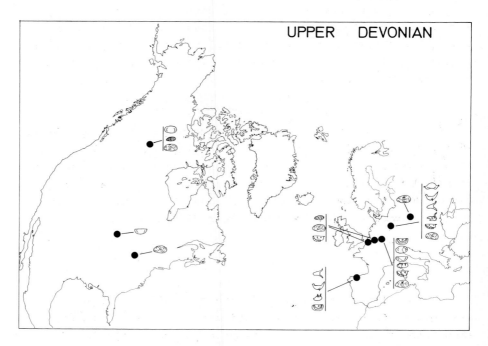

Fig. 3   Schematic distribution of some ostracode
groups in Upper Devonian strata.

ROBINSON (1959) described an ostracod fauna from the
Cowdor Quarry in Derbyshire containing among others Tetrasac-
culus, Kirkbyacea, Discoidella, Paraparchitacea, Monoceratina
and Healdicea.

Germany

KUMMEROW (1939) described 80 species from Lower Carboniferous strata of Germany and Belgium containing among others
Hollinellidae, Kirkbyacea, Bairdiacea, Berounellidae, Glyptopleuridae and Entomozoidae. GRUNDEL (1961) described more than
60 species from the Gattendorfia-stufe. The faunas are mainly
composed of Tricorninidae, Kirkbyacea, Healdiidae, Rectonariidae, Bairdiidae, Cypridinadae, Monoceratina, Triceratina and
Entomozoidae. Ostracod faunas described by GRUNDEL (1962) from
the CU II$_{\beta/\gamma}$ from the southern border of the Ruhr Carboniferous
area contained mainly Rectonariidae, Tricornina, Glyptopleuridae, Healdiopsis and Entomozoidae. Two species, Glyptopleura
elapa and Richterina (R.) aff. latior are related to species
described in 1974 by SANCHEZ DE POSADA & BLESS from Aprath
(Federal Republic of Germany).

BECKER & BLESS (1974) reviewed the stratigraphical range
of the species described from the Lower Carboniferous of Dinant
Basin by DE KONINCK (1840, 1844),JONES & KIRKBY (1874),
TSCHIGOVA (1970), ROME (1971) and ROME & GOREUX (1960). Some
of the species named under open nomenclature are figured in the
above papers; some of them are compared with previously described species, and their stratigraphical range has been proposed. Nine ostracod zones ranging from Fa2a to V3 have been
proposed.

Belgium

ROME & GOREUX (1960) described nine species of Crypto-
phyllus from Strunian strata. ROME (1971) described several
species of Kegelites, Beyrichiopsis, Sansabella, Bairdia,
Silenites, Microcheilinella, and Paraparchites. Some of the
species of Bairdia described by ROME are considered to belong
to a single species (LETHIERS, 1975).

BECKER & BLESS (1974) studied ostracodes in the Dinant
Basin of Germany and Belgium (see above). BECKER & BLESS (1974)
figured and described some species occurring in Fr2 strata in
the Belgian part of the Dinant Basin. Special attention is

paid to Beyrichiopsis glyptopleuroides GREEN and to a group of similar (if not identical) species composed of "Bernix" venulosa KUMMEROW (1939), Pseudoleperditia poolei SOHN (1969) and P. tuberculifera SCHNEIDER (1956) all of which are considered to probably have some value in long-distance correlations. B. glyptopleuroides is known from the uppermost Famennian and lower most Tournaisian strata in the USSR, Europe and Canada. The group of P. poolei, P. tuberculifera and "Bernix" venulosa is known from Dinantian strata in USA, the Federal Republic of Germany and USSR. BLESS & THOREZ (in BECKER et al., 1974) distinguished four ostracodes assemblages, two of which occur in limestone believed to represent a relatively deep subtidal environment, the other occurred in limestones believed to have been deposited in an intertidal and a supratidal environment.

Poland

BLASZYK & NATUSIEWICZ (1974) described 34 ostracode species from Dinantian and Namurian strata in northwestern Poland. Only one species is related to previously described species from Europe, four are identical and two are related to American species.

Spain

JORDAN & BLESS (1970) described two species of entomozoid Ostracoda occurring in Tournaisian strata of northwestern Spain and figured several species of Kirkbya, Acratia and Triplacera.

Conclusions

Entomozoan Ostracoda seem to be useful for stratigraphical purposes in some Lower Carboniferous strata from Europe but a lot of work has still to be done concerning this subject.

BECKER & BLESS (1974) (in BECKER et al.) cited several species of Bernix, Beyrichiopsis and Pseudoleperditia as possible index fossils for long distance correlations.

A most interesting aspect of the ostracode fauna on both sides of the Atlantic is the absence in North America of Tricorninidae and Rectonariidae. On the other hand the most studied American Lower Carboniferous faunas possess a greater variety of species of hollinomorphs, Kirkbyacea, Bairdiacea and Healdiacea.

UPPER CARBONIFEROUS

U.S.A.

Most of the papers on Pennsylvanian ostracodes deal with
faunas restricted either stratigraphically or geographically.
COOPER (1946), BRADFIELD (1935), and KELLETT (1933, 1934, 1935),
however, describe large and diversified faunas from throughout
the Pennsylvanian of Illinois and adjacent areas of Indiana and
Kentucky, the Ardmore Basin in Oklahoma and the Upper Pennsylvanian into Permian of Kansas, all of which are outside the
area of this study. These faunas are mainly composed of Hollinella, Kloedenellacea, Kirkbyacea, Paraparchitacea, Bairdiacea,
and Healdiacea. Little can be said about the usefulness of the
described species for stratigraphical purposes as most of them
seem to have a rather long stratigraphical range and only a
comparatively small number have a restricted range (cf. COOPER,
1946, p. 18; KELLETT, 1933, p. 61).

SHAVER, (in Thomson et al., 1959) described an ostracode
fauna from Kentucky, Bairdiacea and Kirkbyacea being the predominant elements therein. SHAVER & SMITH (1974) described 15
species of Kirkbyacea from Lower and Middle Pennsylvanian strata of Indiana and Kentucky. The A. rothi fauna is found in
rocks belonging to the Profusulinella zone whereas the A.
centronotus fauna is found in rocks of the zones of Fusulinella
and Fusulina.

Lower Pennsylvanian non-marine ostracodes from the S.
Appalachians were described by SCOTT & SUMMERSON (1944). They
included a species, Cypridina radiata (type species of Radii-
cypridina BLESS, 1973), which is common in the Lower Westphalian of Europe.

74

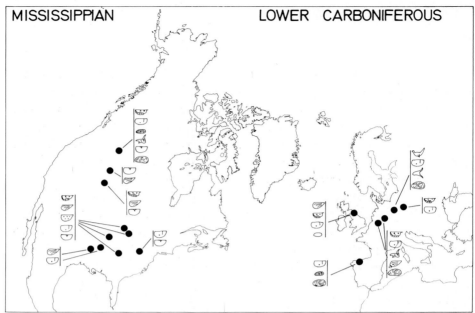

Fig. 4   Schematic distribution of some ostracode groups in Lower Carboniferous and Mississippian strata.

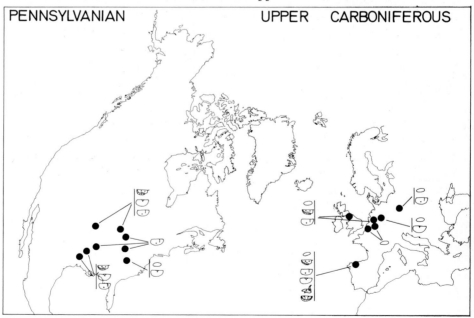

Fig. 5   Schematic distribution of some ostracode groups in Upper Carboniferous and Pennsylvanian strata.

UPPER CARBONIFEROUS OF EUROPE

Great Britain

JONES and JONES & KIRKBY described in the last century
some ostracodes occurring in the Upper Carboniferous of Great
Britain, most of these are species of Geisina and Carbonita.
As stated before species of Carbonita were reviewed by ANDERSON
(1970). POLLARD (1966, 1969) described and noted the occurrence of several Geisina and Carbonita species in non-marine
bands from the Coal Measures of Durham and Northumberland.
BLESS & POLLARD (1973) studied the palaeocology and content of
two ostracode assemblages, one from Upper Westphalian A strata
in The Netherlands, the other one from Lower Westphalian B in
Great Britain.

Only a small number of marine ostracode species are known
from the Upper Carboniferous of Great Britain. RAMSBOTTOM,
1952 described some kirkbyacean and Roundyella species from the
Similis pulchra zone. BLESS (1974) described the occurrence of
several species of Hollinellidae (Hollinella and Jordanites),
Kirkbyacea, Roundyella, Moorites, Cornigella, Pseudoparapar-
chites, Healdia and Asturiella from the Croft's End Marine
Band. Eleven of these species are conspecific or closely related to species of Pennsylvanian age in North America.

Germany

KUMMEROW (1953) described Upper Carboniferous ostracodes
from Germany containing, among others, Paraparchitacea, Kellet-
tina, and Kloedenellacea. Now the fauna needs to be restudied.
KREMP & GREBE (1956), VANGEROW (1958, 1970), described non-
marine ostracodes mainly Kloedenellacea and Carbonita species.

The Netherlands

VAN DER HEIDE (1951) describes species of Carbonita,
Geisina, and Cypridina in the Upper Carboniferous of The Netherlands. BLESS, JORDAN & MICHEL (1969) described a rather diversified ostracode fauna from the base of Westphalian C. This
included: Hollinellidae, Kirkbyacea, Cornigella, Roundyella,
Morrites, Paraparchitacea and Healdiacea. Twelve of these species are conspecific with species described by BLESS (1974)
from the same horizon in Great Britain. BLESS (1974) described
the new genus Cypridelliforma from the Lower Westphalian A of
the Netherlands.

Spain

BLESS (1965, 1967, 1968, 1970 - in VAN AMEROM et al.),
SANCHEZ DE POSADA & BLESS (1971), BECKER & BLESS (in BECKER et
al.,1975) described more than 40 species of Upper Carboniferous ostracodes in northwestern Spain (mostly in Upper Westphalian strata with only a small number from Namurian and Stephanian rocks). Most of the species which had been previously
described from elsewhere are identical or closely related to
Upper Pennsylvanian species from North America. The faunas are
mainly composed of Hollinellidae (Hollinella and Jordanites),
Kirkbyacea, Roundyella, Bairdiacea and Healdiacea. However two
genera, Jordanites and Asturiella do not occur in North America.
Unpublished data of several localities in northwestern Spain
show that some species which occur at the base of Westphalian C
in The Netherlands and England are now known from the Westphalian A-B of northwestern Spain: Healdia sp. D (BLESS, JORDAN
& MICHEL 1969), Asturiella limburgensis BLESS, Pseudobythocy-
pris pediformis (KNIGHT) and Roundyella simplicissima (KNIGHT)
Cavellina sp. cf. cumingsi described by BLESS AND by SANCHEZ
DE POSADA & BLESS (1971) previously described from Upper Westphalian strata only in Spain has now also been found in Lower
Westphalian rocks.

A most interesting aspect of the Spanish Upper Carboniferous ostracode fauna is the presence of some ostracodes previously not known from so young strata as Upper Carboniferous, mainly Tricornidae (Tricornina (Tricornina) and T. (Bohemina) and
Rectonariidae (at least one probable species of Rectoplacera
occurs), and Berounella. The occurrence of such ostracodes is
believed to have a clear environmental control. All these ostracods were found in limestones with silicified fauna or in
shales (as internal and external molds) in sections without
coal bearing strata. The same holds true for a silicified fauna composed of Myodocopida in limestones of Westphalian D age
and for the entomozoan ostracode Truyolsina truyolsi described
by BECKER & BLESS (in BECKER et al. 1974) from rocks of
Namurian A age.

Czechoslovakia

PRIBYL (1958, 1962) described 43 ostracode species from
the Namurian A of Czechoslovakia, containing among others

Carbonita, Hollinella, Healdiacea, Kloedenellidae, and
Kirkbyacea. These bear a close relationship to American faunas
from Chester and Morrow strata (cf. PRIBYL, 1962, p. 6).

Conclusions

Possibly due to the greater development of marine sequences in North America, Upper Carboniferous ostracodes in USA
are more varied than in western Europe where only a relatively
small number of marine faunas has been studied. The marine
ostracodes of Europe have frequently been considered to have
relationships with American ones (cf. PRIBYL, 1962, BLESS,
1974). Both American and European marine bands are mainly composed of Hollinellidae, Kirkbyacea, Kloedenellacea, Paraparchitacea, Bairdiacea, and Healdiacea. Some of the species are
identical or closely related on both sides of the Atlantic. In
Europe work was mainly done with non-marine ostracodes.
Carbonita and Geisina are the more frequent genera.

The presence in Upper Carboniferous strata of northwestern
Spain of some ostracodes belonging to the Families Tricorninidae and Rectonariidae is remarkable.

PERMIAN

U.S.A.

HARRIS & LALICKER (1932) described Wolfcampian ostracodes
from Oklahoma and Arkansas containing Paraparchitacea, Healdiacea and Kloedenellacea. Diversified assemblages of Hollinellidae, Kirkbyacea, Kloedenellacea and Bairdiacea were described
from Wolfcampian strata of Kansas by KELLETT (1933, 1934, 1935)
Wolfcampian and Lower Leonardian from Nebraska by UPSON (1935).
Other Wolfcampian faunas were described by DELO (1930) and
HARRIS & WORELL (1936). All of these are outside the present
study area. HOLLAND (1934) and SCOTT (1942) studied unique
non-marine species from West Virginia.

78

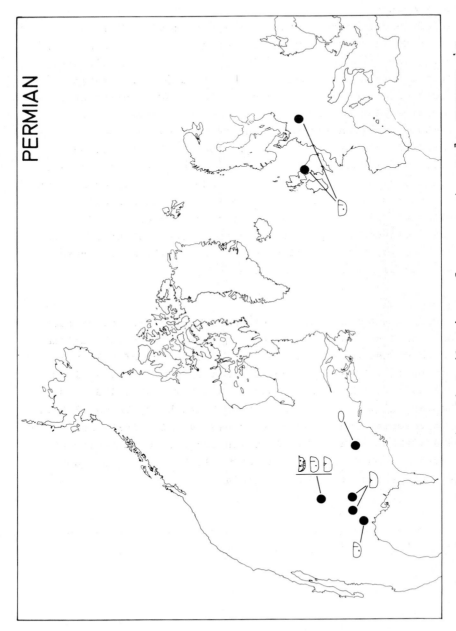

Fig. 6   Schematic distribution of some ostracode groups in Permian strata.

England

Permian ostracodes from England were described in the last century by KIRKBY. Kirkbyacea are the most remarkable elements.

Germany & Poland

Lower Zechstein ostracodes from Pommern have been described by KRÖMMELBEIN (1958). The fauna contains several species of Monoceratina, Roundyella, Healdiacea and Bairdiacea, most of which had not been previously described. RICHTER and REUSS described in the 19th century ostracod faunas from the Zechstein of Germany and in 1968 JORDAN reviewed the Permian ostracodes described from the Zechstein in central Europe.

REFERENCES

Due to reason of space the reader is referred to BECKER & BLESS, 1974 and ECHOLS & CREATH, 1959 for most references dealing with ostracodes described from Devonian and Dinantian strata of Central and Western Europe as well as Mississippian, Pennsylvanian and Permian strata of U.S.A.

Amerom, H. W. J. van, Bless, M. J. M., and Winkler, Prins C. F., 1970. Some paleontological and stratigraphical aspects of the Upper Carboniferous Sama Formation (Asturias, Spain). Med. Rijks Geol. Dienst, N.S., n. 21, pp. 9-79, pls. 1-10.

Anderson, F. W., 1970. Carboniferous Ostracoda. The genus Carbonita Strand. Bull. Geol. Surv. Great Britain, n. 32, pp. 69-121, pls. XII-XIX.

Bandel, D., and Becker, G., 1975. Ostracoden aus paläozoischen pelagischen Kalken der Karnischen Alpen (Silurium bis Unterkarbon). Senck. leth., Band 56, pp. 1-83, pls. 1-8.

Bassler, R. S., 1941. Ostracoda from the Devonian (Onondaga) chert of West Virginia. Washington Acad. Sc. Jour., v. 31, pp. 21-27.

Becker, F., Bless, M. J. M., Streel, M., and Thorez, J., 1974. Palynology and ostracode distribution in the Upper Devonian and basal Dinantian of Belgium and their dependence on sedimentary facies. Med. Rijks Geol. Dienst, N.S., n. 25, pp. 9-98, 30 pls.

Becker, F., and Bless, M. J. M., 1974. Ostracode stratigraphy of the Ardenno-Rhenisch Devonian and Dinantian. Inter. Symp. Belgian Microp. limits, Namur, 1974, Publication n. 1, 52 pp., 50 pls.

Becker, F., Bless, M. J. M., and Kullmann, J., 1975. Oberkarbonische Entomozoen-Schiefer im Kantabrischen Gebirge (Nordspanien). N. Jb. Geol. Palaont. Abh. 150, pp. 92-110.

Becker, G., Mendez Bedia, I., and Sanchez de Posada, L. C., (in press); Una fauna de ostrácodos de la formación Moniello (Devonico, Asturias, NW de Espana). Nota preliminar. Trab. Geol., n. 7.

Becker G., and Sanchez de Posada, L. C., (in press). Ostracoda aus der Moniello Formation Asturiens (Devon; NW Spain) Palaeontographica.

80

Bell, W. A., 1960. Mississippian Horton Group of type Windsor-Horton district, Nova Scotia. Geol. Surv. Canada, Mem. 314, 58 pp., 24 pls.

Berdan, J. M., and Copeland, M. J., 1973. Ostracodes from Lower Devonian formations in Alaska and Yukon Territory. U.S. Geol. Surv. Prof., Paper 825, 47 pp., 14 pls.

Bless, M. J. M., 1965. On two new species of marine ostracodes in the Carboniferous of Asturias, Spain. Leid. Geol. Med., v. 33, pp. 177-182.

————, 1966. Carbonita agnes (JONES) from the Coal Mine "Sabero", (León (Spain). Nots. Coms. Inst. Geol. Min. Espana, n. 90, pp. 93-98.

————, 1967. On the marine beds of some cyclothems in the Central Carboniferous Basin of Asturias with species reference to their ostracode fauna. Id., n. 99-100, pp. 91-134.

————, 1968. On two Hollinid Ostracode genera from the Upper Carboniferous of Northwestern Spain. Leid. Geol. Med., v. 43, pp. 157-212, pls. 1-10.

————, 1974. On a new genus and species of Cypridinacea (Ostracoda) from the Upper Carboniferous of The Netherlands. Med. Rijks Geol. Dienst, N.S., n. 22, pp. 21-23, pls. I.

————, 1974. Ostracodes from Croft's End Marine Band (base of Westphalian C) of the Bristol District. Bull. Geol. Surv. Great Britain, n. 47, pp. 39-53, pls. 4-6.

Bless, M. J. M., and Jordan, H., 1971. The new genus Copelandella from the Lower Carboniferous - the youngest known beyrichiacean ostracode. Lethaia, v. 4, pp. 185-190.

————, 1972. Ostracodes of the family Hollinellidae. Med. Rijks Geol. Dienst, ser. C, v. 3, 83 pp., 35 pls.

Bless, M. J. M., Jordan, H., and Michel, M. Ph., 1969. Ostracods from the Aegir Marine Band (basis Westphalian C) of South Limburg (The Netherlands). Id. n. 20, pp. 19-49, pls. 1-7.

Bless, M. J. M., and Michel, M. Ph., 1967. An ostracode fauna from the Upper Devonian of the Gildar-Monto Region (NW Spain). Leid. Geol. Med. v. 39, pp. 269-271.

Bless, M. J. M., and Pollard, J., 1973. Paleoecology and Ostracode faunas of Westphalian Ostracode Bands from Limburg, The Netherlands and Lancashire, Great Britain. Med. Rijks Geol. Dienst, N.S., n. 24, pp. 1-33, pls. 1-5.

Braun, W. K., 1966. Stratigraphy and microfauna of Middle and Upper Devonian Formations, Norman Wells area, Northwest Territories. N. Jb. Geol. Paläont., Abh. 125, pp. 247-264.

————, 1968. Upper Devonian Ostracod faunas of Great Slave Lake and Northeastern Alberta, Canada. Internat. Symp. Devonian System, Calgary, 1967, Alberta Assoc. Petro. Geol., v. I, pp. 617-652, 9 pls.

Burk, C. F., 1965. Silurian stratigraphy of Gaspé Peninsula. A Reply. Bull. Am. Assoc. Petr. Geol., v. 49, pp. 2305-2316.

Copeland, M. J., 1957. The Arthropod fauna in the Legger Carboniferous rocks of the Maritime Provinces. Geol. Surv. Canada, Mem. 286, 110 pp., 21 pls.

————, 1962. Ostracoda from the Lower Devonian Dalhousie beds, northern New Brunswick. Geol. Surv. Canada, Bull. 91, pp. 18-51.

————, 1974. Biostratigraphic zonation of Devonian and Mississippian Ostracoda from Canada. A summary account. Inter-

nat. Symp. Belgian Microp. limits, Namur, 1974. Publication
n. 7, 9 pp.

Echols, D. J., and Creath, W. B., 1959. Survey of Mississippian, Pennsylvanian and Permian ostracoda recorded in the
United States. Micropaleont., v. 5, pp. 389-414.

Gibson, L. B., 1955. Upper Devonian Ostracoda from the Cerro
Gordo Formation of Iowa. Bull. Am. Paleont., v. 35, pp. 5-36,
pls. 1-2.

Green, R., 1963. Lower Mississippian ostracodes from the Banff
Formation, Alberta. Research Council Alberta, Bull. 11, 273
pp., 17 pls.

Heide, S., van der, 1951. Les Arthropodes du terrain Houillier
du Limburg Méridional (excepté les scorpions et les insectes).
Med. Geol. Stichting, ser. C-IV-3, n. 5, 84 pp., 10 pls.

Jones, T. R., and Kirkby, J. W., 1874. A monograph of the
British fossil Entomostraca from the Carboniferous Formations.
Part I. The Cypridinidae and allied groups. Paleont. Soc.
Monographs, pp. 1-56, pls. 1-5.

————, 1887. A list of the genera and species of bivalved
Entomostraca found in the Carboniferous Formations of Great
Britain and Ireland, with notes on the genera and their dis-
tribution. Geol. Ass. Proc., v. 9, pp. 495-515.

Jordan, H., 1968. Neue taxinomische und biostratigraphische
Ergebnisse mikropaläontologischer Untersuchungen im german-
ischen Zechsteinbacken unter besonderer Berücksichtigung der
Ostracoden. Geol. und Paläont., Band 13, pp. 199-213.

Jordan, H., and Bless, M. J. M., 1970. Nota preliminar sobre
los ostrácodos de la formación Vegamián. Brev. Geol. Astur-
ica, Año XIV, pp. 37-44.

Kesling, R. V., 1952. Ostracodes of the Families Leperditellidae, Primitiidae, Drepanellidae, Aechminidae and Kirkbyidae
from the Middle Devonian Bell Shale of Michigan. Paleont.
Contr. Michigan Univ. Museum, v. 10, pp. 21-44, 5 pls.

————, 1953. Ostracods of the family Hollinidae from the
Arkona Shale of Ontario. Id., v. 10, pp. 203-219, pls. 1-4.

————, 1954. Ostracods from the Middle Devonian Dundee Limestone in Northwestern Ohio. Id., v. 11, pp. 167-186, 3 pls.

————, 1955. Two new species of ostracods from the Center-
field Limestone of Western New York. Id., v. 12, pp. 273-284,
3 pls.

Kesling, R. V., and Kilgore, J. E., 1952. Ostracods of the
families Leperditellidae, Drepanellidae, Glyptopleuridae,
Kloedenellidae, Bairdiidae, Barychilinidae and Thlipsuridae
from the Genshaw formation of Michigan. Id., v. 10, pp. 1-19,
pls. 104.

Kesling, R. V., and McMilland, G. V., 1951. Ostracodes of the
Family Hollinidae from the Bell Shale of Michigan. Id., v. 9,
pp. 45-81, pls. 1-7.

Kesling, R. V., and Peterson, R. M., 1958. Middle Devonian
ostracods from the Falls of Ohio. Micropaleont., v. 4, pp.
129-148, pls. 1, 2.

Kesling, R. V., and Ploch, R. A., 1960. New Upper Devonian
cypridinacean ostracod from Southern Indiana. Paleont. Contr.
Michigan Univ. Museum, v. 15, pp. 281-292, 3 pls.

Kesling, R. V., and Tabor, N. L., 1952. Two new species of ostracods from the Genshaw Formation (Middle Devonian) of Michigan. Jour. Paleont., v. 26, pp. 761-763, pl. 111.

82

Kesling, R. V., and Weiss, M., 1953. Ostracoda from the Norway
    Point Formation of Michigan. Paleont. Contr. Michigan Univ.
    Museum, v. 11, pp. 33-76, 5 pls.
Kremp, G., and Grebe, H., 1956. Beschreibung und stratigraph-
    ischer Wert einiger Ostracodenformen aus dem Rhurkarbon.
    Geol. Jahrb., Band 71, pp. 145-170, pl. 16.
Krömmelbein, K., 1958. Ostracoden aus aus dem Unteren Zech-
    stein der Bohrung Leba in Pommern. Id., Band 75, pp. 115-134,
    pls. 1-3.
Kummerow, E., 1953. Ueber Oberkarbonische und Devonische Os-
    tracoden in Deutschland und in der Bolkrepublik Polen. Geolo-
    gie Beih. zur Zeitch, Beih. 7, 75 pp., 7 pls.
Latham, M., 1932. Scottish Carboniferous Ostracoda. Roy. Soc.
    Edinburgh Trans., v. 57, pp. 351-395.
Lethiers, F., 1972. Ostracodes famenniens dans l'Ouest du
    Bassin du Dinant (Ardenne). Ann. Soc. Geol. Nord., t. XCII,
    pp. 155-169, pls. 23-25.
————, 1974. Ostracodes du passage Frasnien-Famennien de
    Senzeilles (Ardenne). Palaeontograph., A. Band 147, pp. 36-
    69, pls. 7-9.
————, 1974. Ostracodes de la limite Dévonien-Carbonifère
    dans l'Asvenois. C. R. Acad. Sc. Paris, ser. D, t.278, pp.
    1015-1017.
————, 1974. Biostratigraphie des Ostracodes dans le Dévo-
    nien supérieur du Nord de la France et Belgique. Newls. Stra-
    tigraph., v. 3, pp. 73-79.
————, 1975. Les entomozoidés (Ostracodes) du Facies Matagne
    dans le Frasnien Sud-Occidental de l'Ardenne. Geobios, n. 8,
    pp. 135-138.
Loranger, D. M., 1954. Ireton microfossils zones of central
    and northeastern Alberta; in Clark, L. M. (ed.): Western Can-
    ada Sedimentary basin. Am. Ass. Petr. Geol., Ralph Leslie
    Rutherford Mem. vol., pp. 182-203.
————, 1963. Devonian Microfauna of northeastern Alberta.
    Pt. I. Orders Leperditocopida and Palaeocopida. Published by
    the author, 55 pp., 4 pls. Pt. II. Order Podocopida. Id., 53
    pp., 3 pls.
————, 1965. Devonian Paleoecology of northeastern Alberta.
    Jour. Sed. Petr., v. 35, pp. 818-837.
————, 1971. Ostracods, trace elements and Frasnian reefs in
    Sturgeon Lake area. Bull. Centre Rech. SNPA, Colloque Paleoec.
    ostrac., Pau, France, v. 5, supl., pp. 769-768.
Lundin, R., 1968. Haragan Ostracodes. Oklahoma Geol. Surv.
    Bull. 116, 121 pp., 22 pls.
McGill, P., 1963. Upper and Middle Devonian ostracodes from
    the Beaverhill Lake Formation, Alberta, Canada. Bull. Can.
    Petr. Geol., v. 11, pp. 1-26.
————, 1966. Ostracods of probable Late Givetian age from
    Slave Point Formation, Alberta. Id., v. 14, pp. 104-133.
————, 1967. Comparison of a Middle Givetian ostracode fauna
    from Carcajou Ridge, Northwest Territories, Canada, with sim-
    ilar faunas from Europe. Internat. Symp. Devonian System,
    Calgary, 1967. Alberta Soc. Petr. Geol., v. II, pp. 1069-
    1085, 4 pls.
McGuire, O. S., 1966. Population studies in the ostracode ge-
    nus Polytylites from the Chester Series. Jour. Paleont., v.
    40, pp. 883-910, pls. 103-104.

Michel, M. Ph., 1972. Polyzygia Gurich (Ostracoda) in the Devonian of Asturias and León (Spain). Leid. Geol. Med., v. 48,
pp. 207-273, pls. I-XV.
Oswald, D. H. (ed.), 1968. International Symposium on the Devonian System. Alberta Assoc. Petr. Geol., v. I, 1055 pp.,
v. II, 1377 pp.
Pollard, J., 1966. A non-marine ostracod fauna from the Coal
Measures of Durhan and Northumberland. Palaeontology, v. 9,
pp. 667-697.
————————, 1969. Three Ostracod-Mussel Bands in the Coal-Meas-
ures (Westphalian) of Northumberland and Durhan. Proc. Yorkshire Geol. Soc., v. 37, pp. 239-276, pl. 8.
Pribyl, A., 1958. The ostracodes of the Upper Carboniferous
(Nam. A) of Czechoslovakia (Poruba beds) and its importance
for the Ostrava-Karaniva coal district (in Czech with English
summary). Sbornik Ustred Ustav, Geologie, v. 24 (1957), odd.
Paleont., pp. 7-92.
————————, 1961. Biostratigraphical significance of the Carboniferous Ostracoda and their distribution in the Ostrava-
Karviná Coal District (Upper Silesian Basin) of Czechoslovakia. Compt. Rend. Quatriem Congr. pour l'avancement des
etud. de Stratigr. et Géolog. du Carbonif., t. II, pp. 553-
557.
————————, 1961. Upper Carboniferous ostracodes of the Hrusov
and Petrkovice beds of Ostrava-Karviná Coal District (Czechoslovakia). Rozbravy Esl. Akad. ved. r. 71, pp. 1-54, 9 pls.
Ramsbottom, W. H. C., 1952. The fauna of the Cefn Coed Marine
Band in the Coal Measures at Aberbaiden near Tondu, Glamor-
gan. Bull. Geol. Surv. Great Britain, v. 4, pp. 8-30, pls.
II, III.
Robinson, J. E., 1959. The ostracode fauna of the shale facies
of the Cowdor Limestones, noryh end of Cawdor Quarry. Quart.
Jour. Geol. Soc. London, v. 114, pp. 435-448.
Sanchez de Posada, L. C., and Bless, M. J. M., 1971. Una microfauna del Westphaliense C de Asturias. Rev. Esp. de Micropaleont., v. III, pp. 193-204, pls. I, II.
————————, 1974. Preliminary note on the Lower Carboniferous Ostracods from Aprath (Federal Republic of Germany). Internat.
Symp. on Belgian Micropaleont. limits, Namur, 1974. Publication n. 2, 5 pp., 2 pls.
Shaver, R. H., and Smith, S., 1974. Some Pennsylvanian kirkbyacean ostracodes of Indiana and Mid-continent Series terminology. Rep. of Progress, n. 31, 59 pp., 3 pls.
Sohn, I. G., 1960. Paleozoic species of Bairdia and related
genera. U.S. Geol. Surv. Profess. Paper 330A, 115 pp., 6 pls.
————————. 1961. Aechminella, Amphissites, Kirkbyella and related genera. U.S. Geol. Surv. Profess. Paper 330B, pp. 107-
160, pls. 7-12.
————————, (in Gordon et al.), 1969. Revision of some Girty's
invertebrate fossils from the Fayettville Shale (Mississippian) of Arkansas and Oklahoma. Ostracodes. Id. 606-F, pp.
41-55, pls. 6-8.
Sohn, I. G., 1969. Pseudoleperditia Schneider, 1956 (Ostracoda, Crustacea) an early Mississippian genus from Southwestern
Nevada. Id., 643C, pp. 1-6, 1 pl.
————————, 1971. New late Mississippian Ostracode genera and
species from Northern Alaska. Id., 711A, pp. 1-24, pls. 1-9.

84

————, 1972. Late Paleozoic Ostracode species from the Con-
terminous United States. Id., 711B, pp. 1-15, pls. 1, 2.
————, 1975. Mississippian Ostracoda of the Amsdem Formation
(Mississippian and Pennsylvanian) of Wyoming. Id. 848G, pp.
1-22, pls. 1-3.
Stewart, G. A., and Hendrix, W. E., 1945. Ostracoda of the
Plum Broo Shale, Erie County, Ohio. Jour. Paleont., v. 19,
pp. 87-95, pl. 10.
Stumm, E. C., and Wright, J. D., 1958. Check list of fossil
invertebrates described from the Middle Devonian rocks of the
Thedford-Arkona Region of Southeastern Ohio. Paleont. Contr.
Michigan Univ. Museum, v. 14, pp. 81-132.
Swartz, F., 1932. Revision of the ostracode family Thlipsuri-
dae, with description of a new species from the Lower Devon-
ian of Pennsylvania. Jour. Paleont. v. 6, pp. 36-58, pls. 10,
11.
Swartz, F., and Oriel, S. S., 1948. Ostracoda from the Middle
Devonian Windom Beds in Western New York. Pennsylvania St.
Coll. Tech., Paper 142, pp. 541-566, pls. 79-81.
Swartz, F., and Swain, F. M., 1941. Ostracodes of the Middle
Devonian Onondaga Beds of Central Pennsylvania. Geol. Soc.
America Bull., v. 52, pp. 381-458, pls. 1-8.
Sweet, W. C., and Bergström, S. M., 1970. Symposium on Cono-
dont Biostratigraphy. Geol. Soc. America Mem. 127, 499 pp.
Thompson, M. J., and Shaver, R. H., 1964. Early Pennsylvanian
Microfaunas of the Illinois Basin. Trans. Illinois State
Acad. Sc., v. 57, pp. 4-23, 1 pl.
Thompson, M. J., Shaver, R. H., and Riggs, E. A., 1959. Early
Pennsylvanian fusulinids and ostracods of the Illinois Basin.
Jour. Paleont., v. 33, pp. 770-792, pls. 104-107.
Ulrich, E. O., and Bassler, R. S., 1913. Systematic Paleonto-
logy, Lower Devonian Ostracoda Maryland Geol. Surv. Lower
Devonian volume, pp. 513-542, pls. 95-98.
Vangerow, E. F., 1957. Mikropaläontologische Untersuchungen in
den Kohlscheider Schichten im Wurmrevier bei Aachen. Geol.
Jb., Band 73, pp. 457-506, pls. 20-23.
————, 1970. Die fauna des Westdeutschen Oberkarbons. Pa-
laeontogr. Band 134, pp. 133-152, pl. 13.
Weyant, M., 1971. Recherches micropaléontologiques sur le
Paléozoic inférieur et moyen de l'Archipel Arctique Canadien.
Unpublish. Ph.D. theses Univ. Caen.
Wilson, Ch. W., 1935. The ostracode fauna of the Birdsong
Shale, Helderbeg, of Western Tennessee. Jour. Paleont., v. 9,
pp. 629-646, pls. 77-78.

EARLY PALEOZOIC CONODONT BIOSTRATIGRAPHY IN THE ATLANTIC BORDERLANDS

Stig M. Bergström

Department of Geology and Mineralogy, The Ohio State University

Columbus, Ohio 43210

ABSTRACT

Cambrian conodonts, still relatively poorly known but apparently widespread in the Atlantic Borderlands, include some 15 multielement genera. Described faunas are mainly from Scandinavia, Poland, and Germany (erratics) whereas little information is available from, for instance, eastern North America and the British Isles. Although no succession of formally defined conodont zones has as yet been established throughout the system, Cambrian conodonts have considerable potential as guide fossils, particularly in the Upper Cambrian, where many forms show limited vertical ranges and very wide horizontal distributions.

Ordovician conodonts, which include about 80 multielement genera, are much better known than the Cambrian ones, and show striking provincial differentiation throughout the period. In the North American Midcontinent Province, some 17 biostratigraphic units of zonal type have been recognized and in the North Atlantic Province, some 15 zones, and more than 10 subzones, have been formally defined. Ordovician conodonts have proved very useful biostratigraphically, in many instances providing a stratigraphic resolution superior to that of any other fossil group.

Silurian conodont faunas exhibit far less taxonomic diversity (about 15 multielement genera) and provincial differentiation than Ordovician ones. The Silurian conodont succession, best known in the Atlantic Borderlands from Great Britain, Scandinavia, Virginia, and the eastern part of the North American Midcontinent, form the basis for about 12 zones. Many of these zones have been widely recognized not only in Europe and North America but also in Australia, Asia, and northern Africa.

Although published reports on Early Paleozoic conodonts of Africa and South America include only a few papers, the available data suggest that the conodonts faunas from those continents are, by and large, very similar to those known from the northern hemisphere.

Co-occurrence of stratigraphically diagnostic conodonts and

graptolites has made it possible to tie together conodont and graptolite zonal units in the Ordovician and Silurian at a relatively large
number of stratigraphic levels.

Geographic distance apparently was of relatively minor importance in comparison with ecologic factors for the establishment of
patterns in the Early Paleozoic conodont biogeography. This conclusion, along with the fact that very similar conodont faunas were
present on both sides of the Proto-Atlantic, make conodonts of little
use for evaluations of the size and development of the Proto-Atlantic
Ocean, at least at the present time.

INTRODUCTION

Conodonts are common microfossils in many types of Lower Paleozoic marine rocks. Due to their rapid evolution, the very wide
horizontal distribution of many taxa, and the fact that numerous
species apparently were not strongly facies-controlled, conodonts
now rank among the most useful fossils biostratigraphically in the
Lower Paleozoic. The group has been known for 120 years but the
period of modern and intense study of these fossils began about 1950
when their great biostratigraphic potential began to be apparent.
Since then, the conodont literature has grown very rapidly and close
to 100 papers, in which conodonts are dealt with in one form or another, are now published annually. Most papers are of biostratigraphic or taxonomic nature but during the last few years, the paleoecology and micromorphology of these fossils have attracted considerable interest, and it is likely that these areas of research will be
increasingly important in the future.

Since the mid-1960's, conodont taxonomy has gone through a
period of fundamental change. On the basis of the fact that several
different types of conodont elements evidently occurred in the
apparatus of a single individual, the taxonomy has developed rapidly from a pure form taxonomy based on external shape of single elements, to a zoologically more sound multielement taxonomy based on
reconstructed assemblages of elements. A great many conodont taxa
have now been re-evaluated on a multielement basis but it will take
a long time to complete this radical taxonomic revision because many
hundred species and more than 300 genera were originally proposed

as form taxa.

The purpose of the present contribution is to summarize briefly
the conodont biostratigraphy so far established for Cambrian through
Silurian rocks in the areas bordering the present Atlantic. In terms
of geographic extent, the area dealt with includes North America east
of 90°W Longitude, western and central Europe, and the Baltoscandic
region. Because little is known about Lower Paleozoic conodonts in
South America (see, for instance, Serpagli, 1974) and Africa (see,
for instance, Ethington and Furnish, 1962), the review will include
data mainly from North America and Europe. Although published several
years ago, and therefore outdated in some respects, a summary volume
entitled "Symposium on Conodont Biostratigraphy" (Sweet and Bergström,
eds., 1971) still provides a useful summary of Lower Paleozoic conodont biostratigraphy, and the reader is referred to that volume for
a more detailed treatment than is possible within the scope of the
present contribution. Papers in that volume also give illustrations
of many of the stratigraphically important conodonts mentioned below.

CAMBRIAN

The oldest conodonts known are from strata in Siberia currently
classified as late Precambrian (Missarzhevsky, 1973; Matthews and
Missarzhevsky, 1975) but the group, which is represented by species
of the order Paraconodontida, exhibited little diversity before Late
Cambrian time. The paraconodonts are weakly phosphatized and differ
also in other respects from the conodontiform conodonts, which include practically all post-Cambrian forms.

Although known from the Baltoscandic area (Müller, 1959; Poulsen, 1966; Bengtsson, 1976), Germany (Müller, 1959, erratic boulders),
Poland (Szaniawski, 1971), the British Isles (Miller and Rushton,
1973), and New York State (Landing, 1974b, 1976), Cambrian conodonts
are still very little studied in the Atlantic Borderlands (Fig.1).
Their potential as biostratigraphic tools was suggested in the
pioneer work by Müller (1959), but studies on Cambrian conodont
biostratigraphy have so far been concerned only with mainly Late
Cambrian successions in areas such as western North America (Miller,
1975), Iran (Müller, 1973), Queensland (Druce and Jones, 1971), and
China (Nogami, 1966, 1967). Müller (1973) and Miller (1975) intro-

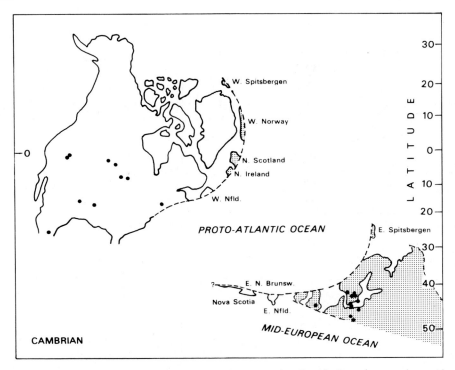

Fig.1. Cambrian conodont occurrences in North America and north-western Europe plotted on a sketch-map showing inferred continent positions during Middle Cambrian time. As in Fig.2 and 4, each dot denotes one or several conodont localities which have produced collections described in the literature or seen by the writer. Latitudinal position of northwestern Europe based on Noltimier and Bergström (in preparation).

duced zonal units based on Late Cambrian species and there are indications that at least some of these zones are recognizeable in the Appalachians (Landing, 1976). No attempt has been made so far to establish formal conodont biostratigraphic units in the Lower and Middle Cambrian and apart from a few papers (Landing, 1974b; Poulsen, 1966; Bengtsson, 1976), conodonts of that age remain virtually unstudied in the Atlantic area. The known Cambrian conodont faunas include some 15 genera of simple forms such as Furnishina, Proconodontus, Prosagittodontus, Muellerina, Hertzina, Prooneotodus, and Proscandodus and have a somewhat monotonous character. However, there are clear indications that some species have a wide geographic distribution combined with a reasonably short vertical range; accordingly, a promising and challenging task would be to explore their

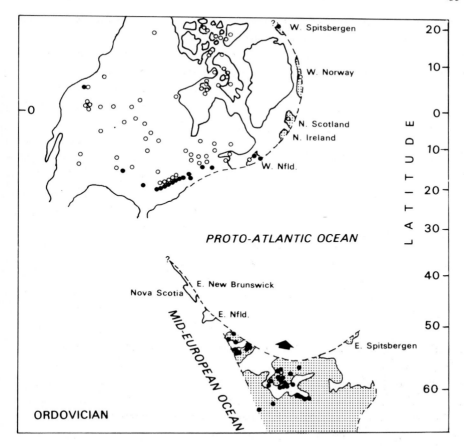

Fig.2. Ordovician conodont occurrences in North America and
northwestern Europe plotted on a sketch-map showing inferred
positions of continents in Early Ordovician (Arenigian) time.
Latitudinal position of Europe according to Noltimier and
Bergström (1976; in preparation), that of North America based on
McElhinny and Opdyke (1973). Dark dots denote North Atlantic
Province faunas, open circles Midcontinent Province faunas. Note
the presence of North Atlantic Province conodont faunas in the
eastern Appalachians and in California (Klamath Mountains).

utility as regionally useful index fossils in the Atlantic area.

ORDOVICIAN

The Early and Middle Ordovician was a time of rapid conodont
evolution that led to a diversification at the generic and specific

level that is apparently greater than that of any other comparable
time interval in the stratigraphic record of conodonts. This is illustrated by the fact that some 150 form genera of conodonts have been
proposed on the basis of Ordovician collections and more than half of
these are distinct as multielement genera. In addition, there are
quite a few characteristic, but still unnamed, genera, and the total
number of Ordovician multielement genera may ultimately prove to be
well in excess of 100. This figure is more than twice as high as that
for any other geologic period.

Ordovician conodonts have been far more intensely studied than
the Cambrian ones in the Atlantic Borderlands but there are, nevertheless, large gaps in our knowledge about faunas of that age. Best
known are faunas from the Baltoscandic area and the North American
Midcontinent but those from several other major regions, for instance,
South America, northern Africa, and Greenland, remain virtually un-
explored. Fig.2 gives the general geographic location of important
Ordovician conodont occurrences in areas bordering the present North
Atlantic.

Taken as a whole, Ordovician conodont faunas probably exhibit a
stronger biogeographic differentiation than those of any other system
(Bergström, 1973a; Barnes et al., 1973; Sweet and Bergström, 1974).
This faunal provincialism, which can be traced back to the Tremadocian, prevailed to the end of the period although modified as to areal
extent and distinctiveness. The two main faunal provinces generally
recognized, the North American Midcontinent Province and the North
Atlantic Province, are characterized by two conodont faunas so
different that separate zonal schemes have been established for each
of them (Sweet et al., 1971; Ethington and Clark, 1971; Bergström,
1971a, 1971b; Lindström, 1971) and there is still considerable uncertainty regarding the precise relations between several units in
these schemes. A provisional correlation between these provincial
schemes is given in Fig.3.

North American Midcontinent Province

Lower Ordovician (Canadian) conodont faunas in this province,
which occupies the central portion of the North American continent as
well as parts of Siberia and Australia, are characterized by "simple-
cone" genera such as Acanthodus, Scolopodus, Ulrichodina, Oneotodus,

91

| SERIES | | | MIDCONTINENT | NORTH ATLANTIC CONODONT | |
N. AM	G. BRITAIN	BALTOSCANDIA	FAUNAS	ZONES	SUBZONES
CININNATIAN	ASHGILLIAN	HARJUAN	12	*Amorphognathus ordovicicus*	NOT YET DISTINGUISHED
	CARADOCIAN		11	*Amorphognathus superbus*	
			10		
			9 8	*Amorphognathus tvaerensis*	*Prioniodus alobatus*
			7		*Prioniodus gerdae*
					Prioniodus variabilis
CHAMPLAINIAN	LLANDEILIAN	VIRUAN		*Pygodus anserinus*	Upper
			6		Lower
	?			*Pygodus serrus*	*Eoplacognathus lindstroemi*
					Eoplacognathus robustus
					Eoplacognathus reclinatus
			?		*Eoplacognathus foliaceus*
	LLANVIRNIAN		5	UNNAMED	*Eoplacognathus suecicus*
				Eoplacognathus variabilis	
			4	*Microzarkodina parva*	
			3	*Paroistodus originalis*	
			2	*Prioniodus navis*	
CANADIAN	ARENIGIAN	OELANDIAN	1	*Prioniodus triangularis*	
			E	*Oepikodus evae*	NOT YET DISTINGUISHED
	?			*Prioniodus elegans*	
			D	*Paroistodus proteus*	
	TREMADOCIAN		C	*Paltodus deltifer*	
			B	*Cordylodus angulatus*	
			A		

Fig.3. Correlation between North American and European Ordovician standard series, Midcontinent Province conodont faunas (Sweet et al., 1971; Ethington and Clark, 1971), and North Atlantic Province conodont zones and subzones (Lindström, 1971; Bergström, 1971a, 1971b). As shown in the diagram, the stratigraphic scope of the three main subdivisions of the Ordovician is, by tradition, not the same in Europe and North America; in the text below, the terms Lower, Middle, and Upper Ordovician are used in their local sense, that is, to denote the Oelandian, Viruan, and Harjuan in Europe, and the Canadian, Champlainian, and Cincinnatian in North America. Top of Canadian is taken as the base of the Whiterockian Stage. As to the un-named intervals between the Tremadocian/Arenigian and the Llanvirnian/Llandeilian, see Bergström et al. (1973, 1974).

"Paltodus", and others. Compound-element genera are less well represented but include Cordylodus, Loxodus, Chosonodina, and Oepikodus. Although earliest Ordovician faunas begin to be well known (Miller, 1969; Druce and Jones, 1971) much remains to be learned about younger Early Ordovician conodont faunas of this province. Fortunately, this serious gap in our knowledge may be at least partially filled by new data from still largely unpublished studies in New Mexico-Texas (Repetski, 1974) and Oklahoma (Mound, 1968; Brand, 1976; Potter, 1975).

Corresponding work in Asia is being carried out by, among others, Lee (1970, 1975).

Descriptions of earliest Middle Ordovician (earliest Champlainian) Midcontinent Province conodont faunas remain largely unpublished in North America but the available published and unpublished data (see, for instance, Mound, 1965; Bradshaw, 1969) show that Midcontinent faunas of this age include, apart from an array of "simple-cone" genera such as Drepanoistodus, Panderodus, Oistodus, and Scolopodus, a distinctive suite of compound-element genera, for instance, Multioistodus, Leptochirognathus, Histiodella, Bergstroemognathus, Erismodus, and Tricladiodus. Typical platform elements are rare or missing in most faunas of this type, but occur (Polyplacognathus, Scyphiodus) in slightly younger Middle Ordovician faunas along with abundant representatives of Phragmodus, Plectodina, and fibrous conodont genera such as Curtognathus, Chirognathus, and Coleodus (Bergström and Sweet, 1966; Schopf, 1966; Webers, 1966). Unfortunately, nowhere in the North American Midcontinent is there a reasonably complete Middle Ordovician (Champlainian) succession and the sequence of faunas proposed by Sweet et al. (1971) is pieced together from sections in several different areas. Although the precise correlation between some units in these areas is still uncertain, the general succession of units in this scheme has proved to be very useful regionally over the North American craton (Sweet and Bergström, in press).

At present the best known Late Ordovician (Cincinnatian) conodont faunal succession in North America is that of the Cincinnati arch area in Ohio and adjacent states (Sweet, in press). The Late Ordovician faunas there include representatives of, among others, Aphelognathus, Belodina, Oulodus, Phragmodus, Plectodina, and Rhipidognathus. In some intervals, there are also occurrences of representatives of Amorphognathus, Icriodella, Periodon, Protopanderodus, and Rhodesognathus, which are generally taken to be basically North Atlantic Province forms. Some of these are important stratigraphically and serve as correlative links with the North Atlantic conodont zonal succession, where they are far more widely distributed. By and large, the indigenous Cincinnatian conodont faunas are of a rather monotonous and conservative type and they are not noted for showing conspicuous evolutionary changes through the sequence.

In the Cincinnati region, as well as elsewhere in eastern USA, the interval within which the Ordovician/Silurian boundary falls is

marked by a striking change in the composition of the conodont
faunas. A majority of the typical Midcontinent province genera apparently became extinct in latest Ordovician time and only Panderodus
and a few other "simple-cone" genera, as well as representatives of
Amorphognathus, Icriodella, Oulodus, and a few other compound-element
genera, survived into Silurian time. Accordingly, Silurian conodont
faunas differ a good deal from Ordovician ones and the general extinction of conodont genera in Late Ordovician time is one of the
most severe during the entire time of existence of the group.

The North Atlantic Province

Early Ordovician (Oelandian) conodont faunas of the North Atlantic Province are currently best known from the Baltoscandic area
(Lindström, 1955, 1960, 1971; Sergeeva, 1964; Viira, 1967, 1975;
Kohut, 1972; van Wamel, 1974) although a few collections have been
described from Scotland (Lamont and Lindström, 1957), eastern North
America (Fåhraeus, 1970; Bergström et al., 1972; Landing, 1974a),
and Argentina (Serpagli, 1974). Early Ordovician faunas of this province are characterized by a wide variety of "simple-cone" genera
such as Drepanodus, Acontiodus, Drepanoistodus, Oistodus, Paltodus,
Paroistodus, Protopanderodus, Scandodus, and Stolodus as well as
compound-element genera such as Cordylodus, Microzarkodina, Oepikodus,
and Periodon. It is of interest to note that true platform conodonts
appeared as early as in earliest Ordovician (Tremadocian) time in the
Baltoscandic area (Lindström, 1955) but such conodonts did not
constitute a substantial part of the conodont faunas until Middle
Ordovician (Viruan) time. Successions of conodont zones proposed by
Sergeeva (1964), Lindström (1971), and Viira (1975), among others,
differ relatively little from each other and the one introduced by
Lindström (1971) has proved to be widely applicable. Van Wamel (1974)
proposed a far more detailed zonal subdivision of Tremadocian and
Early Arenigian strata in eastern Sweden but some of his 20 assemblage zones, which were based on single samples, may be only of local
biostratigraphic significance, if any. On the other hand, units of
Lindström's (1971) zonal succession have been recognized in areas as
far apart as Newfoundland, Texas, Nevada, and Argentina, and it is
clear that conodonts rival graptolites as the biostratigraphically
most useful fossil group in the Lower Ordovician of this province.

In the North Atlantic Province, Middle Ordovician (Viruan) conodonts are best known from the Baltoscandic area (Bergström, 1962, 1971a, 1971b; Hamar, 1964, 1966; Lindström, 1960; Viira, 1975) and eastern North America (Bergström, 1971a, 1973c; Bergström et al., 1974; Bergström and Carnes, in press; Fåhraeus, 1970, 1973; Sweet and Bergström, 1962) but some units in Great Britain (Lamont and Lindström, 1957; Lindström, 1959; Bergström, 1964, 1971a; Rhodes, 1953) and France (Lindström et al., 1974) have also yielded conodonts of this age. Zonal units have been proposed by Bergström (1971a) and Viira (1975) and many of the units in the former's zonal succession have been recognized regionally in the North Atlantic Province. Middle Ordovician conodont faunas of this area are characterized by a succession of stratigraphically important species of Amorphognathus, Eoplacognathus, Prioniodus (Baltoniodus), and Pygodus as well as by the widespread presence of Periodon and "simple-cone" genera such as Protopanderodus. Representatives of Icriodella, although rather common in British faunas, are rare in Baltoscandia but known also from the Appalachians; it should be noted, however, that typical North Atlantic Province faunas of late Middle Ordovician age are currently not known from the Appalachians.

Late Ordovician (Harjuan) faunas of North Atlantic Province type are as yet not documented in eastern North America, but are known from several areas in Europe, including Baltoscandia (Bergström, 1971a; Viira, 1975), Great Britain (Rhodes, 1955; Bergström, 1964, 1971a), the Carnic Alps in Italy and Austria (Serpagli, 1967; Schönlaub, 1971), Thuringia (Knüpfer, 1967), and Spain (Fuganti and Serpagli, 1968). The biostratigraphic zonation of the Upper Ordovician based on conodonts is still preliminary (Bergström, 1971a) and some refinement will no doubt be possible when the pertinent faunas are better known than at the present time. North Atlantic Province faunas of this age are less varied than faunas from older parts of the system; common and widespread elements include Amorphognathus, Dapsilodus, Hamarodus, Periodon, Protopanderodus, Prioniodus, and Strachanognathus whereas forms of Icriodella, Dichodella, and Nordiodus have a more limited distribution. As is also the case in the Midcontinent Province, the interval near the Ordovician/Silurian boundary is marked by a drastic extinction of common and widespread North Atlantic Province taxa; only Amorphognathus, Hindeodella (Ozarkodina), and Icriodella, along with some "simple-cone" genera,

survived into the Silurian. It should be noted, however, that earliest Silurian (early Llandoverian) conodonts are virtually unknown in
the area under discussion and this may make the extinction seem more
abrupt than is really the case. Interestingly, the extinction appears
to have affected tropical-subtropical (Midcontinent) faunas as well
as temperate (North Atlantic) faunas equally severely. Accordingly,
the cause may not be solely a sharp drop in water temperature associated with the Late Ordovician glaciation.It may be significant that
the genera surviving into the Silurian were all among the more widely
distributed, and those not restricted to one province, during the
Ordovician which suggests that these conodonts had the capacity to
adapt to a relatively wide range of environmental conditions.

SILURIAN

In comparison with those from the Ordovician, Silurian conodonts are less well known regionally in the Atlantic Borderlands.
Also, practically nothing has been published on conodont faunas from
the very lowermost part of the system. However, judging from data
currently available, the Middle (Wenlockian) and Upper (Ludlovian
and Pridolian) Silurian conodonts are now relatively well known as
far as their general taxonomy and stratigraphic ranges are concerned.
A conodont-based zonal scheme for most of the Silurian System
was introduced by Walliser (1964). It is based largely on the Cellon
section in the Carnic Alps of Austria. With some modifications,
especially in the Llandoverian, Walliser's zonal system has proved
applicable world-wide. The zonal schemes currently in use in Europe
and North America include about 12 assemblage zones; accordingly,
they are less detailed than that based on graptolites, which includes
about 32 zones (Bulman, 1970).
In the area under discussion, Silurian conodonts are best known
from Great Britain (Aldridge, 1972, 1975), southern and eastern
Sweden (Jeppsson, 1975), central Europe (Walliser, 1964; Schönlaub,
1971; Walmsley et al., 1974), Virginia (Helfrich, 1975), and the
eastern Midcontinent of North America (Cooper, 1975; Nicoll and
Rexroad, 1968; Pollock and Rexroad, 1973; Rexroad, 1967; Rexroad and
Nicoll, 1971). Silurian conodons are also known from North Africa
(Ethington and Furnish, 1962), eastern Canada (Legault, 1968), the

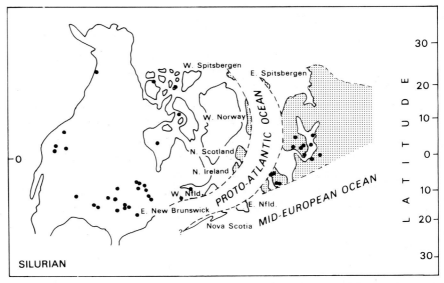

Fig.4. Silurian conodont occurrences in North America and north-western Europe plotted on a sketch-map showing inferred position of continents in Early Silurian (Llandoverian) time. Note that practically all these conodont faunas fall within a belt between 20 degrees South and North latitude.

Iberian Peninsula (Kockel, 1958; van den Boogaard, 1965), France (Feist and Schönlaub, 1974) and scattered localities in a few other areas in the Atlantic Borderlands. A review of Silurian conodont occurrences in the Atlantic Borderlands in North America and northern Europe is given in Fig.4.

Silurian conodont faunas, as now known, include 15-20 multiele-ment genera and are far less varied at both the generic and specific level than the Ordovician ones. Further, they do not show any striking biogeographic differentiation although there are some regional differences in the specific composition of the faunas (Jeppsson, 1975).

Earliest Silurian (early Llandoverian) conodonts are as yet virtually unknown but younger Llandoverian faunas are characterized by Apsidognathus, Carniodus, Distomodus, Icriodella, Llandovery-gnathus, Hindeodella (=Ozarkodina), Icriodina, and Hadrognathus. The considerable number of platform genera in the Llandoverian faunas is a notable feature; in younger Silurian faunas, platform conodonts are uncommon and little varied. Llandoverian strata, as well as younger Silurian beds, contain several widespread "simple-cone"

SERIES	CONODONT ZONES		
	CARNIC ALPS	GREAT BRITAIN	NORTH AMERICA
PRIDOLIAN	*H. s. eosteinhornensis*	*H. s. eosteinhornensis*	*H. s. eosteinhornensis*
LUDLOVIAN	*H. crispa*	NOT YET IDENTIFIED	*H. crispa*
	I. latialatus		*I. latialatus*
	P. siluricus		*P. siluricus*
	A. ploeckensis		*A. ploeckensis*
	H. crassa		*A. bicornuta*
WENLOCKIAN	*H. sagitta*	*H. sagitta*	*H. sagitta*
	K. patula	NOT YET IDENTIFIED	NOT YET IDENTIFIED
LLANDOVERIAN	*P. amorphognathoides*	*P. amorphognathoides*	*P. amorphognathoides*
	P. celloni ?	*I. inconstans* *H. staurognathoides*	*P. celloni*
	?	*I. discreta* *I. deflecta*	*I. irregularis* *P. simplex*
	Bereich 1	?	

Fig.5. Silurian conodont zones in Europe and North America. Note that the precise correlation between some units is somewhat uncertain, particularly in the Lower and Middle Llandoverian.

genera but these are of minor significance biostratigraphically, at least as far as our present knowledge goes.

Wenlockian and younger Silurian faunas are dominated by numerous species of Hindeodella (=Ozarkodina) and Ligonodina but stratigraphically important forms of the platform genera Kockelella, Pelekysgnathus, and Polygnathoides are present in some intervals. A comprehensive and authorative review especially of European Late Silurian faunas has been given recently by Jeppsson (1975) and an excellent summary of the known stratigraphic distribution of British Silurian conodonts has been presented by Aldridge (1975). Their work clearly suggests that not only are conodonts very useful guide fossils in the European Silurian, but also that the presently used system of about 12 zones can be considerably refined after a thorough taxonomic revision of the faunas. The commonly used Silurian conodont zones in Europe and North America are listed and correlated in Fig.5 which also illustrates the relations between conodont zones and European standard series.

RELATIONS BETWEEN CONODONT AND GRAPTOLITE ZONAL UNITS

For a hundred years or more, graptolites have served as the principal guide fossils in regional correlation of Ordovician and Silurian rocks in many parts of the world. Although they are still unsurpassed as biostratigraphic tools in shaly facies, the very scattered occurrence or absence of these fossils in many important carbonate units in America and Eurasia has led to problems in applying the graptolite biostratigraphy to such sequences, especially to those so widely developed on the continental shields. On the other hand, conodonts are present in a wider range of facies types than are graptolites and they may be used for correlations across facies boundaries. The fact that conodonts are most common, and most easily studied, in shelly rocks makes them particularly useful to compliment graptolites in biostratigraphic work.

Considerable efforts have been made in recent years to tie the zonal system based on conodonts into that based on graptolites. Much of this work has been carried out in the Atlantic Borderlands where, at relatively numerous localities, stratigraphically diagnostic conodonts and graptolites have been found together or in such a position that they provide useful information regarding the mutual relations between zonal units based on each of these fossil groups. These relations will be briefly reviewed below.

Cambrian

No standard graptolite and conodont zone systems have been established for the Cambrian and the biostratigraphic significance of the dendroid graptolites present in rocks of that system remains unclear.

Ordovician

The striking provincialism exhibited by both conodonts and graptolites in large parts of the Ordovician has greatly complicated the work aimed at establishing as precisely as possible the relations between conodont and graptolite zones. However, as shown by recent summaries (Bergström, 1971b, 1973b), there is now a large number of ties between North Atlantic Province conodont zones and the standard

| SERIES | | NORTH ATLANTIC CONODONT ZONES | | BRITISH GRAPTOLITE ZONES |
N. AM.	GREAT BRITAIN	AND SUBZONES		ZONES
CINCINNATIAN	ASHGILLIAN	Amorphognathus ordovicicus		Dicellogr. anceps
				Dicellogr. complanatus
				Pleurogr. linearis
	CARADOCIAN	Amorphognathus superbus		Dicranogr. clingani
		Amorphognathus tvaerensis	Prioniodus alobatus	Diplogr. multidens
			Prioniodus gerdae	
			Prioniodus variabilis	
CHAMPLAINIAN	LLANDEILIAN	Pygodus anserinus	Upper	Nemagraptus gracilis
			Lower	Glyptogr. teretiusculus
	?	Pygodus serrus	Eopl. lindstroemi	
			Eopl. robustus	
	LLANVIRNIAN		Eopl. reclinatus	Didymogr. murchisoni
		UNNAMED	Eopl. foliaceus	
			Eopl. suecicus	
		Eoplacognathus variabilis		Didymogr. "bifidus"
	ARENIGIAN	Microzarkodina parva		Didymogr. hirundo
		Paroistodus originalis		
		Prioniodus navis		Didymogr. extensus
		Prioniodus triangularis		
		Oepikodus evae		
		Prioniodus elegans		
	?	Paroistodus proteus		(Tetragr. approximatus)
CANADIAN	TREMADOCIAN	Paltodus deltifer		Anisograptidae
		Cordylodus angulatus		Clonogr. tenellus
				Dictyonema flabelliforme

Fig.6. Correlation between North American and British Ordovician standard series, North Atlantic Province conodont zonal units, and British graptolite zones.

British-Baltic graptolite zones. A review of the relations between these zonal systems, as now known, is given in Fig.6. Some progress has also been made in elucidating the relations between North Atlantic Province conodont zones and North American-Pacific Province graptolite zones (Bergström, 1971b, 1974, 1976b: Landing, 1974a) but much more work along those lines is needed. This applies also to the problem of the precise relations between Midcontinent Province conodont units and graptolite zones, for which relatively little direct evidence is now available. The scarcity of zonal graptolites in rocks with North American Midcontinent Province conodont faunas will doubtless make this work both difficult and time-consuming.

Silurian

The many graptolite zones recognized in this system within the area dealt with herein are, by and large, somewhat less satisfactorily tied directly into the conodont zonal succession than is the case

SERIES	CONODONT ZONES		EUROPEAN GRAPTOLITE ZONES
	GREAT BRITAIN	CENTRAL EUROPE	
PRIDOLIAN	H. s. eosteinhornensis	H. s. eosteinhornensis	Monogr. angustidens
			Pristiogr. transgrediens
			Monogr. perneri
			Monogr. bouceki
			Saetogr. lochkovensis
			Pristiogr. ultimus
LUDLOVIAN	NOT YET IDENTIFIED	H. crispa	Pristiogr. fecundus
		I. latialatus	Saetogr. fritschi linearis
		P. siluricus	Saetogr. leintwardinensis
			Pristiogr. tumescens
		A. ploeckensis	Cucullogr. scanicus
		H. crassa	Neodiversogr. nilssoni
WENLOCKIAN	H. sagitta	H. sagitta	Pristiogr. ludensis
			Cyrtogr. lundgreni
			Cyrtogr. ellesae
			Cyrtogr. linnarssoni
			Cyrtogr. rigidus
	NOT YET IDENTIFIED	K. patula	Monogr. riccartonensis
	P. amorphognathoides	P. amorphognathoides	Cyrtogr. murchisoni
LLANDOVERIAN	I. inconstans	P. celloni	Monocl. crenulata
			Monogr. griestoniensis
	H. staurognathoides	?	Monogr. crispus
			Monogr. turriculatus
	I. discreta I. deflecta	?	Monogr. sedgwicki
			Monogr. convolutus
			Monogr. gregarius
	?	Bereich I	Monogr. cyphus
			Cystogr. vesiculosus
			Akidogr. acuminatus
			Glyptogr. persculptus

Fig.7. Correlation between European Silurian standard series, conodont zones, and graptolite zones.

in the Ordovician. In his standard work, Walliser (1964) presented a preliminary correlation between his conodont zones and standard graptolite zones that is still valid in most respects although subsequent research has made it possible to refine it to a certain extent. However, there is still some uncertainty regarding the precise relation between some conodont and graptolite zones, especially in the Llandoverian, and more detailed conodont work is clearly needed in graptolite-bearing sections. Fig.7 is an attempt to summarize the currently known relations between Silurian conodont and graptolite zones.

In the combined graptolite-conodont zonal systems we have a powerful tool to recognize biostratigraphic units across facies and provincial boundaries, and to correlate units over large distances with unusual precision. Indeed, there are now intervals within which trans-Atlantic correlations can be achieved with a stratigraphic resolution of a few feet, and it is to be expected that further refinement will be possible in many intervals when additional data become available from critical areas and key sections. Obviously, one future goal may be to try to establish one detailed zonal framework based on all stratigraphically important fossil groups but much remains to be learned about the mutual relations between the ranges of representatives of these groups before a meaningful attempt at such a synthesis can be made.

CONODONTS AND THE EVOLUTION OF THE PROTO-ATLANTIC

Fossils played a significant role in the early development of the concept of a Proto-Atlantic Ocean (Wilson, 1966) and indications from especially benthic megafossils such as brachiopods and trilobites have also been widely used in subsequent discussions regarding Paleozoic plate tectonics and the evolution of the Proto-Atlantic (Williams, 1973; Whittington and Hughes, 1972, 1973; Burrett, 1973). The distribution of predominantly planktic or epiplanktic organisms such as graptolites and conodonts has attracted somewhat less interest in these discussions (Bergström, 1971a, 1973a, 1976a; Skevington, 1976; Bergström et al., 1972, 1974). This might be related to the fact that the very wide horizontal distribution of many taxa of these groups certainly suggests that such forms were capable of crossing water bodies that were large enough to serve as migrational barriers to most benthic organisms and, accordingly, it might have been assumed that these groups would be less likely to provide useful data regarding the mutual positions of the continental plates. This idea is reenforced by the fact that, as has been noted by Cook and Taylor (1975), the degree of faunal resemblance between two areas is not necessarily directly related to the geographic distance between them, and presumably, this is likely to apply to planktic and epiplanktic forms even more than to benthic organisms. However, this study is concerned with a time interval and a geographic region characterized by very active seafloor spreading and it is therefore appropriate to consider briefly the relations between the conodont faunas in Europe and North America and the general model of the development of the Proto-Atlantic.

Even a casual look at the trans-Atlantic conodont faunal relations during Cambro-Silurian time shows that there is apparently no simple relation between varying faunal resemblance and geographic

proximity of the North-European and North American plates. During
the period of maximum expansion of the Proto-Atlantic in the Cambrian,
conodont faunas show a high degree of similarity regionally with
numerous cosmopolitan taxa. In Early Ordovician time, a striking
provincial differentiation developed, which lasted through virtually
the entire period. This resulted in the establishment of Midcontinent
Province faunas over the North American Platform and the strikingly
different North Atlantic Province faunas in northwestern Europe and,
interestingly, in regions along the eastern and western margins of the
North American craton. Due to the fact that most of the pertinent
faunas have not yet been revised on multielement basis, it is currently impossible to give meaningful numerical values of faunal resemblance between the two provinces during different time intervals in
the Ordovician; however, it appears that the differences were
especially conspicuous during Middle Ordovician (Champlainian) time.
This period of maximum conodont provincialism is not in phase with
commonly proposed models for the evolution of the Proto-Atlantic,
which suggest maximum continental separation in the Late Cambrian
(when conodont faunas are largely cosmopolitan) and successive
closing during the Ordovician, that is, during the period of most
pronounced conodont provincialism. It should be noted, however, that
the cosmopolitan faunas in the Silurian are in accord with a model
of a then largely subducted Proto-Atlantic (Fig.4).

The important fact that North Atlantic Province conodont faunas
are present not only in Europe but also on the American side of the
Proto-Atlantic in regions clearly belonging to the North American
plate (Fig.2) indicates that the Ordovician Proto-Atlantic was an
inefficient migration barrier to these conodonts. Also, it is clear
that the prime factors that served to prevent migration of these
faunas into the Midcontinent Province area further on to the craton
evidently did not include geographic distance. Rather, it seems likely that they were of ecologic character. The nature of this ecological
control has been dealt with repeatedly in recent years (Barnes et al.,
1973; Barnes and Fåhraeus, 1975; Bergström, 1971a, 1973a; Bergström
and Carnes, in press; Sweet and Bergström, 1974) and a discussion of
it is beyond the scope of the present paper. Yet it is of interest
to note that these ecological factors (among which water temperature
no doubt was important) apparently also strongly influenced the
distributional patterns of several megafossil groups, including

graptolites and brachiopods.

In view of the apparently insignificant role played by geographic distance alone vis-a-vis ecological factors for the changing conodont distributional patterns particularly in the Cambro-Ordovician, along with the fact that very similar conodont faunas were present on both sides of the Proto-Atlantic, it appears that conodonts are not likely to provide critical data toward an interpretation of the relative width of the Proto-Atlantic during different time intervals. Accordingly, one has to be content with the fact that although conodonts are superior tools for trans-Atlantic correlations, at least at the present time they give little evidence regarding the size and development of the Proto-Atlantic Ocean.

ACKNOWLEDGEMENTS

I am indebted to Dr. Walter C. Sweet for reading the manuscript and offering valuable comments and to Mrs. Helen Jones and Mrs. Karen Tayler for technical assistance.

REFERENCES

Aldridge, R.J., 1972. Llandovery conodonts from the Welsh Borderland: Brit. Mus. (Nat. Hist.) Bull. Geol., 22:2, 125-131.

Aldridge, R.J., 1975. The stratigraphic distribution of conodonts in the British Silurian: Geol. Soc. London Jour., 131, 607-618.

Barnes, C.R., and Fåhraeus, L.F., 1975. Provinces, communities, and the proposed nectobenthic habit of Ordovician conodontophorids: Lethaia, 8, 133-149.

Barnes, C.R., Rexroad, C.B., and Miller, J.F., 1973. Lower Paleozoic conodont provincialism: Geol. Soc. America Spec. Paper 141, 157-190.

Bengtsson, S., 1976. The structure of some Middle Cambrian conodonts, and the early evolution of conodont structure and function: Lethaia, 9, 185-206.

Bergström, S.M., 1962. Conodonts from the Ludibundus Limestone (Middle Ordovician) of the Tvären area (S.E. Sweden): Ark. f. Mineralogi o. Geol., 3:1, 1-61.

Bergström, S.M., 1964. Remarks on some Ordovician conodont faunas from Wales: Acta Univ. Lundensis, II:3, 1-66.

Bergström, S.M., 1971a. Conodont biostratigraphy of the Middle and Upper Ordovician of Europe and eastern North America: Geol. Soc. America Mem. 127, 83-157.

Bergström, S.M., 1971b. Correlation of the North Atlantic Middle and Upper Ordovician conodont zonation with the graptolite succession: Mém. Rech. Géol. et Min., 73, 177-187.

Bergström, S.M., 1973a. Ordovician conodonts. In Hallam, A. (Ed.), Atlas of Palaeobiogeography, 47-58. Elsevier Publ. Company.

Bergström, S.M., 1973b. Correlation of the late Lasnamägian Stage (Middle Ordovician) with the graptolite succession: Geol. Fören. Stockholm Förhandl., 95, 9-18.

Bergström, S.M., 1973c. Biostratigraphy and facies relations in the lower Middle Ordovician of easternmost Tennessee: Amer. Jour. Science, 273-A, 261-293.

Bergström, S.M., 1974. Ordovician correlations by means of non-benthonic fossils: On the relations between North American standard graptolite zones and North Atlantic Province conodont zones: Geol. Soc. America, Abstr. with Progr., 6:7, 652-653.

Bergström, S.M., 1976a. Ordovician conodonts and the "Proto-Atlantic" Ocean: Geol. Soc. America, Abstr. with Progr., 8:2, 133.

Bergström, S.M., 1976b. The Marathon Middle and Upper Ordovician succession reconsidered: Conodont and graptolite biostratigraphy of the Woods Hollow and Maravillas Formations: Geol. Soc. America, Abstr. with Progr., 8:4, 463-464.

Bergström, S.M., and Carnes, J.B., in press. Conodont biostratigraphy and paleoecology of the Holston Formation (Middle Ordovician) and associated strata in eastern Tennessee. In Barnes, C. R. (Ed.), Conodont Paleoecology: Geol. Assoc. Canada Spec. Pap.15.

Bergström, S.M., Epstein, A.G., and Epstein, J.B., 1972. Early Ordovician North Atlantic province conodonts in eastern Pennsylvania: U.S. Geol. Surv. Prof. Pap. 800D, D37-D44.

Bergström, S.M., Riva, J., and Kay, M., 1974. Significance of conodonts, graptolites, and shelly faunas from the Ordovician of western and north-central Newfoundland: Can. Jour. Earth Sci., 11:12, 1625-1660.

Bergström, S.M., and Sweet, W.C., 1966. Conodonts from the Lexington Limestone (Middle Ordovician) of Kentucky and its lateral equivalents in Ohio and Indiana: Bull. Amer. Pal., 50:229, 271-441.

van den Boogaard, M., 1965. Two conodont faunas from the Paleozoic of the Betic of Malaga near Vélez Rubio, S.E. Spain: Koninkl. Nederl. Akad. Wetensch. Proc. Ser. B, 68, 33-37.

Bradshaw, L.E., 1969. Conodonts from the Fort Peña Formation (Middle Ordovician), Marathon Basin, Texas: Jour. Paleont., 43, 1137-1168.

Brand, U., 1976. Lower Ordovician conodonts from the Kindblade Formation, Arbuckle Mountains, Oklahoma: Unpubl. M.A. thesis, Univ. of Missouri, Columbia, 110 pp.

Bulman, O.M.B., 1970. Graptolithina with sections on Enterepneusta and Pterobranchia. In Teichert, C. (Ed.), Treatise on Invertebrate Paleontology, V (2nd ed.), 163 pp.

Burrett, C., 1973. Ordovician biogeography and continental drift: Palaeogeogr., Palaeoclim., Palaeoecol., 13, 161-201.

Cook, H.E., and Taylor, M.E., 1975. Early Paleozoic continental margin sedimentation, trilobite biofacies, and the thermocline, western United States: Geology, 3:10, 559-562.

Cooper, B.J., 1975. Multielement conodonts from the Brassfield Limestone (Silurian) of southern Ohio: Jour. Paleont., 49, 984-1008.

Druce, E.C., and Jones, P.J., 1971. Cambro-Ordovician conodonts from the Burke River structural belt, Queensland: Bur. Min. Resour. Geol. Geophys. Bull. 110, 118 pp.

Ethington, R.L., and Clark, D.L., 1971. Lower Ordovician conodonts in North America: Geol. Soc. America Mem. 127, 63-82.

Ethington, R.L., and Furnish, W., 1962. Silurian and Devonian cono-
     donts from Spanish Sahara: Jour. Paleont., 36, 1253-1290.
Fåhraeus, L.E., 1970. Conodont-based correlations of Lower and Middle
     Ordovician strata in western Newfoundland: Geol. Soc. America
     Bull., 81, 2061-2076.
Fåhraeus, L.E., 1973. Depositional environments and conodont-based
     correlations of the Long Point Formation (Middle Ordovician),
     western Newfoundland: Can. Jour. Earth Sci., 10, 1822-1833.
Feist, R., and Schönlaub, H.-P., 1974. Zur Silur/Devon-Grenze in
     östlichen Montagne Noire Süd-Frankreichs: Neues Jahrb. Geol.
     Paläont. Monatsh., Jahrg. 1974, 4, 200-219.
Fuganti, A., and Serpagli, E., 1968. Geological remarks on Urbana
     Limestone and evidence for its Upper Ordovician age by means
     of conodonts - eastern Sierra Morena, South Spain: Geol. Soc.
     Ital. Boll., 87, 511-521.
Hamar, G., 1964. The Middle Ordovician of the Oslo Region, Norway.
     17. Conodonts from the lower Middle Ordovician of Ringerike:
     Norsk Geol. Tidsskr., 44, 243-292.
Hamar, G., 1966. The Middle Ordovician of the Oslo Region, Norway.
     22. Preliminary report on conodonts from the Oslo-Asker and
     Ringerike districts: Norsk Geol. Tidsskr., 46, 27-83.
Helfrich, C.T., 1975. Silurian conodonts from Wills Mountain anti-
     cline, Virginia, West Virginia, and Maryland: Geol. Soc. Amer.
     Spec. Paper 161, 82 pp. + appendix.
Jeppsson, L., 1975. Aspects of Silurian conodonts: Fossils and Strata
     6, 54 pp.
Knüpfer, J., 1967. Zur Fauna und Biostratigraphie des Ordoviciums
     (Gräfenthaler Schichten) in Thüringen: Freiberger Forschungs-
     Hefte, C220, 119 pp.
Kockel, F., 1958. Conodonten aus dem Paläozoikum von Malaga (Süd-
     Spanien): Neues Jahrb. Geol. Paläont. Monatsh., Jahrg., 225-263.
Kohut, J.J., 1972. Conodont biostratigraphy of the Lower Ordovician
     Orthoceras and Stein Limestones (3c), Norway: Norsk Geol.
     Tidsskr., 52, 427-445.
Lamont, A., and Lindström, M., 1957. Arenigian and Llandeilian
     cherts identified in the Southern Uplands of Scotland by means
     of conodonts, etc.: Edinburgh Geol. Soc. Trans., 17:1, 60-70.
Landing, E., 1974a. Lower Ordovician conodont and graptolite bio-
     stratigraphy of the Taconic Province, eastern New York: Geol.
     Soc. America, Abstr. with Progr., 6:6, 525-526.
Landing, E., 1974b. Early and Middle Cambrian conodonts from the
     Taconic Allochthon, eastern New York: Jour. Paleont., 48, 1241-
     1248.
Landing, E., 1976. Late Cambrian Prooneotodus tenuis (Müller) appara-
     tuses and associated conodonts from the Taconic Allochthon,
     eastern New York: Geol. Soc. America, Abstr. with Progr., 8:4,
     487-488.
Lee, H.-Y., 1970. Conodonten aus der Choson-Gruppe (Unteres Ordoviz-
     ium) von Korea: Neues Jahrb. Geol. Paläont. Abh., 136:3, 303-
     344.
Lee, H.-Y., 1975. Conodonten aus dem unteren und mittleren Ordoviz-
     ium von Nordkorea: Palaeontographica, Abt. A, 150, 161-186.
Legault, J.A., 1968. Conodonts and fish remains from the Stonehouse
     Formation, Arisaig, Nova Scotia: Canada Geol. Surv. Bull. 165,
     30 pp.
Lindström, M., 1955. Conodonts from the lowermost Ordovician strata
     of south-central Sweden: Geol. Fören. Stockholm Förhandl., 76,
     517-614.

Lindström, M., 1959. Conodonts from the Crûg Limestone (Ordovician, Wales): Micropaleontology, 5, 427-452.

Lindström, M., 1960. A Lower-Middle Ordovician succession of conodont faunas: Internat. Geol. Congr., XXI Sess., Repts., 7, 88-96.

Lindström, M., 1971. Lower Ordovician conodonts of Europe: Geol. Soc. America Mem. 127, 21-61.

Lindström, M., Racheboef, P.R., and Henry, J.-L., 1974. Ordovician conodonts from the Postolonnec Formation (Crozon Peninsula, Massif Armoricain) and their stratigraphic significance: Geol. et Palaeont., 8, 15-28.

Matthews, S.C., and Missarzhevsky, V.V., 1975. Small shelly fossils of late Precambrian and early Cambrian age: a review of recent work: Geol. Soc. London Jour., 131, 289-304.

McElhinny, M.W., and Opdyke, N.D., 1973. Remagnetization hypothesis discounted: a paleomagnetic study of the Trenton Limestone, New York State: Geol. Soc. America Bull., 84, 3697-3708.

Miller, J.F., 1969. Conodont fauna of the Notch Peak Limestone (Cambro-Ordovician), House Range, Utah: Jour. Paleont., 43, 413-439.

Miller, J.F., 1975. Conodont faunas from the Cambrian and lowest Ordovician of western North America: Geol. Soc. America, Abstr. with Progr., 7:7, 1200-1201.

Miller, J.F., and Rushton, A.W.A., 1973. Natural conodont assemblages from the Upper Cambrian of Warwickshire, Great Britain: Geol. Soc. America, Abstr. with Progr., 5:4, 338-339.

Missarzhevsky, V.V., 1973. Conodontiform organisms from beds close to the Precambrian-Cambrian boundary on the Siberian Platform and in Kazakhstan. In Zhuraleva, I.T., (Ed.), Paleontological and biostratigraphic problems in the Lower Cambrian of Sibiria and the Far East, 57-59. "Nauka" Publ. House, Novosibirsk. (in Russian).

Mound, M., 1965. A conodont fauna from the Joins Formation (Ordovician), Oklahoma: Tulane Studies Geol., 4, 1-46.

Mound, M., 1968. Conodonts and biostratigraphy of the Lower Arbuckle Group (Ordovician), Arbuckle Mountains, Oklahoma: Micropaleont. 14:4, 393-434.

Müller, K.J., 1959. Kambrische Conodonten: Zeitschr. Deutsch. Geol. Ges., 111, 435-485.

Müller, K.J., 1973. Late Cambrian and Early Ordovician conodonts from northern Iran: Geol. Surv. Iran, Repts., 30, 77 pp.

Nicoll, R.S., and Rexroad, C.B., 1968. Stratigraphy and conodont paleontology of the Salamonie Dolomite and Lee Creek Member of the Brassfield Limestone (Silurian) in southeastern Indiana and adjacent Kentucky: Indiana Geol. Surv. Bull. 40, 73 pp.

Nogami, Y., 1966. Kambrische Conodonten von China. Teil 1. Conodonten aus den oberkambrischen Kushan-Schichten: Coll. Sci. Univ. Kyoto, Mem., 32, 351-367.

Nogami, Y., 1967. Kambrische Conodonten von China. Teil 2. Conodonten aus den hoch oberkambrischen Yencho-Schichten: Coll. Sci. Univ. Kyoto, Mem., 33, 211-218.

Noltimier, H.C., and Bergström, S.M., 1976. Paleomagnetic studies of Early and Middle Ordovician limestones from the Baltic Shield: Geol. Soc. America, Abstr. with Progr., 8:4, 501.

Pollock, C.A., and Rexroad, C.B., 1973. Conodonts from the Salina Formation and the upper part of the Wabash Formation (Silurian) in North-Central Indiana: Geol. et Palaeont., 7, 77-92.

Potter, C.A., 1975. Lower Ordovician conodonts of the upper West Spring Creek Formation, Arbuckle Mountains, Oklahoma: Unpubl. M.A. thesis, Univ. of Missouri, Columbia, 133 pp.

Poulsen, V., 1966. Early Cambrian distacodontid conodonts from Born-
     holm: Biol. Medd. Kongl. Dan. Vidensk. Selsk., 23:15, 1-12.
Repetski, J.E., 1974. Conodonts from the Lower Ordovician El Paso
     Group of West Texas: Geol. Soc. America, Abstr. with Program.,
     6:6, 540.
Rexroad, C.B., 1967. Stratigraphy and conodont paleontology of the
     Brassfield (Silurian) in the Cincinnati Arch area: Indiana
     Geol. Surv. Bull. 36, 64 pp.
Rexroad, C.B., and Nicoll, R.S., 1971. Summary of conodont biostrati-
     graphy of the Silurian System of North America: Geol. Soc.
     America Mem. 127, 207-225.
Rhodes, F.H.T., 1953. Some British Lower Paleozoic conodont faunas:
     Roy. Soc. London Phil. Trans., Ser. B, 647, 237, 261-334.
Rhodes, F.H.T., 1955. The conodont fauna of the Keisley Limestone:
     Geol. Soc. London, Quart. Jour., 11, 117-142.
Schönlaub, H.-P., 1971. Zur Problematik der Conodonten-Chronologie
     an der Wende Ordoviz/Silur mit besonderer Berücksichtigung der
     Verhältnisse im Llandovery: Geol. et Paleont., 5, 35-57.
Schopf, T.J.M., 1966. Conodonts of the Trenton Group (Ordovician) in
     New York, southern Ontario, and Quebec: New York State Mus.
     Bull., 405, 105 pp.
Sergeeva, S.P., 1964. On the significance of the Lower Ordovician
     conodonts in the Leningrad region: Leningrad Univ. Vestnik,
     Ser. Geologiya i Geografii, 12:2, 56-60. (in Russian).
Serpagli, E., 1967. I conodonti dell'Ordoviciano Superiore (Ashgill-
     iano) delle Alpi Carniche: Soc. Paleont. Ital. Boll.,6, 30-111.
Serpagli, E., 1974. Lower Ordovician conodonts from Precordilleran
     Argentina (Province of San Juan): Soc. Paleont. Ital. Boll.,
     13, 17-98.
Skevington, D., 1976. Protochordates, palaeolatitudes and the Proto-
     Atlantic Ocean: Geol. Soc. America, Abstr. with Progr., 8:2,
     269.
Sweet, W.C., in press. Conodonts and conodont biostratigraphy of
     post-Tyrone Ordovician rocks of the Cincinnati region: U. S.
     Geol. Surv. Prof. Paper.
Sweet, W.C., and Bergström, S.M., 1962.Conodonts from the Pratt
     Ferry Formation (Middle Ordovician) of Alabama: Jour. Paleont.,
     36, 1214-1252.
Sweet, W.C., and Bergström, S.M., 1971. Symposium on conodont bio-
     stratigraphy: Geol. Soc. America Mem. 127, 499 pp.
Sweet, W.C., and Bergström, S.M., 1974. Provincialism exhibited by
     Ordovician conodont faunas: Soc. Econ. Paleont. Min. Spec.
     Publ. 21, 189-202.
Sweet, W.C., and Bergström, S.M., in press. Conodont biostratigraphy
     of the Middle and Upper Ordovician of the midcontinent of the
     United States: Palaeontol. Assoc. Spec. Paper.
Sweet, W.C., Ethington, R.L., and Barnes, C.R., 1971. North American
     Middle and Upper Ordovician conodont faunas: Geol. Soc. America
     Mem. 127, 163-193.
Szaniawski, H., 1971. New species of Upper Cambrian conodonts from
     Poland: Acta Geol. Polonica, 16, 401-413.
Viira, V., 1967. Ordovician conodont succession in the Ohesaare core:
     Eesti NSV Tead. Akad. Toimetised, 16, 319-329.
Viira, V., 1975. Ordovician conodonts of the East Baltic: Eesti NSV
     Tead. Akad. Geol. Inst. Tallinn, 142 pp. (in Russian).
Walliser, O.H., 1964. Conodonten des Silurs: Hess. L.-Amt Bodenforsch.
     Abh., 41, 106 pp.
Walmsley, V.G., Aldridge, R.J., and Austin, R.L., 1974. Brachiopod

108

and conodont faunas from the Silurian and Lower Devonian of Bohemia: Geol. et Paleont., 8, 39-47.

van Wamel, W.A., 1974. Conodont biostratigraphy of the Upper Cambrian and Lower Ordovician of north-western Öland, south-eastern Sweden: Utrecht Micropaleont. Bull. 10, 126 pp.

Webers, G.F., 1966. The Middle and Upper Ordovician conodont faunas of Minnesota: Minnesota Geol. Surv. Spec. Publ. SP-4, 123 pp.

Whittington, H.B., and Hughes, C.P., 1972. Ordovician geography and faunal provinces deduced from trilobite distribution: Royal Soc. London Phil. Trans., B263, 235-278.

Whittington, H.B., and Hughes, C.P., 1973. Ordovician trilobite distribution and geography: Palaeontol. Assoc. Spec. Pap., 12, 235-240.

Williams, A., 1973. Distribution of brachiopod assemblages in relation to Ordovician palaeogeography: Palaeont. Assoc. Spec. Pap., 12, 241-269.

Wilson, J.T., 1966. Did the Atlantic close and then re-open? Nature, London, 211, 676-681.

Discussion

Dr. I. G. Sohn: Do you consider the diversity of conodonts to be greater in deep than in shallow water?

Dr. S. M. Bergström: It is difficult to give a definite answer to this question but as a rule, faunas were more diversified in somewhat deeper than in very shallow water. In, for instance, the North American Middle Ordovician, the diversity in conodont faunas increases considerably from the central part of the continent to the Appalachians. It is especially striking that Ordovician platform conodonts were much more abundant and varied in the presumably somewhat deeper water along the continental margins than in the shallow epicontinental seas on the platform itself.

Dr. B. K. Holdsworth: 1. What evidence is there for considering the Dalradian part of the American plate in Cambrian time, as the only known trilobites do not necessarily seem to be of the American province?

2. What evidence exists for an Ordovician ocean extending northwards into S.W. England, all France and well into central Europe?

Bergström: As far as the first question is concerned, I am aware of the weaknesses of some of the data behind the Cambrian reconstruction. As you point out, faunal and geologic evidence for placing parts of the British Isles, etc. on one or the other side of the Proto-Atlantic is not very strong and at the present time, it is certainly not sufficient for making any firm judgements as to the former geographic positions of these areas. The position of the Baltic Shield in my map is based on paleomagnetic measurements on Middle Cambrian rocks that we have carried out recently, and I have adopted the conventional view and kept part of the British Isles -- The English Midlands, etc. -- in about the same geographic position to the Baltic Shield as today. But I have no strong feelings regarding where Scotland is likely to have been in Cambrian time. I believe, however, that the marine Lower Paleozoic sediments on the Hebridean Platform in north-westernmost Scotland, for instance, the Durness Limestone, show so close affinities with equivalent sediments in easternmost North America, Greenland, Spitsbergen, etc. that it is reasonable to suggest that they all belong to the same general province.

As far as the second question is concerned, I have preferred, in the model of the Early Ordovician Proto-Atlantic, to leave out from the map Cornwall and the other Hercynian areas in the SW part of the British Isles. I personally believe that there was a mid-European sea extending into that area as suggested by other authors, but I do not know if Cornwall was a part of Great Britain or not during Early Ordovician time.

Dr. J. W. Neale: If I can follow up what Dr. Holdsworth said, I was rather worried about some of your plate tectonic reconstructions. In terms of the conodonts, I noticed this particularly in your Llandeilian reconstruction where you had your American and North Atlantic faunas sharply separated somewhere in the region of the Appalachians. At the same time, the proto-Atlantic Ocean apparently had no effect on the ease of dissemination of conodonts from the European areas to the east American area. In this instance if your proto-ocean is no barrier to your North Atlantic fauna, what controls the sharp division between your North Atlantic fauna and your American fauna in the Appalachian area? In other words, would it not be better to attach the east American part of your plate to the European area at that time?

Bergström: This is a very interesting question and a problem on which, as you know, I have been working for several years. It is striking how similar the Lower-Middle Ordovician conodont faunas in the eastern part of the Appalachians are to those in Europe. In many cases one can simply match faunas species by species. On the other hand, if one goes to the western Appalachians twenty-thirty miles to the west, one encounters Midcontinent-type conodont faunas which are virtually completely different. For a while, I was inclined to believe that maybe, there was a plate boundary between these two conodont faunal provinces. However, if one studies the local Ordovician geology in Tennessee, Virginia, etc., one finds that even if these rocks have been thrusted some distance – probably a rather short distance – to the northwest, one can trace many formations from the western Appalachians into at least the central part of the Appalachian Valley; accordingly, there appears to be no reason to regard the conodont faunal provincial boundary as coinciding with a plate boundary. Actually, there is a zone of transition that may be one or two thrust belts wide in which some North Atlantic province elements are mixed with representatives of the Midcontinent province, and this rather narrow belt has been noted locally from Alabama all the way up to Maryland in the central part of the Appalachian Valley.

I think this faunal differentiation in the Appalachian Valley is likely to have been caused by ecological control, not by geographic separation. I believe that basic controlling factors were water temperature and water depth. According to recently published reconstructions by Walker and others of sedimentary environments in the Middle Ordovician of eastern Tennessee, the critical area was not on the continental slope but on the outer part of the platform. The North Atlantic conodonts preferred the somewhat deeper water along the continental margin, and with few exceptions, they did not inhabit the shallower water farther onto the platform. Although now partly obscured by thrust faults, the transition zone may have been five miles or so wide where one can go from one type of fauna into another. So I do not think that the present provincial differentiation of conodonts in that area can serve for identification of platform boundaries.

LATE PALEOZOIC AND TRIASSIC CONODONT BIOSTRATIGRAPHY:

CORRELATIONS AROUND THE EXPANDING ATLANTIC OCEAN

David L. Clark

Department of Geology and Geophysics
University of Wisconsin, Madison

ABSTRACT

The best Late Devonian configuration for the present Atlantic borderland area includes North America and northern Europe in contact or close proximity and a "proto-Atlantic Ocean" separating these continents from Africa and South America. During the Carboniferous, the Africa-South America block approached the North America-northern Europe block, destroying the "proto-Atlantic" and during the Permo-Triassic, there was no major separation of a tenuously fused South America, Africa, North America, Europe continent. By Late Triassic, the present North Atlantic began forming as the European and African continents began the shift to their present positions.

The relationship of this continental movement to conodont biostratigraphy on either side of the present Atlantic Basin can be compared using the similarity index, $2w/ a + b$, for conodont form-species. Conodont similarity between Europe and North America was greatest during the Late Devonian and Permian, with a gradual decrease in the Triassic that might be related to formation of the present Atlantic Basin. Similarity indices among other Atlantic borderland continents are somewhat ambiguous, as far as documenting continental proximity. This is because of limited taxonomic data. Also, such things as low Carboniferous Atlantic borderland conodont similarity may be explained by other factors such as the ecologic idiosyncrasies of the ancient Pacific and Tethyan water masses, or different continental elevations during the time interval. Regardless, some 100 conodont zones (43 Devonian, 23 Carboniferous, 12 Permian, 22 Triassic) provide an important biostratigraphic framework for the Atlantic borderlands.

A calculated zonal similarity index mimics that of the form-species index. The greatest zonal similarity is during the Late Devonian, much less during the Carboniferous, but high again in the Triassic before dropping in the Late Triassic, perhaps, in part, reflecting Late Triassic separation of North America and Europe.

RÉSUMÉ

La meilleure configuration du dévonien récent pour la région de bord actuelle de l'Atlantique comprend l'Amérique du Nord et l'Europe du nord, soit en contacte l'un avec l'autre soit en proche proximité, et une "Océan proto-Atlantique" qui sépare ces continents de l'Afrique et l'Amérique du Sud. Pendant le carbonifère, le bloc comprenant l'Amérique du Sud et l'Afrique se rapprochait du bloc comprenant l'Amérique du Nord et l'Europe du nord. La "proto-Atlantique" était détruite et pendant le permo-trias, il n'y avait aucune division dans le continent tentativement lié de l'Amérique du Sud, l'Afrique, l'Amérique du Nord et l'Europe. Par le trias récent, l'Atlantique du Nord actuelle avait commencé à prendre forme pendant que les continents européen et africain prenaient leurs positions actuelles.

Le rapport de ce mouvement continental à la biostratigraphie conodonte des deux côtés du Bassin Atlantique actuel peut être comparé en utilisant l'indice de similitude, 2w/a+b, pour le genre-forme conodont. La similitude conodonte entre l'Europe et l'Amérique du Nord était la plus grande pendant le dévonien récent et le permien, avec une lente réduction pendant le trias qui pourrait se rapporter à la formation du Bassin Atlantique actuel. Les indices de similitude parmi d'autres régions de bord de l'Atlantique sont assez ambiguës, quant à la documentation de la proximité continentale. Ceci est dû aux données taxonomiques limitées. Aussi, des choses comme une basse similitude conodonte de la région de bord de l'Atlantique carbonifère peuvent être expliquées par d'autres facteurs comme les idiosyncrasies des anciennes masses d'eaux de la Pacifique et Tethyan ou des différentes élévations continentales pendant ces périodes de temps.

Néanmoins, il y a une centaine de zones conodontes (43 dévoniennes, 23 carbonifères, 12 permiennes, 22 trias) qui comprennent un cadre biostratigraphique important pour les régions de bord de l'Atlantique. Une indice de similitude zonale calculée copie celle du genre-forme. La plus grande similitude zonale arrive pendant le dévonien récent, encore moins pendant le carbonifère, mais beaucoup plus encore pendant le trias avant de se rabaisser pendant le trias récent, peut-être ainsi réfléchissant en partie la séparation de l'Amérique du Nord de l'Europe pendant le trias récent.

INTRODUCTION

Worldwide conodont biostratigraphy was summarized in 1971 (Sweet and Bergstrom, 1971). Since then advances in conodont research primarily have been in areas of biology, paleoecology and paleogeography. Because significant new ideas have developed concerning continental placement during the geologic past, it is now possible to relate conodont biostratigraphy more closely to paleogeography. New Phanerozoic world maps have provided a base for Atlantic borderland biostratigraphy, in particular (Smith and others, 1973).

This paper, while reviewing and in some instances updating Late Paleozoic and Triassic conodont biostratigraphy, includes an attempt at relating the Atlantic Basin development of the time to conodont similarities in the borderlands. Atlantic borderlands geology includes little marine Pennsylvanian, Permian or Triassic. The portion of this review related to conodonts of these systems is based on western North American and European occurrences that, in the strictest sense, are not Atlantic borderlands. The similarities and differences among these conodont faunas are related directly to the Atlantic Basin development, however, and have a legitimate place in any discussion of Atlantic Basin borderland biostratigraphy.

DEVONIAN

Introduction

Devonian conodonts are well known in the British Isles, Belgium, Germany, France, Spain, Portugal and western North Africa (Morocco and Spanish Sahara) for the eastern Atlantic borderlands, but less well known along the Atlantic seaboard of North America and relatively unknown in South America. The most comprehensive biostratigraphy is based on work in central Europe (Germany, Austria, Poland, Bulgaria, and Italy) and western and central North America (Iowa, Illinois, Nevada, Utah, Ontario, Alberta, and the Arctic Islands). The precision of Upper Devonian correlation across the present Atlantic Basin is unexcelled in any other system. Details of Lower and Middle Devonian faunas are not as comprehensive.

114

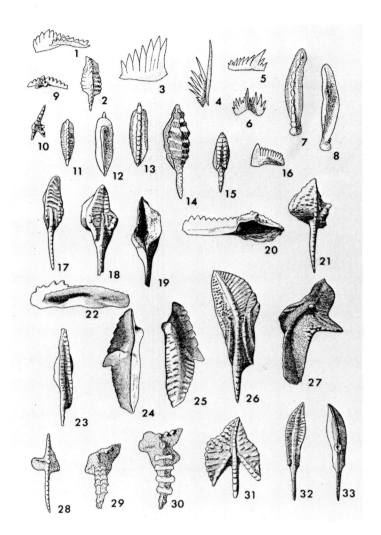

Fig. 1. Devonian to Triassic "index" species. All figures
except 14, adapted from Lindstrom, 1964. Figures, X28, except
as indicated.

1 - 2   Epigondolella mungoensis (Diebel); 1, lateral; 2, up-
          per view; Upper Middle Triassic species; genus impor-
          tant in Upper Triassic, as well.
3   Neospathodus cristagalli (Huckriede); Lower Triassic
          species; genus important in Upper Permian and through-
          out Triassic.
4 - 6   Ellisonia sp.; multielement species important in Up-
          per Permian and Triassic; 4, LC-element; 5, ozarkodi-
          nid element; 6, U-element. (continued on next page)

Lower Devonian

Gedinnian

The most important Gedinnian conodont is <u>Icriodus wo-</u>
<u>schmidti</u> (Fig. 1). It marks the base of the Devonian on both
sides of the Atlantic and ranges through much of the Gedinnian,
at least as recognized in New York (Klapper <u>et al</u>., 1971, p.
288). Other species of <u>Icriodus</u> (<u>I</u>. <u>post-woschmidti</u>, <u>I</u>. <u>pesa-</u>
<u>vis</u> in Europe and <u>I</u>. <u>latericrescens</u> in North America) plus
species of <u>Spathognathodus</u> comprise the important Upper
Gedinnian conodonts (Fig. 2).

Figure 1 (continued)

7 - 8	<u>Gladiogondolella tethydis</u> (Huckriede); 7, upper; 8, lower view; Middle Triassic species.
9 - 10	<u>Furnishius triserratus</u> Clark; 9, lateral; 10, upper view; Lower Triassic species.
11	<u>Neogondolella serrata</u> (Clark and Ethington); Upper Permian species; genus important throughout Permian and Lower and Middle Triassic.
12 - 13	<u>Neogondolella</u> sp.; 12, lower; 13, upper view; important Permian and Lower and Middle Triassic genus.
14	<u>Sweetognathus whitei</u> (Rhodes); X60; Lower Permian species.
15	<u>Neostreptognathodus sulcoplicatus</u> (Youngquist, Hawley and Miller); Middle Permian species.
16	<u>Anchignathodus</u> sp.; X23; Permian and Lower Triassic genus.
17	<u>Idiognathodus acutus</u> Ellison; X29; Upper Carboniferous genus.
18 - 19	<u>Streptognathodus eccentricus</u> Ellison; X29; 18, upper; 19, lower view; Carboniferous genus.
20 - 21	<u>Gnathodus punctatus</u> (Cooper); X29; 20, lateral; 21, upper view; Carboniferous genus.
22 - 23	<u>Cavusgnathus convexa</u> Rexroad; 22, lateral; 23, upper view; Lower Carboniferous species.
24 - 25	<u>Mestognathus beckmanni</u> Bischoff; 24, upper; 25, lower view; Lower Carboniferous species.
26	<u>Siphonodella quadruplicata</u> (Bronson and Mehl); Lower Carboniferous species.
27	<u>Palmatolepis perlobata schindewolfi</u> Muller; Upper Devonian species.
28	<u>Spathognathodus steinhornensis</u> Ziegler; Lower Devonian species.
29	<u>Icriodus woschmidti</u> Ziegler; Lower Devonian species.
30	<u>Icriodus latericrescens bilatericresens</u> Ziegler; Lower Devonian species; genus important throughout Devonian.
31	<u>Ancyrodella lobata</u> (Branson and Mehl); Upper Devonian species.
32 - 33	<u>Polygnathus pennata</u> Hinde; 32, upper; 33, lower view; genus important throughout Devonian.

Siegenian

Species of Icriodus are the important Siegenian indicators, including I. pesavis and I. latericrescens in North America and I. huddlei in Europe. Spathognathodus sulcatus is
important in North American faunas but has not been found in
Europe (Ziegler, 1971).

Emsian

As in the other Lower Devonian units, species of Icriodus
and Spathognathodus are the most important elements of the
Emsian fauna. In addition, Polygnathus appears in the Emsian
in both Europe and North America and is particularly important
in the latter. Polygnathus dehiscens, Spathognathodus stein-
hornensis s. s. and Icriodus species including I. huddlei and
I. bilatericrescens are most important for the Emsian. In
general, Emsian correlations of the Atlantic borderlands area
(e.g., New York, Spain and France) are better understood than
those of the earlier Lower Devonian units.

Middle Devonian

Eifelian

Conodonts of this interval are well defined in Belgium,
Spain, and Germany and can be differentiated into three zones
(Ziegler, 1971). Icriodus coringer marks the base of the
Eifelian on both sides of the Atlantic (New York and Belgium)
but is succeeded upward by Icriodus latericrescens n. subsp. A
(Klapper, et al., 1971) in North America, a form that has not
been found in Europe. Polygnathus species dominate the Atlan-
tic borderland faunas and correlations are fairly secure.

Givetian

Icriodus and Polygnathus dominate the biostratigraphically
important Givetian faunas. Specific differences are common between the European and North American section but sufficient
common species (e.g., Polygnathus varcus) as well as associated megafossils allow some confidence in correlation. The
sections in Germany, Belgium and Spain and the New York section
have become standrads. Some taxonomic problems with potentially significant Icriodus species are yet to be resolved
(Ziegler, 1971, p. 258).

	Western Europe	North America
Givetian	Polygnathus varcus	Polygnathus spp.
	Icriodus obliquimarginatus	Icriodus spp.
Eifelian	Polygnathus kockelianus	Polygnathus spp.
	Spathognathodus bidentatus	
	Icriodus corniger	
Emsian	Non-latericrescid – Icriodus-Polygnathus	Polygnathus foveolatus Spathognathodus exiguus exiguus Spathognathodus steinhornensis
		Polygnathus dehiscens lenzi Spathognathodus exiguus exiguus Spathognathodus steinhornensis
	I.b. bilatericrescens – S. steinhornensis – Polygnathus	Polygnathus dehiscens lenzi
Siegenian	I. huddlei curvicauda	Spathognathodus sulcatus (late forms)
	I. huddlei huddlei	Icriodus latericrescens n.subsp.B Icriodus cf. I. n.sp.A. Spathognathodus sulcatus
	Icriodus huddlei curvicaudata – rectangularis s.l.– angustoides	Icriodus pesavis – Spathognathodus johnsoni – S.transitans
Gedinnian	Ancyrodelloides – Icriodus pesavis	Spathognathodus n. sp.C
	Icriodus w.postwoschmidti	Spathognathodus n. sp. Q
	Icriodus w. woschmidti	Icriodus w.woschmidti

Fig. 2. Conodont zonation of the Lower and Middle Devonian.

Upper Devonian

By almost any standard of measure, the high point of conodont evolution was reached during the Late Devonian. More than
1000 names have been proposed for Late Devonian form taxa,
twice as many as for any other similar interval and evolution
was more rapid than at any other time (Clark, 1972). At least
120 "index species" have been recognized for the Upper Devonian
and the detailed zonation recognized first in Germany has become an ideal model for biostratigraphy and a world standard
for conodonts (Ziegler, 1971) (Fig. 3).

Subdivision of the Upper Devonian into Frasnian and Famennian stages or into ammonoid units for classification is un-
necessary now because the conodont zones have provided a detailed standard that can stand alone. Species of Polygnathus,
Ancyrodella and Ancyrognathus are the important early Upper
Devonian species but these taxa are replaced by a literal flood
of Palmatolepis species that form the structural unity of the
Upper Devonian zonation. Ancestor-descendant relationships can
be determined in much of the zonation. A few species of Spathognathodus and Protognathodus are important for the late Upper
Devonian faunas.

CARBONIFEROUS

Introduction

The most complete Lower Carboniferous conodont faunas are
those recognized in Germany and the Upper Mississippi Valley.
The Atlantic borderlands Lower Carboniferous sequences contain
faunas that, in part, correlate with the German and Mississippi
Valley sections, but which have differences, as well (Rhodes
and Austin, 1971). These latter faunas are from France,
Belgium, Spain, Italy, North Africa, Britain and Ireland, for
the eastern Atlantic borderlands and the central Appalachians
for the western borderlands.

Upper Carboniferous conodonts are poorly known. North
American faunas in the central Appalachians (Pennsylvania, Ohio
and West Virginia) have been obtained from approximately 5% of
these predominantly non-marine sequences (Lane et al., 1971).
Central and western North American sequences are more marine
and, hence, more complete. The European section is quite incomplete. Conodont faunas from the early Upper Carboniferous

Protognathodus	
Siphonodella praesulcata	
Spathognathodus costatus	UPPER
	MIDDLE
	LOWER
Polygnathus styriacus	UPPER
	MIDDLE
	LOWER
Scaphignathus velifer	UPPER
	MIDDLE
	LOWER
Palmatolepis marginifera	UPPER
	LOWER
Palmatolepis rhomboidea	UPPER
	LOWER
Palmatolepis crepida	UPPER
	MIDDLE
	LOWER
Palmatolepis triangularis	UPPER
	MIDDLE
	LOWER
Palmatolepis gigas	UPPER
	MIDDLE
	LOWER
A. triangularis	
Polygnathus asymmetricus	UPPER
	MIDDLE
	LOWER
	LOWERMOST
Schmidtognathus hermani – Polygnathus cristatus	UPPER
	LOWER

Figure 3. Conodont zonation of the Upper Devonian.

of Belgium, France, Britain and Ireland, comprise the record. The relatively poor conodont record for the Upper Carboniferous is related to ecologic constraints that may have as their source the shifting continental plates that closed the "proto-Atlantic" Ocean during this period of different continental elevations.

Lower Carboniferous

Tournaisian—Kinderhookian

These roughly correlatable Lower Carboniferous time-rock units are characterized by similar to identical species of Siphonodella in both Europe and North America. The multiple Siphonodella zones in North America are not as well represented in Europe but are correlated with Spathognathodus and Polygnathus species that are present in both sections (Collinson et al., 1971) (Fig. 4).

Visean—Valmeyeran

These intervals are distinguished by a diverse fauna of Gnathodus, Bactrognathus, Taphrognathus, Apatognathus, and Cavusgnathus species in North America and many of the same species plus important Polygnathus, Mestognathus and Scaliognathus in Europe. Correlation among the various European and North American sections is good. The British Avonian sequence, in particular, is easily correlatable with the Upper Mississippi Valley section (Rhodes and Austin, 1971; Collinson et al., 1971).

Namurian—Chesterian

Although the late Lower Carboniferous interval is incomplete in the British Avonian, the German and Mississippi Valley sections appear complete. All are characterized by Gnathodus, the German and North American sections by Paragnathodus, Kladognathus and Streptognathodus.

Upper Carboniferous

Namurian—Morrowan

The Belgium Namurian sequence contains the best eastern Atlantic fauna (Higgins and Bouckaert, 1968). Gnathodus and Idiognathodus dominate the faunas in Belgium and species are at least similar to those in the Morrowan of North America. Because of the rather poor state of knowledge concerning Upper Pennsylvanian conodont biostratigraphy, few correlations exist

(Rhodes and Austin, 1971). Morrowan faunas are better known than others from the Upper Carboniferous, and Lane et al. (1971) have described Midcontinent species of Spathognathodus, Idiognathodus, and Gnathodus that define seven biostratigraphic units. The same species are recognized in western North America and the central Appalachians.

	British Avonian	Mississippi Valley Region
Namurian		Streptognathodus unicornis
		Kladognathus-Cavusgnathus naviculus
		Kladognathus primus
Visean	Gnathodus girtyi collinsoni	G. bilineatus-Kladognathus mehli
	Gnathodus mononodosus	G. bilineatus-Cavusgnathus altus
	Mestognath. beckmanni-G. bilineatus	G. bilineatus-Cavusgnathus charactus
	Apatognathus scalenus-Cavusgnathus	Apatognathus scalenus-Cavusgnathus
	Taphro. varians-Cavusgnath.-Apatognath	
	Cavusgnath. unicornis-Apatognath.	Taphrognathus varians-Apatognathus
	no conodonts	Gnathodus texanus-Taphrognathus
	Mestognath. beckmanni-Poly. bischoffi	Bactrognathus-Taphrognathus
	P. lacinatus-Mestognathus beckmanni	Bactrognathus-Polygnathus communis
Tournaisian	G. antetexanus-Polygnathus lacinatus	Gnathodus semiglaber-Pseudopolygnath. multistriatus
	G. semiglaber-Pseudo. multistriatus	
	Spath. costatus costatus-Gnathodus delicatus	Siphonodella isosticha-S. cooperi
	Spath. robustus-S. tridentatus	S. quadruplicata-S. crenulata
	Siphonodella-Polygnathus inornatus	
	Patrognathus variabilis-Spathognathodus plumulus	Siphonodella duplicata
		Siphonodella sulcata

Figure 4. Conodont zonation of the Lower Carboniferous.

Post-Morrowan Carboniferous

This Upper Carboniferous interval is one of the more poorly defined units based on conodonts. Post-Morrowan faunas are known in Belgium (Westphalian) and in central and western North America. For the most part, Gnathodus, Adetognathus, Idiognathodus and Anchignathodus dominate the fauna and many of the species range throughout the Upper Carboniferous section (Fig. 5).

Rates of conodont evolution during most of the Paleozoic apparently were higher than during the Late Carboniferous and the uniform conodont taxa of this interval reflect this. An alternative idea to this is that our concept of Late Carboniferous time, particularly in the post-Morrowan interval, may be in error and this interval may represent only a part of what is presently thought to be included. Whatever the explanation, Upper Carboniferous conodonts appear to be less diverse than those of any other similar Paleozoic or Triassic interval.

PERMIAN

Introduction

Marine Permian strata is rare in the Atlantic borderlands. There is a limited section in Greenland and considerably more in the Arctic Islands and western North America. There is little known of the section in South America and Africa but fairly good, if incomplete, sections are found in southern and northern Europe. The Dutch-German Permian Basins have Lower Permian, at least some of which is marine, and Upper Permian (Zechstein) in some abundance. The Zechstein Sea may have had considerable extent in northern Europe (Thomas, 1975) and the German surface sections contain Upper Permian faunas.

Earliest Permian conodonts are known only in western North America (Clark and Behnken, 1971) but latest Permian faunas are known now in Kashmir-Pakistan and parts of the mid-East as well as western North America (Fig. 6).

Lower Permian

Sakmarian-Wolfcampian

Earliest Permian conodonts, corresponding to early Wolfcampian determinations in western North America, essentially are the same as those of the Late Carboniferous, i.e., Gnathodus, Idiognathodus, Adetognathus, etc. (Clark, 1974).

post-Morrowan	Gnathodus, Adetognathus, Idiognathodus, Anchignathodus spp.
Morrowan	Idiognathoides convexus
	Idiognathodus n.sp. A
	Idiognathodus humerus
	Gnathodus bassleri bassleri
	Gnathodus bassleri symmetricus
	Idiognathodus noduliferus
	Spathognathodus muricatus

Fig. 5. Conodont zonation of the Upper Carboniferous.

UPPER	Dzhulfian	Anchignathodus typicalis
		? Neogondolella orientalis
		? Neogondolella rosenkrantzi
	Guadalupian	Neogondolella serrata postserrata
		Neospathodus arcucristatus
		Neogondolella serrata serrata
LOWER	Artinskian-Leonardian	Neostreptognathodus clinei
		Neostreptognathodus prayi
		Neostreptognathodus sulcoplicatus
		Neostreptognathodus pequopensis
	Sakmarian-Wolfcampian	Neogondolella bisselli–Sweetognathus whitei
		Idiognathodus ellisoni

Fig. 6. Conodont zonation of the Permian.

A crisis in conodont evolution during the Early Permian produced a dramatic change in all younger faunas (Clark, 1972). To date, the pre-crisis Early Permian fauna has only been reported from western North America. Younger Early Permian conodonts are known from North and South America (Clark. 1974) and include *Sweetognathus*, *Neogondolella*, *Anchignathodus*, and *Ellisonia*.

Artinskian—Leonardian

Conodonts of this interval are dominated by species of *Neostreptognathodus*, a descendant of the older Permian *Sweetognathus*. The best faunas of this interval are from western North America but some southern European material is known. *Neogondolella* species showing good relationships with older Wolfcampian material are useful fossils in this interval (Behnken, 1975).

Upper Permian

Guadalupian

Species of *Neogondolella*, *Ellisonia* and *Neospathodus* are the important conodonts of this interval. Good sequences in western North America form the standard of reference which has been correlated with northern European and Greenland faunas. The emergence of *Neospathodus* and *Neogondolella* as excellent index forms in the Guadalupian is the beginning of a trend that continues until the Triassic extinction.

Dzhulfian

Youngest marine Permian strata are extremely rare and are unknown in the Atlantic borderlands area (Furnish, 1973). Greenland conodonts appear to be Guadalupian as do Zechstein specimens from northern Europe. Kashmir, Pakistan and Trans-Caucasus areas have produced the only published data on latest Permian conodonts (Sweet, 1970, 1973). Latest Permian taxa probably are present in undescribed sections in the western United States, as well.

Latest Permian faunas consist of Neogondolella species intermediate between those of the Guadalupian and Early Trias-sic, Anchignathodus, and Ellisonia species similar to both Permian and Triassic taxa. In fact, Sweet (1973) concluded that the latest Permian taxa did not experience the Permo-Triassic crisis and are distinguished only because of their similarity to Early Triassic forms.

TRIASSIC
Lower Triassic

Earliest Triassic conodonts are not presently known from the Atlantic borderlands but should be present in Greenland, at least. The faunas from Lower Triassic strata are well known in western North America and in the Salt Range and Trans-Indus Ranges of West Pakistan. The detailed zonation of the Lower Triassic includes species of Neogondolella and Neospatho-dus, evolutionary descendants of Permian species, as well as new forms such as Parachirognathus-Furnishius and Platyvillo-sus (Sweet et al., 1971). The earliest fauna is the same as that found in latest Permian strata but by the middle part of the Lower Triassic the faunas are uniquely Triassic. Species of Ellisonia are most common but are not as stratigraphically restricted as the Neogondolella and Neospathodus species (Fig. 7).

Middle Triassic

Mosher (1973) recognized three faunal associations for the Middle and Upper Triassic. The first, the gladiogondolel-lid group, has only been recognized in the Alpine-Tethyas belt.

UPPER	Rhaetian	Conodonts present but not diagnostic
	Norian	Epigondolella bidentata
		Epigondolella multidentata
		Epigondolella abneptis
	Karnian	Paragondolella polygnathiformis
		Neospathodus newpassensis
MIDDLE	Ladinian	Epigondolella mungoensis
		Neogondolella mombergensis
	Anisian	Neogondolella constricta
		Neogondolella regale
LOWER	Spathian	Neospathodus timorensis
		Neogondolella jubata
		unnamed zone
		Platyvillosus
	Smithian	Neogondolella milleri
		Neospathodus conservativus
		Parachirognathus −Furnishius
	Dienerian	Neospathodus pakistanensis
		Neospathodus cristagalli
		Neospathodus dieneri
		Neospathodus kummeli
	Griesbachian	Neogondolella carinata
		Anchignathodus typicalis

Figure 7.   Conodont zonation of the Triassic.

The second, a group dominated by Neogondolella, has been recognized on both sides of the Atlantic as has the third, the Epigondolella group. Mosher (1973) suggested that, in spite of Triassic continental separation, with the exception of group 1, the Middle and Upper Triassic conodonts were not seriously affected.

Species of Neogondolella, (N. constricta and N. mombergensis) dominate the Middle Triassic faunas. The German, Spanish and northern Italy faunas are well known as well as those from western North America. In addition, species described from the Cameroons in West Africa apparently are widespread (e.g., Epigondolella mungoensis).

Upper Triassic

Upper Triassic faunas include genera whose stratigraphic value is well documented in earlier rocks, e.g., Neospathodus, Epigondolella, and Paragondolella. Distinctive species are widespread in Europe and North America, although the best faunas have been obtained from the Alpine region of Europe. These species represent the end of the conodont lineage and their extinction at the close of the Triassic marks the end of conodont biostratigraphy.

UPPER PALEOZOIC AND TRIASSIC CONODONT EVOLUTION

The biostratigraphic application of conodonts through this ∼150 million year interval is based largely on morphologic modifications of specimens that are interpreted as ancestors and descendants. Thus, approximately a dozen genera and several times that number of species provide the framework of a very sound biostratigraphy (Fig. 8).

Species of Icriodus mark the base of the Devonian and their evolution throughout the Devonian provide nine or ten zones of great importance. Their range is overlapped by that of Polygnathus, a taxon that appeared in the late Early Devonian and provided a dozen or so important Middle and Upper Devonian to Lower Carboniferous zones. The stratigraphic significance of these two genera is overshadowed by species of Palmatolepis that dominate the Upper Devonian and provided more than a dozen zones. The Carboniferous began with new genera, including Siphonodella, a definite Polygnathus descendant,

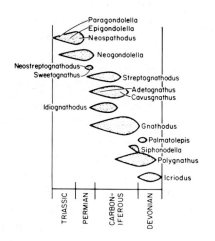

Fig. 8. Evolution and distribution of biostratigraphically important conodont genera from the Devonian-Triassic.

whose species provide four or more Lower Carboniferous (and one Late Devonian) zones. Thus, three genera and their descendants (Icriodus,Polygnathus-Siphonodella, and Palmatolepis) are the basis of the Devonian-earliest Carboniferous biostratigraphy (40-50 million years).

This homogeneous biostratigraphy breaks down in the remainder of the Carboniferous and earliest Permian, and species of a number of genera such as Gnathodus, Cavusgnathus, Adetognathus, Idiognathodus, and Streptognathodus are the basis of the biostratigraphy. Polygnathus, one of the important Devonian genera, is also important for the Carboniferous, and may be ancestral to some of the other important Carboniferous forms.

A marked change, recently described as a conodont "crisis", occurred in the Early Permian (Clark, 1972). All of the important Carboniferous species became extinct and all post-Early Permian conodonts can be traced to a handful of species that were obscure, to some extent, prior to the crisis. Thus, the Sweetognathus-Neostreptognathodus line provided five or so Lower Permian zones, and evolved from Anchignathodus, a less important Carboniferous genus that survived until Early Triassic. Also important was Neogondolella, a genus that provides approximately a dozen Permian and Triassic zones, and may be related to the relatively unimportant Carboniferous genus Gondolella. Neospathodus was new in the Permian and provided species for eight or so Permian and Triassic zones. Neospathodus was ancestral to Epigondolella and Paragondolella plus several other important Triassic forms. This post-Early Permian interval apparently was a return to the homogeneous biostratigraphy of the Devonian-earliest Carboniferous, and three genera, <u>Sweetognathus</u>, <u>Neogondolella</u> and <u>Neospathodus</u> were the ancestors of most of the later Permian and Triassic biostratigraphically important species.

ATLANTIC BORDERLANDS ON A DRIFTING CRUST

The present North Atlantic Ocean began forming ∼180-200 m.y. ago as part of the Russian platform rotation (Ostenso and Wold, 1973). This produced the opening of the North-Atlantic-Labrador Sea that has resulted in the present separation of North America from most of Europe. North America and most of central and northern Europe had been in fairly close positions since at least the Early Paleozoic when the Russian platform collided with the North American-Greenland plate, causing the Caledonian orogeny. The proximity of North America and Europe during the duration from Early Paleozoic to the Early Triassic has been portrayed in a series of maps by Smith <u>et al</u>. (1973).

Unfortunately, details necessary to unambiguously define precise paleogeographic details of North America and Europe during this interval are lacking. Relationship of the Southern Atlantic borderlands continents is better known and during this interval, South America and Africa formed a single mass, separation of which during the Early Mesozoic produced the present South Atlantic. Thus, the Late Devonian configuration of the present Atlantic borderland areas included a North American-Europe proximity and a "proto-Atlantic" separating North America-Europe from Africa and South America. During the Carboniferous the African-South American block approached the North American-Middle and Northern Europe block, destroying the "proto-Atlantic" and by the Permian and Triassic there was no major oceanic area separating a tenuously fused South America-Africa-North America-Europe. The Triassic collision of the Russian platform with the Siberian platform formed the Eurasian continent and produced rotation that opened the North Atlantic-Labrador Sea.

If this can be accepted as a factual description of the Late Paleozoic-Triassic paleogeographic framework, Devonian to Early Triassic conodont faunas in North America and Europe might be expected to be more similar than faunas of the Late

Triassic because of the formation of the modern Atlantic Ocean
that separated the areas during the Triassic. Similarly,
Devonian and Carboniferous conodont faunas of North America-
Europe and those of southern Europe-Africa and South America
should be less similar when they were separated by a "proto-
Atlantic" during the Devonian and Carboniferous than during the
Permian and Early Triassic when the continents were together.
Late Triassic faunas might be expected to reflect the separa-
tion related to formation of the modern Atlantic.

This might be considered a working hypothesis in spite of
the many difficulties of reconstructing continental positions
and the relatively unknown ecologic requirements of conodonts.

In order to test this hypothesis, at least in part, the
similarity index for faunas at various times and geographic
positions has been calculated. This has been done using the
similarity coefficient 2w/a+b, where w = number of conodont
taxa common to two continents, and a + b = total number of
species on both continents. This has been based on form taxa
recorded in the Wisconsin IBM conodont catalogue for all intervals indicated, except the Permian. The Permian figures are
based on natural or multielement species. Resulting figures
should be comparable, at least. These comparisons have been
made with the warning of Cook and Taylor (1975) in mind. These
students have emphasized the danger of unequivocal acceptance
of the idea that faunal similarity = geographic proximity.
Other factors (e.g., thermocline barriers) must be considered.
The results are tabulated in Table 1. Apparent immediately is
the incompleteness of the data matrix for many of the areas
that could be expected to test the hypothesis. That is, South
America and Africa have a unique roll in Late Paleozoic and
Triassic continental patterns but there is no conodont data for
most of this time interval that can be used. Indeed, the lack
of data may be taken to indicate adverse ecology during this
interval. Nicoll (1975) has suggested that conodonts could not
tolerate the cold water associated with Permian glaciation, at
least for western Australia. A Permian paleolatitude of 50° -
65° for this part of Australia may suggest temperature toler-
ances for conodonts of that time interval. The more complete
European-North American data is not inconsistent with

130

continental patterns (Fig. 9) but the drop in similarity index of .290 - .194 - .191 as the Atlantic formed during the Triassic cannot by itself be considered of extraordinary significance. The European-North American data reflects Carboniferous ecologic problems that are not necessarily related to continental position (e.g., Vallentine and Moores, 1972). Indices for other areas (Table 1) may reflect the need for more data more than they do the confirmation of shifting position of the Late Paleozoic and Triassic Atlantic borderlands.

In addition to the form-taxa similarity index, a similarity index based on name zonal species was calculated (Table 2). This similarity matrix has many built in biases, not the least important of which is the fact that the reason a taxon is the name-giver for a zone, is based on its widespread occurrence. Hence, any calculation based on zonal species could be expected to give a higher degree of similarity than the comparison of similarity indices based on total taxa (Table 1). This is confirmed, but the much higher index numbers can be attributed to the bias in selection of zonal species.

This zonal similarity index is plotted in Figure 10 for North America and Europe. Whether or not the Late Triassic dropoff is due to the separation of the continents during that interval is uncertain. Certainly, it is not inconsistent with the plate-tectonic model now proposed. Carboniferous low values probably reflect the well known Carboniferous ecologic changes more than continental proximity.

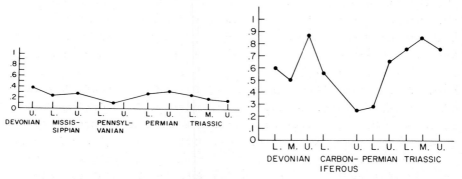

Fig. 9. Plot of diversity index of form-species for Europe and North America.

Fig. 10. Plot of zonal similarity index of form-species for Europe and North America.

ACKNOWLEDGMENTS

This study was made possible by NSF Grant GA-40454. Jim Gamber, University of Wisconsin graduate student, aided in similarity index calculation and updating of the IBM form taxa catalogue. Ed Landing, University of Michigan graduate student, maintained the IBM catalogue and made some preliminary calculations for this study. Paul Dombrowski drafted the figures and the manuscript was typed by Catherine M. Ward.

REFERENCES

Austin, R. L., 1973, Modification of the British Avonian conodont zonation and a reappraisal of European Dinantian conodont zonation and correlation. Ann. Soc. Géol. Belgique, v. 96, p. 523-532.

Behnken, F. H., 1975, Leonardian and Guadalupian (Permian) conodont biostratigraphy in western and southwestern United States. Jour. Paleont. v. 49, p. 284-315.

Clark, D. L., 1972, Early Permian crisis and its bearing on Permo-Triassic conodont taxonomy. Geol. Palaeont., Sp. Volume 1, p. 147-158.

_____, 1974, Factors of Early Permian conodont paleoecology in Nevada. Jour. Paleont. v. 48, p. 710-720.

_____, and Behnken, F. H., 1971, Conodonts and biostratigraphy of the Permian. Geol. Soc. America Mem. 127, p. 415-439.

Collinson, Charles, Rexroad, C. B. and Thompson, T. L., 1971, Conodont zonation of the North American Mississippian. Geol. Soc. America Mem. 127, p. 353-394.

Cook, H. E., and Taylor, M. E., 1975, Early Paleozoic continental margin sedimentation, trilobite biofacies, and the thermocline, western United States. Geol., v. 3, p. 559-562.

Furnish, W. M., 1973, Permian Stage Names, in, The Permian and Triassic Systems and their mutual boundary, ed. Logan and Hills. Canadian Soc. Petrol. Geol. Mem. 2, p. 522-548.

Higgins, A. C. and Bouckaert, J., 1968, Conodont stratigraphy and palaeontology of the Namurian of Belgium. Expl. Cartes Geol. et Mineres Belgique, Mem. 10, 64 p.

Klapper, G., Sandberg, C. A., Collinson, C., Huddle, J. W., Orr, W. N., Rickard, L. V., Schumacher, D., Seddon, G. and

Uyeno, T. T., 1971, North American Devonian conodont biostratigraphy. Geol. Soc. America Mem. 127, p. 285-316.

Lane, H. R., Merrill, G. K., Straka, J. J., III, and Webster,
G. D., 1971, North American Pennsylvanian conodont stratigraphy. Geol. Soc. America Mem. 127, p. 395-414.

Nicoll, R. S., 1975, The effect of Late Carboniferous-Early
Permian glaciation on the distribution of Permian conodonts in Australasia. Geol. Soc. America Abstracts,
v. 7, p. 828-829.

Ostenso, N. A. and Wold, R. J., 1973, Aeromagnetic evidence for
origin of Arctic Ocean Basin, in, Arctic Geology, ed.
M. G. Pitcher. Amer. Assoc. Petrol. Geol. Mem. 19, p.
506-516.

Rhodes, F. H. T. and Austin, R. L., 1971, Carboniferous conodont faunas of Europe. Geol. Soc. America Mem. 127,
p. 317-352.

Sandberg, C. A. and Ziegler, W., 1973, Refinement of standard
Upper Devonian conodont zonation based on sections in
Nevada and West Germany. Geol. Palaeont., v. 7, p. 97-
122.

Smith, A. G., Briden, J. C. and Drewry, G. E., 1973, Phanerozoic world maps, in Organisms and continents through time,
ed. N. F. Hughes. Palaeont. Assoc. Spec. Paper 12,
p. 1-42.

Sweet, W. C., 1970, Uppermost Permian and Lower Triassic
conodonts of the Salt Range and Trans-Indus Ranges, West
Pakistan. Univ. Kansas Spec. Pub. 4, p. 207-275.

_____, 1973, Late Permian and Early Triassic conodont
faunas, in, The Permian and Triassic Systems and their
mutual boundary, ed. Logan and Hills. Canadian Soc.
Petrol. Geol. Mem. 2, p. 630-646.

Sweet, W. C., and Bergstrom, S. M., 1971, eds. Symposium on
conodont biostratigraphy. Geol. Soc. America Mem. 127,
449 p.

_____, Mosher, L. C., Clark, D. L., Collinson, J. W., and
Hasenmueller, W. A., 1971, Conodont biostratigraphy of the
Triassic. Geol. Soc. America Mem. 127, p. 441-465.

Thomas, T. M., 1975, Search for hydrocarbons in shelf areas of
Northwest Europe: Progress and prospects. Amer. Assoc.

Petrol. Geol. Bull., v. 59, p. 573-617.

Vallentine, J. W. and Moores, E. M., 1972, Global tectonics and the fossil record. Jour. Geol., v. 80, p. 167-184.

Ziegler, W., 1971, Conodont stratigraphy of the European Devonian. Geol. Soc. America Mem. 127, p. 227-284.

Table 1.

Similarity index for conodont form species during indicated time and for designated geography.

	U. Dev		U. Miss		L. Penn		U. Perm		M. Tri		Dev		Perm	
		L. Miss		L. Penn		L. Perm		L. Tri		U. Tri		Carb		Tri
Eu & NA	.348	.136	.231	0	.019	.34	.37	.290	.194	.191	.348	.204	.355	.225
Eu & Af	.268	.017	x	x	x	x	x	x	.088	.021	.268	.004	x	.036
Eu & Aus	.249	.312	x	x	x	x	x	0	x	x	.249	.078	x	x
Eu & Asia	.109	.071	.239	0	0	.57	.53	.065	.239	.270	.109	.077	.55	.191

x = insufficient data

Table 2.

Zonal similarity index for North America and Europe.

Upper Triassic	.75
Middle Triassic	.85
Lower Triassic	.76
Upper Permian	.67
Lower Permian	.29
Upper Carboniferous	.25
Lower Carboniferous	.57
Upper Devonian	.89
Middle Devonian	.50
Lower Devonian	.60

Discussion

Dr. F. M. Swain: To what extent may facies control have played an important role in late Paleozoic-Triassic conodont distribution in contrast to continental separation or proximity?

Dr. D. L. Clark: Much of our research during the past 7 years has been related to conodont paleoecology. There is good evidence that conodonts behaved much like most other groups of organisms; some were very closely related to facies factors, but many were part of the pelagic realm. The similarity index calculations did not consider facies control. The simplified assumption was that a greater similarity may indicate proximity of environments. It was recognized that geographically adjacent but different environments may support completely different taxa. The calculations for this report were made simply to see if the similarity index for conodonts (most of which were nekton or plankton) correlated with the best guesses for continental separations).

I would answer the Holdsworth question more simply -- differential solubility of apatite during diagenesis of carbonates (and clastics) has not been recognized. Only reworked conodonts show preservational problems and this can be explained more easily by things other than solution of calcium phosphate.

Dr. M. C. Keen: Could you comment on the status of the conodonts reported from the Cretaceous of W. Africa?

Clark: The "Cretaceous" conodonts from Africa are part of the upper Middle Triassic fauna recognized worldwide. Further, most of the conodont elements from the African sample became extinct (in normal sequences) before the close of the Triassic. The same is true for the Japanese "Jurassic" material. There are no unambiguous post-Triassic conodonts.

Dr. B. K. Holdsworth: Is there any possibility that apparent facies restriction of conodonts could be connected with differences in susceptibility to solution of particular forms during diagenesis?

Clark: To this I would say that as far as I know, there are no notable differences in relative solubility of conodont specimens in the sediments we generally work with. Under normal circumstances, the conodonts are very well preserved and show no indications of being partially dissolved. Accordingly, I do not think that selective preservation for that reason, if at all present, is likely to be a problem. One big problem has, however, appeared and it involves the laboratory preparation. Up to recently, some laboratories were using monochloric acid for preparation of conodont samples but unfortunately, it has turned out that when treated with this acid for some time, some forms were dissolved and other unaffected, thereby producing biased faunas as laboratory artifacts. Fortunately, as far as I know, these early-dissolved elements have not played an important role in the discussions of conodont biogeography and biostratigraphy.

NOTES ON PALEOZOIC ACRITARCHS FROM THE ATLANTIC MARGINS

F. M. Swain

University of Minnesota, Minneapolis, Minnesota;
and University of Delaware, Newark, Delaware

Abstract

Stratigraphically useful acritarch assemblages of Atlantic margins have been described from Ordovician rocks of Europe and North Africa; Silurian of Europe, North Africa, and North America; Devonian of Europe, North and South America; Carboniferous of North America, Europe, and North Africa; and Permian of Europe, North Africa, and North America.

Zusammenfassung

Formationskundig, nützliche Versammlungen der Acritarchen von Atlantischen Ränder sind von Ordovizischen Felsen des Europas und Nordafrikas; dem Silur des Europas, Nordafrikas und Nordamerikas; Devon des Europas, Nord- und Südamerikas; Karbon des Nordamerikas, Europas, und Nordafrikas; und dem Perm des Europas, Nordafrikas, und Nordamerikas beschreibt.

Introduction

The following brief discussion will deal with a few selected acritarch assemblages from localities bordering the Atlantic Ocean, together with references that will hopefully guide the reader to additional studies on these useful microfossils. A specialist was not available to prepare an acritarch article for this volume and the present writer has depended completely on publications from which a few notes are presented herein.

Cambrian

A few acritarch assemblages have been recorded from Cambrian rocks but the writer has not reviewed the literature on them.

Ordovician

Acritarchs of Ordovician rocks of Europe and North Africa have been studied by Burmann (1968), Cramer (1964), Cramer et al. (1974a, 1974b), Deunff (1961, 1968), Martin (1968), Vavrdová (1965, 1972), and Eisenack, et al. (1973).

Cramer et al. (1974b) recorded 10 species which they be-

lieve have short stratigraphic ranges from the late Arenigian
shales of the Tadla Basin of Morocco: Aureotesta clathrata
Vavrdová 1972 (Text-fig. 1-8), Morrocanium simplex Cramer,
Kanes, Diez, and Christopher, 1974, Neoveryhachium carminae
(Cramer, 1964) and related forms, Priscotheca siempreplicata
Cramer et al., 1974, Rugulidium microrugulatum (Martin, 1968),
R. varirugulatum Cramer, et al., 1974, R. rugulatum Cramer et
al., 1974 (Text-fig. 1-9), R. scabratum (Cramer, 1964), R. tri-
angulatum Cramer et al., 1974, and Veryhachium lairdi (Deflandre, 1946). Most of the species are late Arenigian to early
Llanvirnian in range but P. siempreplicata is of Tremadocian to
late Arenigian age. Several of the forms also occur in Belgium,
northern Florida subsurface, Spain, Libya, Tunisia, Algeria,
and Saudi Arabia.

In a comparative study Cramer et al. (1974a), seven late
Arenigian to early Llanvirnian acritarchs also are believed to
have restricted ranges: Pirea dubia Vavrdová 1972 (Text-fig.
1-1), Multiplicisphaeridium hoffmanensis Cramer et al. (Text-
fig. 1-2), M. moroquense Cramer et al. 1974 (Text-fig. 1-3),
M. rayii Cramer et al. 1974 (Text-fig. 1-4), Oodium mordidum
Cramer et al. 1974 (Text-fig. 1-5), Coryphidium elegans Cramer
et al. 1974 (Text-fig. 1-6). P. dubia has also been recorded
from Bohemia, Libya and in presumably reworked Silurian deposits of NW Spain.

Other than the occurrences in the relatively undisturbed
Ordovician rocks of northern Florida subsurface referred to
above, little is known of Ordovician acritarchs of the western
margin of the Atlantic Ocean.

Silurian

Studies of Silurian "hystrichospheres" (acritarchs) of the
European Atlantic margins include those by Eisenack (1934,
1954, 1955, 1959, 1962, 1965, 1970, 1971, 1972), Downie (1959,
1963), Deflandre (1942, 1945), Deunff (1954), Cramer (1964,
1966a, 1966b), Bachmann and Schmid (1964), Martin (1965, 1966),
Stockmans and Willière (1963), and Lister (1970).

The Llandoverian Visby Marl of the Baltic (Gotland) contains a rich acritarch assemblage with Multiplicisphaeridium
spp. and other forms, many of which also range into the Wenlockian elsewhere (Eisenack, op. cit.).

An assemblage of acritarchs from the type Wenlockian Wenlock Shales of Wenlock, England comprises, among other longer ranging forms, 38 species or subspecies that Downie (1963) believed are confined to the Wenlockian. Several of the forms he cited, however, are now known to occur in older or younger deposits. Correlation of the Wenlockian assemblage with Wenlockian beds in France (Montagne Noire, Deflandre, 1945) and in the Baltic region (Eisenack, op. cit.) is shown by certain of the species. Three assemblages are recognized by Downie in the type Wenlockian: Assemblage I of the Buildwas Beds, lower Wenlock, typified by Deunffia and Domasia, Assemblage II of the middle Wenlock Coalbrookdale Beds with "Veryhachium" bulbiferum and "V." elongatum as typical species; and Assemblage III of the upper Coalbrookdale and Tickhill Beds, upper Wenlock, have abundant Micrhystridium, leiofusid acritarchs and Multiplicisphaeridium granulatispinosum. The attempt has been made here to justify Downie's generic assignments with those of the acritarch catalogue of Eisenack et al. (1973) but some of Downie's species have not yet been listed in the published volume of the catalogue.

Silurian acritarch floras are also known from the San Pedro and equivalent formations of northern Spain (Cramer 1966, 1967, 1968, 1970; Cramer and Diez 1968), Tarannonian and late Wenlockian? of Belgium (Martin, 1965, 1968); Tunisia and Libya (Hoffmeister, 1959; Cramer, 1970) and Shropshire England, (Lister, 1970).

Silurian acritarchs of the western Atlantic margins have been recorded by Fisher (1953), and especially by Cramer and Diez de Cramer in a series of papers.

A comprehensive examination of Silurian acritarchs of eastern North America was made by Cramer and Diez de Cramer (1972). Localities pertinent to the present volume are those in Maine, New York, Ontario, Pennsylvania, Virginia, Georgia, and Florida. The authors recognized five acritarch biofacies which they believe to be time-transgressive in the Silurian. The five, probably temperature-controlled biofacies are: (1) Neoveryhachium carminae (Text-fig. 12) ; (2) Domasia (Text-fig. 18) ; (3) Deunffia (Text-fig. 15) (with D. eisenacki subfacies); (4) Gloecocapsamorpha prisca, (5) Pulvinosphaeridium-

Estastra biofacies (tolerant of warm climate); when the distribution of the biofacies is plotted on a Pangaea arrangement
of the Silurian continents. The authors recognize several acritarch realms all in the Silurian southern hemisphere. The
proposed time-transgressive distribution of the biofacies
through the Silurian permits the authors to suggest a rate of
continental drift of 3 cm. per year during the Wenlockian. The
Gondwana pole, assumed to be the South Pole of that time, probably lay in northern South Africa. An interesting suggestion
is made that the long-problematical early Paleozoic block of
northern Florida subsurface originally lay between Brazil and
Africa as a small sliver of continental crust, and subsequently
drifted to its present position.

Cramer and Diez de Cramer list the total chronostratigraphic ranges of species or species-groups that occur abundantly and more or less world-wide in the Silurian, as follows:

Neoveryhachium carminae (Text-fig. 12), middle early
  Llandoverian (graptolite zone 18 or 19) to basal early
  Gedinnian.

Domasia trispinosa, middle early Llandoverian to top Wenlockian (g. z. 31).

Domasia elongata (Text-fig. 18), early late Llandoverian
  (g. z. 22 or 23) to top Wenlockian.

Deunffia (D. monospinosa, D. eoramusculosa, D. furcata,
  (Text-fig. 15), middle late Llandoverian (g. z. 23) to
  top Wenlockian.

Deunffia eisenacki, early Ludlovian (g. z. 33 top) to end
  Ludlovian (post - g. z. 36)

Quadraditium fantasticum, late early Llandoverian to top
  Silurian. (Text-fig. 13)

Duvernaysphaera aranaides (Text-fig. 14), late early
  Llandoverian to early Emsian.

Baltisphaeridium denticulatum (Text-fig. 16) (s.l.) base of
  Silurian (g. z. 16) to top Gedinnian.

Eupoikilofusa striatifera (Text-fig. 17), (well sculptured-
  forms), base of Silurian to basal early Gedinnian.

Devonian

A collection of acritarchs from green shales of the San
Pedro Formation, early Gedinnian? at Oblanca, Cantabric Moun-

tains, northern Spain was recorded by Cramer (1946b). Twenty-
five species occur in frequencies greater than 1%, and the following are present in numbers greater than 10% of the fauna:
Veryhachium trisulcum Deunff, V. reductum Deunff, V. carminae
Cramer, Micrhystridium stellatum Deflandre, and M. fragile
Deflandre. There are five new species of Leiofusa. One of the
five species, L. bernesga Cramer is stated to be widely distri-
buted in the San Pedro Formation of the Cantabric Mountains;
the other four are apparently more restricted in distribution.
Some additional references to studies of Devonian acritarchs in
Spain and other parts of western Europe and North Africa in-
clude those of Cramer (1966a, 1967), Jardiné et al. (1972),
Lanzoni and Magloire (1969), Rauscher (1969), and Stockmans and
Willière (1960, 1962a, b, 1969).

A small collection of acritarchs from the early Devonian
of Uruguay, Department of Durazno (Martinez-Macchiavello,
1968), comprises 11 species, including a new species of Leoni-
ella and several new subspecies of Baltisphaeridium simplex
Stockmans and Willière, Veryhachium legrandi St. and W., and
V. exasperatum Deunff. The collection was obtained from shales
that have also yielded sporomorphs.

Brito (1967) described both Silurian and Devonian acritarchs from the Maranhao Basin of northern Brazil. The acri-
tarch-bearing sequence occurs within about 2,500 meters of
Paleozoic rocks ranging from Cambro-Ordovician to Permian in
age and littoral-to bathyal-to continental in lithofacies. The
Silurian is represented by the Itaim Formation (middle and upper parts). Two palynological zones termed S (poorly developed)
and T lie in the Itaim Formation; Zone T contains Dactylofusca
maranhensis Brito and Santos (Text-fig. 19) , and Leiofusa
mulleri Brito and Santos. The overlying Picos Shale is early
to middle Devonian. Zones Lower Q and R are within this forma-
tion; R has Veliferites tenuimarginatus Brito (Text-fig. 21),
Evittia sommeri Brito (Text-fig. 18) , and Triangulina alargada
Cramer; Lower Q has Tasmanites sp. aff. T. mourai Sommer,
Leiofusa bacillum Deunff and L. brazilensis Brito and Santos
(Text-fig. 20) . The lower and middle parts of the overlying
Cabeças Formation, middle Devonian are represented by Upper
Zone Q and Zone P; Upper Zone Q has Pseudolunulidea imperatri-

zensis Brito and Santos (Text-fig. 23), and Zone P has Maranhites sp., Duvernaysphaera radiata Brito (Text-fig. 22) as
well as Polyedrixum sp., Netromorphitae and Chitinozoa. The
upper Cabeças Formation, middle to late Devonian has Maranhites
brazilensis Brito, M. mosesi (Sommer) (Text-fig. 24), and
Pterospermopsis brazilensis Brito (Text-fig. 25), but not
Chitinozoa or Netromorphitae.

Other western Atlantic Devonian acritarch floras were described from Canada by Deunff (1955, 1961).

Devonian sporomorphs are well developed around the Atlantic margins but are not dealt with herein.

Carboniferous

A major assemblage of acritarchs and prasinophycean algae
was recorded from late Devonian and early Mississippian rocks
in a bore hole near Barberton, northeastern Ohio, by Wicander
(1974). He described 15 new genera and 56 new species from the
late Devonian Cleveland and Chagrin Shales and the early Mississippian Bedford Shale. Of these the following were only
recorded from the Mississippian: Cymatiosphaera velicarina
Wicander (Text-fig. 26), Conradidium firmamentum W. (Text-fig.
27), Diexallophasis absona W. (Text-fig. 28), D. cuspidus W.,
Exilisphaeridium simplex W. (Text-fig. 29), Multiplicisphaeridium verrucarum W. (Text-fig. 30), Navifusa drosera W. (Text-
fig. 31), and Stellinium cristatum W. (Text-fig. 32).

Carboniferous acritarchs were described from north Africa
(Algerian Sahara) by Lanzoni and Magloire (1969) and from Belgium by Stockmans and Willière (1966).

There is also extensive literature on Carboniferous sporomorphs of the Atlantic margins not considered herein.

Permian

Acritarchs of particularly small size and included in
Baltisphaeridium (Text-fig. 33), Micrhystidium (Text-fig. 34),
Verhyachium (Text-fig. 35), and Leiofusa (Text-fig. 36), were
found in Lower Permian Marls of Yorkshire England (Wall and
Downie 1962). A striking feature of the assemblages is a variable plexus of forms ranging between those of Veryhachium type
and those of Micrhystridium type. The former are polygonal
tests with as few as four processes and represent a Veryhachium?
irregulare complex. The latter are small spherical forms with

as many as 20 spines of _Micrhystridium_ type.

Other Permian acritarchs are known from the Permian of Yugoslavia and from the Sahara (Jekhowsky, 1961), but in western Atlantic, only from Oklahoma (Wilson, 1960).

References

Bachmann, A., and Schmid, M., 1964. Mikrofossilien aus dem österreichischen Silur., Verh. Geol. Bundesanst., H. l, p. 63, 64.

Brito, I. M., 1967. Silurian and Devonian acritarchs from Maranhão Basin, Brazil, Micropaleontology, v. 13, p. 473-482.

Burmann, G., 1968. Diachrodien aus dem unteren Ordovizium, Paläont Abh., v. 2, no. 4, p. 639-652.

Cramer, F. H., 1964a. Microplankton from three Paleozoic Formations in the Province of León, NW Spain, Leidse Geol. Meded., v. 30, p. 253-361, pls. 1-24.

_____, 1964b. Some acritarchs from the San Pedro Formation (Gedinnien) of the Cantabric Mountains in Spain, Bull. Soc. belge Géol., v. 73, p. 33-38.

_____, 1966a. Palynomorphs from the Siluro-Devonian boundary in NW Spain, Notas y Communs. I. G. M. España, v. 85, p. 71-82.

_____, 1966b. Hoegispheres and other microfossils incertae sedis of the San Pedro Formation (Siluro-Devonian boundary) near Valporquero, León, Spain, Notas y Communas, I. G. M. España, v. 86, p. 75-94.

_____, 1967. Palynology of Silurian and Devonian rocks of Northwest Spain, Bol., I. G. M. España, no. 77, p. 225-286.

_____, 1968. Silurian palynologic microfossils and paleolatitudes, Neues Jahrb. Geol. Palaont., Jg. 1968/10, p. 591-597.

_____, 1970. Distribution of selected Silurian acritarchs, Rev. Española Micropaleont., Num. Extraord. VI, p. 1-203.

Cramer, F. H., Allan, B., Kanes, W. H., and Diez, M. d. C. R., 1974a. Upper Arenigian to lower Llanvirnian acritarchs from the subsurface of the Tadla Basin in Morocco. Palaeontographia Abt. B., v. 145, Lfg. 5, 6, p. 182-190.

Cramer, F. H., and Diez de Cramer, M. d. C. R., 1968, Considerationes taxonomicas sobre las acritarcas del Silurico Medio y Superior del Norte de España. Primera Parte: Las acritarcas acantomorfiticas, Bol. I. G. M. España, nr 79/6, p. 541-574.

_____, 1972. North American Silurian palynofacies and their spatial arrangement: acritarchs, Palaontogr. Abh. B., v. 138, Lfg. 5-6, p. 107-180.

Cramer, F. H., Kanes, W. H., Diez, M. d. C. R., and Christopher, R. A., 1974b. Early Ordovician acritarchs from the Tadla Basin of Morroco. Palaeontographica, Abt. B., v. 146, Lft. 3-6, p. 57-64.

Deflandre, G., 1942. Sur les Hystrichosphères des calcaires siluriens de la Montagne Noire, Acad. Sci. Paris, C. R., 215, p. 475-476.

_____, 1945. Microfossiles des calcaires siluriens de la Montagne Noire, Ann. Paléont., v. 31, p. 41-76.

Deunff, J., 1954. Sur le microplancton du Gothlandien armoricain, Soc. Géol. France, C. R., Somm. v. 3, p. 54, 55.

_____, 1955. Un microplancton fossile Dévonien a Hystrichosphères du continent Nord-Américain, Bull. Microscopie Appliquée, ser. 2, v. 5, p. 138-149, pls. 1-4.

_____, 1961a. Un microplancton à Hystrichosphères dans le Trémadoc du Sahara, Rev. Micropaleont., v. 4, p. 37-42.

_____, 1961b. Quelques précisions concernant les hystrichosphaeridées du Dévonien du Canada, C. R. Soc. Geol. France, v. 8, p. 216-218.

_____, 1968. Arbusculidium, genre noveau d'Acritarche du Tremadocien marocain, C. R. Somm., Seances Soc. Geol. France, 1968, p. 101, 102.

Downie, C., 1959. Hystrichospheres from the Silurian Wenlock Shale of England, Palaeontology, v. 2, p. 56-71.

_____, 1963. "Hystrichospheres" (Acritarchs) and spores of the Wenlock Shales (Silurian) of Wenlock, England. Palaeontology, v. 6, pt. 4, p. 625-652, pls. 91, 92.

Eisenack, A., 1934. Neue Mikrofossilien des baltischen Silurs III und neue Mikrofossilien des böhmischen Silurs I, Paläont 2, v. 16, p. 52-76.

_____, 1954. Hystrichosphären aus dem baltischen gotlandium, Senckenbergiana, v. 34, p. 205-211.

_____, 1959. Neotypen baltischen Silur-Hystrichosphären und neue Arten, Palaeontographica (A), v. 112, p. 193-211.

_____, 1962. Einige Bemerkungen zu neuen Arbeiten über Hystrichosphären, Neues Jahrb. Geol. Paläont., m.n., 1962, p. 92-101.

_____. 1965. Mikrofossilien aus dem Silur Gotlands, Hystrichosphären, Problematika, Neues Jahrb. Geol. Palaont. Abh. 122, p. 257-274.

_____, 1970. Mikrofossilien aus dem Silur Estlands und der Insel Ösel, Geol. Foren., Stockholm. Forh., v. 92, p. 302-322.

_____, 1971. Weitere Mikrofossilien aus dem Beyrichienkalk (Silur), Neues Jahrb. Geol. Paläont. mn., p. 449-460.

_____, 1972. Chitinozoen und andere Mikrofossilien aus der Bohrung Leba, Pommern, Palaeontogr. Abt. A., v. 139, p. 64-87.

Eisenack, A., Cramer, F. H. and Diez, M. d. C. R., 1973. Katalog der fossilen Dinoflagellaten, Hystrichosphären und verwandten Mikrofossilien., v. 3, Acritarcha (I), E. Schweizbart'sche Verlagbuchlandlung, Nagele und Obermiller, Stuttgart, 1973, p. 1-1104.

Fisher, D. W., 1953. A microflora of the Maplewood and Neagher Shales, Buffalo Soc. Nat. Sci., Bull. 21, p. 13-18.

Hoffmeister, W. S., 1959. Silurian plant spores from Libya, Micropaleontology, v. 5, p. 331-334.

Jardiné, S., Combaz, A., Magloire, L., Peniguel, G., and Vachey, G., 1972. Acritarches du Silurien terminal et du Dévonien du Sahara Algérien, C. R. Sept. Congr. internat. Stratigr. Géol. Carbonifère, Krefeld 1971, v. 1, p. 295-310.

Jekhowsky, B. de, 1961. Sur quelques hystricosphères Permo-Triasiques d'Europe et d'Afrique, Rev. Micropaléontologie, v. 3, p. 207-212, pls. 1, 2.

Lanzoni, E., and Magliore, L., 1969. Associations palynolo-
    giques et leurs applications stratigraphiques dans le Dévo-
    nien supérieur et Carbonifère inférieur de Grand erg Occi-
    dental (Sahara Algerien), Rev. de L'Institut Français du
    Petrole, v. 24, p. 441-468, pls. 1-8.
Lister, T. R., 1970. A monograph of the Acritarchs and Chiti-
    nozoa from the Wenlock and Ludlow Series of the Ludlow and
    Millichope areas, Shropshire, Palaeontogr. Soc. Monographs,
    1, p. 1-100.
Martin, F., 1965. Les acritarchs du sondage de la brasserie
    Lust à Kortrijk (Courtrai) (Silurien belge), Bull. Soc. belge
    Geol. Pal. Hydr., t. 74, p. 351-400.
_____, 1968. Les Acritarches de l'Ordovicien et du Sil-
    urien belges. Détermination et valeur stratigraphique, Inst.
    Roy. Sc. Nat. (1969) Mem. 160, p. 1-175.
Martinez-Machiavello, J. C., 1968. Quelques Acritarches d'un
    échantillon du Dévonien inférieur (Cordobés) de Blanquillo,
    Departement de Durazno, Uruguay, Rev. Micropaléontologie
    v. 11, no. 2, p. 77-84.
Rauscher, R., 1969. Presence d'une forme nouvelle d'Acritarchs
    dans le Dévonien de Normandie, C. R. Acad. Sci. Paris, v. 268,
    ser. D., p. 34-36, pl. 1.
Stockmans, F., and Willière, Y., 1960. Hystrichosphères du
    Dévonien belge (Sondage de l'Asile d'alienes à Tournai),
    Senck. Leth., v. 4, p. 1-11, pls. 1, 2.
_____, 1962a. Hystrichosphères du
    Dévonien belge (Sondage de l'Asile d'alienes à Tournai),
    Bull. Soc. Belge Géol. Paléont. Hydrol., v. 71, p. 41-77,
    pls. 1, 2.
_____, 1962b. Hystrichosphères du
    Dévonien belge (Sondage de Wepion, Bull. Soc. Belge Géol.
    Paléont. Hydrol., v. 71, p. 83-89, pls. 1, 2.
_____, 1963. Hystrichosphères ou
    mieux les Acritarches du Silurien belge, Sondage de la bras-
    serie Lust à Courtrai (Kortrijk), Bull. Soc. Belg. de Géol.,
    t. 71, fasc. 3, p. 450-481, pl. 1-3 (Brussels).
_____, 1966. Les acritarches du
    Dinantien du Sondage de l'asile d!alienes à Tournai (Bel-
    gique), Bull. Soc. Belge Géol. v. 74, p. 462-477, pl. 1.
_____, 1969. Acritarches du Fammeni-
    en Inférieur, Mem. Acad. R. Belgiques, Cl. Sci., v. 38, p.
    1-63, pls. 1-5.
Vavrdová, m., 1965. Ordovician acritarches from central Bohe-
    mia, Vestnik U.U.G., v. 40, p. 351-357.
Vavrdová, M., 1972. Acritarchs from Klabava Shales (Arenigian),
    Vestnik U.U.G., v. 47, p. 79-86.
Wall, D., and Downie, C., 1962. Permian hystrichospheres from
    Britain, Paleontology, v. 5, pt. 4, p. 770-784, pls. 112-114.
Wicander, E. R., 1974. Upper Devonian-Lower Mississippian acri-
    tarchs and prasinophycean algae from Ohio, U.S.A., Palaeonto-
    graphica, Abt. B., v. 148, Lfg. 1-3, p. 9-43, 19 pls.
Wilson, L. R., 1960. A Permian hystrichosphaerid from Oklahoma,
    Okla. Geol. notes, v. 20, p. 7, 170.

Text-figures

1. Pirea dubia Vavrdova, X495, late Arenigian to early Llanvirnian shales, Tadla Basin, Morocco (after Cramer et al., 1974)
2. Multiplicisphaeridium hoffmanensis Cramer, et al., X825, late Arenigian to early Llanvirnian, Tadla Basin, Morocco (after Cramer et al., 1974)
3. Multiplicisphaeridium maroquense Cramer, et al., X450, late Arenigian to early Llanvirnian, Tadla Basin, Morocco, (after Cramer et al., 1974)
4. Multiplicisphaeridium rayii Cramer et al., X495, late Arenigian to early Llanvirnian, Tadla Basin, Morocco (after Cramer et al., 1974)
5. Oodium mordidum Cramer et al., X900, late Arenigian to early Llanvirnian, Tadla Basin, Morocco (after Cramer et al., 1974)
6. Coryphidium elegans Cramer et al., X825, late Arenigian to early Llanvirnian, Tadla Basin, Morocco (after Cramer et al., 1974)
7. Picostella perforata Cramer et al., X495, late Arenigian to early Llanvirnian, Tadla Basin, Morocco (after Cramer et al., 1974)
8. Aureotesta clathrata Vavrdova, X495, late Arenigian, Tadla Basin, Morocco (after Cramer et al., 1974)
9. Rugulidium rugulatum Cramer et al., X495, late Arenigian, Tadla Basin, Morocco (after Cramer, et al., 1974)
10. Acanthodiacrodium sp., X495, late Arenigian, Tadla Basin, Morocco (after Cramer et al., 1974)
11. Polygonium gracile Vavrdova, X825, late Arenigian, Tadla Basin, Morocco (after Cramer et al., 1974)
12. Neoveryhachium neocarminae Cramer, two different specimens, X495, middle early Llandoverian to basal early Gedinnian, eastern U.S.A. (after Cramer and Diez de Cramer, 1972)
13. Quadriditium fantasticum Cramer, X495, late early Llandoverian to top Silurian, eastern U.S.A. (after Cramer and Diez de Cramer, 1972)
14. Duvernayosphaera aranaides Cramer, X495, late early Llandoverian to early Emsian, eastern U.S.A. (after Cramer and Diez de Cramer, 1972)
15. Deunffia furcata Cramer, X495, middle late Llandoverian to top Wenlockian, eastern U.S.A. (after Cramer and Diez de Cramer, 1972)
16. Eupoikilofusa striatifera stericula Cramer, X495, species ranges from base of Silurian to basal early Gedinnian, eastern U.S.A. (after Cramer and Diez de Cramer, 1972)
17. Dactylofusa maranhensis Brito and Santos, X240, Itaim Formation, Silurian, Maranhao Basin, Brazil (after Brito, 1967)
18. Evittia sommeri Brito, X240, Picos Shale, early to middle Devonian, Maranhao Basin, Brazil (after Brito, 1967)
19. Baltisphaeridium denticulata indianae Cramer, X495, base Silurian to top Gedinnian (range of species) eastern U.S.A., (after Cramer and Diez de Cramer, 1972)
20. Leiofusa brazilensis Brito and Santos, X300, Picos Shale, early to middle Devonian, Maranhao Basin, Brazil (after Brito, 1967)
21. Veliferites tenuimarginatus Brito, X750, Picos Shale, early to middle Devonian, Maranhao Basin, Brazil (after Brito, 1967)

22. <u>Duvernayosphaera</u> <u>radiata</u> Brito, X300, Cabecas Formation, middle Devonian, Maranhao Basin, Brazil (after Brito, 1967)
23. <u>Pseudolunulidea</u> <u>imperatrizensis</u> Brito and Santos, X240, Cabecas Formation, middle Devonian, Maranhao Basin, Brazil (after Brito, 1967)
24. <u>Maranhites</u> <u>mosesi</u> (Sommer), X375, Cabecas Formation, middle Devonian, Maranhao Basin, Brazil (after Brito, 1967)
25. <u>Pterospermopsis</u> <u>brazilensis</u> Brito, X240, Cabecas Formation, middle Devonian, Maranhao Basin, Brazil (after Brito, 1967)
26. <u>Cymatiosphaera</u> <u>velicarina</u> Wicander, X825, Bedford Shale, early Mississippian, Ohio (after Wicander, 1974)
27. <u>Conradidium</u> <u>firmamentum</u> Wicander, X525, Bedford Shale, early Mississippian, Ohio (after Wicander, 1974)
28. <u>Diexallophasis</u> <u>absona</u> Wicander, X750, Bedford Shale, early Mississippian, Ohio (after Wicander, 1974)
29. <u>Exilisphaeridium</u> <u>simplex</u> Wicander, X750, Bedford Shale, early Mississippian, Ohio (after Wicander, 1974)
30. <u>Multiplicisphaeridium</u> <u>verrucanum</u> Wicander, X712, Bedford Shale, early Mississippian, Ohio (after Wicander, 1974)
31. <u>Navifusa</u> <u>drosera</u> Wicander, X450, Bedford Shale, early Mississippian, Ohio (after Wicander, 1974)
32. <u>Stellinium</u> <u>cristatum</u> Wicander, X637, Bedford Shale, early Mississippian, Ohio (after Wicander, 1974)
33. <u>Baltisphaeridium</u> <u>debilispinum</u> Wall and Downie, X750, Lower Permian Marl, Yorkshire, England (after Wall and Downie, 1962)
34. <u>Micrhystridium</u> <u>stellatum</u> Deflandre, X750, Lower Permian Marl, Yorkshire, England (after Wall and Downie, 1962)
35. <u>Veryhachium?</u> <u>irregulare</u> Jekhowsky, X750, Lower Permian Marls, Yorkshire, England (after Wall and Downie, 1962)
36. <u>Leiofusa</u> <u>jurassica</u> Cookson and Eisenack, X750, Lower Permian Marls, Yorkshire, England (after Wall and Downie, 1962)
37. <u>Domasia</u> <u>elongata</u> Downie, X1050, Buildwas Beds, early Wenlockian, Wenlock, England (after Eisenack <u>et</u> <u>al</u>., 1973)

148

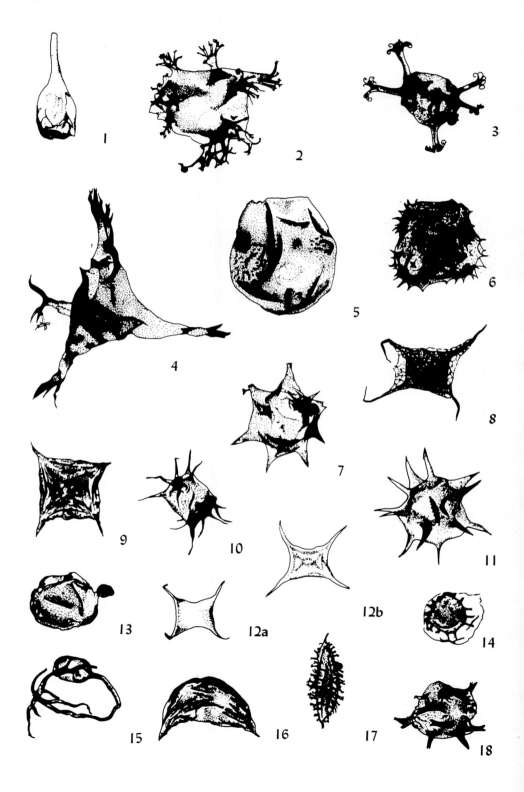

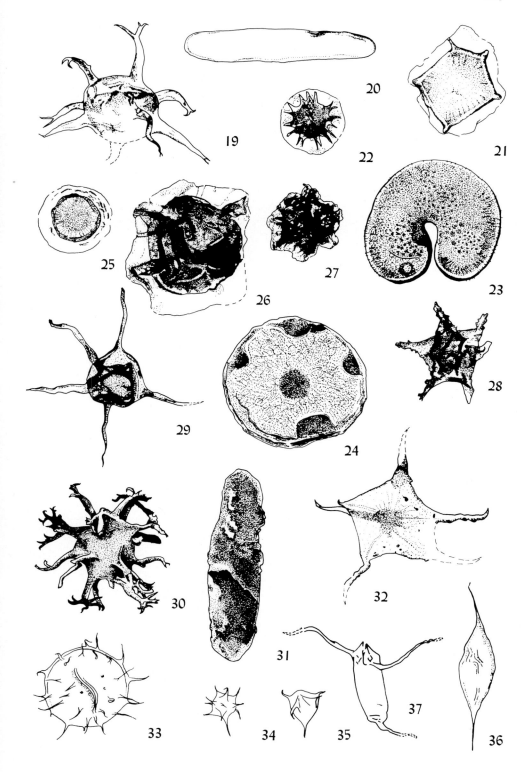

PALEOBIOGEOGRAPHY OF CHITINOZOA

By
Donald W. Zalusky
Glassboro State College
Glassboro, New Jersey 08028

Abstract

Biogeographic distribution of Ordovician to Silurian Chitinozoa of the ancient Atlantic borderlands is discussed. The classification system followed is that of Jansonius (1970). The genus Conochitina is restricted to forms with a mucro or copula; some forms designated Conochitina by previous authors are assigned to Euconochitina, therefore, Conochitina, sensu stricto, has rarely been reported from North America. The genus Cyathochitina is subdivided into two informal groups: Cyathochitina I which includes the conical forms, and Cyathochitina II which includes the cylindrical forms. Cosmopolitan or provincial genera and species are noted.

Résumé

La distribution biogéographique des Chitinozoaires Ordoviciens aux Chitinozoaires Siluriens des anciens confins Atlantiques est discutée. Le système de classification suivi est celui de Jansonius (1970). Le genre Conochitina est limité aux formes contenant un mucro ou un copula; certains formes attribuées auparavant à Conochitina par quelques auteurs, sont attribuées à Euconochitina, donc, Conochitina, sensu stricto, ont été rarement rapporté de l'Amérique du Nord. Le genre Cyathochitina se subdivise en deux groupes informels: Cyathochitina I qui comprend les formes coniques, et Cyathochitina II qui comprend les formes cylindriques. Des genres et des especes cosmopolites ou provinciaux sont notés.

Introduction

Chitinozoa, discovered in Silurian Baltic glacial erratics by Eisenack in 1929, are now documented from all continents except Antartica in the Ordovician through the Devonian interval. Initial finds were generally black and opaque and were illustrated only as silhouettes. Subsequent discoveries of well-preserved material provided some information on internal structures and additional information on surface ornamentation. It is now recognized that spines and other ornamentation originate from an external tegument or periderm which is frequently degraded and lost more readily than the principle wall layer or ectoderm.

The primary morphological features utilized in chitinozoan taxonomic assignment are general overall shape, presence or absence of spines, and other ornamentation. Loss of the periderm generally results in the loss of all smaller spines, although some genera possess larger periaboral spines which may be retained. Undoubtedly this has caused some confusion as many degraded spinous forms then mimic smooth forms of similar overall shape.

Chitinozoan investigations have demonstrated that Chitinozoa were geographically widespread and evolved rapidly in the Ordovician through Devonian Periods thereby providing the potential for an extremely useful correlation tool. Work in progress and recent investigations will undoubtedly clarify much of the taxonomic and biostratigraphic confusion. Recommended is the two volume work of the Commission Internationale de Microflore du Paleozoique (cited in bibliography) for all Chitinozoa research prior to 1965. Subsequent to this date, important publications (Eisenack, 1968, 1970, 1971, 1972a, 1972b, 1973; Boneham and Masters, 1973; Cramer, 1966, 1968, 1969, 1973; Cramer and Cramer, 1972a, 1972b; Jenkins, 1967,

1969, 1970a, 1970b; Laufeld, 1967, 1974; Lange, 1967, 1974; Urban, 1972; Urban and Kline, 1970; Urban and Newport, 1973; and others) have provided a substantial degree of maturity to Chitinozoa knowledge. Early workers reporting first occurrences from a particular geographic entity contributed much to the extension of the paleobiogeography of Chitinozoa but undoubtedly forced some identifications due to the reasons discussed earlier. It is hoped that investigators in other specialities encountering chitinozoans in their samples will request the assistance of chitinozoan workers to avoid invalid taxonomic assignments.

The stratigraphic range of Chitinozoa is confined to the Paleozoic but is only known with some degree of certainty in the Ordovician through Devonian interval. Isolated Cambrian occurrences have been reported but are (1) doubtful chitinozoans, (2) found in rocks whose Cambrian age is questionable, or (3) possibly represent stratigraphic leaks. Chitinozoans have been reported present in Paleozoic rocks younger than Devonian but generally only their presence was noted--forms were not identified or illustrated. Tasch (1973), however, has illustrated well-preserved Chitinozoa representing two new genera from the Permian Fort Riley Fm. of Kansas (USA), but has not published further details.

Taxonomic Considerations

The genus Conochitina was recognized as so broadly defined that many similar though distinct forms had been assigned therein. As emended by Taugourdeau (1966), Conochitina
was restricted to exclude forms without a mucro or conspicuous
callus. For the excluded forms, Taugourdeau erected the genus
Euconochitina.

It is generally considered that most if not all Chitinozoa formed chains although chain length must have varied between and within genera. Desmochitinids and related forms are
frequently found in short to long chains whereas other forms
are generally found isolated or not infrequently in chains of
only two or three tests attached oral pole to aboral pole.
Frequently more or less distinct concentric circles, the callus, are found centered on the aboral pole of many genera,
probably the result of chain forming. Therefore the "conspicuous" callus becomes a matter of individual judgement in assigning a form to Conochitina or to Euconochitina. This auth-
or relies primarily on the presence of a mucro to differentiate forms for assignment. On this basis, most North American
forms previously assigned to Conochitina are considered to be
members of Euconochitina. Jansonius (1970, p. 803) stated,
"...Conochitina sensu stricto is common in the Baltic area,
but so far not recorded from North America." A review of the

published illustration of North American "conochitinids" discloses only a few forms with a mucro. Illustrations of C. micrancantha (Collinson and Schwalb, 1955, Pl. 2, Fig. 20) and C. dactylus (Pl. 2, Figs. 16-19) each show more or less clearly a mucro; further, in the diagnosis they describe the presence of a small papilla (mucro) for each species. Boneham and Masters (1973) illustrated a specimen of C. dactylus which possesses a mucro (Fig. 3, no. 8). C. minnesotensis is shown to have a "basal process" by Jenkins (1969, Pl. 3, Figs. 11-12). Jenkins (p. 15) in reference to C. robusta noted, "A few specimens in each population bear a short cylindrical or distally tapered rod-like basal process (copula, Eisenack, 1959; 'mucro,' Laufeld, 1967)...."

With these exceptions as noted above, Conochitina sensu stricto is notably scarce in North America whereas Euconochitina is rather common. Possibly continued work will shed more light on this anomaly.

It is not the author's intent to substitute informal nomenclature, however, a useful purpose would be served if attention is called to the generic description of Cyathochitina, Eisenack, 1955. Eisenack described this genus as having a more or less cylindrical neck and a conical chamber. Such is the overall shape found in the designated type species, C. campanulaeformis. The distinguishing characteristic of Cya-

thochitina is the periaboral carina in association with the a-
bove mentioned shape. Eisenack and others have subsequently
included within Cyathochitina forms with little or no flexure
to distinguish the neck, or, that is, essentially cylindrical
forms. Without desiring to change the affiliation to this genus, the author proposes that investigators designate those
cyathochitinids possessing essentially a conical chamber and
cylindrical neck as Cyathochitina I and those essentially cylindrical in overall shape as Cyathochitina II; for example,
C. cylindrica II. This author believes such a procedure may
serve a useful purpose in both phylogenetic and paleobiogeo-
graphical distribution studies particularly in those cases
where illustrations do not accompany chitinozoan faunal lists.

Perhaps the best example of how this may be applied is
found on Plate VII, Vol. 1, The Chitinozoan (Combaz, et al.,
1967) wherein all species of Cyathochitina known through 1965
are illustrated. These species are catagorized as Type I or
II in Table 1 of this report. Of the 23 species and subspecies illustrated, 11 are Type I and 12 are Type II. Ten forms
are provincial to the Northwest Africa--2 Type I, 8 Type II.
One of the former, C. infundibuliformis, Taugourdeau and de
Jekhowsky, 1960, was questionably assigned to Cyathochitina
by Taugourdeau and de Jekhowsky because it did not possess an
observable carina.

European provincial forms include 4 Type I and 4 Type II.
The latter include C. conica, Taugourdeau, 1961, from the
Aquitane (S.W. France) and C. elenitae, Cramer, 1964 from the
adjacent area of Northwest Spain. The acritarch and chitinozoan fauna of Northwest Spain have been described by Cramer
(1966) as more closely allied with the Saharan Fauna than the
Baltic. C. stentor and C. striata, the other two species of
Type II, are unusually large species from the Baltic area
originally described by Eisenack (1937) in the genus Conochitina and subsequently transferred to Cyathochitina as included
species by Eisenack in 1955.

Cosmopolitan species occurring in two or more continents
consist of 5 Type I including the rather ubiquitous calix-
campanulaeformis-kuckersiana complex.

It would thus appear that Cyathochitina II is restricted
to the present southern latitudes if the atypical forms C.
stentor and C. striata are excluded. Cyathochitina I appear
to have a more cosmopolitan distribution with some species restricted to a transitional zone between the Saharan and the
Baltic faunas. In particular, forms closely related to C.
campanulaeformis are present in Baltic, Saharan, and North
American strata.

This somewhat restricted example utilizing only cyathochitinids known prior to 1965 appears to hold with respect to

post-1965 reports of cyathochitinid occurrences. Excluding a
possibly premature restriction of this genus a notation of
type would serve to elicit useful paleobiogeographic and phylogenetic information.

Table 1. Paleobiogeographic distribution of
pre-1965 cyathochitinids, Type I and II.

Cyathochitina Eisenack, 1955

Provincial Forms

NW Africa

Type	Species
II	alata
II	cylindrica
II	djadoensis
II	elongata
II	fistulosa
II	fusiformis
I	hymenophora nigarica
I	?infundibuliformis
II	koumeidaensis
II	obtusa

Europe

Type	Species
II	conica
II	elenitae
I	granulata
I	hymenophora
I	novempopulanica
I	regnelli
II	stentor
II	striata

Cosmopolitan Forms

Type	Species
I	calix (NW Africa, Baltic)
I	campanulaeformis (NW Africa, Baltic, N. America)
I	dispar (NW Africa, Baltic)
I	dispar verrucata (Aquitane, N. America)
I	kuckersiana (NW Africa, Baltic, N. America)

Ordovician and Silurian Biogeography

The analysis of the paleobiogeography of Chitinozoa is at
present complicated by several factors. A decided tendency
for chitinozoans to react to as yet unknown environmental factors exacerbates the picture. The sudden disappearance and

reappearance at higher intervals by some species has been discussed by Mannil (1972). Cramer (1970) noted the sudden dramatic increases in narrow zones, $\pm$ 2 mm, characterized by only one or two species, a phenomenon which he compared to algal blooms. Rapid decreases in abundance without a concomitant change in lithology have been described by Laufeld (1967).

Certain areas of the world have emerged as quasi-reference sections for chitinozoans due to the intensive work of one or more individuals. Foremost among these is the Baltic area as the result of Eisenack's activity spanning more than four decades. Material obtained from areas greatly removed from the Baltic have of necessity have been compared to the host of species described by Eisenack.

A southern reference fauna came into being due primarily to the efforts of Taugourdeau, and Taugourdeau and de Jekhowsky. Material from core samples spanning Ordovician through Devonian strata permitted Taugourdeau and de Jekhowsky (1960) to identify eighty-four species (forty-nine new) in two new genera and ten known genera. At the time of this work the precise age of the strata was indefinite.

Cramer (1964), 1966) reported on the occurrence of Chitinozoa in Silurian strata of Northwest Spain adding another assemblage for comparative purposes. Laufeld (1967) reported on the Caradocian Chitinozoa of Shropeshire, Wales. Andress,

Cramer, and Goldstein (1969) reported on Ordovician and
Silurian chitinozoans obtained from three well cores in Florida.

Other significant works of interest here include Cramer
(1968), chitinozoan fauna of the Red Mountain Fm. (Silurian)
of Alabama and Laufeld's (1974) extensive report on the
Silurian Chitinozoa of Gotland. Although Oklahoma is geographically removed from the present Atlantic Ocean, reports
by Jenkins (1967, 1969) on Chitinozoa of the Ordovician formations. Viola Ls. Fernvale Ls. and the Sylvan Shale of Oklahoma present important considerations for paleobiogeographic
distribution. Jenkins' contributions are enhanced by his previous extensive experience with Baltic faunas.

The foregoing constitute the primary reports which have
contributed most to the present knowledge of the Ordovician-
Silurian chitinozoan biogeographic distribution of the peri-
pheral areas of the Atlantic Ocean. South American chitinozoan faunas are presently little known and do not constitute
a reference in the sense used here. Many additional reports
(over 300) covering this region and others supplement those
previously mentioned but for various reasons are not germane
to this discussion.

The paleobiogeographic pattern of cyathochitinids previously discussed indicates a northern or "Baltic" chitinozoan

province and a southern or "Saharan" province with some degree of both provincialism and cosmopolitanism. The intervening boundary is blurred and the boundary itself may be the result of a limiting environmental factor rather than a physical barrier. It is likely that direct communication between the two provinces existed in some time intervals during the long time span under consideration.

The chitinozoan fauna of the Ordovician Sylvan Shale was considered by Jenkins (1970) to be related to the Baltic fauna on the basis of species in common which represented twenty-five per cent of the twelve species identified in the Sylvan Shale. Species in common were Conochitina cactacea, C. elegans, Desmochitina minor and D. scabiosa. Previously, Jenkins (1969) had found the Viola and Fernvale Limestones (Ordovician, Oklahoma) to contain cosmopolitan elements consisting of eleven previously described species, and noted a striking resemblance to faunas from the Cardoc-Ashgill of Estonia, Sweden, Britain, and the Osterkalk of the Baltic region. The cosmopolitan species consisted of Angochitina capillata, Conochitina miracantha, C. wesenbergensis, C. hirsuta, C. robusta, C. minnesotensis, Cyathochitina kuckersiana, Desmochitina minor, D. lata, Rhabdochitina usitata, and R. turgida.

Laufeld (1967) concluded that the Caradocian Chitinozoa from Dalarna Sweden show a striking resemblance to Estonian faunas of equivalent age (p. 292, "...as might be expected.") and suggested similarities, limited by sparse information in the literature, with faunas from the eastern United States. Study of the Shropshire Caradocian chitinozoan fauna by Jenkins (1967) led him to conclude that Baltic faunas were closely comparable whereas Saharan faunas were only remotely comparable, containing only two species recognized in Shropshire (Rhabdochitina magna and Siphonochitina pellucida). However, as Jenkins pointed out (p. 485), three additional species known from the Ordovician of Britian and the Baltic are included in a Range Chart of Saharan forms by Benoit and Taugourdeau (1961) although not figured or described by them. The presence of Clathrochitina in several Saharan assemblages and its absence in Shropshire is believed by Jenkins (p. 485) to emphasize the differences between these two faunas. Somewhat anomalous is the occurrence of chitinozoans belonging to the genus Siphonochitina in Britian and the Sahara but absent in the Baltic.

In a comparative study of Gotlandic and Saharan Silurian chitinozoans. Taugourdeau and de Jekhowsky (1964) found a high degree of affinity between the two but noted a time lag in certain cases in which some forms appeared much earlier in the Sahara. This time lag was attributed to the large migration distance involved. The Baltic chitinozoans were found by

Taugourdeau and de Jekhowsky to generally be larger than Saharan forms particularly Conochitina. Desmochitinids were remarkably similar and ancryochitinids appeared to follow parallel evolution although they were much more diversified in the Saharan region. Finely pilose angochitinids were noted to be absent in Africa. Saharan assemblages were considered by Taugourdeau and de Jekhowsky to be more varied and evolved than their European counterparts.

Cramer (1973) considered Florida Silurian chitinozoans to be quite similar to equivalent material from Portuguese Guinea.

Laufeld (1974) included within the genus Gotlandochitina, Laufeld, 1974, three previously described Chitinozoa (Ancyrochitina aequoris, Angochitina callawayensis, and A. milanensis) peculiar to the northern tier of states of the Central United States. This genus has not been recognized from the "Saharan" province.

From the foregoing, it is evident some degree of provincialism exists in Ordovician-Silurian chitinozoan faunas in spite of some cosmopolitan elements which gives rise to northern and southern provinces as has been demonstrated for other organisms. The northern or "Baltic" province encompasses lands bordering on the present Baltic Sea, the British Isles, Anticosti Island, Ontario, the northern states of the Central

United States and into Oklahoma. The southern or "Saharan"
province includes the Northwest African region, Portuguese
Guinea, Northwest Spain, and Florida. The intervening border
between the two provinces may be transitional, however,
"Baltic" and "Saharan" assemblages are distinguishable based
on the character of the fauna as well as composition.

References Cited

Andress, N. E., Cramer, F. H., and Goldstein, R. F., 1969.
Ordovician chitinozoans from Florida well samples: Transact.
Gulf Coast Assoc. Geol. Soc., v. XIX.

Benoit, A., and Taugourdea, P., 1961. Sur quelques chitinozoaires de l'ordovicien du Sahara: inst. Français Pétrole
Rev., v. 16, p. 1403-1421, 2 figs., 2 pls.

Collinson, C., and Schwalb, H., 1955. North American Paleozoic Chitinozoa: Ill. Geol. Survey, Rpt. Invest. 186, 31 p.,
12 figs., 2 pls.

Combaz, A., et al., 1967. Microfossiles organique du paléozoique; pt. 2, les chitinozoaires, morphographie: Centre
National Recherche Sci. Ed., 42., 8 figs., 5 pls.

Cramer, F. H., 1964. Microplankton from three Paleozoic formations in the Province of Leon (NW-Spain): Leidse Geol.
Mededel., v. 30, p. 253-261, 56 figs., 24 pls.

_____, 1966. Chitinozoans of a composite section of
upper Llandoverian to basal lower Gedinnian sediments in
northern Leon, Spain; a preliminary report: Soc. Belg.
Geologie Bull., v. 75, p. 69-146, 7 figs., 5 pls.

_____, 1968a. Considerations paléogéographiques à
propos d'une association de microplanctontes de la série
gothlandienne de Birmingham (Ala., USA): Soc. Geol. France
Bull., ser. 7, v. 10, p. 126-131, 1 fig.

_____, 1969. Possible implications for Silurian pa-
leogeography from phytoplancton assemblages of the Rose Hill
and Tuscarora Formations of Pa.: Jour. Paleon., v. 43, p.
485-491, 2 figs., pl. 70.

_____, 1973. Middle and Upper Silurian chitinozoan
succession in Florida subsurface: Jour. Paleon., v. 47, p.
279-288, 2 figs., 2 pls.

Cramer, F. H., and Diez de Cramer, C. R., 1972. Subsurface
section from Portuguese Guinea dated by palynomorphs as Middle Silurian: Am. Assoc. Petroleum Geologists Bull., v. 56,

p. 2271-2272, 1 fig.

Eisenack, A., 1937. Neue Mikrofossilien des baltischen Silurs, IV: Palaeont. Zeitschr., v. 19, p. 217-243, 22 figs., pls. 15, 16.

_____, 1955. Neue Chitinozoen aus dem Silur des Baltikums und dem Devon der Eifel: Senckenbergiana Lethaea, v. 36, p. 311-319, 3 figs., 1 pl.

_____, 1968. Über Chitinozoen des baltischen Gebietes: Palaeontographica, Abt. A, v. 131, p. 137-198, 13 figs., pls. 24-32.

_____, 1970. Mikrofossilien aus dem Silur Estlands und der Insel Oesel: Geol. Foren. Stockholm Forh., v. 92, p. 302-322, 7 figs.

_____, 1971. Weitere Mikrofossilien aus dem Beyrichienkalk (Silur): Neues Jahrb. Geologie u. Palaeontologie Monatsh., p. 449-460, 34 figs.

_____, 1972a. Chitinozoen und andere Mikrofossilien aus der Bohrung Leba, Pommern: Palaeontographica, Abt. A, v. 139, p. 64-87, 6 figs., pls. 16-20.

_____, 1972b. Beitrage zur Chitinozoen-Forschung: Palaeontographica, Abt. A. v. 140, p. 117-130, 1 fig., pls. 32-37.

_____, 1973. Kleinorganismen als Zerstoerer saeurefester organischer Substanzen und von Biophosphaten: Paleont. Zeitsch., v. 47, p. 8-16, 2 pls.

Jansonius, Jan, 1964. Morphology and classification of some Chitinozoa: Canadian Petroleum Geology Bull., v. 12, p. 901-918, 2 figs., 2 pls.

_____, 1967. Systematics of the Chitinozoa: Rev. Paleobotany and Palynology, v. 1, p. 345-360, 2 figs., 1 pl.

_____, 1970. Classification and stratigraphic application of Chitinozoa: Jour. Paleontology, v. 43, p. 889.

Jenkins, W. A. M., 1967, Ordovician Chitinozoa from Shropshire: Paleontology, v. 10, p. 436-488, 12 figs., pls. 68-75.

_____, 1969. Chitinozoa from the Ordovician: Viola and Fernvale Limestones of the Arbuckle Mountains Oklahoma: Palaeont. Assoc. Spec. Papers Palaeontology 5, 44 p., 10 figs., 9 pls.

_____, 1970. Chitinozoa from the Ordovician Sylvan Shale of the Arbuckle Mountains, Oklahoma: Paleontology, v. 13, p. 261-288, 7 figs., pls. 47-51.

Lange, F. W., 1967. Subdivisao bioestratigrafica e rivisao da coluna Siluro-Devoniana da basia do baixo Amazonas: simposio sobre a Biota Amazonica Atlas, v. 1, p. 215-326, 4 figs., 12 pls.

Laufeld, Sven, 1967. Caradocian Chitinozoa from Dalarna,
    Sweden: Geol. Foren. Stockholm Forh., v. 89, p. 275-349,
    34 figs.

_____, 1974. Silurian Chitinozoa from Gotland, Fos-
    sils and Strata, no. 5, p. 1-128.

Mannil, R., 1972. The zonal distribution of chitinozoans in
    the Ordovician of the East Baltic area. Int. Geol. Congr.
    24th Sess., Sect. 7, 569-571.

Tasch, Paul, 1973. Paleobiology of the Invertebrates: John
    Wiley and Sons, Iec., New York, 923 p.

Taugourdeau, Philippe, 1961. Chitinozoaires du silurien
    d'Aquitaine: Rev. Micropaléontologie, v. 4, p. 135-154,
    8 figs., 6 pls.

_____, 1966. Les chitinozoaires: techniques
    d'études, morphologie et classification: Soc. Geol. France
    Mem., n. ser., v. 45, mem. 104, 64 p., 4 pls.

Tougourdeau, Philippe, and Jekhowsky, Benjamen, de, 1960.
    Repartition et description des chitinozoaires silur-devonien
    de quelques sondages de la C.R.E.P.S., de la C.F.P.A. et de
    la S.N. Repal au Sahara: Inst. Français Pétrole Rev., v. 15,
    p. 1199-1260, 19 figs., 13 pls.

_____, 1964.
    Chitinozoaires Siluriens de Gotland; comparison avec les
    Formes Sahariennes, Inst. Français Pétrole, v. 19, p. 845-
    870.

Taugourdeau, Philippe, et al., 1967. Microfossiles organiques
    de paléozoique, pt. 1; les chitinozoaires; analyse biblio-
    graphique illustrée: Centre National Recherche Sci. Ed.,
    96 p., 4 fig., 11 pls.

Umnova, N., 1973. Methods of Investigation, Use of Infared
    Light for the Study of Chitinozoa: Paleont. Jor., 1973,
    no. 3, p. 394-400.

Urban, J. B., 1972. A reexamination of Chitinozoa from the
    Cedar Valley Formation of Iowa with observations on their
    morphology and distribution: Bulls. Am. Paleontology, v. 63,
    no. 275, p. 1-43, 9 figs., 8 pls.

Urban, J. B., and Kline, J. E., 1970. Chitinozoa of the Cedar
    City Formation, Middle Devonian of Missouri: Jour. Paleonto-
    logy, v. 44, p. 69-76, 2 figs., pl. 18.

Urban, J. B., and Newport, R. L., 1973. Chitinozoa of the
    Wapsipinicon (Middle Devonian) of Iowa: Micropaleontology,
    v. 19, p. 239-246, 2 figs., 2 pls.

PALEOZOIC RADIOLARIA : STRATIGRAPHIC DISTRIBUTION IN ATLANTIC BORDERLANDS

B.K. HOLDSWORTH, Department of Geology, The University,
Keele ST5 5BG, England

ABSTRACT: A few well-preserved, well-dated faunas suggest that Ordovician,
Silurian, Devonian and Carboniferous Systems may each possess
characteristic Radiolaria populations, with a major change in fauna
occurring in the Upper Silurian or Lower Devonian. Known Paleozoic and
post-Paleozoic taxa appear to be wholly unrelated.

RÉSUMÉ: Quelques faunes bien préservées et bien datées donnent l'impression
que chacun des Systèmes de l'Ordovicien, du Silurien, du Dévonien
et du Carbonifère peut avoir des populations Radiolaires caractéristiques,
un changement radical de faune ayant lieu au Silurien Supérieur ou au
Dévonien Inférieur. Les taxa connus du Paléozoïque et du Postpaléozoïque
paraissent complètement sans rapport.

INTRODUCTION: For reliable information on stratigraphic distribution of
Radiolaria in Paleozoic rocks we depend on a few well-
preserved, well-dated faunas. Nevertheless, understanding of Paleozoic
evolution in the Subclass has undergone a major revolution in the last 25
years. The new information cannot easily be summarized. Some important
genera and even families remain unnamed and the majority of important
assemblages are but partly described or wholly undescribed. What follows
is an extended abstract of relatively recent work, much of it unpublished.
Space allows only inadequate illustration and priority is given to
important, hitherto unfigured forms.

SYSTEMATIC FRAMEWORK: The systematic framework adopted in this account,
outlined below, differs in important respects from
those of previous authors, but space precludes full discussion.

Subclass RADIOLARIA Müller
Order POLYCYSTINA Ehrenberg (emended Riedel)
Suborder SPUMELLARIA Ehrenberg

I. "PALAEOACTINOMMIDS" new informal grouping.
All the Paleozoic Spumellaria with single latticed shell or two or more
concentric shells, sometimes pylomate, lacking the internal spicular
system of Entactiniacea and the characteristic lattice structure of
Rotasphaerids (below).
A large, diverse group, previously included in the Actinommidae
Haeckel (emended Riedel 1967). Though Palaeoactinommids show clear
morphologic similarities to post-Paleozoic Actinommidae, known Paleozoic
faunas seem to show a marked decline in Palaeoactinommids by Late
Paleozoic, and it is not completely clear that they are, in fact, present
in Late Paleozoic rocks. Stratigraphic distribution suggests, therefore,
that Palaeoactinommids and Actinommidae could be phylogenetically separate.[1]

II. "ROTASPHAERIDS" new informal grouping.
Paleozoic Spumellaria with single, spherical, latticed shell with angular
meshes, lacking the internal spicular system of the Entactiniacea and
displaying a point or points upon the shell from which radiate 5 or more
strong, straight lattice bars.

1. F'note: Suggestions of hollow spines amongst some Ordovician
Palaeoactinommids tend to reinforce this view, but the "grouping"
is almost certainly polyphyletic and stratigraphic ranges and
phyletic relationships are very unclear at present.

Superfamily ENTACTINIACEA Riedel
(Family Entactiniidae Riedel 1967)

Paleozoic Radiolaria, conventionally - though perhaps incorrectly - allocated to Spumellaria, with a spicule of four or more rays, point- or bar-centred, constituting the main part of the skeleton or incorporated within a latticed or spongy shell, essentially spherical, with or without pylome. Shell very rarely tubular, pylomate, and rarely (and doubtfully) conical, pylomate.

III. Family ENTACTINIIDAE Riedel 1967 (emended herein). Entactiniacea with unlatticed skeletons, the spicule rays not differentiated to define apical and basal hemispheres. More commonly, skeletons latticed, essentially spherical, lacking pylome. Type Genus: Entactinia Foreman.

The Entactiniidae as defined here includes the whole large group of previously recognized "non-pylomate Entactiniidae" (cf. Holdsworth 1973). Constituent genera are interpreted in the senses of Foreman (1963) and Fortey and Holdsworth (1972).

IV. Family PYLENTONEMIDAE Deflandre 1963 (emended herein). Entactiniacea with single essentially spherical, latticed (extremely rarely spongy) shell, rarely more than one shell, the outermost shell displaying an actual or incipient pylome with differentiated rim. Lacking tubular or conical extension associated with pylome. Type Genus: Pylentonema Deflandre.

? Superfamily ENTACTINIACEA

V. Family PALAEOSCENIDIIDAE Riedel 1967 (emended herein). Paleozoic spicular, siliceous skeletons comprising 6 to 8 divergent rays, arising most commonly from the ends of a very short median bar. 3 to 4 rays stronger and/or more elaborately ornamented define a "basal hemisphere" : 2 to 4 weaker, usually unornamented rays define an "apical hemisphere". Type Genus: Palaeoscenidium Deflandre.

VI. Family POPOFSKYELLIDAE Deflandre 1964

Lattice shell perforate, tubular or conical, with open aperture. Spicules confined to adapertural interior of shell, but with portions of some rays incorporated in shell wall. Type Genus: Popofskyellum Deflandre.

Of the three constituent genera, Cyrtentactinia Foreman - now construed as a monospecific genus represented by Cyrtentactinia primotica Foreman - is an unquestionable entactiniaceid. The inclusion of Popofskyellum and Tuscaritellum Deflandre within the Entactiniacea is less certain, and amongst species of these genera, one species of Popofskyellum only is definitely known to possess internal spicules.

Suborder ALBAILLELLARIA Deflandre 1953 (emended Holdsworth 1969a). Bilaterally symmetrical Paleozoic Radiolaria with closed or open frames in which can be detected a fundamental pattern of three rods arranged in triangular form.

VII. Family CERATOIKISCIDAE Holdsworth 1969a

Albaillellaria in which the skeletal frame is closed and typically consists of 3 rods, a-rod, b-rod and intersector, the a-rod commonly bearing paired spines (caveal ribs) developed in a plane normal to that of the frame. Ideal frame with both ends of all principal rods produced as extratriangular extensions, but some extensions may be suppressed. Type Genus: Ceratoikiscum Deflandre.

VIII. Family ALBAILLELLIDAE Deflandre 1952 (em. herein). Albaillellaria with a-rod and intersector longer than b-rod, in part developed as columellae within largely imperforate shell, not joining

distally or with junction tenuous (see Holdsworth 1969). Type Genus:
Albaillella Deflandre.

RADIOLARIA INCERTAE SEDIS

IX. RADIOLARIA FAMILIA NOVA Fortey and Holdsworth 1972
Paleozoic latticed or spongy conical shells, open at base, with spirally
disposed lateral spines and also apical spines. Lacking internal spicule.

X. Genus CORYTHOECIA Foreman 1963
Paleozoic tubular, imperforate, siliceous shells. Structure probably
variable and imperfectly known.

XI. SPICULAR FORMS
A diverse, probably polyphyletic, group of forms with skeletons consisting
of thread-like spicule rays, not obviously related to any family or genus
listed above.

Note: Implicit in the above treatment is a rejection at the moment of
      arguments for phyletic connections of the Entactiniacea with
Nassellaria and extant spiculate spheroids, such connections appearing
doubtful in the light of presently available morphologic and stratigraphic
data. Of post-Paleozoic families considered by Riedel (1967) to have
Paleozoic representatives, Phacodiscidae, Spongodiscidae and Lithelidae are
omitted, as no completely convincing evidence of their existence has been
found in well-preserved Paleozoic faunas.

BRIEF DESCRIPTIONS OF TAXA AND OUTLINE OF STRATIGRAPHIC DISTRIBUTION

CAMBRIAN: No adequate faunas are known from the Atlantic Borderlands or
elsewhere, but Palaeoactinommids are probably present and Nazarov (1975)
reported Entactiniidae (sensu this paper).

ORDOVICIAN: Excellent late Arenigian Radiolaria have been described from
the Valhallfonna Formation, Spitsbergen (Fortey and Holdsworth 1972). A
second fauna is known from the early Arenigian of this Formation (Bruton
et al., in preparation) and a similar occurrence is reported from the
Table Head Formation, Newfoundland (Bergström 1974).

Palaeoactinommids dominate both Spitsbergen horizons ("sphaeroids"
of Fortey and Holdsworth 1972). Architecture is particularly clear in the
earlier assemblage and entactiniaceid spicules are unquestionably absent.
Commonly the innermost structure is a subspherical microsphere with
straight lattice bars, from the junctions of which arise the primary
spines. In some cases this is the only complete lattice, but more
commonly it is enclosed by a very much larger spherical shell with
circular pores, less commonly by two such shells. A second broad group
lacks innermost microspheres, having 1 to 3 essentially spherical shells
with pores of irregular size and shape, shells being sometimes ill-defined
so that a sub-spongy structure results. Palaeoactinommid spines are
invariably delicate, unbladed, needle-like, sometimes slightly curved.

Contemporaneous Entactiniidae show various architectures, all

radically different from Palaeoactinommids. Important are Entactiniid
Genus Novum Fortey and Holdsworth, with subspherical shell developed from
lateral spinules of 3 of 6 near orthogonal, point-centred spicule rays, and
aff. Entactinia Foreman (Fortey and Holdsworth) spp. These latter possess
crude, sometimes incomplete single lattice shells generated either by
terminal ramification of 4 of 6 very strong, bar-centred spicule rays, or
by lateral spinules developed from 1 or more rays of a similar spicule, all
rays continuing as main spines, frequently terminating with simple
bifurcation. Examples of other architectures are very rare and include aff.
Stigmosphaera Rüst Fortey and Holdsworth and Entactiniid Genus Novum 2.
This last is distinctive, the sub-spherical skeleton developing from a bar-
centred, 6-ray spicule by repeated bifurcations of all rays, resulting
spinules not fusing. Entactiniidae are also probably represented by 6-
ray, point- and bar-centred spicules unassociated with any true lattice.
Conical Radiolaria Familia Nova spp. are confined to the higher
Spitsbergen horizon.

Arenigian Entactiniidae have some similarities to Silurian
forms (as have Palaeoactinommids) but differ completely from Late Paleozoic
forms. Rotasphaerids, Pylentonemidae, all Albaillellaria and probably all
Palaeoscenidiidae are apparently absent in the earliest Ordovician.
Adequate younger Ordovician material is unknown from the Atlantic Border-
lands, but Dunham and Murphy (in press : in litt.) show Arenigian-like
Palaeoactinommids and Entactiniidae to persist in the Caradocian of
Nevada, together with some new ?Palaeoactinommids and possibly Entactinia
Foreman s.s. sp.

SILURIAN: Well-preserved, hitherto undescribed Radiolaria of Llandoverian
through Wenlockian age are known from graptolitic concretions of the Cape
Phillips Formation, Cornwallis Islands. Preliminary study of assemblages
from 12 horizons shows Palaeoactinommids, Rotasphaerids, Palaeoscenidiidae
and Ceratoikiscum spp. to be the numerically dominant groups, with
latticed Entactiniidae less common than in Late Paleozoic faunas and with
bladed-spine forms absent. Ceratoikiscum spp. are undetected below the
Cyrtograptus rigidus Zone (Wenlockian). The relative importances of
environmental change, selective dissolution and evolutionary change in
determining the wide variation in assemblage make-up cannot yet be assessed.

Amongst Palaeoactinommids, smaller single and double shelled
species appear closely comparable with Arenigian forms, but far greater
diversity is now found in the group. Maximum lattice shell diameter varies

from 0.15mm. to 0.85mm. The most distinctive forms are the larger, with outermost simple, reticulate lattice widely separated from one or more much smaller, reticulate inner lattices. Investigation of innermost structure is more difficult than at Ordovician horizons, but in no instance so far where it has been possible to determine the innermost details of a Silurian form with more than one spheroidal lattice has any trace been found of an entactinaceid spicule. Presently it can only be concluded that the common, multishelled and rarer spongy Silurian spheroids are Palaeoactinommids. Number and spacing of lattice shells is variable in the group, as is main and subsidiary spine count, but spines are invariably unbladed rods - except in the instance of a relatively rare group with delicate, sub-spongy shells which possess 4 to probably 6 strong, three-bladed major spines. These are the earliest examples of bladed spines definitely known to the writer in the Paleozoic record.

The Rotasphaerids are totally distinct from the Palaeoactinommids in possessing only one spherical lattice shell which invariably shows at least part of the shell incorporating notably straight bars which radiate from a centre, often associated with a spine base. Most commonly, perhaps invariably, more than one radiation point is present. Interiors are clearly seen and there is no trace of structure within the single shell. Spines are never less than 6, commonly more, invariably rod-like but in very rare instances with a very slight suggestion of incipient blading. The commonly robust shells are likely to have been solution-resistant, so that the presence of this group in the Cornwallis Silurian assemblages, but apparently in no definitely older or definitely younger faunas, may be of particular biostratigraphic significance.

Entactiniidae occur in rather insignificant numbers. A single species of Entactiniid Genus Novum is very scarce, differing from the Arenigian forms in having rounder pores and more delicate shell. In material so far examined Entactinia s.s. is probably represented by no more than 4 small, rare species, all with unbladed spines and delicate or very delicate internal spicules, and Entactinosphaera s.s. has not been definitely recognized. Cf. Entactinia is represented by even scarcer specimens with very strong, 6-ray, bar-centred spicule and highly eccentric, crude and probably incomplete reticulate shell generated by lateral spinules from spicule rays which continue as rod-like spines. Entactiniidae lacking true, reticulate shell are slightly more abundant as a group, but individual species equally rare. Amongst these, Entactiniid Genus Novum 2 is closely comparable with the form from the lowest Arenigian. A form with bar centre and 7 very strong rays shows spinules arising laterally at more than one level from rays, preferentially developed on 4 rays. Most abundant is a group with 6 rays and short or very short bar centre having lateral spinules arising from one or more levels, equally developed on each ray - Haplentactinia Foreman s.l. spp.

Aff. Entactinia spp. of Arenigian type appear absent. The Silurian Entactinia spp. and Haplentactinia-like forms seem to represent an "advance" in development relative to the Arenigian, but it is noticeable that the Cornwallis reticulate Entactiniidae still appear to be scarcer and less diverse than in known Late Paleozoic assemblages, and it must be concluded that the major entactiniid radiation was post-Wenlockian.

Palaeoscenidiidae are represented by common Palaeoscenidium s.l. spp. with 3 or 4 basal spines and 2 or 3 apicals. They differ from Palaeoscenidium s.s. in lacking the imperforate, tent-like shell surrounding the proximal ends of the basals, but basals are ornamented by pendant spinules, sometimes amalgamating to form an irregular, perforate "tent". The entactiniid-like bar from which basals and apicals arise is clear and can be detected in Late Paleozoic Palaeoscenidium s.s. spp. This bar is also clear in a further, widely diverse group of spicular forms here referred to as "aff. Palaeoscenidium spp.". Such forms differ from Haplentactinia s.l. in the markedly differential development of the primary spicule rays, and there is a clear approach to the basal/apical polarisation characteristic of Palaeoscenidium s.l. and s.s. Single and commonly multiple bifurcation tends to mark the stronger rays : lesser length, absence of bifurcation or weak single bifurcation marks the less-developed, apical analogues. Pendant lateral spinules, reminiscent of Palaeoscenidium s.l. are found on some rays. It is tempting to view such forms as intermediates between Haplentactinia s.l. (Entactiniidae) and Palaeoscenidium s.l. (Palaeoscenidiidae) - and hence the tentative assignment of Palaeoscenidiidae to the Superfamily Entactiniacea (see Systematic Framework).

Ceratoikiscum spp. and excessively scarce Radiolaria Genus B Foreman sp. represent the first known appearance of Ceratoikiscidae and of Albaillellaria outside Asia. (Nazarov et al. (1975) described the new, possibly doubtful Ceratoikiscum acatangulatum from the Llandeilo-Lower Caradoc of Kazakhstan.) Absence of both genera in the lower part of the Cape Phillips Formation (Llandoverian levels) may well be due to solution. Ceratoikiscum spp. show ideal, rather delicate frames with no or weak patagium and slender caveal ribs - a few commonly developed from the dorsal end of the b-rod - ornamented only with minute serrations. A very few "mutant" specimens show doubling of one or more extra-triangular rods, or gross distortion of the usual, essentially equilateral frame. All forms are totally distinct from known Late Paleozoic species.

Of rare, Spicular Forms, most significant is a type with 8 curving spines radiating from the end of a short "stem", each spine ornamented with regularly spaced spinules. Similar but simpler forms are often abundant at Upper Paleozoic levels.

Forms not present in Cornwallis assemblages but known in European Silurian faunas are apparent pylomate Palaeoactinommids of Germany (Stürmer 1966) and Archocyrtium Deflandre spp. (Deflandre 1972), probably Wenlockian, and from Brétignolles, France. These latter are, in the writer's view, the earliest convincing Pylentonemidae known.[1]

LATE SILURIAN, EARLY AND MIDDLE DEVONIAN: Adequate assemblages from this considerable interval are unknown to the writer from the Atlantic Border-lands, so that, to identify the time of major assemblage change which, it appears, took place between Wenlockian and Frasnian, it is necessary to look elsewhere.

Most significant is Hinde's (1899) account of "Tamworth Series" forms (New South Wales) which, it would appear, must be from rocks of Early-Middle Devonian age. Though it is impossible to pin-point the horizon of any described species, the entire assemblage is clearly more closely related to well-dated Late Devonian faunas than to Cornwallis Silurian assemblages. Nazarov (1975) also indicates typical Late Devonian forms extending into the Middle Devonian of Kazakhstan. In both regions, essentially spheroidal forms with bladed spines predominate - a significant percentage, at least, being Entactiniidae.

Thus Aberdeen's (1940) "Devonian" fauna from the Santiago Member of the Caballos Novaculite, Texas, appears anomalous in apparently lacking Entactiniacea completely (Riedel and Foreman 1961). The age of the Novaculite and the position within it of the Santiago Member are both questionable (cf. Nigrini and Nitecki 1968). Lower Devonian conodonts from the "upper" part of the Novaculite (Graves 1952 : Glenister and Klapper (in litt.)) raise the possibility that some part of the unit might possibly be pre-Devonian, and the gross aspect of the Santiago assemblage, as described, seems to invite some comparison with the less diverse of Cornwallis pre-Ludlovian faunas. Santiago "Hexastylus" and "Heliosphaera" spp. as refigured by Riedel and Foreman (1961) appear to be possible Rotasphaerids and the larger multishelled Santiago spheroids compare with Cornwallis Palaeoactinommids. Of Santiago forms, only "Xiphostylus" and "Stylosphaera" are obviously foreign to the Cornwallis Silurian and closely comparable with Tamworth forms. But no reliable date exists for the first appearance of these and closely similar forms with spheroidal to slightly ellipsoidal shells and only two prominent spines, and it is quite possible that they are of late Silurian origin.

The unquestionable contrasts between proven Wenlockian and proven Frasnian assemblages (see below) could have been initiated prior to the

1. F'note: Mid. Ordovician "Pylentonema" spp. (Nazarov et al. 1975) appear from their descriptions to be pylomate Palaeoactinommids.

Middle Devonian and quite possibly in latest Silurian time.

LATE DEVONIAN: Most completely described and illustrated of known Paleozoic assemblages are those of the Huron Member of the Ohio Shale (Foreman 1963) - Famennian in age on conodont evidence (Hass 1947 : Glenister and Klapper (in litt.)).   Entactiniidae dominate the three assemblages studied, represented by the genera Haplentactinia, Entactinia, Entactinosphaera Foreman, Polyentactinia and Tetrentactinia Foreman. Of 44 essentially spherical, latticed species, 3 only may possibly lack an entactiniid spicule and be referable to the Palaeoactinommids. Rotasphaerids are absent. Contrast with Cornwallis Silurian assemblages is thus marked and further underlined by the prevalence of Entactiniidae with bladed spines and concentric lattices - unknown in the Lower Paleozoic. Palaeoscenidiidae show a marked decline, now represented by only two species, Palaeoscenidium cladophorum Deflandre and Palaeoscenidium ? quadriramosum Foreman - the latter apparently a remnant of the Silurian aff. Palaeoscenidium stocks. Ceratoikiscidae are represented by 4 species of Ceratoikiscum and one Radiolaria Genus B species, all very distinct from Silurian forms. The most significant innovation in this family is the first appearance of a shelled genus - Holoeciscus Foreman - but Albaillellidae are absent. Cyrtentactinia primotica Foreman is here considered the earliest representative of the Popofskyellidae, and Formaniella cibdelosphaera (Foreman) is the earliest known of the Pylentonemidae apart from Archocyrtium spp. figured by Deflandre (1972) from the "Silurian" of Brétignolles (see above). The problematic Corythoecia is first known from the Huron Member, accompanied by several Spicular Forms.

An older, early Frasnian (Polygnathus asymmetricus Zone) assemblage in excellent preservation is known from the Canol Shale, North West Territories (MacKenzie and Holdsworth, in preparation). Pylentonemidae, Popofskyellidae and Holoeciscus are unrepresented, Haplentactinia excessively scarce. In other respects the assemblage compares closely with the Huron faunas, certain entactiniid species being identical, others closely related. Palaeoactinommids have not been detected and Rotasphaerids and Albaillellidae are absent. Palaeoscenidiidae are represented by P. cladophorum and P. aff. cladophorum Ceratoikiscidae are richer than in the Huron assemblages, Ceratoikiscum spp more diverse, so that the absence of Holoeciscus strongly suggests that the genus is of post-early Frasnian origin. 3 of the 4 Huron Ceratoikiscum spp are clearly related to Canol stocks, but minor specific differences are detectable and probably of biostratigraphic utility. A distinctive Canol variety of Ceratoikiscum bujugum Foreman is recognizable in low Frasnian faunas of Nevada and Western Australia. Caveal rib vane development in Ceratoikiscum spinosiarcuatum Foreman s.l. Group is greater than at Huron horizons and in marked contrast to Wenlockian development. Within Ceratoikiscum planistellare Foreman Group exists a Canol form close to Ceratoikiscum avimexpectans Deflandre (Tournaisian), the Canol form being unseen at Huron levels but detectable in the low Frasnian of Western Australia. Certain Ceratoikiscum stocks, at least, appear to have had a world-wide distribution at Late Devonian tropical latitudes and are recognizable in the Urals (Nazarov 1973, 1975).

CARBONIFEROUS: The earliest adequate fauna is of the Montagne Noire, Southern France (eg. Deflandre 1960), probably Tournaisian and closely comparable with a proven Tournaisian Turkish assemblage (Holdsworth 1973). Most important is the first appearance of the Albaillellidae, represented in France by at least 3 Albaillella spp. and 1 species of Lapidopiscum Deflandre. Holoeciscus is now absent, but 5 and perhaps 6 Ceratoikiscum spp. were figured by Deflandre (1960). One may be identical with the Famennian C. spinosiarcuatum s.s., but both French and Turkish Tournaisian forms are mostly clearly distinguishable from Late Devonian species, though related in some cases. Particularly significant is the appearance of a Turkish species with the ideal 6 extratriangular spines reduced to 4 (possibly "C. (?) apertum" Deflandre of France).

Very striking in French and Turkish faunas is the abundance and diversity of Pylentonemidae - vastly greater than in the North American latest Devonian. Deflandre gave the names Pylentonema, Cerarchocyrtium Cyrtisphaeractenium, Cyrtisphaeronemium, Paraarchocyrtium and Archocyrtium to encompass the variation, but the classification remains unsatisfactory. Formaniella s.s. appears to be absent. Popofskyellidae are represented by the earliest known species of Popofskyellum and Tuscaritellum, but the Famennian Cyrtentactinia s.s. is undetected. Palaeoscenidium cladophorum persists from the Famennian, but the only other representative of the Palaeoscenidiidae is the new species Palaeoscenidium bicorne Deflandre. Non-pylomate Tournaisian spheroids are very poorly known. A number of species are Entactiniidae, but internal structure is seldom clear in material presently available.[1]

Viséan Radiolaria remain very poorly known. Re-examination of G.J. Hinde's late Viséan chert slides from S.W. England (Hinde and Fox 1895) reveals highly dissolved, partly crushed assemblages in which the

1. F'note: Triaenosphaera Deflandre may be a true Palaeoactinommid.

only forms able to be reliably identified are a few Entactiniidae.
Hinde's "Lithocampe" and "Stichocapsa" spp. are probably Albaillella spp.
outlines, and "Porodiscus" spp. are fairly clearly crushed and partially
dissolved spheroids, in some cases Entactiniidae.   A late Viséan (P2b)
horizon in black, goniatite-bearing, shales of County Leitrim, Ireland,
contains distinctive, new Radiolaria Genus B spp. accompanied by the
earliest known inflated, bi-winged Albaillella sp. of Albaillella pennata
Holdsworth Group.

A. pennata Group is also known from a very low Namurian (E1)
goniatite horizon of the Fayetteville Shale, Arkansas, associated with a
second spinose Albaillella sp. and spheroids with external characteristics
typical of Late Paleozoic Entactiniidae (Nigrini and Nitecki 1968).
The Namurian record is continued by often exquisitely preserved Radiolaria
in the black shale goniatite bands of Staffordshire and Derbyshire,
England - E2b through early R2a in age (Holdsworth 1964, 1966). In these
bands, Entactiniidae of the genera Entactinia, Entactinosphaera and
Polyentactinia predominate. Tetrentactinia is present, but shows less
diversity than in the Famennian. Haplentactinia and all Palaeoscenidiidae
are absent. Convincing evidence of Palaeoactinommids has not yet been
found, and there is a striking absence of forms with only two, major
spines, known in Devonian faunas (see above, Late Silurian, Early and
Middle Devonian). Pylentonemidae are known only by a single specimen of
Archocyrtium. Albaillellidae are represented most commonly by A. pennata
and A. cf. pennata (Holdsworth 1966a), spinose species being excessively
rare : Lapidopiscum is absent. Ceratoikiscidae include Radiolaria Genus B
spp., quite distinct from all older forms, and Ceratoikiscum spp., most
commonly of Ceratoikiscum bicancellatum Holdsworth and Ceratoikiscum
tricancellatum Holdsworth Groups (Holdsworth 1969a) - again, quite
distinct from all known pre-Namurian species. Popofskyellidae are
represented only by extremely rare, sporadic Popofskyellum spp., and the
last known occurrences of both Popofskyellidae and Ceratoikiscidae are in
Zone R1a. All Albaillellaria are restricted to relatively few of the
Radiolarian horizons, and it is at a few horizons where the rare,
Popofskyellum and sometimes cf. Corythoecia specimens are found. Spicular
Forms are frequent at many levels and have Late Devonian affinities.

Youngest known of adequate Paleozoic assemblages is that from
Eoasianites-bearing, sponge-rich concretions of the Intararé Formation,
Uruguay (see eg. Kling and Reif 1969). Age is questionable but apparently
within the interval Westphalian (Late Carboniferous) (Ramsbottom pers.

·comm.) to Early Permian (Bharadwaj 1969). Though of lesser age and apparently significantly higher latitude (see below) than English Namurian faunas, the gross aspect of the Intararé assemblage is rather similar. Preliminary study reveals at least 24 species of Entactiniidae, mainly Entactinia and Entactinosphaera spp., but at least 1 species of Polyentactinia is present - close and perhaps identical to a common English Namurian species - and an unnamed English Namurian genus probably occurs also. Presence of Palaeoactinommids has not been established. Pylentonemidae, Palaeoscenidiidae, Popofskyellidae and Ceratoikiscidae are absent. Albaillellidae are represented by at least two Albaillella spp., one a markedly spinose form, but A. pennata Group is probably absent. Spicular Forms are closely comparable to types known in the English Namurian, and there occurs also a remarkable scalariform Incertae Sedis, first seen at English Namurian horizons.

PERMIAN: Other than the Intararé assemblage, which could be of Early Permian age, no adequate faunas from this System are yet known to the writer. That a major revolution in Radiolaria history occurred somewhere between Late Carboniferous and Jurassic is unquestionable, but the present paucity of reliable information from the Permo-Trias leaves the time of this major change in doubt. No Albaillellaria, Entactiniacea, Popofskyellidae or Palaeoscenidiidae have, to the writer's knowledge, been reliably reported from Mesozoic rocks, and it has already been pointed out that the Actinommid-like forms of the Paleozoic (Palaeoactinommids) may be unrelated to the post-Paleozoic Actinommidae. But the Nassellaria - in the writer's view unrepresented by any reliably known Paleozoic form - were in existence by Triassic (Anisian-Rhaetian) time, represented by the Theoperidae (Riedel 1967). Thus a fundamental step in the evolution of the Radiolaria had occurred as early, perhaps, as Middle Trias, and the record of the striking pre-Jurassic extinctions and innovations is probably to be sought in Late Permian and Early Triassic assemblages.

PALEOLATITUDE AND PALEOOCEANOGRAPHY: It is unlikely that the few faunas discussed above, widely separated both in space and time, provide a fully representative record of Paleozoic Radiolaria evolution.

All but one fauna apparently belonged to a belt within 25° of the equator. High latitude forms are represented only in the ? Late Carboniferous Uruguayan assemblage which could have originated at latitude greater than 50°S. It is important to note, however, that in gross aspect this fauna is comparable with some tropical English Namurian assemblages.

Paleoenvironment could have been a more important factor in determining assemblage make-up. None of the faunas described above can be considered to occur in truly oceanic sediments. Modern Radiolaria appear to show maximum diversity in open oceans. Shallow, ensialic marginal seas appear to have restricted, endemic populations, seasonally enriched by ingress of open-ocean forms in some cases (Björklund 1974). Portions of extensive, comparatively shallow epicontinental seas, relatively distant from connection with contemporaneous oceans might be expected to have contained Radiolaria populations of somewhat lower diversity than the oceans themselves, and diversity could well have been closely related to ocean proximity. Slight but intriguing hints of such a relationship are detectable in the known Paleozoic record.

The profusion of earliest Carboniferous Pylentonemidae in Southern France and Turkey contrasts starkly with the virtual absence of the family in the English early to mid Namurian. Evolutionary decline may be the explanation, but from Early Paleozoic time Southern France and Turkey can be interpreted as having directly bordered a major "Tethyan" ocean (Briden et al. 1974), whereas the English Namurian accumulated in an epicontinental embayment (Ramsbottom 1971) to which oceanic water may well have been denied easy access. Conceivably, location at the "Tethys" margin also explains the Silurian appearance of Pylentonemidae at Brétignolles - anomalously early relative to the known first occurrence in North America. Within the shallower and more enclosed of Paleozoic epicontinental seas, salinity and temperature may have favoured specialised populations, depths being insufficient for the establishment of lower water layers capable of accommodating more open-sea forms. Comparatively few English Namurian marine bands contain significant numbers of Albaillellaria, and there is some independent geologic evidence to suggest that appearances of the sub-order correspond with periods of relatively high sea level within the Central England Basin.

The possibility cannot be ignored that other, apparently age-diagnostic, contrasts between the few Paleozoic faunas available are in some cases no more than a reflection of differences in the environments of the assemblages concerned.

Aberdeen, E., 1940. Radiolarian fauna of the Caballos formation, Marathon Basin, Texas: Jour. Paleont., v.14, p.127-139.

Bergström, S.M., 1974. The oldest known well-preserved Radiolarians from North America: Abst. Prog. Geol. Soc. Am., v.6, p.491.

Bharadwaj, D.C., 1969. Lower Gondwana Formations: Compte Rendue Sixième Cong. Internat. de Strat. et de Géol. du Carbonifère, v.1, p.255-278.

Björklund, K.R., 1974. The seasonal occurrence and depth zonation of Radiolarians in Korsfjorden, Western Norway: Sarsia, v.56, p.13-42.

Briden, J.C., Drewry, G.E. and Smith, A.G., 1974. Phanerozoic equal-area world maps: Jour. Geol., v.82 p.555-574.

Deflandre, G., 1960. A propos du développment des recherches sur les
Radiolaires fossiles: Rev. Micropaléont., v.2, p.212-218.
------------ 1972. Le système trabéculaire interne chez les
Pylentonémides et les Popofskyellidés, Radiolaires du Paléozoique.
Phylogenese des Nassellaires: C. R. Acad. Sci. Paris, t.274, p.3535-3540.
Dunham, J.B. and Murphy, M.A., in press. An occurrence of well-preserved
Radiolaria from the Upper Ordovician (Caradocian), Eureka County,
Nevada: Jour. Paleont.
Foreman, H.P., 1963. Upper Devonian Radiolaria from the Huron member of
the Ohio Shale: Micropaleont., v.9, p.267-304.
Fortey, R.A. and Holdsworth, B.K., 1972. The oldest known well-preserved
Radiolaria: Boll. della Soc. Paleont. Italiana, v.10, p.35-41.
Graves, R.W., Jr., 1952. Devonian conodonts from the Caballos novaculite:
Jour. Paleont., v.26, p.610-612.
Hass, W.H., 1947. Conodont zones in Upper Devonian and Lower Mississippian
formations of Ohio: Jour. Paleont., v.27, p.131-141.
Hinde, G.J., 1899. On the Radiolaria in the Devonian rocks of New South
Wales: Quart. Jour. Geol. Soc. London, v.55, p.38-64.
---------- and Fox, H., 1895. On a well-marked horizon of Radiolarian
rocks in the Lower Culm Measures of Devon, Cornwall and West Somerset:
Quart. Jour. Geol. Soc. London, v.51, p.609-688.
Holdsworth, B.K., 1964. Radiolarian nature of the thicker-shelled
goniatite faunal phase in some Namurian limestone "bullions": Nature
London, v.201, p.697-699.
--------------- 1966. A preliminary study of the palaeontology and
palaeoenvironment of some Namurian limestone "bullions": Mercian
Geologist, v.1, p.315-337.
--------------- 1966a. Radiolaria from the Namurian of Derbyshire.
Palaeontology, v.9, p.319-329.
--------------- 1969. The relationship between the genus Albaillella
Deflandre and the ceratoikiscid Radiolaria: Micropaleont., v.15,
p.230-236.
--------------- 1969a. Namurian Radiolaria of the genus Ceratoikiscum
from Staffordshire and Derbyshire, England: Micropaleont., v.15,
p.221-229.
--------------- 1973. The Radiolaria of the Baltalimani Formation, Lower
Carboniferous, Istanbul: in Kaya, O., Editor, Paleozoic of Istanbul: Ege
Universitesi Fen Fakultesi Kitaplar Seresi No.40, p.117-134.
Kling, S.A. and Reif, W.-E., 1969. The Paleozoic history of amphidisc and
hemidisc sponges: new evidence from the Carboniferous of Uruguay: Jour.
Paleont., v.43, p.1429-1434.
Nazarov, B.B., 1973. First discovery of Radiolaria Entactiniidae and
Ceratoikiscidae in the Upper Devonian of the Southern Urals: Dokl. Akad.
Nauk S.S.S.R., v.210, p.696-699. (In Russian).
------------ 1975. Lower and Middle Paleozoic Radiolarians of Kazakhstan:
Trudy Geol. Inst. Akad. Nauk S.S.S.R., v.275, 202pp. (In Russian).
------------, Popov, L.E. and Apollonov, M.K., 1975. Radiolaria from the
Lower Paleozoic of Kazakhstan: Izv. Akad. Nauk S.S.S.R. Ser. Geol. for
1975, no.10, p.96-111. (In Russian).
Nigrini, C. and Nitecki, M.H., 1968. Occurrence of Radiolaria in the
Mississippian of Arkansas: Fieldiana. Geology, v.16, p.255-268.
Ramsbottom, W.H.C., 1971. Palaeogeography and goniatite distribution in
the Namurian and Early Westphalian: Compte Rendue Sixième Cong.
Internat. de Strat. et de Géol. du Carbonifère, v.4, p.1394-1399.
Riedel, W.R., 1967. Class Actinopoda: in Harland, W.B. et al., Editors,
The Fossil Record: London (Geological Society), p.291-298.
------------ and Foreman, H.P., 1961. Type specimens of North American
Paleozoic Radiolaria: Jour. Paleont., v.35, p.628-632.
Stürmer, W., 1966. Das Wachstum silurischer Sphaerellarien und ihre
späteren chemischen Umwandlungen: Päleont. Z., v.40, p.257-261.

180

SPECIMENS FIGURED: Listed below are the slide numbers of all figured specimens with England Finder co-ordinates (slide label to right) in brackets. In the cases of Silurian specimens the numbers (GSC) assigned to the specimens in the National Type Collection of Invertebrate Fossils, Geological Survey of Canada, Ottawa, are also indicated. All specimens are presently stored in the author's collection.

Plate 1. 1. PMO/P/24. (K43.4); 2. PMO/P/15. (J40.4); 3. PMO/P/17. (H37); 4. CWI.CN/P/2. (K43.2). GSC 48235; 5. CWI.CB/P/13. (S40.3). GSC 48236; 6. CWI.CC/P/13. (P35.1). GSC 48237.
Plate 2. 1. CWI.CS/P/8. (M43.2). GSC 48238; 2. CWI.CN/P/39. (S35). GSC 48239; 3. CWI.CN/P/39. (S35.3). GSC 48240; 4. CWI.CN/P/2. (H37). GSC 48241; 5. CWI.CE/P/8. (N44). GSC 48242; 6. CWI.CN/P/11. (O34.3). GSC 48243; 7. CWI.CC/P/12. (R35.1). GSC 48244; 8. PMO/P/17. (K34).

ACKNOWLEDGEMENTS: For their co-operation in production of this account I am indebted to the Paleontologisk Museum Oslo, the Geological Survey of Canada and to the following individuals who have contributed invaluable material and paleontologic and stratigraphic information: S.M. Bergström, A. Brandon, D.L. Bruton, the late G. Deflandre, J. Dunham, R.L. Ethington, H.P. Foreman, R.A. Fortey, B.F. Glenister, W.S. MacKenzie, M.A. Murphy, O. Kaya, G.L. Klapper, S.A. Kling, W.H.C. Ramsbottom, W.R. Riedel. The Cornwallis Island material was discovered and collected by R. Thorsteinsson.

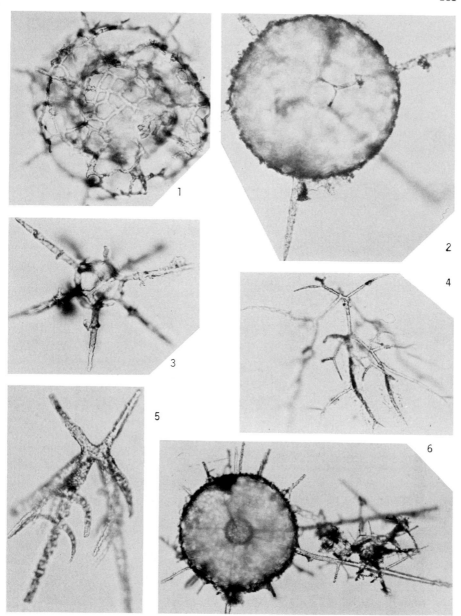

Plate 1. 1-3 Early Ordovician Palaeoactinommids (1. Paleontologisk Museum
Oslo NF.3280 x262, 2. PMO NF.3281 x262, 3. PMO NF.3282 x285).
4-5 Silurian Palaeoscenidiidae (4. Aff. Palaeoscenidium sp. x262,
5. Palaeoscenidium s.l. sp. x285). 6. Silurian Palaeoactinommid
x102.

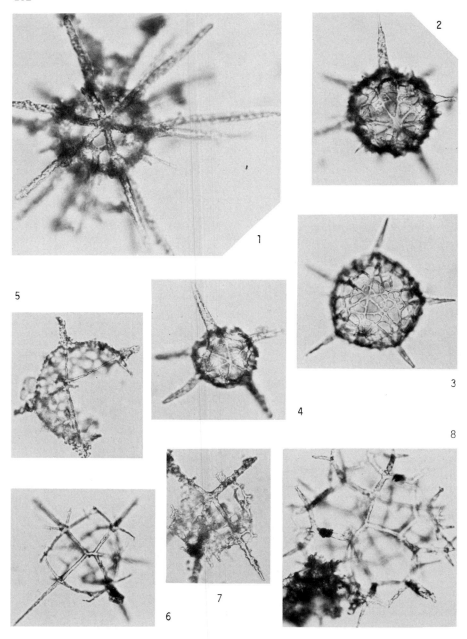

PLATE 2. All figs. x285. 1-4 Silurian Rotasphaerids; 5-7 Silurian
        Entactiniidae (5. Entactinia sp., 6. Haplentactinia s.l. sp.,
        7. Cf. Entactinia sp.); 8. Entactiniid Genus Novum 2 sp., Early
        Ordovician (Paleontologisk Museum Oslo N.F.3279).

Discussion

Dr. S. M. Bergstrom: Is there any morphological feature by which entactiniid radiolarians without shell can be safely distinguished from sponge
spicules? I have spicule-like fossils in my collections from the upper
Arenigian of western Newfoundland which are very similar indeed to some
of the radiolarians you illustrated, but they lack shell. I am not sure
if they are radiolarians or sponge spicules and wonder if you can advise
me how to classify these forms?

Dr. B. K. Holdsworth: In the Lower Ordovician the spicules of the Entactiniidae that I know are remarkably strong. All might well be mistaken for
some kind of sponge skeletal element, but for the fact that in many cases
the rays of the spicules give rise to a more or less complete lattice
shell, sometimes reticulate. But I admit that a real problem exists.

The presumed entactiniid, Haplentactinia arrhinia Foreman, is one of
relatively very few Late Paleozoic Entactiniacea lacking a true lattice,
the spicule rays being ornamented by circlets of short spinules. Super-
ficially it does look rather as though it could be a sponge element. On
the other hand, it is quite different from any kind of spicule that I know
to have been found definitely forming part of a Late Paleozoic sponge
skeleton. I agree with Helen Foreman that this spicular form really is an
entactiniid radiolarian.

But in the Early Paleozoic I share your difficulty. There are a lot of
often robust spicular fossils showing the bar-centre similar to that so
often found in unquestionable Entactiniacea, but with no true latticed
shell - only short spinules. Should we consider these to be haplentacti-
niidae in the very broadest sense - or are they completely unrelated?
Possibly important clues are their similarity in size and spicule strength
to unquestionable Early Paleozoic Entactiniidae; the broadly continuous
spectrum existing between truly latticed and completely unlatticed forms,
and the common failure of obvious sponge megascleres to appear abundantly
in the same rocks. I confess, however, to considerable ignorance regarding Early Paleozoic sponges, and I really cannot say at present that I
know of any infallible test which can differentiate between an unlatticed
Early Paleozoic entactiniid and all kinds of sponge spicules. I don't
know whether Helen has any views on your problem.

Dr. H. P. Foreman: No, I really have nothing to add.

Dr. F. Gradstein: You stated that you hope that Paleozoic Radiolaria may be
of use as indicators of 'blue', deep oceanic water masses. Do you think
that it is depth rather than distance from a shore which you hope to delineate, or both? Would the occurrence of Radiolaria-rich deposits not
be rather due to a "sediment starved" situation with little or no detrital
sediment influx instead of the presence of deep oceanic conditions?

Holdsworth: I do not think that my supposed 'blue' water assemblages necessarily indicate either particularly deep water sediments or particularly
land-remote sediments. In the cited cases of southern France and northwest Turkey I intended to suggest that the presence of unusually diverse
Early Carboniferous Radiolaria assemblages might be due to an oceanic influence induced by subsidence of French and Turkish shallow water, cratonic platforms immediately bordering a deep, 'Tethyan' ocean. To what
depth that depositional surface would have needed to have been submerged
to allow ingress of little-modified open ocean assemblages, and the distance of the depositional site from contemporaneous shorelines after such
submergence I would not like to say. What I do suggest is that the influence on Radiolaria populations of such 'overspill' of open ocean water
is likely to have declined rapidly in the direction of the continent
interiors. Certainly, sediments which accumulated well within major Late
Paleozoic oceans, presumably at very considerable depth and distance from
shore, should exhibit similar, high-diversity assemblages. But the stra-

tigraphic contexts of the French and Turkish occurrences do not, I think, indicate convincingly that they fall into this category, and as yet I know with certainty of no occurrences which might.

Sediment starvation in both deep and shallow water will often increase the bulk of Radiolaria in unit volume of rock. It alone is unlikely to markedly increase diversity - and I stress that it is marked difference in diversity of assemblages in Paleozoic rocks with large Radiolaria fractions that I believe to be of importance in paleooceanography. I did not mean to imply that simple abundance of Radiolaria in Paleozoic sediments is an indication of oceanic influence or of deep water origin. On the contrary, I am sure that some Paleozoic Radiolaria could occur profusely in distinctly shallow epicontinental seas far removed from the deep oceans, and given certain conditions - including slow sediment accumulation - they were preserved in considerable numbers.

Dr. W. Manspeizer: What independent criteria are available to support your inference that the Late Paleozoic Radiolaria are 'blue' water fauna?

Holdsworth: The 'independent criteria' I mentioned do not relate to my inference that extremely diverse Late Paleozoic assemblages with a diverse pylentonemid element are 'blue' water faunas. The 'criteria' relate to my suspicion that in quite shallow epicontinental seas, relatively far removed from deep oceans, comparatively slight increase in water depth tended to result in increase in Radiolaria diversity whilst not permitting the extreme diversity of the supposed 'blue' water assemblages.

In the 'basin facies' of the European Namurian it is common to find the predominant goniatite-pelecypod macrofauna confined to relatively thin beds of very fine black shale or mudstone, a few inches to a few feet thick, separated by very variable thicknesses of unfossiliferous strata. The fossiliferous beds are the so-called 'marine bands', and it is exclusively within these bands that Radiolaria are known to occur. The conventional explanation of 'marine bands' is that the rock record reflects successive rises and falls of sea level, probably eustatic, with 'marine bands' marking eustatic 'highs'. Falling sea level might be expected to have increased basin margin erosion, resulting in the interpolation of sands and silts between 'marine bands'. Often such interpolations occur. But at the southern end of the Central England Basin it is noticeable that certain parts of the succession display 'groups' were periods during which average sea level in the southern area was sufficiently high to prevent significant local erosion even during eustatic 'lows'. (Enhanced local subsidence, higher average eustatic sea level or a combination of these factors could be invoked in explanation.) Concomitantly, eustatic 'highs' bringing in the macrofauna during these periods should have resulted in local relatively deep seas. In fact, presently known occurrences of Albaillellaria, Ceratoikiscidae, Popofskyellidae and Corythoecia spp. are confined to 'marine bands' of those 'groups' which lack interpolated sands and silts. Such bands tend to have rather low goniatite and sponge contents - further indicators, it can be argued, of relatively deep water conditions. But the Pylentonemidae are virtually absent from all 'marine bands'.

Autunian and Carnian Palynoflorules

Contribution to the Chronology and Tectonic History

of the Moroccan Pre-Atlantic Borderland

Harold L. Cousminer
The American Museum of Natural History
New York, New York

Warren Manspeizer
Newark College of Arts and Sciences
Rutgers University
Newark, New Jersey

Abstract

A palynoflorule from the Moroccan Meseta north of Khenifra closely matches fossil pollen assemblages from the European Autunian Series, as described by Daubinger (1974) and from the Pictou Group of Nova Scotia (Barss and Haquebard,1967). The age of these assemblages is latest Carboniferous to earliest Permian.

A second palynoflorule from the Central High Atlas south of Marrakech is of mid-Carnian age based upon the concurrence of age-diagnostic pollen species that have been reported from the Swiss and English middle Keuper, type Carnian of Austria, and North American Triassic beds in Virginia, North Carolina, Pennsylvania, New Jersey, Texas, New Mexico, and Arizona.

These age determinations serve to document a large-scale unconformity that encompasses virtually all the Permo-Triassic on the Moroccan Meseta and High Atlas. This episode of pronounced crustal thinning preceded Triassic rifting, and ultimately led to the fragmentation of Pangaea.

Introduction

Late Paleozoic and Early Mesozoic reconstructions of Pangaea are inferred largely on paleomagnetic, geophysical and lithologic data (see Pitman and Talwani, 1972; Phillips and Forsythe, 1972; Ballard and Uchupi, 1975; Schenk, 1971; Olsen and Leydon, 1973; Dewey, Pitman and Ryan, 1973; and many others). Paleontologic data bearing on this question and the time of opening of the Atlantic, however, are sparse. This paper presents new palynologic data from the Late Carboniferous and Late Triassic of Morocco that bear on the sequence of events leading up to the opening of the North Atlantic Ocean.

Geologic Setting

Late Carboniferous sediments of Morocco were deposited in small isolated basins on the Moroccan-Oranian Meseta, and in large Paleozoic troughs of the Tindouf and Colomb Bechar Basins (Fig. 1). According to Michard and Sougy (1974, cited in Van Houten, 1976), the Meseta was formed by the Late Paleozoic Hercynian (Variscan) Orogeny. The main deformational phases occurred from the Namurian to the end of the Westphalian, and increased in intensity towards the margin of the Atlantic Ocean.

A new extensional phase of crustal deformation occurred in the Early Mesozoic that fragmented the Meseta. Late Triassic continental sediments of Middle Carnian and younger age were laid down in grabens in the western High Atlas as the Meseta began to separate from the African Plate (Cousminer and Manspeizer, 1976). In the Early Jurassic (Sinemurian), as North America and Africa began to move apart, the Oranian and Moroccan Plates separated along the line of the Middle Atlas Mountains (Manspeizer, Puffer, and Cousminer, 1976b). Early Jurassic seas, emanating from the Tethys Sea to the north and east, transgressed the Meseta along the High Atlas and Middle Atlas troughs.

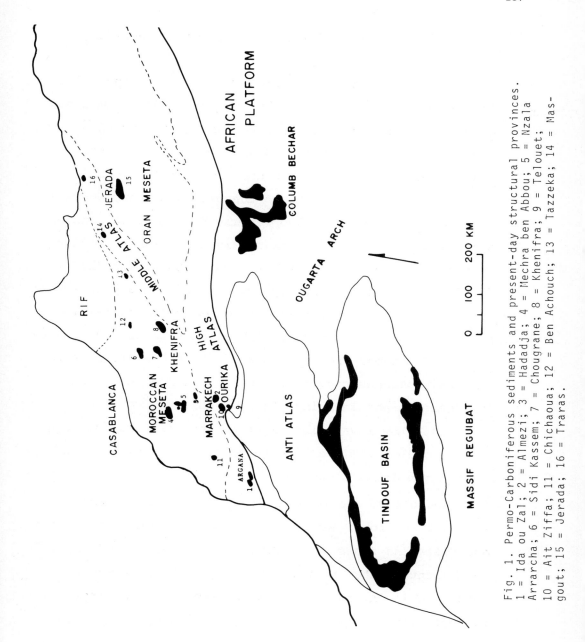

Fig. 1. Permo-Carboniferous sediments and present-day structural provinces.
1 = Ida ou Zal; 2 = Almezi; 3 = Hadadja; 4 = Mechra ben Abbou; 5 = Nzala
Arrarcha; 6 = Sidi Kassem; 7 = Chougrane; 8 = Khenifra; 9 = Telouet;
10 = Ait Ziffa; 11 = Chichaoua; 12 = Ben Achouch; 13 = Tazzeka; 14 = Mas-
gout; 15 = Jerada; 16 = Traras.

Autunian Palynoflorule: Moroccan Meseta, Khenifra Basin

A diversified palynoflorule was recently recovered from samples collected by John Lorenz in 1974 from a conglomeratic sandstone on the Moroccan Meseta, about 3.5 km. north of the city of Khenifra and along the west bank of the Oued Oum er Rbia. The productive samples were collected about 100 meters above the base of a conglomeratic sandstone that is about 500 meters thick, folded, and unconformably overlying Namurian (Mississippian to Pennsylvanian) metamorphic rocks. According to Lorenz (1974), the formation is principally composed of sandstones intercalated with massively bedded conglomerates, containing poorly sorted, well rounded cobbles of vein quartz, quartzite, and metamorphic rock in a sandy and pebbly matrix.

Palynomorphs recovered from the samples are listed on Figure 2, and fall into four groups:

> I. Lower vascular plant spores; II. monosaccate
> pollen; III. disaccate pollen; IV. miscellaneous
> gymnospermous pollen.

Because of the occurrence in the Moroccan samples of Pictou Group zonal markers from both the *Potonieisporites* and *Vittatina* zones, a stratigraphic level is indicated equivalent to that near the boundary between these two zones. This interpretation signifies that the sample is of uppermost Stephanian to Lower Permian age.

Based on this work a direct comparison can be made between the palynoflorules from the type Autunian sediments of Europe and the Khenifra Basin. In Figure 3 are histograms showing the percentages of spores, monosaccates, disaccates, monocolpates and *Vittatina* species

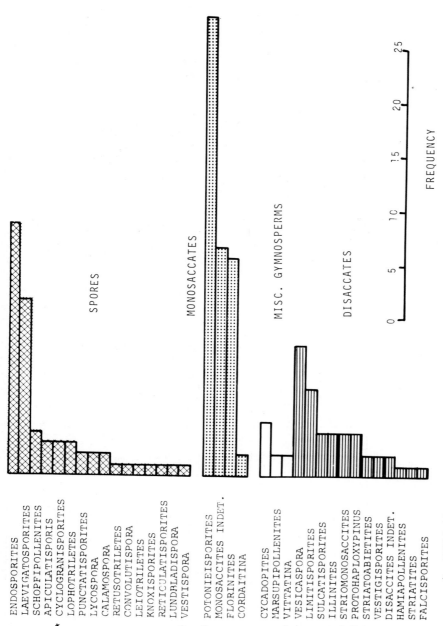

ENDOSPORITES
LAEVIGATOSPORITES
SCHOPFIPOLLENITES
APICULATISPORIS
* CYCLOGRANISPORITES
LOPHOTRILETES
PUNCTATISPORITES
LYCOSPORA
CALAMOSPORA
RETUSOTRILETES
CONVOLUTISPORA
LEIOTRILETES
KNOXISPORITES
RETICULATISPORITES
LUNDBLADISPORA
* VESTISPORA

* POTONIEISPORITES
MONOSACCITES INDET.
FLORINITES
CORDAITINA

CYCADOPITES
MARSUPIPOLLENITES
* VITTATINA
VESICASPORA
LIMITISPORITES
SULCATISPORITES
ILLINITES
* STRIOMONOSACCITES
* PROTOHAPLOXYPINUS
* STRIATOABIETITES
VESTIGISPORITES
DISACCITES INDET.
HAMIAPOLLENITES
STRIATITES
FALCISPORITES

SPORES

MONOSACCATES

MISC. GYMNOSPERMS

DISACCATES

FREQUENCY

0   5   10   15   20   25

* ZONAL MARKERS IN PICTOU GROUP, NOVA SCOTIA

Fig. 2. Frequency histogram of palynomorphs from the Morrocan Meseta.

190

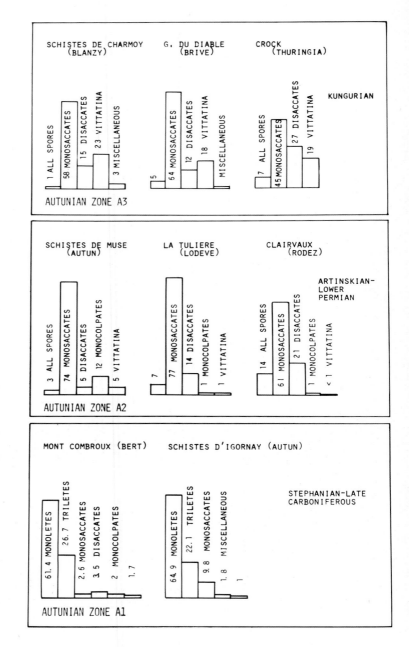

Palynomorphs in
sample from
Khenifra area
(sample K²)

Fig. 3. Histograms of pollen from European and Moroccan Autunian
sediments (European data from Doubinger, 1974).

characteristic of Autunian 1 (lower series), Autunian 2
(intermediate series) and Autunian 3 (upper series) sediments from the type Autunian deposits of Europe, (from
data tabulated by Daubinger, 1974). Included for comparison is a similar histogram of the Khenifra palynoflorule.
The Morroccan sample is clearly intermediate in character
between the Autunian 1 (Stephanian or Late Carboniferous)
and the Autunian 2 (Lower Permian) histograms. Thus, on
the bases of comparisons with both the Pictou Group pollen
sequence (Barss and Hacquebard, 1967) and the type Autunian of Central Europe (Daubinger, 1974) the Moroccan
palynoflorule from Khenifra is of latest Carboniferous to
earliest Permian age (Fig. 4).

A Triassic Palynoflorule from the Central High Atlas

     Recently (Cousminer and Manspeizer, 1976a) we reported
on the biostratigraphic significance of fossil pollen recovered from samples of the Oukaimeden Sandstone, collected
in the Ourika Valley south of Marrekech. According to
Mattis (1975), the Oukaimeden sandstone is about 200 meters
thick, and is composed of fine-grained, well-sorted, and
well-rounded quartz sands that are cross-bedded, ripple-
marked and lensoid. The productive samples occur about 400
meters above the base of a clastic sequence that overlies
Precambrian rock, and about 200 meters below the High Atlas
basalts which have an average radiometric age of about 195
m.y. The base of the Triassic section, in the northern
part of the Ourika Valley, is compsed of an andesitic lava
flow and volcanic clasts that rest unconformably (Choubert
and Faure-Muret, 1962) on fossil-bearing red beds con-

192

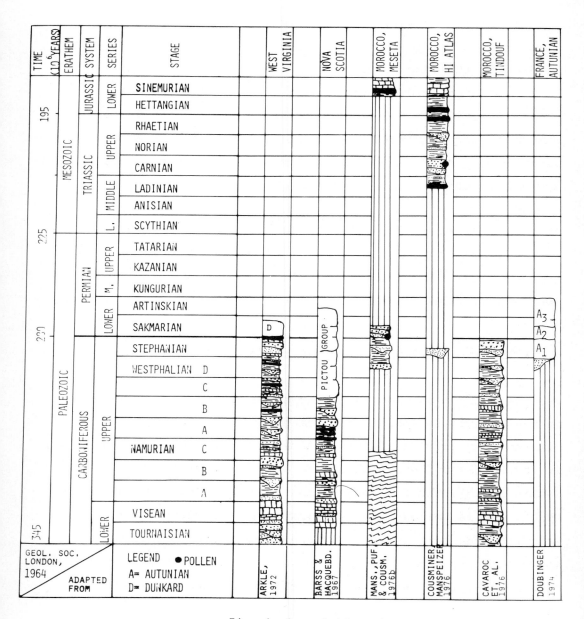

Fig. 4. Correlation chart.

sidered of Stephanian age because they contain *Walchia linearifolia, Mixoneura neuropteroides* (Clariond and Leca, 1934), and *Walchia piniformis* (Greber and Proust, 1958).

Figure 6 is modified from our previous publication (Cousminer and Manspeizer, 1976a) and gives all cited stratigraphic and geographic occurrences of age-diagnostic palynomorphs, recovered from the Central High Atlas of Morocco. They include the following comparatively long-ranging forms that are either entirely restricted to Triassic sediments or have their maximum distribution in sediments of Triassic age:

Aratrisporites species, *Chordasporites* species, *Triadispora* species, *Ovalipollis* species, *Porcellispora longdonensis, Enzonalasporites vigens.*

The Moroccan sample from the Central High Atlas is dated as of mid-Carnian age by the concurrent range of three fossil pollen species (Groups 1 and 5 on the Range Chart) that have overlapping and partially concurrent ranges in several European localities.

Conclusions

1) Two palynoflorule datums have been established in Morocco. These date the youngest Paleozoic rocks on the Meseta (Khenifra Basin) as of Autunian age (Stephanian-Sakmarian); and the Triassic rocks on the High Atlas (Ourika Valley) as of mid-Carnian age. Based on the palynoflorules these Moroccan strata are correlated with beds of equivalent age in many localities in both North America and Europe.

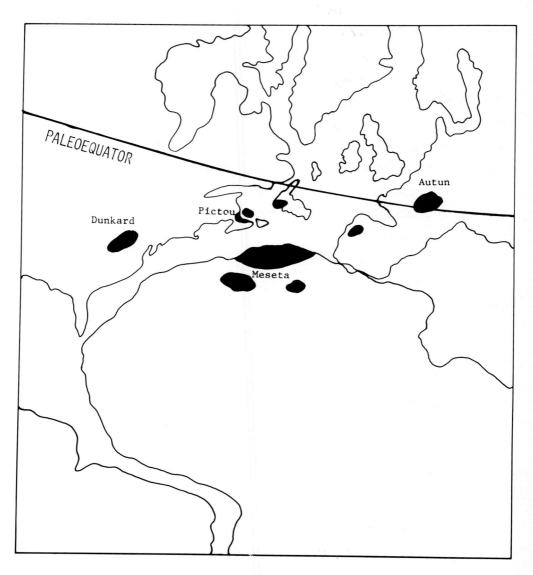

Fig.5. Stephanian - Autunian depositional domaines and paleogeography.

2) The Autunian palynoflorule from the Moroccan Meseta closely resembles those from the upper members of the Pictou Group in Nova Scotia, as described by Barss and Hacquebard (1967). This gives additional support to the Bullard reconstruction of Pangaea (Bullard *et al*, 1965) which places Nova Scotia opposite the Moroccan Meseta during pre-drift time.

3) The pollen data document a major unconformity in Morocco that extends from: (a) the Autunian-Sakmarian to the overlying Sinemurian on the Moroccan Meseta; and (b) from the Stephanian-Autunian to the middle Carnian of the High Atlas. The hiatus indicates that marked uplift and erosion of this region may have been dominant processes for for over 75 million years. This episode of extensive crustal thinning is considered to be of great significance in that it led to isostatic uplift, rifting, volcanism and the ultimate fragmentation of Pangaea.

4) The great increase in gymnospermous pollen near the Permo-Carboniferous boundary, which to a great extent replaced the dominant lower vascular plant spore assemblages so characteristic of the Carboniferous swamp lowlands, is connected with climatic changes that profoundly affected the distribution of Late Paleozoic plant communities. Since paleomagnetic data indicate that the position of the equator remained fairly stable through the Permo-Carboniferous interval, this climatic change is primarily correlated with orographic uplift.

5) Palynofloristic, paleobotanical, and lithostratigraphic records indicate that the degree of uplift and its duration in the Late Paleozoic-Triassic interval varied markedly at different geographic localities. Rapid change upwards from typically Carboniferous spore assemblages to

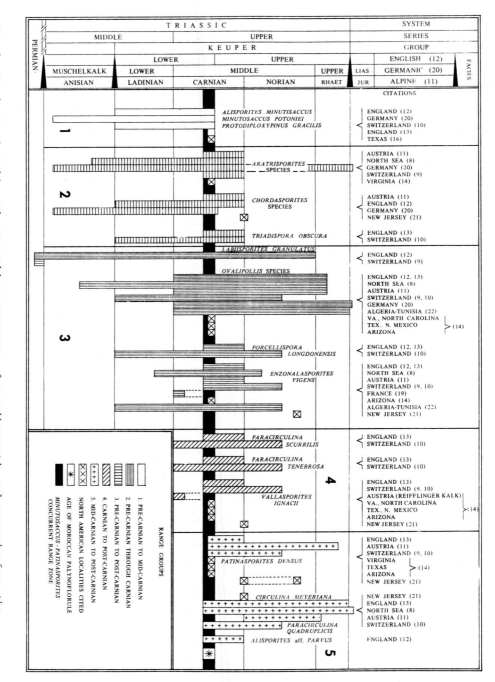

Fig. 6. Cited stratigraphic and geographic occurrences of age-diagnostic palynomorphs from the Central High Atlas of Morocco (modified from Science Volume 191, p. 944).

dominantly gymnospermous saccate pollen associations occurs in both the type Autunian Basin of France and the
Pictou Group sediments of Nova Scotia. Dunkard Group sediments of West Virginia are presumed to be of equivalent
age based on paleobotanical and paleontological evidence;
however, spore assemblages of Carboniferous aspect remain
dominant throughout the Dunkard sequence, indicating the
persistence of a marginal lowland basin in this region extending into the Lower Permian.

References

Ballard, R.D. and Uchupi, E., 1975. Carboniferous and
    Triassic rifting: a preliminary outline of the tecto-
    nic history of the Gulf of Maine. Geol. Soc. America,
    Bull., 83: 2285-2302.

Barlow, J.A., (editor) 1975. Proceedings of the First I.
    C. White Symposium-"The Age of the Dunkard". West Vir-
    ginia Geological and Economic Survey, 352pp.

Barss, M.S., 1967. Carboniferous and Permian spores of
    Canada. Geol. Surv. Can. Pap. 67-11 , 17pp.

Barss, M.S., and Hacquebard,P.A., 1967. Age and the stra-
    tigraphy of the Pictou Group in the Maritime Provinces
    as revealed by fossil spores. Geol. Assoc. Can., Spec.
    Paper 4: 267-282.

Barss, M.S., 1972. A problem in Pennsylvanian-Permian
    palynology of Yukon Territory. Geoscience and Man, 14:
    67-71.

Bode,H.H., 1975. The stratigraphic position of the
    Dunkard; In: J.A. Barlow (editor), Proceedings of the
    First I.C. White Symposium-"The Age of the Dunkard".
    West Virginia Geological and Economic Survey, pp.
    143-154.

Brown, R.H., 1974. The Argana Basin: A Triassic model
    for early rifting. M.S. thesis, Univ. of S. Carolina;
    Columbia, S.C.

Bullard, E.C., Everett, J.C. and Smith, A.G., 1965. The
    fit of the continents around the Atlantic; In: A sym-
    posium on continental drift. Royal Soc. London Phil.
    Trans. Sec. A; 258: 41-51.

Cavaroc, V.V., Padgett, G., Stephens, D.G., Kanes, W.H., Boudda-Ahmed and Wollen, I.D., 1976. Late Paleozoic of the Tindouf Basin-North Africa. Jour. Sed. Pet. 46(1): 77-88.

Chamot, G.A., 1965. Permian section at Apillapampa, Bolivia and its fossil content. Jour. Paleontology 39(6), 1112-1124.

Choubert, G., and Faure-Muret, A., 1962. Évolution du domaine Atlasique-Marocain depuis les temps Paleozoiques. In: Durand-Delga, M. ed. Livre à la mémoire du Prof. Paul Fallot: Soc. Géol. France, Mém. 1: 447-527.

Clariond, L. and Leca, F., 1934. Études sur le Stephanien du versant nord de l'Atlas de Marrakech: Soc. Géol. France, Comptes Rendus Sommaire 10: 210-212.

Clarke, R.F.A., 1965. Keuper miospores from Worcestershire, England. Palaeont. 8: 294-321.

Clendening, J.A., 1965. Palynological evidence for a Pennsylvanian age assignment of the Dunkard Group in the Appalachian Basin. Part I. In: J.A. Barlow (editor), Proceedings of the First I.C. White Symposium-"The Age of the Dunkard". West Virginia Geological and Economic Survey, pp. 195-221. Part II. Coal Geology Bulletin 3. West Virginia Geological Survey.

Cornet, B., Traverse, A., and McDonald, N.G., 1973. Fossil spores, pollen, and fishes from Connecticut indicate Early Jurassic Age for part of the Newark Group. Science 21: 1243-1247.

Cornet, B., and Traverse, A., 1975. Palynological Contributions to the chronology and stratigraphy of the Hartford Basin in Connecticut and Massachusetts. Geoscience and Man 11: 1-33.

Cousminer, H.L., 1965. Permian Spores from Apillapampa, Bolivia. Jour. Paleontology 39(6), 1097-1111.

Cousminer, H.L. and Manspeizer, W., 1974. Late Triassic palynoflorules form Morocco: Comparison with eastern North America. 1974 Annual G.S.A. meeting, Miami Beach, Florida, p. 697.

_____, and _____, 1976a. Triassic pollen date the incipient rifting of Pangea as Middle Carnian. Science 191: 943-945.

_____, and _____, 1976b. Autunian and Carnian palynoflorules: Contribution to the chronology and tectonic history of the Moroccan Pre-Atlantic Borderland. In: Swain, F.M. (editor), Symposium on Stratigraphic Micropaleontology of Atlantic Basin and Borderlands, Proceedings Volume (in press), Elsevier Scientific Publishing Company.

Cross, A.T., 1975. The Dunkard in perspective: Geology, sedimentation and life. In: J.A. Barlow (editor), Proceedings of the First I.C. White Symposium-"The Age of the Dunkard". West Virginia Geological and Economic Survey, pp. 297-299.

Darrah, W.C., 1975. Historical aspects of the Permian flora of Fontaine and White, Ibid, pp. 81-101.

Daubinger, J., 1974. Études palynologiques dans l'Autunien. Rev. Palaeobot. Palynol., 17(1/2); 21-38.

Defretin, S. and Fauvelet, E., 1951. Présence de phyllopodes triassiques dans la région d'Argana-Bigoudine (Haut Atlas Occidental). Notes it Mém. Serv. Géol. Maroc 5(85), 129-134.

Dewey, J.F., Pitman, W.C., Ryan, W.B.F., and Connin, J., 1973. Plate tectonics and the evolution of the Alpine system. Geol. Soc. Amer. Bull. 84: 3137-3180.

Dunay, R.E., and Fisher, M.J., 1974. Late Triassic palynoflorules of North America and their European correlatives. Rev. Palaeobot. Palynol., 17(1/2): 179-186.

Durden, C.J., 1975. Age of the Dunkard: Evidence of the insect fauna. In: J.A. Barlow (editor), Proceedings of the First I.C. White Symposium-"The Age of the Dunkard". West Virginia Geological and Economic Survey, p. 295.

Dutuit, J.M., 1964. Découverte de gisements fossilifères dans le Trias du couloir d'Argana (Atlas-occidental Marocain). Comptes Rendus Acad. Sci. Paris 285(4): 1285-1287.

Eager, R.M.C., 1975. Some nonmarine bivalve faunas from the Dunkard Group and underlying measures. In: J.A. Barlow (editor). Proceedings of the First I.C. White Symposium-"The Age of the Dunkard". West Virginia Geological and Economic Survey. pp. 23-67.

Feys, R., and Greber, C., 1963. Stephanien et l'Autunien du Souss dans les Ida ou Zal (Haut Atlas Occidental-Maroc); Notes et Mém. Serv. Geol. Maroc 22(170): 19-35.

Fisher, M.J., 1972. The Triassic palynofloral succession in England. Geoscience and Man 4: 101-109.

Geologic Map of Morocco, 1954-56. Protectorate Rép. Fr. Maroc; Div. Mines, Service Géologique.

Gillespie, W.H., Hennen, G.J., and Belasco, C., 1965. Plant megafossils from Dunkard strata in northwestern West Virginia, and southwestern Pennsylvania. In: J.A. Barlow (editor), Proceedings of the First I.C. White Symposium-"The Age of the Dunkard". West Virtinia Geological and Economic Survey, pp. 223-248.

Greber, C., and Proust, F., 1958. Sur le Permien et le Trias dans le Haut Atlas Occidental. Soc. Géol. France, Comptes Rendus Sommaires, 10: 210-212.

Havlena, V., 1965. European Upper Paleozoic *Callipteris conferta,* and the Permo-Carboniferous Boundary. In: J.A. Barlow (editor), Proceedings of the First I.C. White Symposium-"The Age of the Dunkard". West Virginia Geological and Economic Survey, pp. 7-22.

Jansa, L.F., and Wade, J.A., 1975. Geology of the continental margin off Nova Scotia and Newfoundland, In: Offshore geology of eastern Canada. Geol. Surv. Can. paper 74-30, 2.

Leschik, G., 1955. Die Keuperflora von Neuwelt bei Basel. Schweiz. Paläont. Abh. (Separatdrücke) 72: 1-68 pp.

Lorenz, J., 1974. Triassic sediments and basin structure of the Kerrouchen Basin, central Morocco. M.S. thesis, Univ. S. Carolina, Columbia, S. Carolina.

Lund, R., 1975. Vertebrate fossil zonation and correlation of the Dunkard Basin. In: J.A. Barlow (editor), Proceedings of the First I.C. White Symposium-"The Age of the Dunkard". West Virginia Geological and Economic Survey, pp. 171-182.

Maedler, K., 1964. Die geologisch Verbreitung von Sporen und Pollen in der deutschen Trias.Geol. Jahrb., Beih., 65: 147 pp.

Manspeizer, W., Puffer, J., and Cousminer, H.L., 1975. The Triassic record, a view from northwest Africa (Morocco). Symposium on Triassic Stratigraphy, Wesleyan College, Connecticut. (Oral Presentation only).

_____, _____, and _____, 1976. Subduction, rifting, and sea floor spreading, Eastern Section G.S.A. meeting, Arlington, Virginia, March.

_____, _____, and _____, 1976b. Opening of the North Atlantic Ocean: A Triassic-Jurassic record in Morocco and North America. Section of geological sciences, The New York Academy of Sciences, Monday, April 5, (Oral Presentation only).

Mattis, A., 1975. Nonmarine Triassic sedimentation, central High Atlas Mountains, Morocco. Ph.D. dissertation, Rutgers University, New Brunswick, New Jersey.

Olsen, W., and Leyden, R.J., 1973. North Atlantic rifting in relation to Permian and Triassic systems and their mutual boundary. Lamont-Daugherty Geol. Observ. Contr. #1810, pp. 720-732.

Olson, E.C., 1975. Vertebrates and the biostratigraphic position of the Dunkard. In: J.A. Barlow (editor), Proceedings of the First I.C. White Symposium-"The Age of the Dunkard". West Virginia Geological and Economic Survey, pp. 155-165.

Phillips, J.D., and Forsyth, D.W., 1972. Plate tectonics, paleomagnetism, and the opening of the Atlantic. Geol. Soc. America Bull. 83: 1579-1600.

Pitman, W.C. III, and Talwani, M., 1972. Sea floor spreading in the North Atlantic. Geol. Soc. America Bull. 83: 619-746.

Remy, W., 1965. The floral changes at the Carboniferous-Permian Boundary in Europe and North America. In: J. A. Barlow (editor), Proceedings of the First I.C. White Symposium-"The Age of the Dunkard". West Virginia Geological and Economic Survey. pp. 305-352.

Robb, J.M., 1971. Structure of the continental margin between Cape Rhir and Cape Sim, Morocco, Northwest Africa. Am. Assoc. Petr. Geol. Bull. 5: 643-650.

Schenk, P.E., 1971. Southeastern Atlantic Canada, northwestern Africa, and continental drift. Canadian Journ. Earth Sci., 8: 1218-1251.

Scheuring, B., 1970. Palynologische und palynostratigraphische Untersuchungen des Keupers in Bolchen Tunnel (Solothurner Jura). Schweiz. Palaeont. Abh. 88: 1-119.

Schultz, G., and Hope, R.C., 1973. Late Triassic flora from the Deep River Basin, North Carolina. Palaeontographica B. 141: 63-88.

Sohn, I.G., 1975. Dunkard Ostracoda-An evaluation. In: J.A. Barlow (editor), Proceedings of the First West Virginia Geological and Economic Survey. pp. 265-278.

Tasch, P., 1975. Dunkard estherids as environmental and age indicators. Ibid pp. 281-292.

Van Houten, F.B., 1975. A precocious Atlantic reconstruction. Geology 3: 194-195.

_____, 1976. Late Variscan nonmarine deposits, Northwestern Africa: Implications for pre-drift North Atlantic reconstructions. American Journal of Science, 276: 671-693.

Van de Poll, H.W., 1973. Stratigraphy, sediment dispersal, and facies analysis of the Pennsylvanian Pictou Group in New Brunswick. Maritime Sediments, 9: 72-77.

Yochelson, E.L., 1975. Monongahela and Dunkard nonmarine gastropods. In: J.A. Barlow (editor), Proceedings of the First I.C. White Symposium-"The Age of the Dunkard". West Virginia Geological and Economic Survey. pp. 249-263.

Acknowledgements

This research has been supported by a National Science Foundation Grant (GF325104) comprising a special foreign currency grant from the Office of International Programs, and from the Earth Science Section of the Division of Environmental Science. This grant is administered by the University of South Carolina.

The senior author acknowledges the use of office and laboratory facilities at York College of the City University of New York, where this research was begun. Thanks are also given to Professor Daniel Habib of Queens College, CUNY, for reviewing the manuscript.

Discussion

Dr. B. K. Holdsworth: Has your new work thrown any more light on the dating of the late Carboniferous-early Permian rocks at higher latitudes? Can you say anything more about the position of the Permo-Carboniferous boundary in the Gondwana Formations, for instance?

Cousminer: Concerning that area, I can't think of any data.

Holdsworth: It's simply that I have been trying to find out the true age of the Eoasianites (goniatite) level in the Itarare Formation of Uruguay which seems to compare quite nicely with part of the South African succession. Some people consider this level in the Gondwana as late Carboniferous, others take it as early Permian.

Cousminer: Are there no terrestrial beds associated? What did you say these are?

Holdsworth: These are interglacials. It's the Dwyka in South Africa.

Cousminer: There is some pollen data on the Dwyka Series.

Holdsworth: There is indeed, and I'm simply wondering how you would assess this?

Cousminer: I'll take a look at that. You put us onto something.

Holdsworth: Well, it may be that there is simply no connection at all. It's obviously much higher latitude than the stuff you are talking about.

Cousminer: That's right.

Dr. I. G. Sohn: I assume that you attribute the distribution of these to wind patterns. Is that correct?

Cousminer: In part, yes.

Sohn: Well, what confuses me is that you have a north-south distribution no matter how you turn it, which would sort of contradict a wind distribution for all the areas you are trying to correlate, or am I wrong?

Cousminer: There could have been a seasonal change in wind direction, due to alternation of the thermocline.

Sohn: Yes, but for that long a distance?

Cousminer: All I can do is present my data. The other thing is, of course, these continents were much closer together than they are now. The Dunkard Group was very close to the Moroccan Salient and may have been closer to a Bullard fit if you discount the continental shelf which may be of post-Triassic age.

Dr. D. Habib: Based on the reconstruction of continents, your palynofloral assemblages were probably within an essentially single phytogeographic zone. If this is the case, then you don't have to resort to long distance wind transportation.

Cousminer: I think that a more firm control might be elevation, where you might get a lowland and an upland distribution of plants that are close geographically and might differ quite a bit. You see this in the Dunkard Group in which you have what many people say are Permian (or Lower Permian) or Autunian equivalents in which the spores are of largely Carboniferous character. I think that that can be explained by the fact that the Dunkard basin was still rather low, relative to these other basins.

MESOZOIC FORAMINIFERA - WESTERN ATLANTIC

RICHARD K. OLSSON

GEOLOGY DEPARTMENT, RUTGERS UNIVERSITY, NEW BRUNSWICK, N.J.

ABSTRACT

Planktic foraminiferal zonal control of Upper Cretaceous sequences is provided by species of the genera Globotruncana, Marginotruncana, Rotalipora, Hedbergella, Rugotruncana, Archeoglobigerina, Abathomphalus, and Ventilabrella. Many of these species are tethyan so that recognition of tethyan biostratigraphic zones is apparent for the most part. Other species, however, are strictly boreal. Consequently, a zonal framework for this region reflects tethyan and boreal water as they influenced the region during specific time intervals of the Cretaceous. Due to preservational factors a planktic zonation for the Lower Cretaceous is at present incomplete.

Benthic foraminiferal assemblages are known from a few Lower Cretaceous and Upper Jurassic sections. They are sparse but apparently belong to shelf to abyssal facies. Upper Cretaceous assemblages occur in shelf, slope, and abyssal facies. Shelf assemblages include species of the genera Coryphostoma, Globulina, Guttulina, Gaudryina, and Gavelinella in addition to a number of lagenid species. Slope assemblages are characterized by species of the genera Gyroidinoides, Praebulimina, Pullenia, Osangularia, Gavelinella, Bolivinoides. Dorothia and others. A number of these species also occur in abyssal planktic oozes. Biostratigraphic control is rather general being useful for the recognition of state intervals. The Bolivinoides zonation is the one exception and it complements the planktic zonation in the Santonian, Campanian, and Maestrichtian.

INTRODUCTION

The intent of this paper is to briefly summarize the available data on the Mesozoic foraminifera of the Western Atlantic exclusive of the Canadian Shelf (see papers by Ascole and Bujack, Gradstein and Williams in this volume). Most of the data comes from the Cretaceous but a little comes from the Jurassic. There are no Triassic foraminifera known from the Western Atlantic.

Along the Atlantic margin of the U.S. sediments of Cretaceous age are exposed in the Atlantic Coastal Plain (fig. 1). The Cretaceous strata dip eastwardly beneath the coastal plain and continental shelf. Cretaceous strata also crop out along submarine canyons and the continental slope. In the adjacent Atlantic Basin a number of DSDP sites have penetrated Cretaceous deposits (fig. 1). Jurassic strata are not exposed in the coastal plain but they have been identified in a few coastal plain wells. Two DSDP drill sites have encountered Jurassic deposits (Sites 100 and 105).

The Upper Cretaceous of the Atlantic Coastal Plain consists of an alternating sequence of marine-marginal marine strata deposited during several cycles of transgression and regression. The Lower Cretaceous is entirely nonmarine in the coastal plain. These nonmarine sediments must extend some distance below the continental shelf for they are still encountered in coastal plain wells. Marginal marine Jurassic is suggested in some coastal plain wells. In the Atlantic Basin marine Lower and Upper Cretaceous and marine Jurassic were penetrated in the DSDP drill sites.

Only beds of Campanian-Maestrichtian age have yielded foraminifera in coastal plain outcroppings (Olsson, 1960, 1964 and Mello et al., 1964). In the coastal plain subsurface more complete sequences of foraminifera which range in age from Cenomanian to Maestrichtian have been recovered (Maher and Applin, 1971; O'Grady, 1976; Olsson and O'Grady, 1976; Olsson and Petters, 1975; Perlmutter and Todd, 1965; and Petters, 1975). Foraminifera from DSDP sites have been reported on by Peterson et al.,(1970, Upper Cretaceous), Cita and Gartner (1971), Luterbacher (1972, Lower Cretaceous and Upper Jurassic), Hayes et al.,(1972, Lower and Upper Cretaceous), and Laughton et al. (1972, Lower and Upper Cretaceous). Lower and Upper Cretaceous strata were penetrated in DSDP drill Sites 382, 384, 385, 386, 387, 390, 391, and 392 (Tucholke et al., 1975; Benson et al., 1975) but paleontologic reports are not yet available.

The following discussion attempts to summarize the data from the above sources in terms of planktic foraminiferal biostratigraphy, benthic foraminiferal biostratigraphy, and paleobathymetric distribution.

PLANKTIC FORAMINIFERAL BIOSTRATIGRAPHY

A planktic foraminiferal biostratigraphic framework has recently been presented by Petters (1975, 1976) for the Upper Cretaceous Atlantic Coastal Plain. This zonation (fig. 2) is similar to that proposed for the western Gulf Coast (Pessagno, 1967) and also bears a number of similarities with the idealized zonation for the low latitudes given by Van Hinte (1976). This is undoubtedly due to warm water influences from southern areas in a smaller Atlantic Ocean. The utility of species of the genera Rotaliapora and Marginotruncana in Cenomanian-Santonian sections would seem to indicate a strong Tethyan influence in the Atlantic region during this time interval. On the other hand Campanian and Maestrichtian zones based on species of Archeoglobigerina and Rugotruncana, suggest diminished Tethyan influence in the Atlantic region. Influence of warmer water masses in the Atlantic during late Cretaceous time is apparently related to the development of major current systems during the opening of the Atlantic (Berggren and Hollister, 1974).

Although many of the valuable biostratigraphic markers utilized in tethyan regions can be utilized also in the west-

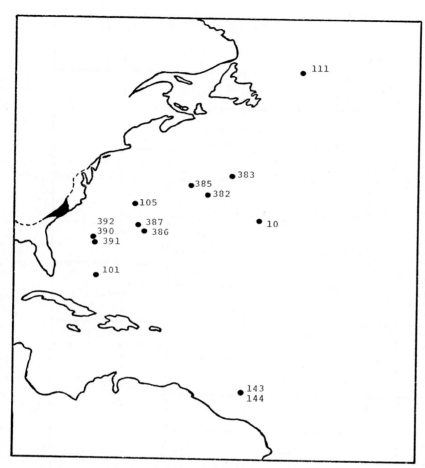

Fig. 1   Outline map of Western Atlantic showing loca-
         tion of DSDP drilling sites where Mesozoic sec-
         tions have been encountered.   Dashed line shows
         western limit of Atlantic Coastal Plain and the
         darkened areas show where Cretaceous sediments
         outcrop.   The numerous coastal plain wells that
         penetrate Mesozoic sections are not shown for
         lack of space.

STAGE	PLANKTIC FORAMINIFERAL ZONATION	BOLIVINOIDES ZONATION
MAESTRICHTIAN	ABATHOMPHALUS MAYAROENSIS	BOLIVINOIDES DRACO
MAESTRICHTIAN	GLOBOTRUNCANA GANSSERI	BOLIVINOIDES DRACO
MAESTRICHTIAN	RUGOTRUNCANA SUBCIRCUMNODIFER	B. MILARIS
CAMPANIAN	G. CALCARATA	B. DECORATUS
CAMPANIAN	G. ELEVATA	B. DECORATUS
CAMPANIAN	VENTILABRELLA GLABRATA	B. CULVERENSIS
CAMPANIAN	ARCHEOGLOBIGERINA BLOWI	B. CULVERENSIS
SANTONIAN	G. FORNICATA	B. STRIGILLATUS
SANTONIAN	MARGINOTRUNCANA CONCAVATA	
CONIACIAN	M. RENZI	
TURONIAN	M. SIGALI	
TURONIAN	M. HELVETICA	
CENO-MANIAN	ROTALIPORA CUSHMANI	
CENO-MANIAN	R. GREENHORNENSIS-PRAEGLOBOTR. DELRIOENSIS	

Fig. 2   Planktic foraminiferal zonation and Bolivinoides zonation for Atlantic Coastal Plain (after Petters, 1976).

ern Atlantic region, in general these elements are much rarer. Upper Cretaceous planktic foraminiferal faunas of the Western Atlantic are dominated more by species of globigerine forms such as Hedbergella, Globigerinella, Rugoglobigerina, Archeoglobigerina and Rugotruncana than are Tethyan faunas. Faunal distributions in the late Cretaceous of Atlantic planktic foraminifera appear to be characterized by broad geographic range of many species with significant changes in abundances from one region to the next. Biogeography would appear to be a fruitful line of study for this fossil group.

Lower Cretaceous planktic foraminifera have been identiin Deep Sea Drilling Sites 101, 105, 111, 143 and 144 but because of preservational factors a zonation is yet to be realized. Nevertheless, the presence of species such as Hedbergella trochoidea, H. infracretacea, H. hauterivica, H. globigerinelloides, Rotalipora appenninica, Globigerinelloides eaglefordensis, G. ultramicrus, and Planomalina buxtorfi attest to the diversity of species in the Atlantic during Early Cretaceous time and assure that in suitable sections planktonic foraminifera will provide a useful biostratigraphic zonation. Deep Sea Drilling Sites 384, 386, 387, 390, 391, and 392 penetrated Lower Cretaceous sections in the Bermuda Rise, J. Anomaly, Blake Nose, and Blake-Bahama Basin but foraminiferal reports are not yet available.

Although planktic foraminifera offer little potential for zonation in the Jurassic where their origin lies, it is useful to note that Luterbacker (in Laughton et al., 1972) has identified the species "Globigerina" helveticojurassica in the Tithonian-Kimmeridgian of site 105.

BENTHIC FORAMINIFERAL BIOSTRATIGRAPHY

CAMPANIAN - MAESTRICHTIAN:

The most useful group of benthic foraminifera for purposes of biostratigraphy are species of Bolivinoides. Petters (1975) has shown that the Bolivinoides zonation (Barr, 1966, 1970) can be utilized in Atlantic Coastal Plain sections (fig. 2). This upper Santonian-Maestrichtian zonation compliments the planktic foraminiferal zonation. Although there is no comparable benthic zonation for older Cretaceous and Jurassic sections the presence of certain species tends to characterize various stratigraphic levels.

CONIACIAN-SANTONIAN:

As noted above the evolution of Bolivinoides is seen in the upper Santonian with the advent of Bolivinoides strigillatus. Benthic species associated with the Coniacian-Santonian include Citharina texana, C. wadei, C. simondsi, Tritaxia capitosa, Planulina texana, and Gaudryina ellisorae. Coniacian foraminifera are absent in updip sections of the Atlantic Coastal Plain due to an important and widely recognized disconformity.

CENOMANIAN - TURONIAN:

Benthic assemblages are characterized by a rich diversity. In coastal plain wells these assemblages are associated with the well-known major transgression of the Albian-Turonian. In the Atlantic Coastal Plain this transgression covered the Early Cretaceous nonmarine sediments with marine sediments of Cenomanian-Turonian age. Deep Sea Drilling has recovered Cenomanian-Turonian sediments at sites 101, 105 (Luterbacker, 1972); 111 (Van Hinte in Laughton et al., 1972); and 143, 144 (Beckman, in Hayes et al., 1972). Site 386 also encountered sediments of this age (Tucholke et al., 1975).

A number of benthic species characterize assemblages of this age. These include Vaginulina debilis, Marginulina siliquina, Citharina kochi, Saracenaria duckcreekensis, Lenticulina gaultina, Eponides moremani, Hoeglundina charlottae, Valvulineria infrequens, V. lotterlie, Gavelinella dakotensis, G. plummerae, G. baltica and Gavelinopsis cenomanica. The latter species appears not to range above the Cenomanian (Van Hinte, 1976)

EARLY CRETACEOUS:

Marine sediments of Early Cretaceous age have not been identified so far in the Atlantic Coastal Plain. Sediments of this age have been recovered at sites 100, 101, 105 (Luterbacker, 1972); 111 (Van Hinte in Laughton et al., 1972); and 143, 144 (Beckman, in Hayes et al., 1972) of the Deep Sea Drilling Project. Sites 384, 386, 387, 390, 391, and 392 have encountered early Cretaceous sediments (Tucholke et al., 1975; Benson et al., 1975) but paleontologic reports are not yet available. Early Cretaceous benthic assemblages are rather sparse and so far do not offer good stratigraphic control. Dorothia praehauteriviana is reported by Luterbacker (1972) as a possible marker species for the Valanginian.

Species identified include Reophax helveticus, Textularia washitensis, T. rioensis, Spiroplectammina alexanderi, Lituola subgoodlandensis, Lenticulina saxocretacea, L. ex gr. muensteri, L. ouachensis ouachensis, L. ouachensis multicella, Astacolus incurvatus, Discorbis minutissima, Lingulina nodosaria, and Gavelinella cf. barremiana.

JURASSIC:

Although strata of Jurassic age have been questionable identified in a few Atlantic Coastal Plain wells no definitive analysis of foraminifera has been published. Luterbacker (1972) has reported on the diverse microfaunas found in Upper Jurassic sediments in sites 100 and 105 of the Deep Sea Drilling Project. He subdivided the benthic foraminifera into three groups, "primitive foraminifera" which predominate and consist of the simply structured arenaceous foraminifera, the lagenids, and other foraminifera not contained in the first two groups and which are minor elements of the assemblages. Some of the more typical species include Brotzenia mosquensis,

Bigenerina jurassica, B. arcuata, Bolivinopsis helveto-
jurassicus, Marssonella doneziana, Trocholina transversarii,
Lenticulina polonica, Marginalina minuta, and "Planularia"
pseudoparallela.

PALEOBATHYMETRIC DISTRIBUTION

Benthic foraminifera are well known for their importance
in delineating environments of deposition and bathymetry in
Cenozoic deposits. Few studies have used foraminifera for
such interpretations in pre-Cenozoic deposits because of the
uncertainty of the environmental significance of extinct
species and genera. Sliter and Baker (1972) are the first to
propose a bathymetric model for Cretaceous benthic foraminifera that defines the generic composition of shelf and slope
assemblages. Their model is based on a study of Campanian-
Maestrichtian shelf and slope deposits of California. It is
based on criteria which includes comparison of depth restriction of genera as deduced from recent distributions, variations in faunal diversity and abundance, recurrent species
associations, and homeomorphic comparisons of Cretaceous and
Recent species.

In the Western Atlantic region, Cretaceous nonmarine,
marginal marine, and shallow-middle shelf deposits are exposed
in the Atlantic Coastal Plain, outer shelf-bathyal deposits
are encountered in some coastal plain wells (Olsson, 1975;
Olsson and O'Grady, 1976), and bathyal-abyssal sections are
encountered in the DSDP drill sites. Some sites such as 111
(Laughton et al., 1972) are drilled in sunken portions of the
continental margin and thus contain shallower-water deposits
than the depth from which the samples were taken. Other sites
must represent locations where bathyal-abyssal deposits of
Cretaceous age are sampled. The following presentation is a
preliminary effort to summarize the known foraminiferal distributions and assemblages as bathymetric models following the
approach of Sliter and Baker (1972).

CAMPANIAN-MAESTRICHTIAN:

Four bathymetric assemblages of Campanian-Maestrichtian
foraminifera can be recognized in the Western Atlantic region;
inner shelf, outer shelf, bathyal, and lower bathyal-abyssal
(figs. 3 and 4). The inner shelf assemblage is characterized
by low species diversity and high dominance. Planktic foraminifera occur rarely in this assemblage. Benthic species
that tend to dominate in this assemblage are Gavelinella
henbesti and Praebulimina carseyae.

Outer shelf assemblages are characterized by increased
diversity of species, a greater variety of lagenid and
rotaliid species. Species of planktic foraminifera become
prominent associates of this assemblage. The more abundant
species of this assemblage include Anomalinoides nelsoni, A.
pinguis, Clavulina trilaterus, Pulsiphonina prima, and Coryphostoma plaitum.

SHELF	INNER	MILIOLIDS LAGENIDS GLOBULINA LACRIMA GUTTULINA NONIONELLA CRETACEA PRAEBULIMINA ASPERA PRAEBULIMINA CARSEYAE	GAVELINELLA HENBESTI CIBICIDES HARPERI GYROIDINOIDES DEPRESSA VALVULINERIA ALLOMORPHINOIDES GAUDRYINA
	OUTER	POLYMORPHINIDS LAGENIDS NEOFLABELLINA RETICULATA PRAEBULIMINA CARSEYAE KYPHOPYXA CHRISTNERI CORYPHOSTOMA PLAITUM PULSIPHONINA PRIMA CLAVULINA TRILATERUS NEOBULIMINA SPINOSA SIPHOGENERINOIDES PLUMMERAE	ANOMALINOIDES PINGUIS ANOMALINOIDES NELSONI GAVELINELLA NACATOCHENSIS ANGULOGAVELINELLA GRACILIS ALLOMORPHINA ALLOMORPHINOIDES HOEGLUNDINA SUPRACRETACEA BOLIVINOIDES TEXTULARIA RIPLEYENSIS

Fig. 3 Tentative bathymetric model of Campanian -
Maestrichtian shelf assemblages based on distribution
of foraminifera in Atlantic Coastal Plain outcrop and
well sections.

BATHYAL	STENSIOEINA	RZEHAKINA EPIGONA
	CORYPHOSTOMA INCRASSATA	DOROTHIA OXYCONA
	PRAEBULIMINA ARKADELPHIANA	DOROTHIA ELLISORAE
	PRAEBULIMINA KICKAPOOENSIS	DOROTHIA BULLETTA
	ARAGONIA VELASCOENSIS	DOROTHIA RETUSA
	GYROIDINOIDES NITIDUS	ARENOBULIMINA FRANKEI
	GYROIDINOIDES GIRARDANUS	ARENOBULIMINA SUBSPHAERICA
	CHILOSTOMELLA TRINIDADENSIS	GAUDRYINA LAEVIGATA
	OSANGULARIA NAVARROANA	CLAVULINA TRILATERUS
	BOLIVINOIDES	BATHYSIPHON
	PULLENIA CRETACEA	SPIROPLECTAMMINA
	STILOSTOMELLA	
	PYRAMIDINA SZAJNOCHAE	
	PLEUROSTOMELLA TORTA	
LOWER BATHYAL- ABYSSAL	ARAGONIA VELASCOENSIS	BATHYSIPHON
	PRAEBULIMINA TAYLORENSIS	PELOSINA COMPLANATA
	LAGENA	SPIROPLECTAMMINA MEXIAENSIS
	OSANGULARIA	TROCHAMMINOIDES IRREGULARIS
	DOROTHIA OXYCONA	GLOMOSPIRA GORDIALIS
	RZEHAKINA EPIGONA	VERNEUILINA

Fig. 4 Tentative bathymetric model of Campanian-
Maestrichtian bathyal-abyssal assemblages based on
distribution of Foraminifera in Atlantic Coastal
Plain wells and DSDP drill sites.

SHELF	CITHARINA TEXANA CITHARINA WADEI CITHARINA SIMONDSI LAGENIDS POLYMORPHINIDS NONIONELLA ROBUSTA PRAEBULIMINA CUSHMANI	TRITAXIA CAPITOSA CLAVULINA TRILATERUS PLANULINA TEXANA GYROIDINOIDES UMBILICATA GYROIDINOIDES DEPRESSUS HOEGLUNDINA SUPRACRETACEA
BATHYAL	LENTICULINA PRAEBULIMINA FABILIS CORYPHOSTOMA INCRASSATA GYROIDINOIDES NITIDUS VALVULINERIA PLUMMERAE ELLIPSOGLANDULINA STILOSTOMELLA	BERTHELINELLA BATHYSIPHON VULVULINA ARENOBULIMINA DOROTHIA TEXTULARIA CLAVULINA TRILATERUS
BATHYAL- ABYSSAL	BATHYSIPHON REOPHAX LITUOTUBA AMMODISCUS	STENSIOEINA BANDYELLA

Fig. 5 Tentative bathymetric model of Coniacian-Santonian shelf-abyssal assemblages based on distribution of foraminifera in Atlantic Coastal Plain wells and DSDP drill sites.

SHELF	**INNER**	LAGENIDS	HAPLOPHRAGMOIDES
		POLYMORPHINIDS	TROCHAMMINA
		VAGINULINA DEBILIS	GAVELINELLA DAKOTENSIS
		MARGINULINA SILIQUINA	GAVELINELLA MINIMA
		NEOBULIMINA ALBERTENSIS	GAVELINELLA PLUMMERAE
		PRAEBULIMINA EXIQUA	CASSIDELLA TEGULATA
		EPONIDES MOREMANI	CERATOBULIMINA PARVA
		QUINQUELOCULINA	HOEGLUNDINA CHARLOTTAE
	OUTER	LAGENIDS	GAVELINELLA BALTICA
		POLYMORPHINIDS	GAVELINELLA DAKOTENSIS
		CITHARINA KOCHI	GAVELINELLA MONTERELENSIS
		VAGINULINA DEBILIS	GAVELINELLA PLUMMERAE
		MARGINULINA SILIQUINA	GAVELINOPSIS CENOMANICA
		SARACENARIA BONONIENSIS	VALVULINERIA INFREQUENS
		SARACENARIA DUCKCREEKENSIS	VALVULINERIA LOTTERLIE
		LENTICULINA GAULTINA	HOEGLUNDINA CHARLOTTAE
		BULIMINELLA FABILIS	GAUDRYINA
		FURSENKOINA CRONEISI	SPIROPLECTAMMINA
		NEOBULIMINA ALBERTENSIS	QUINQUELOCULINA

Fig. 6 Tentative bathymetric model of Turonian-Cenomanian shelf assemblages based on distribution of foraminifera in Atlantic Coastal Plain wells and DSDP drill sites.

Bathyal assemblages contain a greater diversity of arenaceous species than do shelf assemblages. Species of Dorothia, Arenobulimina, Gaudryina, Clavulina, and Bathysiphon are prominent. Typical calcareous genera include Stensioeina, Praebulimina, Gyroidinoides, Osangularia, Pullenia, Chilostomella, and Bolivinoides. Planktic foraminiferal species are well-developed and diverse.

Lower bathyal-abyssal assemblages are distinguished by dominance of planktic foraminifera and low numbers and diversity of benthic species. Benthic species include Aragonia velascoensis, Rzehakina epigona, Pelosina complanata, and Glomospira gordialis.

CONIACIAN-SANTONIAN:

Little information is available on assemblages of this age in part due to the Coniacian disconformity. Shelf assemblages appear to be characterized by a diversity of lagenids of which species of Citharina are prominent; bathyal assemblages contain the genera Praebulimina, Gyroidinoides, Bathysiphon, Arenobulimina, Dorothia, and Clavulina, among others; and a few benthic genera are observed in bathyal-abyssal assemblages (fig. 5)

TURONIAN-CENOMANIAN:

Four bathymetric assemblages, an inner shelf, outer shelf, bathyal, and lower bathyal-abyssal, can be recognized in this interval (figs. 6 and 7). Species which tend to dominate the low diversity inner shelf assemblage are Praebulimina exiqua, Gavelinella dakotensis and G. minima. Hoeglundina charlottae is fairly common. Planktic species are rare.

Outer shelf assemblages are characterized by increased diversity of species, a variety of lagenid species of which a number are typical of the age, and prominent association of planktic species. Abundant species include Gavelinella plummerae, G. monterelensis, Valvulineria lotterlie, and Buliminella fabilis.

The deeper bathymetric assemblages are composed of species of Neobulimina, Gavelinella, Valvulineria, Gyroidinoides, Bathysiphon, Dorothia, Ammobaculites, Ammodiscus, Spirillina, and Glomospira which in addition to a variety of other arenaceous taxa contrasts distinctly with the shelf assemblages.

EARLY CRETACEOUS:

Luterbacker's (1972) study of sites 100, 101 and 105 showed that the lower Cretaceous sections are characterized by sparse assemblages of foraminifera, the composition of which consists of a predominance of simply structured arenaceous foraminifera and rare lagenids. He compared these to similar assemblages that occur in Early Cretaceous strata of the Alpine-Mediterranean region which are interpreted as being de-

BATHYAL	NODOSARIIDS NEOBULIMINA MINIMA NEOBULIMINA ALBERTENSIS BULIMINELLA FABILIS EPISTOMINA LACUNOSA LAGENA SULCATA	GAVELINELLA PLUMMERAE GAVELINELLA MONTERELENSIS VALVULINERIA INFREQUENS VALVULINERIA LOTTERLIE HOEGLUNDINA CHARLOTTAE
LOWER BATHYAL - ABYSSAL	GAVELINELLA LENTICULINA NODOSARIIDS OSANGULARIA NEOBULIMINA GYROIDINOIDES AFF. NITIDUS GYROIDINOIDES OCTACAMERATUS LINGULINA PSEUDONODOSARIA TUBEROSA PSEUDOTEXTULARIIDAE	BATHYSIPHON REOPHAX HELVETICUS DOROTHIA AMMOBACULITES AMMODISCUS TEXTULARIA SPIROPLECTAMMINA SPIRILLINA TROCHAMMINA GLOMOSPIRA

Fig. 7   Tentative bathymetric model of Turonian-Cenomanian
bathyal-abyssal assemblages based on distribution of
foraminifera in Atlantic Coastal Plain wells and DSDP
drill sites.

SHELF	LAGENIDS POLYMORPHINIDS VALVULINERIA	GAVELINOPSIS CENOMANICA GAVELINELLA AMMONOIDES ARENOBULIMINA PRESSLII
BATHYAL	LENTICULINA SAXOCRETACEA LINGULINA NODOSARIA DISCORBIS MINUTISSIMA PATELLINA SUBCRETACEA	QUINQUELOCULINA SABELLA SPIROPLECTAMMINA ALEXANDERI TEXTULARIA RIOENSIS LITUOLA SUBGOODLANDENSIS
BATHYAL-ABYSSAL	LAGENIDS GAVELINELLA GYROIDINOIDES BATHYSIPHON REOPHAX HELVETICUS DOROTHIA PRAEHAUTERIVIANA AMMOBACULITES TROCHAMMINA AMMODISCUS SPIRILLINA	HYPERAMMINA HAPLOPHRAGMOIDES GLOMOSPIRA RHIZAMMINA PROTEONINA LAGENAMMINA TOLYPAMMINA VERNEUILINOIDES AMMOVERTELLA HECHTINA

Fig. 8   Tentative bathymetric model of Lower Cretaceous shelf-abyssal assemblages based on distribution of foraminifera in DSDP drill sites.

posited in bathyal to abyssal water-depths (fig. 8).
This view is strengthened also by analogy with the simply-
structured arenaceous assemblages found at great depth in the
oceans today. The absence for the most part of calcareous
foraminifera also suggests deposition below the carbonate compensation depth.

The Lower Cretaceous in Site 111 penetrated shallow-water
deposits (Van Hinte, in Laughton et. al., 1972)   The assemblages in this section consist of lagenids, polymorphinids,
and species of the genera Gavelinella, Gavelinopsis, Arenobulimina, Valvulineria. This compares well with the shelf
assemblages of the Upper Cretaceous. The analysis of foraminiferal distributions in Sites 384, 386, 387, 390, 391 and
392 will add valuable data to the understanding of Early Cretaceous bathymetric assemblages.

JURASSIC:

The foraminiferal faunas of Sites 100 and 105 are generally rich and consist mostly of simple arenaceous foraminifera
and lagenids (Luterbacher, 1972). They are typical of Juras-
ic assemblages found in many other parts of the world. The
bathymetric interpretation, as noted by Luterbacher, for these
assemblages is difficult because of the much lower diversity
than assemblages which succeed them in time. Jurassic foraminifera were probably very broadly niched. The Sites 100 and
105 assemblages compare with shelf assemblages as defined by
Gordon (1970) but the dominance of "Spirillina" in some sections compare favorably with similar faunas from deposits interpreted as bathyal in origin from the Central Appennines.
Obviously, more data is needed on the bathymetric composition
of Jurassic foraminifera.

ACKNOWLEDGEMENT

"Acknowledgment is made to the Donors of the Petroleum
Research Fund, administered by the American Chemical Society,
for partial support of this research".

REFERENCES

Barr, F.T., 1966. The foraminiferal genus Bolivinoides from
    the Upper Cretaceous of the British Isles: Palaeontology,
    v. 9, p. 220-240.

_____, 1970. The foraminiferal genus Bolivinoides from
    the Upper Cretaceous of Libya: Jour. Palaeontology, v. 44,
    p. 642-654.

Benson, W. E., et al., 1975. Summary of Deep Sea Drilling
    Project: Leg XLIV. Unpublished.

Berggren, W. A., and Hollister, C., 1974. Paleogeography,
    paleobiogeography and the history of circulation of the

Atlantic Ocean: In W. W. Hay (Editor), Studies in Oceanography. Soc. Econ. Paleontol. Mineral., Spec. Publ., v, p. 126-186.

Cita, M. B., and Gartner, S. Jr., 1971. Deep Sea Upper Cretaceous from the Western North Atlantic: In Proceedings of the II Planktonic Conference, Roma 1970 (A. Farinacci,Ed.) p. 287-319.

Gordon, W. A., 1970. Biogeography of Jurassic foraminifera: Bull. Geol. Soc. Am. v. 81, p. 1689-1704.

Hayes, D. E., et al., 1972. Shipboard Site Reports: In Hayes, E. E., Pimm, A. C., 1972, Initial Reports of the Deep Sea Drilling Project, Vol. XIV, Washington (U. S. Government Printing Office) p. 1-338.

Laughton, A. S., et al., 1972. Shipboard Site Reports, Site 111: In Laughton, A. S., Berggren et al., 1972, Initial Reports of the Deep Sea Drilling Project, Vol. XII, Washington (U. S. Government Printing Office) p. 33-160.

Luterbacher, H., 1972. Foraminifera from the Lower Cretaceous and Upper Jurassic of the northwestern Atlantic: In Hollister, C. D., Ewing, J. I., et al., 1972, Initial Reports of the Deep Sea Drilling Project, Vol. XI. Washington (U. S. Government Printing Office) p. 561-594.

Maher, J. C., and Applin, E. R., 1971. Geologic framework and petroleum potential of the Atlantic Coastal Plain and Continental shelf: U. S. Geol. Survey Prof. Paper 659, 98 p.

Mello, J. F., Minard, J. P., and Owens, J. P., 1964. Foraminifera from the Exogyra ponderosa zone of the Marshalltown Formation at Auburn, New Jersey: U. S. Geol. Survey Prof. Paper 501-B, p. 61-63.

Olsson, R. K., 1960. Foraminifera of Latest Cretaceous and Earliest Tertiary Age in the New Jersey Coastal Plain: Jour. Paleontology, v. 34, p. 1-58.

_____, 1964. Late Cretaceous planktonic foraminifera from New Jersey and Delaware: Micropaleontology, v. 10, p. 157-188.

_____, 1975. Upper Cretaceous and Lower Tertiary Stratigraphy of New Jersey Coastal Plain: Second Annual Field Trip Guidebook, Petrol. Exploration Soc. N. Y., p. 1-49.

Olsson, R. K., and O'Grady, M. D., 1976. Cretaceous and early Tertiary Paleobathymetric history of New Jersey Coastal Plain: Annual Meetings, New Orleans, Abstracts, Amer. Assoc. Petrol. Geol., Bull., v. 60, p. 704.

Olsson, R. K., and Ulrich, B. C., 1976. Timing of Transgressions and Regressions in Cretaceous and Tertiary of New Jersey: Abstracts, Annual Meetings New Orleans, Amer.Assoc.

Petrol. Geol., Bull., v. 60, p. 704.

Perlmutter, N. M., and Todd, R., 1965. Correlation and foraminifera of the Monmouth Group (Upper Cretaceous) Long Island, New York: U. S. Geol. Survey Prof. Paper 483-1, p. 1-
24.

Pessagno, E. A., Jr., 1967. Upper Cretaceous planktonic foraminifera from the Western Gulf Coastal Plain: Paleontogra-
phica Americana. v. 5, no. 37, p. 245-445.

Peterson, M. N. A., et al., 1970. Shipboard Site Reports: In
Peterson, M. N. A., et al., 1970, Initial Reports of the
Deep Sea Drilling Project, Vol. II. Washington (U. S. Gov-
ernment Printing Office) p. 1-306.

Petters, S. W., 1975. Subsurface Upper Cretaceous stratigra-
phy and foraminiferal biostratigraphy of the Atlantic
Coastal Plain of New Jersey: Unpublished Ph. D. thesis,
Rutgers University, New Brunswick, N.J., 258 p.

_____, 1976. Upper Cretaceous Subsurface stratigraphy
of Atlantic Coastal Plain of New Jersey: Amer. Assoc.
Petrol. Geol., Bull., v. 60, p. 87-107.

Sliter, W. V., and Baker, R. A., 1972. Cretaceous bathymetric
distribution of benthic foraminifers: Jour. Foram. Res.,
v. 2, no. 4, p. 167-183.

Tucholke, B., et al., 1975. Summary of Deep Sea Drilling Project: Leg XLIII. Unpublished.

Van Hinte, J. E., 1976. A Cretaceous time scale: Amer. Assoc.
Petrol. Geol., Bull., v. 60, p. 498-516.

EXPLANATION OF PLATES

All figured specimens are from the Cretaceous of New Jersey. Stage refers to the stratigraphic level from which the figured specimen was taken.

Plate I

A. *Dorothia ellisorae* (Cushman), X100, Campanian.

B. *Dorothia stephensoni* Cushman, X 110, Santonian.

C. *Dorothia retusa* (Cushman), X 150, Campanian.

D. *Clavulina clavata* Cushman, X 35, Santonian.

E. *Tritaxia capitosa serrulata* (Cushman), X 45, Santonian.

F. *Tritaxia capitosa* (Cushman), X 65, Maestrichtian.

G. *Clavulina trilatera plummerae* (Sandidge), X 35, Campanian.

H. *Clavulina trilatera* Cushman, X 50, Campanian.

I. *Heterostomella faveolata* (Marsson), X 75, Campanian.

J. *Texularia ripleyensis* W. Berry, X 100, Campanian.

K. *Arenobulimina subsphaerica* (Reuss), X 110, Campanian.

L. *Eggerella trochoides* (Reuss), X 160, Campanian.

Plate II

A. *Bolivinoides decoratus* (Jones), X 120, Campanian.

B. *Chilostomella trinidadensis* Cushman and Todd, X 90, Maestrichtian.

C. *Pullenia cretacea* Cushman, X 125, Campanian.

D. *Praebulimina kickapooensis* (Cole) X 95, Maestrichtian.

E. *Siphogenerinoides plummerae* (Cushman), X 90, Campanian.

F. *Praebulimina cushmani* (Sandidge), X 150, Campanian.

G. *Praebulimina carsayae* (Plummer), X 125, Campanian.

H. *Vaginulina debilis* (Berthelin), X 120, Cenomanian.

I. *Citharina kochii* (Roemer), X 90, Turonian.

J. *Saracenaria duckcreekensis* Tappan, X 90, Cenomanian.

K. *Marginulina siliquina* Eicher and Worstell, X 120, Cenomanian.

Plate II (cont'd)

L.   Osangularia navarroana (Cushman), X 150, Campanian.

Plate III

A.   Hoeglundina charlottae (Vieaux), X 150, Turonian.

B - D.   Valvulineria lotterlie (Tappan), B - X 360, C - X 450,
         D - X 300, Turonian.

E,F. Gyroidinoides globosus (Hagenow), E - X 135, F - X 185,
     Campanian.

G,H. Globorotalites michelinianus (D'Orbigny), X 110,
     Campanian.

I - K.   Gavelinella plummerae (Tappan), X 150, Turonian.

L.   Bathysiphon alexanderi Cushman, X 30, Santonian.

Plate IV

A - C.   Valvulineria infrequens Morrow, A,B - X 300,
         C - X 250, Cenomanian.

D,E. Gavelinella ammonoides (Reuss), D - X 120, E - X 140,
     Campanian.

F,G. Anomalinoides pinguis (Jennings), X 150, Maestrichtian.

H,I. Gavelinella nochatochensis (Cushman), X 150, Campanian.

J,K. Anomalinoides nelsoni (W. Berry), X 225, Campanian.

PLATE I

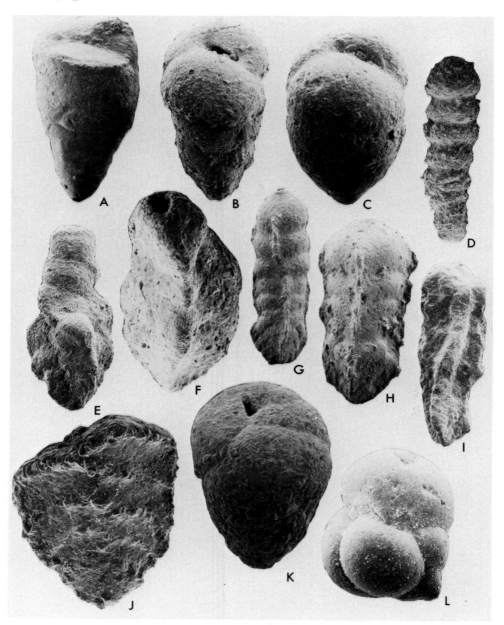

PLATE II

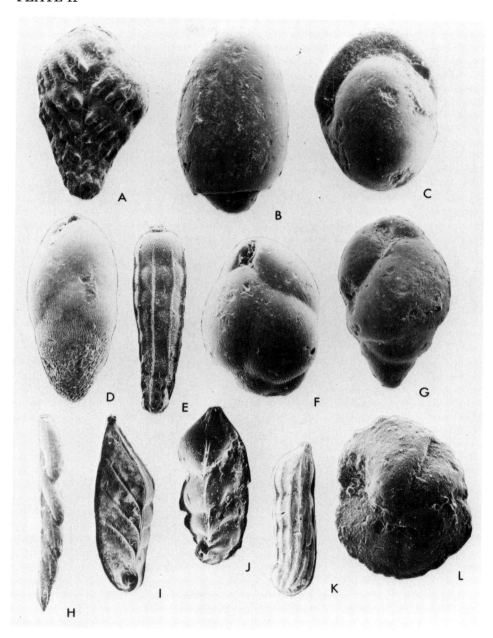

226

PLATE III

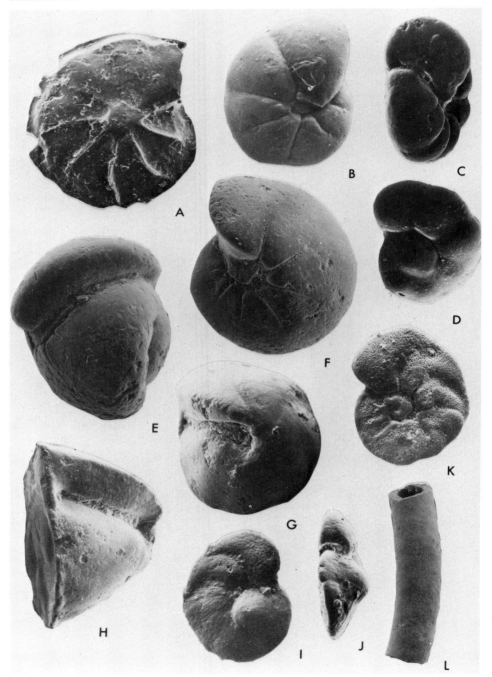

PLATE IV

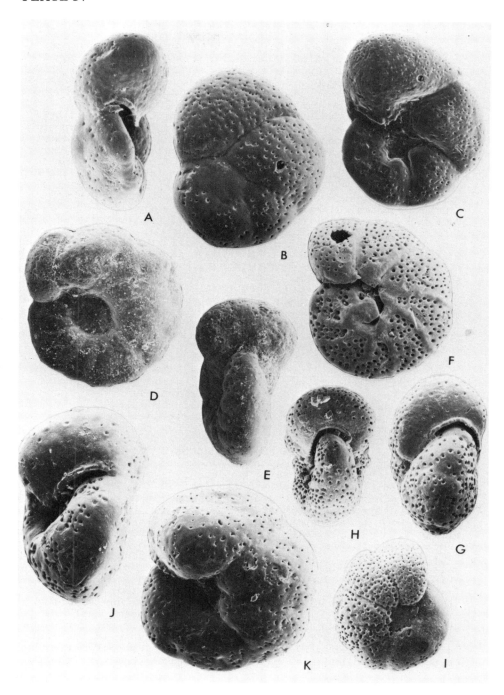

Discussion

Dr. F. M. Gradstein: Do you have indication that the ratio of Globotruncaniids/Hedbergelliids and Heteroheliciids is related to distance from shore? The latter two groups would tend to predominate the more shallow deposits, but are by no means confined to them. Globotruncaniids tend to favour more off-shore conditions. Such a simple distribution has been brought forward earlier by Dr. W. Sliter and Dr. V. Scheibnerova. In my own experience this might be true also for the Upper Cretaceous of the Grand Banks and Scotian Shelf.

Olsson: We know that Recent species of planktic Foraminifera float within a certain depth range in the water column; some are shallow floaters whereas others utilize a deeper water column. As the water column becomes shallower and shallower over the continental margin the ecologic column of planktic species is interfered with because it is intersecting the sea floor. And so those species that are deep floaters in the adult stage cannot occur in shallow waters. They are more strictly oceanic forms. Others which are shallow water floaters can occur farther in over the shelf, particularly if they are very abundant. As an example, rubers in the present seas and other globigerinid species show this relationship. Some of the Hedbergellas seem to have done that and the heterohelicids, in particular, occur in shallow water deposits. Matter of fact, there were a lot of questions on whether some species of heterohelicids were really true planktic forms because they are found in situations which were obviously shallow water. It was only until we began the DSDP where we know we are dealing with very deep water material that we could say for certain that some of them floated in the oceans. But there is this, you'll lose diversity of planktic forms in shallow water and in some cases planktic forms are absent. Such genera as the Archeoglobigerinas, the Rugoglobigerinas and others appear to have been very shallow water dwellers because they occur in shallow water sediments. The globotruncanids apparently preferred open water conditions and perhaps were deep floaters in the adult stage because they are almost absent from shallow water deposits. Thus the criteria of diversity of benthic species, the diversity of planktic species, and the changes in the abundances of certain planktonic taxa such as the heterohelicids all go together in pointing towards either shallow water or deeper water deposition.

Dr. H. Habib: Do the benthonic Foaminifera represent minimum depths, since they can be carried downslope posthumously? Can you use abundances that way?

Olsson: Most transportation of Foraminifera that we are familiar with usually is with turbidites-turbidity current transported Foraminifera; and that becomes quite obvious in a change of sediment texture and in the occurrence of different assemblages. There are some fairly good examples of that kind of thing. But we really don't see much evidence of massive transportation of Foraminifera such that it destroys the whole profile so that you don't know what you're doing. You don't see that. Usually one can spot forams that are transported and out of place. They're usually not very abundant either. Mostly rare forms. Now of course in the Cretaceous we don't really know for sure what the bathymetric ranges of species were, but we can compare them with their modern analogs, in some cases very well and in other cases not so well. Morphotype analysis is a useful approach. It becomes evident when a genus had a shallower range than it does today; for instance species of <u>Osangularia</u>. Back in the Cretaceous and Paleocene times some species of the genus occurred in shallow shelf depths but they are very rare and small in size. They were dwellers of the greater water depths, too and today the genus is confined to the deeper water dwellers.

Dr. J. E. Conkin: Can you give any reasons for agglutinated Foraminifera being found in deep (or deeper) water in the Mesozoic-Cenozoic whereas they are found abundantly in shallow water during the Paleozoic. Could this be partially competition between arenaceous and calcareous forms beginning in middle Devonian and resulting in the gradual diminishing of arenaceous forms from the epicontinental niches, and thus dominance of the calcareous walled forms there?

Olsson: Well the arenaceous forams occur from the lagoons right out into the deep sea

Conkin: Though you usually think of them as being deep water and they are very abundant in deep water.

Olsson: Their diversity in shallow water is very much less than it is in the deeper water.

Conkin: And the abundance of them as well, isn't this true?

Olsson: The abundance? Well it depends, in certain shallow water sediments you can get very high dominance of Textularia, 99% or something like that. It could be very abundant there. With the advent of the Rotaliniids, the calcareous forms, that there was a significant radiation onto the continental shelf. That may in turn have led perhaps to the extinction of many arenaceous forms there or the crowding out of them.

Conkin: Could there be any competition between these calcareous forms and the arenaceous ones?

Olsson: Some deeper water assemblages are composed of mixtures of calcareous and arenaceous types, some assemblages are almost entirely arenaceous and in others calcareous forms are dominant. Faunas below the carbonate compensation depth are entirely arenaceous.

Conkin: Is it not unusual to find abundant arenaceous forams in beach deposits today?

Olsson: That depends where you are but in general perhaps it is true.

Mr. M. Polugar: Did I get the impression awhile ago that you were implying that there was a direct connection between a keeled form and depth rather than temperature? You mentioned something about single-keeled <u>Globotruncana</u> vs. double-keeled ones.

Olsson: In shallow water sediments <u>Globotruncanas</u>, single or double-keeled are either absent or extremely rare; so they look as if they preferred a more oceanic habitat. Whether you're up in mid-latitude or down in the tropics the same thing applies. There may be different species because of temperature control, that is latitudinal control on the distribution of species.

Polugar: In California, in the Miocene, where we have a very deep water abundant planktonic fauna, you'll find no keeled globorotaliids regardless of how deep water you go into. So you can't really relate them to depth.

Olsson: Well, it depends, in California where there was influence of cold water, yes, you couldn't do this. In high latitude areas, say Alaska or the higher latitudes of the Atlantic Globotruncana can be expected to be absent.

Polugar: I don't think you can do the same thing for Globotruncana either. I think single- or double-keeled Globotruncana is a function of possible temperature and time, as you
find double-keeled Globotruncana as you......

Olsson: I don't think I said anything about that. We were
talking about the bathymetric profile and how the distribution of planktonics is related to that profile.

JURASSIC OSTRACODA OF THE ATLANTIC BASIN

by

R.H. BATE

British Museum (Natural History), London SW7 5BD, England

ABSTRACT

Marine Jurassic sediments of the Atlantic Basin are widely developed in the eastern borderland region of the North Atlantic. Here, warm shallow seas supported a rich and varied ostracod fauna – over 200 species being found in the English Bathonian alone. This region is regarded as being one of the major areas in the development and evolution of Post-Palaeozoic Ostracoda. The eastern borderlands may be divided into a northern European (epicontinental sea) Province, a deeper water Tethyan Province and a southern North African province. The South Atlantic Basin was not developed at that time.

On the western borders of the North Atlantic Basin Jurassic sediments are present along the east coast of Greenland and offshore along the North American continent – ostracods having affinities with the epicontinental European faunas are present off Canada, while deep water ostracods, in part comparable with tethyan species from Italy, are present off U.S.A.; shallower water faunas are recorded from the Gulf Coast region (American Province).

Biostratigraphically important ostracod genera of the epicontinental provinces belong to three families: the Progonocytheridae, the Schulerideidae and the Protocytheridae. This contrasts with the Jurassic of the Indian Ocean Basin (East African Province) where only the Progonocytheridae is of equal importance. Ostracod genera belonging to other families play an important but subsidiary role in the biostratigraphy of the North Atlantic Jurassic. Fresh to brackish water ostracods become important in the Middle Jurassic but reach their biostratigraphical acme in the continental deposits of the Upper Jurassic/Lower Cretaceous when the genus *Cypridea* achieved a world-wide distribution.

INTRODUCTION

The Atlantic Ocean was not in existence during Lower Jurassic times as the separation of the continents only commenced about the time of Middle Jurassic. In order, therefore, to present the distribution of the marine Jurassic sediments on a map of that time a base map showing the situation during the Hauterivian has been used. This map [Text-fig. 1], the most accurate one available, has been taken from Owen (1976) and, because of the slow rate of displacement of the continents, shows a position of the continents not too far removed from their position at the close of the Period.

Before the separation of the continents began, marine conditions appear to have been restricted to the north-eastern area of the Atlantic; there is no evidence of marine sediments south of Nova Scotia until Upper Jurassic times. Indeed the southward transgression of the Tethys appears to have been dependent upon the opening of the Atlantic. Europe, on the other hand, was covered by a large epicontinental sea, interspersed with islands, that extended

232

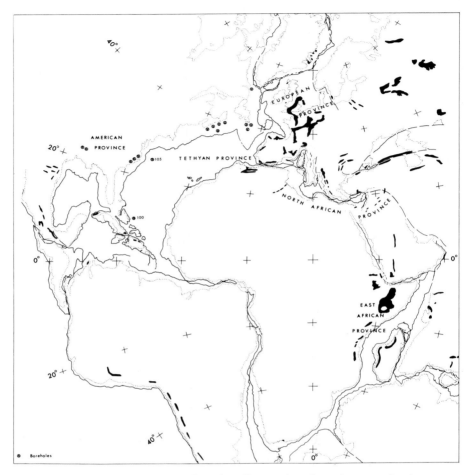

Text-Fig. 1. Outcrop of marine Jurassic sediments plotted on a base map showing
the position of the continents during the Hauterivian [base map taken
from Owen 1976].

eastwards into Asia. In Middle to Upper Jurassic times this linked northwards with the epicontinental sea of north-western America. It was through these large epicontinental seas that migration of the ostracods took place. The east-west orientated Tethys appears to have been a
somewhat narrow but deeper water sea/ocean that acted more as a barrier than as a route for
migration. This is especially noticeable when comparing the ostracod faunas of the north with
those of the southern hemisphere.

The study of the Jurassic ostracods from the Atlantic Basin indicates that four faunal
provinces may be recognised: a European, a Tethyan, a North African and an American. The
North African Province is perhaps rather tenuous at this stage and may have to be merged with the
Tethyan Province at a later date —.

The ostracod fauna of the East African Province is not directly relevant to a study of the Atlantic Basin but is important in the context of being able to identify a southern as distinct from a northern hemisphere fauna — it is in this way that this province will be briefly involved.

Although at the present time the greatest diversity of faunas is generally in the tropics and declines away from the equator this does not appear to have been the case in the Jurassic where a more uniform climate existed. Circumstances at the onset of the Mesozoic, following on after the extinction of most of the Palaeozoic ostracods, were unique in that ecological conditions in certain parts of the world were ideal for speciation and subsequent ostracod evolution. Such conditions existed in Europe from the close of Lower Jurassic times onwards. Thus by the end of the Lower Jurassic four important families were already established : the Schulerideidae, the Progonocytheridae, the Protocytheridae and the Cytheruridae. In the Middle Jurassic the Trachyleberididae and the Limnocytheridae appear while in the Upper Jurassic the Ilyocypridi- dae dominate the continental deposits.

At the present time a zonation of the Jurassic is not available although some authors have initiated a zonal scheme for parts of the succession: Anderson (1971) for the Purbeck; Bate (1965) in the Bajocian; Bate & Coleman (1975) in the Toarcian and Michelsen (1975) for the Lower Jurassic. A zonal scheme for the British Jurassic is currently in preparation by Bate, Kilenyi and Lord (see Bate & Robinson).

FAUNAL PROVINCES

The European Province. The lateral extent of this province ranges from Nova Scotia in the west to the Ukraine in the east. Northwards the extent is not known as the Greenland faunas have not been described, southwards the epicontinental sea of the province merges with the deeper waters of the Tethys. The provincial boundaries do not remain static through the entire period and it will be some time before precise boundaries can be drawn. Spain, for example, is con- sidered to have been a part of the European Province during the Lower Jurassic but with its anit-clockwise rotation through the Jurassic possibly moved into the Tethyan Province towards the close of the period. The maximum diversity of ostracod genera and species exists only at the centre of the province extending through northern France, England, Germany and Poland. This region appears to have been covered by warm, shallow seas in which numerous islands existed. Thus not only are marine ostracods numerically abundant but brackish water lagoonal and estuarine conditions existed for further speciation to take place. Following the extinction of most of the Palaeozoic lineages and a period when continental conditions predominated in the Trias, the marine transgression during the Jurassic entered a region here that was ideally suited to ostracod colonisation and subsequent speciation. Because of this rapid evolution in the ostracods the European Province appears to have been the birth-place of the majority of the important Mesozoic families. Indeed some families which evolved then are important at the present time [e.g. Cytheruridae and Limnocytheridae].

Lower Jurassic — Hettangian to Toarcian. Although sediments of Lower Jurassic age have been found off Newfoundland and Nova Scotia the ostracods have not yet been described. It does not appear to be unreasonable, however, to tie this region in with the European Province par- ticularly as Ascoli has noticed a relationship between the faunas for Middle and Upper Jurassic (pers. comm.).

In general the lower stages of the Lias are not as rich in ostracod species as is the case from the Sinemurian onwards. Initially the faunas are dominated by ogmoconchids and these have been used by Michelsen (1975) as the basis of his zonation of the Danish Lower Jurassic. The genus *Kinkelinella* first appears in the British Trias, evolves rapidly throughout the province and the many stratigraphically short ranging species make it an ideal ostracod for zonation (Bate & Coleman 1975). Indeed the two subgenera, *Kinkelinella (Kinkelinella)* and *Kinkelinella (Ektyphocythere)* are the most important of the cytheracean ostracods [family Protocytheridae] developing at that time. The genus *Micropneumatocythere* is first recorded from the Sinemurian [see *Procytheridea reticulata* Klingler & Neuweiler 1959] and this is currently considered to be the earliest true record of the Progonocytheridae. This family replaces the Protocytheridae in importance in the Middle and Upper Jurassic with the decline of the *Kinkelinella-Ektyphocythere* lineage, although the Protocytheridae still retain an important position in ostracod phylogeny and become important again with respect to their use in stratigraphical correlation. Many new genera appear towards the close of the Lower Jurassic [e.g. *A phelocythere, Trachycythere, Nanacythere & Procytherura*] that are significantly important in correlation during the Toarcian and the Aalenian — there is no clear division between the Lower and the Middle Jurassic at this point with respect to the ostracod faunas. Also by this time the family Cytheruridae was well established [the earliest record being of *Cytheropteron reticulatum* Michelsen (1975) from the Lower Sinemurian] but the fourth important family, the Schulerideidae, was only just developing — *Praeschuleridea pseudokinkelinella* Bate & Coleman (1975) from the Toarcian, the earliest representative, considered to be very close to being the ancestor of the family. The Schulerideidae developed too late to be important in the Lower Jurassic but achieved this role in the Middle and Upper Jurassic.

Middle Jurassic — Aalenian to Callovian. The Aalenian has a fauna that links it closely with the Toarcian below but at the same time contains ostracods having affinities with the Bajocian [e.g. *Pneumatocythere*]. In particular the genus *Praeschuleridea* becomes important all over Europe where several species occur, all characterised by a coarsely pitted ornamentation (Malz 1966). Throughout the other stages of the Middle Jurassic *Praeschuleridea* remains important although its species become smooth-shelled [with a few exceptions — *P. batei* in the Callovian] and, through *Eoschuleridea* [see Bate 1967, p. 40], gives rise to the *Schuleridea* lineage sometime in the early Bathonian. The *Praeschuleridea* lineage dies out in the Callovian with *P. batei* in Britain and *P. caudata* in southern France.

In the Bajocian the Progonocytheridae add the important genus *Glyptocythere* that is so valuable for correlation all over the province. Indeed this genus is known to extend as far east as the Ukraine (Permjakova 1970) and, outside the province, as far east as Uzbekistan (Masumov 1973, see *Macrodentina aspera*). The genus *Micropneumatocythere,* currently the oldest of the family also develops rapidly through the Middle Jurassic where, together with *Glyptocythere* it may be used in zonation. In fact a zonal scheme using species of these two genera is currently being prepared for publication (see Bate & Robinson).

The genus *Progonocythere* appears in the Bathonian of England and Poland more or less at about the same time and once was thought to be restricted to the Middle Jurassic. The many species subsequently placed in this genus from the Upper Jurassic have proven to belong more correctly to other genera, although one species, as yet undescribed, has been found in the Kimmeridgian of Spain. Although giving its name to the family this genus has become relatively unimportant stratigraphically. In the southern hemisphere, however, a very close relative of *Progonocythere,* the genus *Majungaella,* becomes strikingly successful throughout the Middle to Upper Jurassic and through the Cretaceous. The geographical distribution of *Majungaella* and the subgenera/genera that developed from it is from Australia and Africa to South America. This is quite a remarkable achievement considering that both genera certainly arose from a common ancestor and became virtually restricted to their respective hemispheres [only one species of *Progonocythere* – *P. laeviscula* in the Callovian to Oxfordian of India is known from the south while no species of *Majungaella* are known from the north].

Within the Progonocytheridae there are two morphological groups – the first rather quadrate/oval in outline and represented by *Progonocythere* and *Glyptocythere* and the second having a much more elongate carapace as in *Lophocythere* and *Fuhrbergiella.* Both groups are characterised by having many short ranging species important in the correlation of the Jurassic throughout this province.

The *Kinkelinella* – *Ektyphocythere* lineage, so important in the Lower and basal Middle Jurassic apparently dies out in the Bathonian to be continued in the Callovian by the genus *Pseudohutsonia.* Whatley (1970) rightly considers *Pseudohutsonia* Weinholz (1967) to be congeneric with *Balowella* Wienholz (1967) both originally described from the Callovian of East Germany. Thus the record of *Balowella* from the Oxfordian of Canada (Brooke & Braun 1972) and of *Pseudohutsonia* from the Callovian and Oxfordian of Britain (Whatley 1970) extends the geographical range of this important lineage. Interestingly, lineages in the northern hemisphere continue with many more changes at genus level than is the case in the southern hemisphere.

In this consideration of the faunal provinces the important ostracods have all been marine in habit, the diversity of the faunas being measured by the fact that in the Bathonian of England alone there are some 55 genera and over 200 species. Within this Province, however, the variety of ecological niches provides for the development of both brackish and of freshwater ostracods some of which are peculiar to the Middle Jurassic but some, *Theriosynoecum, Bisulcocypris* and *Klieana,* are also important in later parts of the geological column. The genus *Darwinula,* extending through the Palaeozoic to the Recent is, of course, also represented. With the exception of the last named, development of these new ostracods appears to have taken place within the European Province with subsequent migration effecting a world-wide distribution. This also appears to apply to *Timiriasevia* and *Limnocythere* although stratigraphically they are never as important as the first three. The record from Israel by Gerry & Oertli (1967) of *Bisulcocypris* from the Trias is probably misleading as the age of the beds concerned is now considered to be Jurassic (Gerry, pers. comm.).

Upper Jurassic – Oxfordian to Tithonian. Upper Jurassic ostracods of the European Province continue the diversity of forms already encountered. The Progonocytheridae continue with *Lophocythere* s. l. in the Oxfordian and with *Macrodentina* in the Kimmeridgian. *Galliaecytheridea* is characteristic of the Upper Oxfordian and Kimmeridgian of northern France (Oertli 1957) and of the Kimmeridgian of southern England (Kilenyi 1969) but is not well represented in the Upper Jurassic of Switzerland (Oertli 1959) close to the Tethyan Province. It is in Switzerland, as well as in southern France (Donze 1962), on the margins of the Tethyan Province, that the true European representatives of *Procytheridea* are found. Also within the Upper Jurassic the important Cretaceous genus *Protocythere* makes its appearance and is recorded by Oertli (1957) from the Kimmeridgian of the Paris Basin and by Pokorny (1973) from the Tithonian of Czechoslovakia. Some early form of *Protocythere* appears to be present in the Bathonian of England but this still has to be verified. Otherwise the migration route of the genus would be from the east where it has been recorded from Uzbekistan (see *Macrodentina aspera* in Masumov 1973).

Important amongst the late Jurassic marine ostracods is *Paranotacythere* — species of which are stratigraphically short ranging but geographically widely dispersed throughout the epicontinental sea area of the province (Bassiouni 1974).

Large areas of the province became emergent during the final phase of the Jurassic and continental deposits with fresh and brackish-water ostracods stretch from Denmark through England and France and northern Germany into northern Spain and Portugal. Indeed such conditions existed over large areas of Asia, North and West Africa and South America during this time.

Freshwater ostracods first appear in the Upper Jurassic of Portugal where they have been utilised by Helmdach (1971) to subdivide both the Oxfordian and the Kimmeridgian. It is only in the Upper Kimmeridgian that *Cypridea* begins to dominate the freshwater faunas both in Portugal and in the offshore Kimmeridgian of southern Ireland. Henceforth *Cypridea* becomes established world-wide in the Jurassic — Cretaceous interval. Indeed, the use of ostracods in zonation was first employed in the Purbeck by Forbes as early as 1851. Subsequently refinements have been made and the zonal importance of this genus in the British sequence owes much to the work of F. W. Anderson.

The Tethyan Province. In Lower Jurassic times this province extended as far west as the Canadian coast (Nova Scotia and Newfoundland) but expanded in the Upper Jurassic into the southern Atlantic Basin. The north-south extent of the Province takes in western North Africa and the southern part of Europe.

Lower Jurassic — Hettangian to Toarcian. As the Lower Jurassic marine transgression spread north and west over Europe the ancestral stock of the new cytheracean lineages was introduced. The resultant diversity that ensued developed away from the Tethys which was a deeper water environment characterised in the main by smooth-shelled ostracods.

In the Lower Jurassic the genera *Ogmoconcha* and *Ogmoconchella* dominate the Tethyan faunas, especially in the Pliensbachian where highly ornate species of *Ogmoconcha* occur in southern Germany (Lord & Moorley 1974c and Malz 1975), Sicily (Barbieri 1964a) and in the Djebel Zaghouan section in Tunisia. Material collected from the Schwabische Alb [SW-Germany] exhibits both a preponderance of ogmoconchids and some of the diversification more commonly associated with the European Province; as such this region is considered to have been situated on the boundary between the two at that time. The fauna described by Drexler (1958) from Bavaria, south of the Schwabische Alb, is of a deeper water fauna more typical of the Tethys. Similarly the Hettangian fauna in the Ardèche [S. France] described by Donze (1966) is also of this deeper water province and is characterised by species of *Ogmoconchella, Bairdiacypris?, Cytherella* and *Cytherelloidea.*

The ogmoconchids appear to have migrated west through the Tethys and to have moved northwards into the European Province where they survived until the basal Toarcian when possibly they were unable to compete with the rapidly evolving cytheraceans.

A parallel with the ornate tethyan ogmoconchids is the ornate group of bairdiids that were common in the Trias of this province. For example, *Ptychobairdia schaubergeri* is found both in the Kleckenmergel of Austria (Kollmann 1963) and in the Pliensbachian of Djebel Zaghouan,

Tunisia while Lord & Moorley (1974a & b) record highly ornate bairdiids: *B. hahni* and
B. aselfingenensis from the Pliensbachian of SW-Germany. Rarely do these bairdiids occur out-
side the Tethyan Province although fragments of *Ptychobairdia* sp. have been recorded from the
Toarcian of England (Bate & Coleman 1975).

Barbieri (1964b) describes the only ostracods to come from the centre of the province and
this Sicilian fauna is composed entirely of smooth-shelled ostracods.

All the available evidence indicates, therefore, that the evolution of the cytheracean
ostracods was initiated in the warmer epicontinental seas bordering the Tethys.

Middle Jurassic — Aalenian to Callovian. A rather poor fauna is recorded by Barbieri (1964b)
from the Bajocian of Sicily and by Donze (1962) from the Bathonian — Callovian of southern
France where only three species: *Schuleridea caudata*, *Procytheridea martini* and *Oligocythereis
gauthieri* are recorded. The French fauna is not particularly representative of deep water conditions
and is considered to be marginal between the European and Tethyan Provinces. If this is so it
would indicate that the northern boundary of the Tethys had receded since the Lower Jurassic.

Eastwards through Europe the Middle Jurassic faunas described all lie outside the Tethyan
Province — this is true as far east as the Dnieper-Don Depression where Permjakova (1969, 1970)
describes a fauna containing at least 5 species of *Glyptocythere*. Further east in Uzbekistan the
faunas described byMasumov (1973) belong to an epicontinental sea province that must have had
continuity with the European Province and at the same time lay to the north of the Tethys.

Perhaps the Middle to Upper Jurassic faunas of the Middle East should more accurately
belong to the Tethyan Province — in which case the North African Province could be dispensed
with entirely. This will become more apparent as research on these faunas continues.

Upper Jurassic — Oxfordian to Tithonian. In the Upper Jurassic the Tethys moved into the
opening Atlantic Basin and appears to have taken with it the important ostracod family, the
Schulerideidae. The migration route would not be through the deeper water regions but through
the more shallow waters of the margins.

The genus *Procytheridea* sensustricto, first described from the Callovian of western North
America (Peterson 1954), is know to be present in the Callovian of Tanzania (Bate 1975), the
Callovian and Oxfordian of southern France (Donze 1962) and in the Oxfordian of Switzerland
(Oertli 1959) — as well as in western North America and Canada. Because of this wide distribu-
tion pattern a simple westward migration by way of the Tethys appears to have been extremely
unlikely.

Two ostracod faunas described by Oertli (1972) from DSDP wells in the Atlantic [site 100
off the Bahamas and site 105 off Cape Hatteras] have been compared with the deeper water
Tethyan faunas described by Oertli (1967) from Italy and with the Middle to Upper Jurassic
ostracods described from SE. France by Donze (1962). Only *Bairdia (Akidobairdia) farinacciae*
correlates the Upper Jurassic of Italy directly with DSDP site 100 but the faunas are, nevertheless clearly representative of a deeper water Tethyan Province. Similarities between species of
Acrocythere figured by Oertli (1972) and by Pokorny (1973) are not conclusive and in any case
the Czechoslovakian fauna is considered to belong more correctly with the European Province.

The close proximity of DSDP sites 100 & 105 to the American Province is remarkable for
the complete contrast between the faunas. This contrast is considered to be due entirely to
differences in water depth — a clear indication that *Schuleridea* could not have migrated westwards in the deeper water regions of the Tethys.

Ostracods such as *Macrodentina* and *Galliaecytheridea*, so typical of Upper Jurassic sediments of the European Province are absent from deeper waters of the Tethys where Donze (1962) describes a fauna, from southern France, dominated by either smooth-shelled or alate species.

Whereas a zonal scheme for the epicontinental sea facies of Europe and North America is possible it will be some time before this will be so for the deeper water Tethys where smooth-shelled ostracods predominate and where there is a considerable decline in numbers.

The North African Province. the area covered by this province is considered to extend over that part of North Africa lying south of Tunisia and taking in Egypt and the Middle East. There is no published information available and studies on the ostracod faunas are at a very preliminary stage. As a result it may be necessary to revise the decision taken here to separately identify this province.

Lower Jurassic — Hettangian to Toarcian. No Lower Jurassic ostracods have been described as those obtained from Djebel Zaghouán are considered to belong to the Tethyan Province.

Middle to Upper Jurassic — Aalenian to Tithonian. This interval is grouped together simply because the Jurassic material examined has been rather loosely dated as Callovian/Oxfordian. In Jordan, *Amicytheridea* and *Afrocytheridea* and in southern Israel, *Afrocytheridea,* are represented amongst an ostracod fauna typical of a shallow marine environment. Both *Amicytheridea* and *Afrocytheridea* were originally described from the Callovian of Tanzania (Bate 1975) and have since been observed in the Jurassic of Kenya and Somalia. Thus these two genera extend northwards as far as North Africa but do not appear to cross the barrier of the Tethys. It is because of this that a decision was taken to separately identify the North African fauna as being distinct from either the Tethyan to the north or the East African to the south, but having connections with both.

The East African Province. The Southern hemisphere ostracod faunas during the Jurassic and, indeed for the Mesozoic in general, are quite distinct from those of the northern hemisphere. Certainly many ostracod genera are cosmopolitan in their distribution but the essential composition of the faunas is different.

Whereas in the European Province three families, the Progonocytheridae, the Protocytheridae and the Schulerideidae, play an important role in stratigraphical correlation, here only the Progonocytheridae assumes any importance — the Schulerideidae and Protocytheridae are represented but in a subsidiary role. In the East African Province the most important genus is *Majungaella* which is very closely related to *Progonocythere* of the European Province. The latter plays a very insignificant part in Jurassic stratigraphy whereas *Majungaella* may be used for correlation between East Africa, India, Madagascar and South Africa. Indeed, five zonal species of *Majungaella* have been recognised in the Jurassic (Bate 1975): *M. mundula* — Callovian; *M. oxfordiana* — Oxfordian; *M. kimmeridgiana* — Lower Kimmeridge; *M. praeperforata* — Middle or Upper Kimmeridge (precise age uncertain) and *M. perforata* — Tithonian. Additionally *Majungaella* ranges up into the Cretaceous and is known from Africa, South America [see *Novocythere* Rossi de Garcia 1972] and Australia. Thus the presence of this genus, or of one of the very closely related subgenera, readily identifies a southern hemisphere fauna as distinct from one from the north.

The American Province. The buried Upper Jurassic of the Gulf Coast and of the SE-Atlantic coast of the USA together form the American Province which in faunal terms is characterised by the dominance of *Schuleridea, Paraschuleridea* and *Hutsonia*. This is a unique, shallow water marine fauna that is more easily correlated with the western North American Jurassic than with the faunas figured by Oertli (1972) from the DSDP well sites in the Atlantic.

Lower Jurassic — Hettangian to Toarcian. Absent.

Middle Jurassic — Aalenian to Callovian. Absent.

Upper Jurassic — Oxfordian to Tithonian. The age dating of this sequence is not precisely known and probably not all stages are present.

The characteristic feature of this province is the absolute importance of the the the Schulerideidae. Of the large number of rapidly evolving genera that appeared in Europe during the Jurassic only the genus *Schuleridea* in association with *Asciocythere* and *Paraschuleridea* appears to have developed here. European genera such as *Fuhrbergiella* and *Pseudohutsonia* present in the Upper Jurassic of western North America must have arrived there via Greenland and one could postulate that the southward extension of the Tethys with the opening of the Atlantic was initially a deep water transgression that was unsuited to the migration of the European genera. In this context a detailed study of the eastern Canada Jurassic faunas should be very interesting as it would appear that there was little or no continuity between there and the American Province further south.

The absence of Lower and Middle Jurassic marine sediments in the southern North Atlantic is taken to indicate that marine conditions were totally dependent upon the opening of the Atlantic and the introduction of the Tethys. The marine Upper Jurassic ostracods of Louisiana and Arkansas [Swain 1946, and Swartz & Swain 1946] and from North Carolina [Swain 1952 and Swain & Brown 1972] are considered to represent a rather restricted, shallow water marine environment that has little direct correlation with the marine Provinces of Europe and North Africa. Unfortunately we have no information concerning the Jurassic ostracods of Mexico but it is possible that some connection west with a Pacific Province might have been in existence during the Upper Jurassic even though the essential marine transgression of the southern North Atlantic came from the North.

CONCLUSIONS

The distribution of the Jurassic Ostracoda was examined in the context of a re-positioning of the continents as they were in late Jurassic times.

With the Atlantic closed during the Lower Jurassic, marine conditions entering from the east reached only as far west as eastern Canada. The central part of this transgression — the Tethys — has been identified as a deeper water Tethyan Province characterised by a rather restricted fauna of smooth-shelled ostracods.

The primary role of the Tethys appears to have been to act as a north-south barrier to ostracod migration. Possibly this was because the newly evolving ostracods were unable to tolerate the deeper [cooler?] waters of the Tethys.

Bordering the Tethyan Province epicontinental seas spread north over previously barren Triassic landmasses and, following on from the extinction of most of the Palaeozoic lineages, a unique situation arose that led to rapid speciation and subsequent evolution of many new lineages. Although there is some evidence that similar conditions existed eastwards into Asia, the European Province is considered to have been one of the major sites of ostracod evolution and certainly the most important in terms of the present study. Elements of the European fauna migrated north to reach western North America during the Upper Jurassic and likewise migrated east into Asia.

In the Lower Jurassic of the European Province the important ostracods are the ogmoconchids and the *Kinkelinella — Ektyphocythere* lineage of the Protocytheridae. The cytheracean ostracods develop rapidly towards the end of the Lower Jurassic and through the Middle and the Upper Jurassic the stratigraphically important ostracods belong to the Progonocytheridae and to the Schulerideidae. Fresh to brackish-water ostracods of the Limnocytheridae become important in the continental deposits at the close of the Jurassic where the genus *Cypridea* achieves a worldwide distribution. The development of many short ranging species of this genus led to its recognition as a zonal fossil by the English geologist Forbes as early as 1851.

The American Province does not appear to have had an epicontinental sea connection with the European Province and as the Tethys was not an effective migratory route only a very restricted fauna developed — of which the most important is the Schulerideidae. Indeed the common occurence of species of *Schuleridea, Paraschuleridea* and of *Asciocythere* are a characteristic feature of this province.

The North African Province has been little studied to date and is presently recognised as containing elements of the East African fauna associated with the cosmopolitan genus *Cytherella*. It is too early in the investigation of this province to indicate any positive trends in ostracod development.

The East African Province is introduced into this study solely to draw attention to the differences that exist between the northern and the southern hemisphere faunas and to the restricting effect on migration of the Tethys. Although many ostracods such as *Cytherella* and *Bairdia* have a cosmopolitan distribution those ostracods that had their origins in the Jurassic were essentially restricted to a migration route within their respective hemisphere. *Amicytheridea* and *Afrocytheridea* for example extend north into the North African Province but do not extend further. The genus *Majungaella* from East Africa appears to achieve a world-wide distribution in the southern hemisphere but is not known in the north.

The Tethys did, however, permit some migration — *Procytheridea* present in North America and East Africa appears to be associated in Europe with the marginal facies of the Tethys and the Lower Jurassic ogmoconchids were certainly brought into the European Province by the Tethys. Finally, three Lower Jurassic genera of the European Province: *Procytherura, Eucytherura* and *Cytheropteron* appear much later in the Middle to Upper Jurassic of East Africa (Bate 1975) indicating that some north-south migration has been possible.

REFERENCES

ANDERSON, F. W. 1971. In ANDERSON, F. W. & BAZLEY, R. A. B. The Purbeck Beds of the Weald (England). *Bull. Geol. Surv. G. B.* no. 34, 174 pp, 23 pls.

BARBIERI, F. 1964a. *Hungarella hyblea* nuovo Ostracode del Domeriano. *L'Ateneo Parmense*, vol. 35, supl. 1, p. 3-7.

--------1964b. Micropaleontologia del Lias e Dogger del pozzo Ragusa 1 (Sicilia). *Riv. Ital. Paleont.,* vol. 70, p. 709-830, pls. 51-66.

BASSIOUNI, M. E. A. A. 1974. *Paranotacythere* n. g. (Ostracoda) aus dem Zeitraum Oberjura bis Unterkreide (Kimmeridgium bis Albium) von Westeuropa. *Geol. Jb.,* vol. 17, p. 3-111, 13 pls.

BATE, R. H. 1965. Middle Jurassic Ostracoda from the Grey Limestone Series, Yorkshire. *Bull. Br. Mus. nat. Hist. geol.,* vol. 11, p. 73-133, pls. 1-21.

--------1967. The Bathonian Upper Estuarine Series of Eastern England. Pt. 1: Ostracoda. *Bull. Br. Mus. nat. Hist. geol.,* vol. 14, p. 21-66, 22 pls.

--------1975. Ostracods from Callovian to Tithonian sediments of Tanzania, East Africa. *Bull. Br. Mus. nat. Hist. geol.,* vol. 26, p. 161-223, pls. 1-14.

--------& COLEMAN, B. E. 1975. Upper Lias Ostracoda from Rutland and Huntingdonshire. *Bull. Geol. Surv. G. B.* no. 55, p. 1-42, 15 pls.

--------& ROBINSON, E. (eds.). *A Stratigraphical Index of British Ostracoda.* Seel House Press (in prepn.).

BROOKE, M. M. & BRAUN, W. K. 1972. Biostratigraphy and microfaunas of the Jurassic System of Saskatchewan. *Dept. Min. Res. Saskatchewan Report.,* 161, p. 1-83, 26 pls.

DONZE, P. 1962. Contribution a l'étude paléontologique de l'Oxfordien Supérieur de Trept (Isére). *Trav. Lab. Géol. Lyon, N. S.,* no. 8, p. 125-142, pls. 9-11.

--------1966. Ostracodes de l'Hettangien entre Aubenas et Privas (Ardéche). *Trav. Lab. Géol. Lyon, N. S.,* no. 13, p. 121-139, pls. 5-7.

DREXLER, E. 1958. Foraminiferen und Ostracoden aus dem Liascon von Siebeldingen/Pfalz. *Geol. Jb.,* vol. 75, p. 475-554, pls. 20-27.

GERRY, E. & OERTLI, H. J. 1967. *Bisulcocypris ? triassica* n. sp. (Crust., Ostrac.) from Israel. *Bull. Centre Rech. Pau-SNPA.,* vol. 1, p. 375-381, 1 pl.

HEIMDACH, F. F. 1971. Zur Gliederung limnischbrackischer sedimente des portugiesischen Oberjura (ob. Callovien – Kimmeridge) mit Hilfe von Ostrakoden. *N. Jb. Geol. Paläont. Mh.,* vol. 11, p. 645–662.

KILENYI, T. I. 1969. The Ostracoda of the Dorset Kimmeridge Clay. *Palaeontology,* vol. 12, p. 112-160, pls. 23-31.

KLINGLER, W. & NEUWEILER, F. 1959. Leitende Ostracoden aus dem deutschen Lias β. *Geol. Jb.,* vol. 76, p. 373-410, 6 pls.

KOLLMANN, K. 1963. Ostracoden aus der alpinen Trias 2. Weitere Bairdiidae. *Jb. Geol.,* vol. 106, p. 121-203, pls. 1-11.

LORD, A. & MOORLEY, A. 1974a. On *Bairdia hahni* Lord & Moorley sp. nov. *Stereo Atlas Ostracod Shells,* vol. 2, p. 1-4.

--------1974b. On *Bairdia aselfingenensis* Lord & Moorley sp. nov. *Stereo Atlas Ostracod Shells,* vol. 2, p. 5-8.

--------1974c. On *Ogmoconcha ambo* Lord & Moorley sp. nov. *Stereo Atlas Ostracod Shells,* vol. 2, p. 9-16.

MALZ, H. 1966. Zur Kenntnis einiger Ostracoden-Arten der Gattungen *Kinkelinella* und *Praeschuleridea. Senck. leth.,* vol. 47, p. 385-404, pls. 48-49.

MASUMOV, A. S. 1973. *Jurassic ostracods of Uzbekistan.* 157 pp., 14 pls. Tashkent [in Russian].

MICHELSEN, O. 1975. Lower Jurassic biostratigraphy and ostracods of the Danish Embayment. *Danmark. geol. unders.,* no. 104, 287 pp., 41 pls.

OERTLI, H. J. 1957. Ostracodes du Jurassique Supérieur du Bassin de Paris (Sondage Vernon 1). *Rev. Inst. Franc. Pétrole,* vol. 12, p. 647-695, pls. 1-7.

--------1959. Malm-Ostrakoden aus dem schweizerischen Juragebirge. *Mém. Soc. Helv. Sci. Nat.,* vol. 83, p. 1-44, pls. 1-8.

--------1972. 22 Jurassic ostracodes of DSDP Leg 11 (sites 100 and 105) – Preliminary account. p. 645-6, pls. 1-5. In. HOLLISTER, C. D., EWING, J. I., et al., *Initial Reports of the Deep Sea Drilling Project,* vol. 11.

OWEN, H. G. 1976. Continental displacement and expansion of the Earth during the Mesozoic and Cenozoic. *Phil. Trans. roy. Soc.,* Ser. A., vol. 281, p. 223-291.

PERMJAKOVA, M. N. 1969. New species of Ostracoda from the Bajocian Deposits of the Dnieper-Don Depression. *Paleont. Sbornik.,* no. 6, p. 34-38, 1 pl. [in Russian].

--------1970. Ostracoda of the genus *Glyptocythere* from the Middle Jurassic Deposits of the Dnieper-Don Depression. *Paleont. Sbornik.,* no. 7, p. 61-67, 1 pl. [in Russian].

POKORNY, V. 1973. The Ostracoda of the Klentnice Formation (Tithonian?) Czechoslovakia. *Acad. naklad. Ces. Akad. ved.,* vol. 40, p. 1-107, 20 pls.

ROSSI DE GARCIA, E. 1972. Ostracoda. In: MALMUMIAN, N., MASIUK, V. & ROSSI DE GARCIA, E. Microfósiles del Cretácico superior de la perforación SC-1, provincia de Santa Cruz, Argentina. *Revta. Asoc. geol. Argent.,* vol. 27, p. 265-272, 1 pl.

SWAIN, F. M. 1946. Upper Jurassic Ostracoda from the Cotton Valley Group in Northern Louisiana; the genus *Hutsonia. J. Paleont.,* vol. 20, p. 119-129, pls. 20-21.

--------1952. Ostracoda from wells in North Carolina: Part 2. Mesozoic Ostracoda. *U. S. Geol. Surv. Prof. Pap.,* 234-B, p. 59-93, pls. 8, 9.

REFERENCES

SWAIN & BROWN, P. M. 1972. Lower Cretaceous, Jurassic (?) and Triassic Ostracoda from the Atlantic Coastal Region. *U. S. Geol. Surv. Prof. Pap.,* 795, p. 1-55, 9 pls.

SWARTZ, F. M. & SWAIN, F. M. 1946. Ostracoda from the Upper Jurassic Cotton Valley Group of Louisiana and Arkansas. *J. Paleont.,* vol. 20, p. 362-373, pls. 52, 53.

WHATLEY, R. C. 1970. Scottish Callovian and Oxfordian Ostracoda. *Bull. Br. Mus. nat. Hist. geol.,* vol. 19, p. 297-358, 9 pls.

WIENHOLZ, E. 1967. Neue Ostracoden aus dem norddeutschen Callov. *Freiberger. ForschHft.,* vol. 213, p. 23-51, pls. 1-5.

Discussion

Dr. W. Manspeizer: Our (Manspeizer, Cousminer, and Puffer) work over the past few years disagrees, in part, with the reconstruction presented in this paper. Our studies and those of Prof. Ager, Chris Kendell, Ian Evans, Bill Ryan, etc. generally show that the High Atlas contains a marine fauna, and was a deep-water basin that became detached from the African Platform in the early Jurassic. Similar marine conditions prevailed throughout most of the middle and late Jurassic. Our studies in easternmost Morocco show that the Muschelkalk Sea emanating from the Tethys Sea to the north and east, transgressed the region near the Algerian border in the middle Jurassic and probably extended into the Rif Province of western Morocco by Late Triassic and Jurassic time. These waters from Tethys may well have served as the source for the salt in the Argo Salt (early Jurassic) on the Scotian Shelf, and for the marine transgression associated with the acritarchs and tasminids of the Waterstone Formation of the Central Midlands of England. In short, stratigraphic data from Morocco indicates that a marine Tethys transgression extended into the Atlantic region by the late Triassic.

Bate: Dr. Manspeizer quite rightly indicates that Morocco belonged to a Tethyan Province during the Jurassic-this agrees with my paper as I have placed Tunisia firmly in this province for the Jurassic. What must have confused Dr. Manspeizer was the use of the term North African Province. This term was used for that part of North Africa extending from Libya to the Middle East. Even so I have suggested on the second page of my paper that the eastablishment of this province"is rather tenuous and may have to be merged with the Tethyan Province at a later date". With respect to Dr. Mansseizer's comment concerning the Upper Trias in Morocco I see nothing in my paper that would disagree with this. All I would suggest is that Morocco was situated on the most easterly extension of the deeper-water Tethyan transgression with shallower marine conditions extending farther east.

Dr. F. M. Swain: There may be a little too much emphasis on provincialism in this discussion and not enough on the possibilities of local environmental variations and possible endemism. The Upper Jurassic ostracodes from the mid-Atlantic and Gulf of Mexico regions have certain similarities, particularly in the Schulerideas, Hutsonias, and Fabanellas. In the Scotian Shelf area there are also similarities to the preceding, the three areas showing gradational relations. But a difference occurs between these late Jurassic ostracode faunas and those of the Sundance Sea of the western interior U. S. There we find ostracodes that are more like those from western Europe than they are like Gulf and Atlantic.

Bate: I would say that some relationship between the Gulf Coast, Mid-Atlantic, and Scotian areas is to be expected through migration via shallow epicontinental seas. The deeper water Tethys appears to have been less helpful in this respect.

Dr. I. G. Sohn: I believe I found the same ornate bairdiids (ie. Ptychobairdia) in the Triassic of Nevada, and Kolmann's from Austria are Triassic also.

Bate: The intimation from Dr. Sohn that he has found ornate
bairdiids from the Trias of Nevada similar to forms described by Kolmann from Austria is of interest because it para-
llels the sort of distribution I have mentioned for the Jurassic genus Procytheridea sensu stricto.

Cretaceous Ostracoda of the North Atlantic Basin

JOHN W. NEALE

Department of Geology, The University, Hull, Yorkshire, England

Abstract

For most of Lower Cretaceous time, North Atlantic ostracod faunas are only well-developed in the Eastern area and our knowledge is one-sided. Marine transgression and the extension of deposition in the Albian and Upper Cretaceous gave excellent faunas on both sides of the Atlantic and the Basin appears 'two-sided'. Besides being stratigraphically useful, the ostracods provide ecological information on factors such as salinity, temperature and migration.

Introduction

During Cretaceous times the North Atlantic Basin was relatively small, the South Atlantic Rift was only just commencing and most of our information is derived from land areas surrounding the Basin. Lower and Upper Cretaceous both have diverse ostracod faunas which are stratigraphically useful. They also often provide much additional information such as changes from non-marine deltaic/mudflat to marine environments which can have important applications in oil exploration. Besides salinity, these small crustaceans may also give insights into other parameters such as temperature, relative depth and biogeography. In the North Atlantic area the development of ostracod-bearing sediments was noticeably one-sided in the Lower Cretaceous. With the Albian and Upper Cretaceous, a big marine expansion produced a two-sided Basin and faunas of these ages are well known in Western areas. This review deals first with Lower Cretaceous and then with the Upper Cretaceous working from east to west in each case.

Lower Cretaceous Non-Marine Faunas

In late Jurassic-Early Cretaceous times non-marine deposition was a notable feature in many areas. Such 'Wealden' facies is characterised by rapid lateral and vertical variations of muds

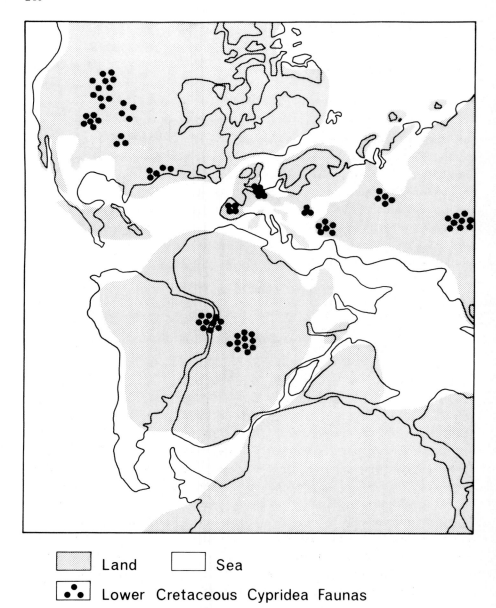

Land   Sea

Lower Cretaceous Cypridea Faunas

Fig.1 EARLY CRETACEOUS – Distribution of Land, Sea and
    Non-marine Ostracod Faunas

and sands and is often of considerable interest to the petroleum geologist and reservoir engineer.   Formerly regarded as
deltaic sedimentation, the most recent theory regards the
classic Wealden environment as a variable salinity mudplain
with overloaded streams periodically providing a braided sandur
environment.   The deposits yield excellent ostracod faunas,
especially the genus Cypridea, and the fullest development
occurs in southern Britain where the beds range from Late
Jurassic up to and including the Barremian.   Based on these
'beaked' species, Anderson (1973) in a masterly review has been
able to recognise two assemblages of Jurassic age and eight of
Cretaceous age which enables correlations to be made in the NW
European area (Fig. 2).   In N. Germany the 'Wealden' Beds only
include the first two Cretaceous assemblages and are overlain
by marine Middle Valanginian.   The situation in Holland is
similar.   In the Swiss Jura the first Cretaceous assemblage is
present but the bulk of the non-marine beds are Jurassic, as in
the case of the French Jura and the Paris Basin where Cretaceous non-marine faunas have not been attested.   The Danish
Island of Bornholm yields both Jurassic and Cretaceous assemblages of which the highest from the Jydegaard Formation suggests the second Cretaceous assemblage.   Outside Britain the
most complete developments appear to be those of the Iberian
peninsula where the earlier Jurassic and Cretaceous non-marine
assemblages have been found at a number of places.   West of
Logrona, north of Burgos and also near Cuenca, the highest
Cretaceous assemblage has been found.

   The genus Cypridea is surprisingly widespread and Anderson
(1973) has suggested that it was climatically controlled and
occupied a zone on either side of the Cretaceous equator whose
climate was similar to that of the present day Mediterranean.
The means of dispersal of this group of ostracods, which occur
in disjunct basins, has so far not been satisfactorily explained.   Although the forms show a general similarity, doubtless due to the limited number of usable taxonomic characters,
indigenous species are developed.   Thus while correlation can
be made between Gabon and NE Brazil, and emphasises the close
juxtaposition of those two areas in early Cretaceous times (see
Bertels, this symposium), there is not a single species in

248

	ENGLAND			GERMANY HOLLAND DENMARK	SPAIN	JURA	APTIAN
8	WEALD CLAY	C.tenuis C.insulae C.comptonensis C.vectae C.hamata C.caudata C.cuckmerensis C.warlinghamensis C.rotundata	S.cornigera M.mantelli		Burgos ··· Cuenca		Upper Barremian
7	WEALD CLAY	C.spinigera C.pseudomarina C.fasciata C.rotundata C.valdensis C.clavata	T.fittoni		Logroño ·· Zaragoza · Empasa		Lower Barremian
6	WEALD CLAY	C.clavata C.bogdenensis C.valdensis C.rotundata	T.fittoni				Upper Hauterivian
5	WEALD CLAY	C.dorsispinata C.tuberculata C.marina C.pumila	M.henfieldensis				Lower Hauterivian
4	TUNBRIDGE WELLS SAND & WADHURST CLAY	C.aculeata C.r.tillsdenensis C.arenosa C.melvillei C.bispinosa C.laevigata	R.jurassica	Germany & Holland	Sandstone?		Upper Valanginian
3	WADHURST CLAY	C.paulsgrovensis C.menevensis C.tuberculata C.frithwaldi	T.alleni				Middle Valanginian
2	ASHDOWN SANDS / UPPER PURBECK	C.setina C.dolabrata C.brevirostrata C.propunctata C.alta formosa C.wolburgi C.lata latissima C.wicheri	R.jurassica	W6 / W3 Denmark Orgaño			Lower Valanginian
1	MIDDLE PURBECK	C.g.fasciculata C.vidrana C.bimammata C.altissima C.lata senilis C.misia	M.mediostricta	W2 / W1		French · Swiss	Berriasian

Cinder Beds

Fig. 2 Europe-Lower Cretaceous Non-Marine Ostracod Faunas

(Anderson 1973 [pars] redrawn)

Plate 1. Paired stereoscopic photographs of non-marine Lower Cretaceous ostracods from southern England. 1. Cypridea granulata fasciculata Jones. LV x 35. 2. Cypridea setina Anderson. RV x 50. 3. Cypridea paulsgrovensis (Anderson). LV x 52. 4. Cypridea aculeata Jones. LV x 50. 5. Cypridea dorsispinata (Anderson). LV x 60. 6. Cypridea clavata Anderson. LV x 52. 7. Cypridea spinigera (J. de C. Sowerby). LV x 57. 8. Cypridea tenuis Anderson. Carapace from right x 50.

PLATE 1

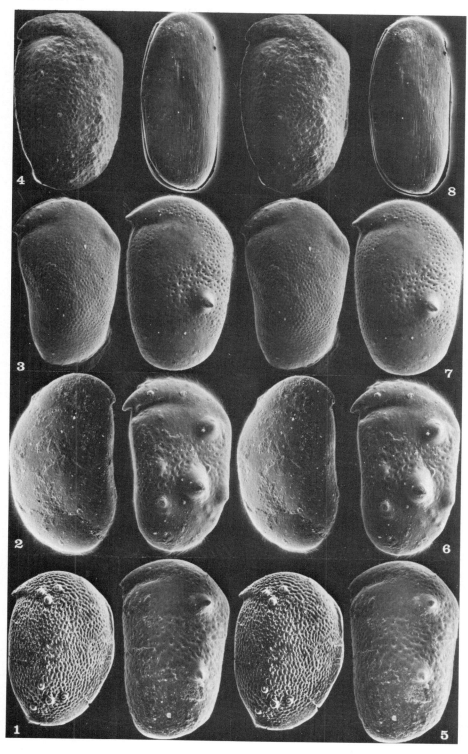

common with the European area.

In America, species of Cypridea and Hutsonia, have been found in the Middle Atlantic States in Unit H of Swain and Brown (1972) which is regarded as possibly Jurassic at the base ranging up to early Comanchean (? Aptian).   Marine species also occur in Unit H, which is predominantly marine in eastern North Carolina, becoming increasingly non-marine in Virginia and New Jersey except in the extreme east.   In the same area Unit G also contains Cypridea but is essentially marine (see under ALBIAN). Cypridea faunas are also found in Oklahoma, Utah, Idaho, Wyoming, Montana, South Dakota and Alberta but are too far west to be legitimately considered in this contribution.

From time to time, intercalations occur of ostracods suggestive of higher salinities than those of the normal Cypridea faunas.   Anderson calls these 'S' Phase and they have been variously called marine or quasi-marine bands.   The close association of species such as Fabanella boloniensis, F.ansata and Mantelliana purbeckensis with evaporite deposits has led to the suggestion that they may have flourished in hypersaline waters.   Kilenyi and Allen (1968) who made a detailed study of bands in the lower part of the Weald Clay of Sussex and Surrey concluded on the basis of Schuleridea (Eoschuleridea) wealdensis, Hutsonia capelensis, Ammobaculites and Cirripedes that the salinity in that sediment was polyhaline/euhaline on occasion. The specialised faunas of these 'S' phase bands has so far proved of little use in correlation.

Lower Cretaceous Marine Faunas

Marine ostracod faunas in northern Europe have recently been reviewed by Neale (1973) who gives a full bibliography.

1. BERRIASIAN
Provincialism was most marked during this stage with clear separation of the colder, northerly faunas of Britain and Denmark and those of the warmer areas further south.   The British fauna is characterised by Galliaecytheridea teres (used as the stage index fossil by Christensen 1974), Mandelstamia sexti, Schuleridea juddi, Paracypris caerulea, Cytheropterina triebeli and Pontocypris felix.   Notable absentees are representatives

of the warm water genus Cytherelloidea and Cytherella, Protocythere and Cythereis, whilst Galliaecytheridea and Mandelstamia linger on from the Jurassic.

In the Vocontian Trough of S. France some 10° further south,
which includes the type area of Berrias on the west, Berriasian
faunas have a very different aspect with Cytherella, Cytherelloidea, Protocythere, Cythereis and Acrocythere already established, different species of Orthonotacythere, Paracypris and
Schuleridea, and indigenous genera such as Kentrodictyocythere
and Raymoorea. Donze (1971) has been able to recognise differences in passing from the margins (facies semi-emersif)
towards the deeper parts of the basin (facies neritique). The
latter is characterised by the disappearance of genera such as
Euryitycythere, Exophthalmocythere, Orthonotacythere, Quasiher-
manites, Kentrodictyocythere, Schuleridea, 'Clithrocytheridea'
and others, a marked increase in the proportion of Paracypris,
Cytherelloidea, Cytherella and Bairdia, and the presence of
Cypridina, Polycope, Cardobairdia, Annosacythere, Raymoorea and
Hemicytherura. The marginal facies includes occasional
Cypridea, Fabanella, Limnocythere and Scabriculocypris. In
Portugal the Berriasian starts with a regressive phase contain-
ing both non-marine and marine elements, followed by a transgressive phase. The marine ostracods suggest a melange of
both northern and southern European forms.

North Africa, after separation for a long period of Mesozoic
time, shows links with Europe to the north during the Berriasian and migration of the fauna northwards, probably due to
shallowing of Tethys in between (Donze 1975). In North
Tunisia, the Djebel Ouest sections provide ostracods from the
Tithonian through the Berriasian. The Berriasian genera
include Amphicythere, Eucytherura, Hemicytherura, Paracypris
and Pontocyprella. Tethysia, which appears earlier in N.
Africa, also occurs in the Berriasian of both there and the
Vocontian Trough and confirms the affinities between the two
areas. In Algeria Protocythere mazenoti, P. paquieri and P.
cf. P. revili also emphasise the close links with the south of
France in Berriasian times.

2. VALANGINIAN

Marine conditions become much more widely established and in the British area the relict Jurassic genera disappear and Cytherella and Protocythere make their appearance. Faunas of this age appear in Germany above non-marine deposits and are widely known from Heligoland and the north German oilfield across to Poland. Particularly characteristic are Protocythere hannoverana, Stravia crossata, Dolocytheridea wolburgi and Schuleridea praethoerenensis. Northerly migration from Tethys can again be seen (Donze 1973) and by the end of Valanginian times Euryitycythere and Parexophthalmocythere had already reached NW Germany and the latter genus and Kentrodictyocythere had reached Poland. Moving southward into warmer seas, different species of many genera are present as clearly seen in the area of Neuchâtel and the Alpes de Haute Provence where Protocythere is represented by P. praetriplicata, P. divisa, P. helvetica and P. reicheli.

3. HAUTERIVIAN

In the Hauterivian the sharp distinctions between northern and southern European faunas disappear. Cytherelloidea, Euryitycythere, Parexophthalmocythere and Rehacythereis reach the British area and the ostracods are abundant and widespread. In Britain and N. Germany, Paranotacythere diglypta, Protocythere hechti and Rehacythereis senckenbergi are characteristic as well as elements of the continuously evolving Protocythere triplicata and Cytherelloidea ovata lineages. Other important genera include Apatocythere, Dolocytheridea, Eucytherura and Schuleridea. In the Lower Saxony Basin, Kemper (1971) recognised three ostracod facies in the Upper Valanginian and Lower Hauterivian. Associated with a typical suspension feeder community was a fauna rich in species but poor in individuals with Protocythere frankei, Bairdia, Pontocyprella and many cytherurid species. The shallow neritic to sub-littoral crinoid facies provided the richest faunas which included Haplocytheridea kummi, Protocythere triplicata, Rehacythereis senckenbergi, Schuleridea thoerenensis, Dolocytheridea spp., and many others. In the deeper parts transitional to the bathyal areas H. kummi and Protocythere hechti might be present. Cytherelloidea was considered indicative of warm, shallow water.

Further south in the Paris Basin the genera are similar as well as some of the species but new species such as Protocythere pumila, P. cancellata and Schuleridea extranea appear.

4. BARREMIAN

In the British area characteristic species such as Cytherelloidea dalbyensis, Amphicytherura bartensteini, Eucytherura nuda, Paranotacythere blanda, P. inversa inversa and others are present although the genera remain much the same as before. Pseudobythocythere is recorded for the first time in Britain, Platycythereis and Metacytheropteron occur in France and Schulapacythere in Romania. In the Paris Basin a characteristic Protocythere fauna includes P. strigosa and P. villierensis. Paranotacythere damottae damottae occurs, as well as P. inversa inversa, Schuleridea virginis and S. bernouilensis. In the South of France, Donze (1971) has again drawn attention to the difference between the margins of the Vocontian Trough where 25 genera occurred in one sample and the deeper part of the Trough where only the four genera Cythereis, Protocythere, Cytherella and Pontocyprella were present in the Gisement d'Angles (Basses-Alpes).

5. APTIAN

Aptian is well-known and yields a great variety of ostracods. In the general British, N. German and Paris Basin area typical species are Centrocythere bordeti, Cythereis bekumensis, C. geometrica, Dolocythere rara, Dolocytheridea intermedia Eocytheropteron stchepinskyi, Paranotacythere inversa tuberculata, Protocythere croutesensis, P. derooi, Schuleridea derooi and Veenia florentinensis. Developments of Protocythere give rise to the genera Saxocythere and Batavocythere of Kemper (1971) which make their appearance in this stage and Saxocythere tricostata tricostata is found in Britain and N. Germany. In N. Germany Batavocythere hiltermanni appears in the Upper Aptian. In S. France, Moullade (1963) found Protocythere bedoulensis, P. oertli and P. alexanderi characteristic of the Bedoulian (L. Aptian) although the first two range down into the Barremian and P. bedoulensis is also found in the Lower Gargasian. In the Gargasian (Upper Aptian) at the type locality of Apt, Oertli (1958) recorded amongst others, Dolocythe-

ridea intermedia (found also in the north), Schuleridea jonesiana (which does not apparently reach the Paris Basin and northern Europe until the Albian), and a number of new species
including Cythereis bartensteini, C. buchlerae, Neocythere mertensi and Platycythereis rectangularis.

6. ALBIAN

Hitherto, our information has been confined to the eastern side
of the North Atlantic Basin. With the Albian came a striking
extension of the shallow shelf seas into the eastern and southern United States so that deposition in the Atlantic Basin was
no longer one-sided. The faunas of these beds are comparatively well-known.

In Europe, typical Albian forms include Batavocythere gaultina, Isocythereis fissicostatus, I. fortinodis, Neocythere
vanveeni, Protocythere mertensi, P. nodigera, P. speetonensis,
Saxocythere dividera, S. notera, Schuleridea jonesiana and
Veenia harrisiana. Dolocytheridea intermedia intermedia gets
into the Lower Albian and D. bosquetiana is characteristic of
the Middle and Upper Albian and just gets into the Cenomanian.
In addition there is a wide variety of species of Acrocythere,
Alatacythere, Argilloecia, Clithrocytheridea, Cythereis, Cytherelloidea, Dolocytheridea, Eucythere, Eucytherura, Habrocythere, Hemicytherura, Isocythereis, Krausella, Macrocypris, Neocythere, Platycythereis, Protocythere, Saxocythere, Schuleridea,
and Veenia. The genera Bairdia, Conchoecia and Polycope also
occur. In the Paris Basin Matronella appears in the Albian
and has been recorded from the Northern Ireland Cenomanian.
Although intermediates are not known it is thought to be the

Plate 2. Paired stereoscopic photographs of European Lower
Cretaceous marine Ostracoda. All specimens from England
except where stated.
1. Galliaecytheridea teres (Neale), RV x 52, Berriasian; 2.Phi-
lomedes donzei Neale, ♀ carapace x 35, Basal Valanginian,France;
3. Mandelstamia sexti Neale, LV x 70, Berriasian; 4.Protocyth-
ere hannoverana Bartenstein and Brand, LV x 64; Valanginian;
5. Paranotacythere diglypta (Triebel), LV x 99, Hauterivian;
6. Cythereis acuticostata Triebel, LV x 65, Barremian; 7.Schu-
leridea jonesiana (Bosquet), ♀ LV x 47, Albian-Cenomanian; 8.
Platycythereis gaultina (Jones), LV x 88, Albian.

PLATE 2

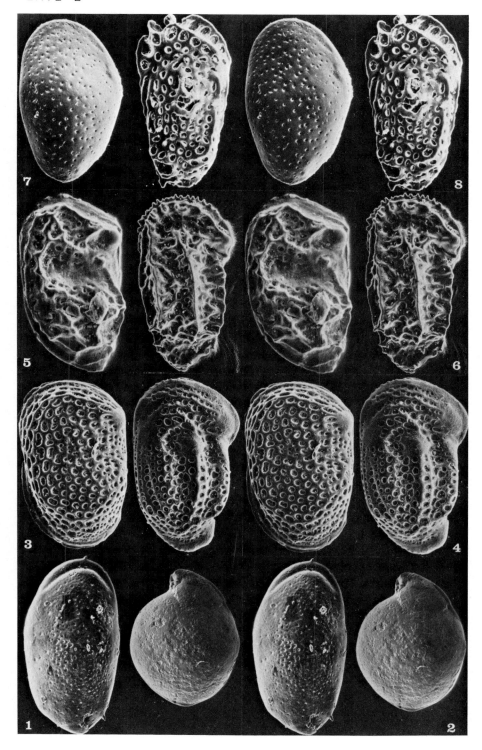

ancestor of Spinoleberis in the Campanian.   Also very characteristic are Cornicythereis, Cythereis, Isocythereis and Platycythereis.   The genus Paranotacythere appears restricted to
Britain in the Albian where the single species P. fordensis
occurs (Bassiouni 1974).   Chapmanicythereis which first
appeared in the Aptian and continued on to the Turonian is also
important and is separated from the allied Platycythereis by
the presence of an eye tubercle and longitudinal ribbing.   The
European fauna is fairly cosmopolitan during this (and subsequent stages) although some regional differences in species are
apparent.

Swain recorded species of Cytherella, Asciocythere, Schuleridea?, Cythereis and Protocythere from a core taken in the
Eastern Atlantic at 25°55.53'N, 27°03.64'W approximately 1000
kms west of Cap Blanc on the African coast and concluded that
the age of the fauna was probably Albian.   A Western Atlantic
core from 09°27.23'N, 54°20.52'W, some 400 kms north of the
Guianas yielded a larger fauna of 25 species including three
species found in the core above and Swain again concluded that
an Albian age was most likely.

Information on pre-Albian beds in the United States is
Scarce.   Swain and Brown (1972) suggest that some of the
material from the Atlantic Coastal Plain (Unit H) is early
Cretaceous in age but that the base of the unit may be Jurassic.
Vanderpool (1928, 1933) described forms from the Upper Trinity
Group of SW Arkansas, SE Oklahoma, N Texas and NW Louisiana
which included Asciocythere perforata, A. rotunda, Eocytheropteron trinitiensis, Paracypris weatherfordensis and Schuleridea
dorsoventrus.   Swain and Brown (1964) added the new name Cythereis praeornata and Cypridea dequeenensis and two other Cypri-
deas, C. diminuta and C. wyomingensis occur.   The age is usually assumed to be late Aptian? but more probably early Albian in
the case of the Upper Trinity material.   Swain and Brown (1964)
describe a nearshore Albian fauna in Central and NE North Carolina of 12 species (almost all new) which were assigned to Clithrocytheridea, Cythereis, ?Dolocytheridea, Eocytheropteron,
Eucythere, Eucytheroides, Fossocytheridea, Haplocytheridea, Orthonotacythere and Perissocytheridea.   The Middle and Upper
Trinity fauna from the Atlantic Coastal Plain of Swain and

Brown (1972) has no species in common with the previous fauna
and includes mainly species of Asciocythere, Dolocytheridea,
Schuleridea, Paraschuleridea and Hutsonia.

The higher Lower and Middle Albian (Fredericksburg Group)
and Upper Albian (Washita Group) have been covered in detail by
Alexander (1929, 1932-34) for N. Texas and Swain and Brown
(1972) correlate their Unit F of the Atlantic Coastal Plain
with these two groups.   Alexander describes many species assigned to Bairdia, Bythocypris, Cytherella, Cytherelloidea, Cytheridea, Cytheropteron, Cytherura, Eocytheropteron, Paracypris
and others whose generic taxonomy could do with considerable
updating.   Swain and Brown's (1972) Unit F of the Atlantic
Coastal Plain contains nothing in common with the Texas fauna
and consists of new species, together with principally species
of Swain and Brown (1964) and Swain (1952) assigned to Asciocythere, Dolocytheridea, Cythereis, Eocytheropteron, Clithrocytheridea, Eucythere, Fossocytheridea, Haplocytheridea, Orthonotacythere and Perissocytheridea.

Upper Cretaceous

With the continuing Cretaceous transgression and widening Atlantic Rift, chalk facies became widely established in the northern hemisphere as time progressed.   Only in the earliest and
latest stages are brackish water ostracods known.   Many Lower
Cretaceous genera such as Apatocythere, Cytheropterina, Dolocytheridea, Habrocythere, Paranotacythere, and Protocythere died
out or only survived a short distance above the base of the
Cenomanian.   Chapmanicythereis replaced Platycythereis and
during the Upper Cretaceous there was a great expansion and
proliferation of trachyleberids.   The Platycopa, Bairdia,
Krithe, Paracypris and many other genera continue and the faunas begin to take on a much more modern aspect.

1. CENOMANIAN

In the Eastern Atlantic, faunas are known from Northern Ireland
to the Atlas Mountains.   In the north Dolocytheridea bosquetiana, Neocythere vanveeni and Schuleridea jonesiana linger on
from the Albian and Matronella matronae also crosses the boundary.   Veenia replaces Protocythere as the dominant protocyth-

erinid.   Schuleridea continues as S. tumescens in the Dordogne
and Touraine and Dolocytheridea in both the Dordogne and Touraine and Algeria as D. crassa and D. atlasica respectively.
In the Paris Basin the new genera Dordoniella and Risaltina
appear.   Many new trachyleberids make their appearance, such
as Dumontina, Mauritsina, Oertliella and probably Limburgina
although the latter is better known from the Campanian and Maastrichtian.   In southern France, Amphicytherura, Annosocythere,
Curfsina, Metacytheropteron, Opimocythere, and Bairdia are
important.   Particularly notable is the large number and variety of species of Cythereis in all areas.   Cytherella is ubiquitous but Cytherelloidea is rarely recorded.   Only in the
Dordogne (SW France) has a brackish water fauna of Metacypris
and Theriosynoecum been noted (Colin 1974).

In the Atlantic Coastal Plain comparable faunas have been
described by Swain and Brown (1972) in Unit E from N. Carolina
and Virginia and Swain and Brown (1964) from the Lower Atkinson
of S. Alabama, N. Florida and Georgia.   A wide variety of Cythereis with Cytherella occurs throughout.   The more northerly
area with Dolocytheridea and Schuleridea? recalls the position
in Europe.   The more southerly area with Cytherelloidea suggests perhaps some temperature zoning.   Both areas yield the
well-known Cythereis eaglefordensis, regarded as a reliable Upper Cenomanian index further west by Hazel (1969).   The southern areas show an approach to the Gulf Cretaceous described by
Alexander (1929 et al.) with the occurrence of Cytherella austinensis, Bairdia comanchensis, Paracypris alta and Schuleridea
washitaensis.   The Woodbine of N. Texas yields Cythereis burle-
sonensis, C. roanokensis, C. worthensis, Paracyprideis grayson-
ensis and Pontocyprella alexanderi.   In the Hammond Well, Maryland (Swain 1948), Leguminocythereis? pustulosa*from 1588-98'
came from beds thought to be of approximately this age.

2. TURONIAN
The Turonian is essentially a continuation and consolidation of
the position in the Cenomanian.   Eastern Atlantic faunas are
dominated by the trachyleberids and important new species of
Cythereis, Curfsina, Pterygocythereis and Spinoleberis, together with Asciocythere, Dordoniella, Pterygocythere and other
genera.   In S. France, Dumontina, Mauritsina and Trachyleberi-
*This species is type of Eucytheroides Swain and Brown 1964 (ed.)

dea occur amongst others.   Saida is established in middle Europe and produced a stratigraphically useful sequence of species
up through into the Tertiary (Herrig 1968).   In Algeria the
typically African Ovocytheridea and Brachycythere sp. gr. ekpo
suggest links with the marine Cretaceous of the Gulf of Guinea.

In the W. Atlantic the Upper Atkinson of S. Alabama, N. Florida and Georgia has yielded only four species belonging to the
genera Brachycythere, Cytherella, Cythereis and Haplocytheridea.
The development of ostracod faunas seems equally restricted in
N. Texas where the Eagle Ford Shale yielded only Cytherella
'munsteri', Bairdia alexandrina, Pterygocythere saratogana and
Cythereis eaglefordensis.   How far the Eagle Ford in Texas
should be regarded as Turonian in age and how far as Cenomanian
is arguable.

3. CONIACIAN

Least well-known of the stages as regards its ostracod fauna,
in Europe Pokorny (1964) has shown evolution in the subspecies
of Cythereis marssoni from the U. Turonian to the Coniacian.
The species is not known from the Santonian.   A similar evolution is seen in the subspecies of C. ornatissima.   There is a
wide variety of Cythereis and other genera include Cytherelloidea Eucythere, Idiocythere. Karsteneis, Oertliella, Phacorhabdotus, Pterygocythereis, Spinicythereis and Spinoleberis.   In
Algeria Ovocytheridea producta and O. brevis along with Brachycythere sp. gr. ekpo, Veenia, Mauritsina?, and Protobuntonia
again indicate links with the West African area.

In the W. Atlantic the Coniacian includes the Tokio Formation of Arkansas and N. Louisiana and the Austin Chalk of Texas
but the upper part of these is almost certainly Santonian and
the faunas are best regarded as 'Lower Senonian'.   Cythereis
bicornis is a very typical species.   In Hammond Well, Maryland
the fauna from 1470-80' of Cythereis cf. bicornis, C. parausti-
nensis and Haplocytheridea parvasulcata probably belongs in
this stage.   In the Gulf, Phacorhabdotus pokornyi, Schuleridea
travisensis and Veenia reticulata are recorded in E. Texas
(Hazel & Paulson 1964).   In Arkansas the Tokio Formation yielded Cythereis bicornis, C. hannai, Pterygocythereis tokiana and
Cytheropteron sp. (Israelsky 1929).   From the Austin Formation
of Texas come Bairdippilata rotunda, Brachycythere sphenoides,

260

FIG. 3   UPPER CRETACEOUS - DISTRIBUTION OF LAND AND SEA

Plate 3.   Paired stereoscopic photographs of American Cretaceous
Marine Ostracoda.   1. Eocytheropteron tumidum (Alexander) LV x 76.
2. Rehacythereis fredericksburgensis (Alexander). LV x 57.   3.Hap-
locytheridea? globosa (Alexander). LV x 63.   4. Haplocytheridea?
cf. plummeri (Alexander) ♂ RV x 70.   5. Bradleya hazardi (Israelsky). LV x 62.Cytherella tuberculifera Alexander.LV x 52.   7.
Sphaeroleberis pseudoconcentrica (Butler & Jones) RV x 139.   8.
Curfsina communis (Israelsky). RV x 73.   1,2. Alexander Station
5, Texas.   Kiamichi Fm., U. Albian.   3-6. Alexander Station 60,
Texas.   Navarro Fm., Maastrichtian.   7,8. Ca. 2 miles S. of Lucky,
Louisiana.   Butler & Jones Locality 1957.   Saratoga Fm., Campanian

PLATE 3

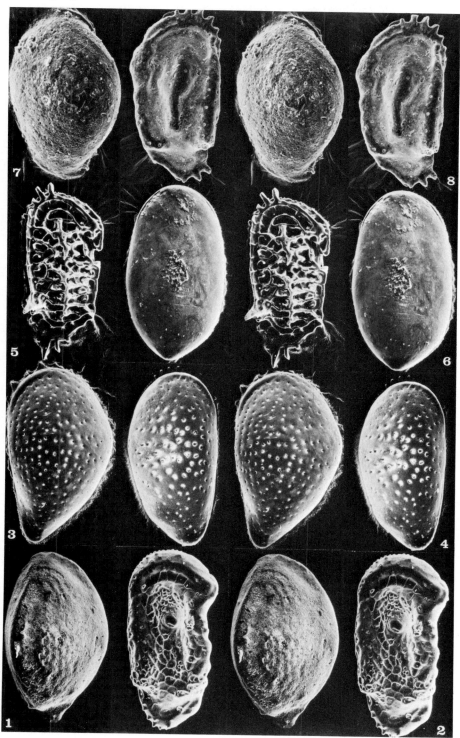

B. taylorensis, Cythere cornuta var. gulfensis, C. foersteriana,
Cythereis austinensis, C. bicornis, C. dallasensis, Cytherella
austinensis, Krithe cushmani, Paracypris tenuicula and Veenia
ozanana.

4. SANTONIAN

A large number of species is known from this stage in northern
and central Europe.   Imhotepia? similis and Idiocythere defin-
ita go right through the stage and a little above the base of
the Campanian in the Baltic Island of Rügen.   Herrig (1967 et
al.), Grundel (1968 et al.) and Ohmert (1973) have described
faunas from Rügen and north Europe and the appearance of Golco-
cythere, Mosaeleberis and Kikliocythere is notable in these
beds, together with various species of Costaveenia, Cytherella,
Cytherelloidea, Cytheropteron, Eucytherura, Trachyleberidea,
Veenia etc.   Ohmert described the new genera Deroocythere and
Kikliopterygion which occur in Santonian beds.   Phacorhabdotus
semiplicatus seems fairly widespread and this, or a closely related species, is found in the Alpes Maritimes of France where
Donze and Porthault (1970) have also noted species of Cythereis,
Karsteneis, Oertliella, Pterygocythereis and Trachyleberidea.

In the W. Atlantic the Austin Chalk fauna has been consider-
ed for convenience with the Coniacian.   Similarly, the Taylor
fauna which may be in part U. Santonian is considered under
Campanian.   Hazel and Paulson (1964) consider the Brownstown
Marl in which Israelsky (1929) found Veenia spoori to be L.
Campanian.

5. CAMPANIAN

A number of Jones andHinde's classic species such as Paracypris
siliqua, Pariceratina tricuspidata, Curfsina nuda, and Cytherella obliquirugata come from deposits of this age in Britain,
but more recent work on the Island of Rügen has better stratigraphic control (Herrig 1967 et al.).   Here Idiocythere repli-
cata, Amphicytherura chelodon, Argilloecia decussata and Bairdia denticulata all make their appearance at the base of the
Campanian, the latter three continuing into the Maastrichtian.
Generally in Europe Curfsina, Dumontina, Limburgina, Mosaeleberis, Planileberis and Spinoleberis are important genera together with Amphicytherura, Phacorhabdotus, Physocythere and Saida

as well as many others and the ubiquitous Platycopa.   In the
Aquitaine Basin of SW France the fauna has considerable affinities with that of the Maastrichtian of Holland.   Deroo (1966)
has suggested migration from south to north but Colin (1973)
favours an ecological explanation.   The full significance of
the similarities has still to be worked out.   A fauna from Algeria includes the genera Acanthocythereis, Bradleya, Cythereis,
Cytheropteron, Veenia and Veenidea but probably the most interesting genus is Buntonia which is so well developed in the Nigerian Cretaceous.

In the west the Matawan Group of Maryland, Delaware and Virginia has yielded 27 species (Schmidt 1948).   Genera represented include Bairdoppilata, Brachycythere, Cythereis, Cytherelloidea, Krithe and Xestoleberis.   The generic status of many of
the species referred to Cythereis needs reassessment in terms
of modern taxonomy.   Beds assigned to the same Group in Hammond Well, Maryland (Swain 1948) yielded a fauna of Bairdia,
Cythereis, Haplocytheridea and Paracypris.   It is from the
Gulf Coast area, however, that the fauna is best known.   According to the distribution tables in Alexander (1929), Cythereis austinensis, C. bicornis and C. dallasensis die out between
the Austin and Taylor Formations while in the latter C. rugosis-
sima, Haplocytheridea? grangerensis, H? plummeri and Paracypris
angusta make their appearance.   Cytherella navarroensis,
Veenia ozanana and V. paratriplicata become common for the first
time.   Detailed studies of the Campanian have also been made
by Israelsky (1929) and Benson and Tatro (1964) in Arkansas and
Butler and Jones (1957) in Louisiana.   Important genera include Alatacythere, Amphicytherura, Bairdoppilata, Brachycythere,
Bythocypris, Cytherella, Cytherelloidea, Haplocytheridea?,
Krithe, Loxoconcha, Orthonotacythere, Paracypris, Pterygocythere, Veenia, and Xestoleberis.   From E. Texas Hazel and Paulson
described a fauna regarded as of this age which included species of Brachycythere, Bradleya, Pterygocythereis and Veenia
(Nigeria) but was chiefly notable for Schuleridea travisensis.
This is the highest Cretaceous record of Schuleridea (also
noted by them in the Coniacian) and is separated from other
records by a considerable interval.   They also thought that
Brachycythere durhami, which occurs in both the Austin Group

and the Gober Chalk, was facies controlled.

6. MAASTRICHTIAN

In the type area the fauna is known in considerable detail from
the work of Deroo (1966) who described 163 Maastrichtian species.   Further, he was able to recognise seven divisions in
the stage and separate it satisfactorily from the underlying
Campanian and overlying Danian.   Species of Alatacythere, Asciocythere, Brachycythere, Curfsina, Dumontina, Eucytherura,
Hemicytherura, Kikliocythere, Kingmaina, Limburgina, Mauritsina,
Mosaeleberis, Netrocytheridea, Phacorhabdotus, Sphaeroleberis,
Veenia, Veenidea and Xestoleberis are important.   Acuticythe-
retta, Globoleberis, Protojonesia and Tumidoleberis are recorded for the first time.   At least ten species are referred to
Bythoceratina and according to Deroo the Tertiary genera Murr-
ayina and Paleomonsmirabilia and the Tertiary and Recent Orionina are also present.   The Rügen fauna (Herrig 1965 et al.)
contains a number of species in common with Holland as well as
new species of Argilloecia, Cytherura, Polycope and two new
species of Saida.   In SW France, Blanc & Colin (1975) have
described an Upper Maastrichtian fauna which includes Asciocythere, Bairdia, Eucythere, Krithe, Limburgina?, Paleomonsmirabilia, and Xestoleberis as well as rather doubtful Paracypris,
Centrocytheridea and Cytheromorpha.   In Algeria in rocks of
possible Maastrichtian age Bellion et al. (1973) noted Protobuntonia numidica (ranging to the local base of the Senonian),
Brachycythere sp. gr. ekpo, Acanthocythereis sp., and Mauritsina sp. Evidence of brackish and lagoonal conditions is forth-;
coming for the first time since the Cenomanian.   Damotte and
Fourcade (1971) found Neocyprideis murciensis in SE Spain and
interpreted the conditions as a lagoon or marginal environment
with variable salinity.   Also in Spain in the S. Pyrenees,
Liebau (1971) was able to recognise four different salinity
controlled facies.   The two more brackish facies contained
Bisulcocypris, Cyprideis, Cypridopsis?, Cytheromorpha?, Neocyprideis, Parakrithe? and Paleomonsmirabilia.

    In the west our knowledge is based on faunas from the Navar-
ro Formation described by Alexander (1929) and the fauna from
the Arkadelphia Marl described by Israelsky (1929) consisting
of Brachycythere ledaforma, B. sphenoides, Cythereis costatana,

Bradleya hazardi, Phacorhabdotus tridentatus, Curfsina communis and two other species of which Cythereis ivii is probably Veenia arachoides. Schmidt's fauna from the Monmouth Group of Maryland is possibly in large part Campanian. It contains Brachycythere rhomboidalis, Haplocytheridea? fabaformis, H? f.multilira, H? macropora, H? amygdaloides brevis, H? plummeri, H? ulrichi, Cythereis pidgeoni, Cytheropteron coryelli, Loxoconcha cretacea, Paracypris monmouthensis and Veenia arachoides.

Conclusions

During the Cretaceous, Ostracoda are well developed on the Eastern side of the Atlantic in the Berriasian to Aptian Stages, and on both sides of the Atlantic from the Albian onwards. General faunal studies have shown that they are useful in recognising the Stages and sometimes finer divisions. Phylogenetic studies of single genera such as those of Bassiouni (1974), Bettenstaedt (1958), Herrig (1968), Neale (1973) and others hold out possibilities of finer discrimination. Much remains to be done in this field and a lot of the older work needs revising and updating taxonomically. Often, the ostracods can give a lot of useful information about the nature of the environment at the time they lived.

Acknowledgements

I wish to thank Dr. R. H. Bate, Dr. R. G. Clements and Dr. T.I. Kilenyi who kindly loaned me specimens of Cypridea.

References

Alexander, C. I. 1929. Ostracoda of the Cretaceous of North Texas. Univ. Texas Bull. 2907, 137 pp., 10 pls.

Alexander, C. I. 1934. Ostracoda of the genera Monoceratina and Orthonotacythere from the Cretaceous of Texas. J. Paleont. v. 8, pp. 57-67, pl. 8.

Anderson, F. W. 1973. The Jurassic-Cretaceous transition: the non-marine ostracod faunas. Geol. J. Special Issue No. 5, pp. 101-111, 1 text-fig.

Bassiouni, M. el A. A. 1974. Paranotacythere n.g. (Ostracoda aus dem Zeitraum Oberjura bis unterkreide (Kimmeridgium bis Albium) von Westeuropa. Geol. Jahrb. v.17A, pp. 1-112, pls. 1-13, 1 table, 5 text-figs.

Bellion, Y., Donze, P, and Guiraud, R. 1973. Repartition Stratigraphique des principaux ostracodes (Cytheracea) dans le Crétacé Supérieur du Sud-Ouest Constantinois (Confins Hodna - Aurès, Algérie du Nord). Publ. Serv. géol. Algérie (N.S.) No. 44, pp. 7-44, pls. 1-7, 2 tables, 2 text-figs.

Benson, R. H., and Tatro, J. O. 1964. Faunal description of Ostracoda of the Marlbrook Marl (Campanian), Arkansas. Univ. Kansas Paleont. Contr. 7, pp. 1-32, pls. 1-6, 15 text-figs.

Bettenstaedt, F. 1958. Phylogenetische Beobachtungen in der Mikropaläontologie. Palaont. Z. v. 32, pp. 115-140.

Blanc, P. L., and Colin, J. P. 1975. Étude Micropaléontologique et Paléoécologique du Maastrichtien de Cézan-Lavardens (Gers, S.O. France). Palaeontographica, v. 148A, pp. 109-131, pls. 1-4, 13 text-figs.

Butler, E. A., and Jones, D. E. 1957. Cretaceous Ostracoda of Prothro and Rayburns Salt Domes Bienville Parish, Louisiana. La. Geol. Surv. Bull. No. 32, pp.1-65, pls. 1-6, 5 text-figs.

Christensen, O. B. 1974. Marine Communications through the Danish Embayment during Uppermost Jurassic and Lowermost Cretaceous. Geoscience and Man, v. 6, pp. 99-115, 1 pl., 5 text-figs., 5 tables.

Colin, J. P. 1973. Nouvelle Contribution a l'étude des ostracodes du Crétacé supérieur de Dordogne (S.O. France). Palaeontographica, v. 143, pp. 1-38, pls. 1-6, 3 text-figs.

Colin, J. P. 1974. Nouvelles espèces des genres Metacypris et Theriosynoecum (Ostracodes lacustres) dans le Cénomanien de Dordogne (S.O. France). Rev. Espan. Micropal. v. 6, pp. 183-189, 1 pl.

Damotte, R, and Fourcade, E. 1971. Neocyprideis murciensis n. sp., Ostracode nouveau du Maestrichtien de la province de Murcie (Sud-Est de l'Espagne). Bull. Soc. geol. France. v. 13, pp. 169-173, pl. 6, 1 text-fig.

Deroo, G. 1966. Cytheracea (Ostracodes) du Maastrichtien de
Maastricht (Pays-Bas) et des régions voisines; résultats stratigraphiques et paléontologiques de leur
étude. Med. Geol. Sticht. C. V. 2, pp. 1-197, pls.
1-27, 9 tables, 20 text-figs.

Donze, P. 1971. Rapports entre les faciès et la répartition
générique des ostracodes dansquatre gisements-types,
deux à deux synchroniques, du Berriasien et du Bar-
rémien du sud-est de la France. Bull. Centre Rech.
Pau - SNPA, v. 5 suppl., pp. 651-661, 2 tables, 2
text-figs.

Donze, P. 1973. Ostracod migrations from the Mesogean to Boreal Provinces in the European Lower Cretaceous.
Geol. J. Spec. Issue, Liverpool, v. 5, pp. 155-160,
4 text-figs.

Donze, P. 1975. Paléobiogéographie des populations d'ostracodes
de part et d'autre de la Téthys (Afrique du Nord et
Europe Occidentale), au Jurassique supérieur et au
Crétacé inférieur. Bull. Soc. Géol. France. (7)
v. 27, pp. 843-849, 3 text-figs.

Donze, P., Porthault, B., Thomel, G., and Villoutreys, O. de.
1970. Le Sénonien inférieur de Puget-Théniers
(Alpes-Maritimes) et sa microfaune. Geobios v. 3,
pp. 41-106, pls. 8-13, text-figs. 1-4.

Grundel, J. 1968. Neue Ostracoden aus der Salzbergmergel-
Fazies (Santon) im westlichen Teil der Deutschen
Demokratischen Republik. Geologie, 17, pp. 947-963,
pls. 1-3

Hazel, J. E. 1969. Cythereis eaglefordensis Alexander 1929 -
a guide fossil for deposits of latest Cenomanian age
in the western interior and Gulf Coast regions of
the United States. Geol. Surv. Res. 1969. D155-
D158, 2 figs.

Hazel, J. E., and Paulson, O.L., Jr. 1964. Some new ostracode
species from the Austinian and Tayloran (Coniacian
and Campanian) rocks of the East Texas Embayment.
J. Paleont. v. 38, pp. 1047-1064, pls. 157-159, 2
text-figs.

268

Herrig, E. 1965. Cythereis reticulata varia ssp.n., eine neue Ostracoden-Unterart aus der Rügener Schreibkreide (Unter-Maastricht). Ber. geol. Ges. DDR. v. 10, pp. 403-419, pls. 1-4, 5 text-figs.

Herrig, E. 1967. Möglichkeiten einer Feinstratigraphie der höheren Oberkreide in Nordostdeutschland mit Hilfe von Ostracoden. Ber. deutsch. Ges. Wiss. A. Geol. Palaont. v. 12, pp. 557-574, pls. 1,2, 8 text-figs.

Herrig, E. 1968. Zur Gattung Saida Hornibrook (Ostracoda, Crustacea) in der Oberkreide. Geologie, 17, pp. 964-981, pl. 1, 7 text-figs.

Israelsky, M.C. 1929. Upper Cretaceous Ostracoda of Arkansas. Ark. Geol. Surv. Bull. No. 2, pp. 1-29, pls. 1-4, 1 table.

Kemper, E. 1971a. Die Paläoökologische Verbreitung der Ostra-koden im Obervalanginium und Unterhauterivium des Niedersachsischen Beckens (NW-Deutschland). Bull. Centre Rech. Pau -SNPA, v. 5 suppl., pp. 631-649, pls. 1,2, 2 text-figs.

Kemper, E. 1971b. Batavocythere und Saxocythere, zwei neue Protocytherinae-Gattungen (Ostracoda der Unterkreide). Senck. Leth. v. 52, pp. 385-431, pls. 1-8, 1 text-fig.

Kilenyi, T. I., and Allen, N.W. 1968. Marine brackish bands and their microfauna from the lower part of the Weald Clay of Surrey and Sussex. Palaeontology, v.11, pp. 141-162, pls. 29, 30, 9 text-figs.

Liebau, A. 1971. Die Ableitung der Palökologischen systematik Oberkretazischen Lagune. Bull. Centre Rech. Pau - SNPA. v.5 suppl., pp. 577-599, 2 pls, 2 tables, 1 text-fig.

Moullade, M. 1963. Principaux représentants du genre Protocy-there (Ostracodes) dans le Crétacé inférieur du Sud-Est de la France. Rev. Micropal., v. 6, pp. 102-108, pls. 1,2, 2 tables.

Neale, J. W. 1973. Ostracoda as means of correlation in the Boreal Lower Cretaceous with special reference to the British marine Ostracoda. Geol. J. Spec. Issue, Liverpool, v. 5, pp. 169-184, 2 text-figs.

Oertli, H. J. 1968. Les Ostracodes de l'Aptien-Albien d'Apt.
        Revue Inst. fr. Pétrole 13, pp. 1499-1537, pls. 1-9,
        1 table, 3 text-figs.

Ohmert, W. 1973. Ostracoden aus dem Santon der Gehrdener Berge.
        Ber. Naturhist. Ges. v. 117, pp. 163-194, pls. 15-18.

Pokorný, V. 1964. The Phylogenetic lines of Cythereis marssoni
        Bonnema. 1941 (Ostracoda, Crustacea) in the Upper
        Cretaceous of Bohemia, Czechoslovakia. Acta Univ.
        Carol. Geol. No. 3, pp. 255-274, pls. 1,2, 16 text-
        figs.

Schmidt, R. A. M. 1948. Ostracoda from the Upper Cretaceous
        and Lower Eocene of Maryland, Delaware, and Virginia.
        J. Paleont. v. 22, pp. 389-431, pls. 61-64.

Swain, F. M. 1948. Ostracoda from the Hammond Well. Maryland
        Board Nat. Res., Cretaceous and Tertiary Subsurface
        Geology, pp. 187-213, pls. 12-14.

Swain, F. M. 1952, Ostracoda from wells in North Carolina.
        Part 2. Mesozoic Ostracoda. U.S.G.S. Prof. Pap.
        243-B.

Swain, F. M., and Brown, P. M. 1972. Lower Cretaceous Jurassic
        (?), and Triassic Ostracoda from the Atlantic
        Coastal Region. Prof. Pap. U.S. geol. Surv. 795,
        pp. 1-55, 10 pls., 32 figs.

Vanderpool, H. C. 1928. Fossils from the Trinity Group (Lower
        Comanchean), J. Paleont., v. 2, pp. 95-107, pls.
        12-14.

Vanderpool, H. C. 1933. Upper Trinity Microfossils from south-
        ern Oklahoma. J. Paleont., v. 7, pp. 406-411, pl.
        49.

Discussion

Dr. M. C. Keen: Is it possible to relate the _Cypridea_ zones of the North Atlantic with those of West Africa and South America?

Neale: The difference in the faunas of the two areas is so great that any direct correlation of the zones is impossible. Attempts to relate the zones in the respective areas to the standard stages are speculative and can only suggest very tentative broad equivalences.

Dr. B. K. Holdsworth: Can one recognise bathyal assemblages in the Cretaceous yet?

Neale: Depth in itself, as distinct from its affect on temperature, light intensity and to some extent substrate, is unimportant as an ecological factor and very difficult to assess accurately. In consequence it depends how accurately you want to define bathyal. Certainly it is possible to recognise inshore and offshore faunas as seen in the work of Kemper in the North German Basin and Donze in Vocontian Trough. Unfortunately, information obtained by oil exploration in the European shelf areas has remained confidential but I understand that the shelf faunas pass outwards into faunas that contain a high proportion of _Bairdia_ which may indicate an approach to a tru bathyal environment.

Keen: Do you think ostracods will be of any help in determining the depth of deposition of the Chalk (including chalk marls, glauconitic layers chalk rocks)? I ask this because Hancock has recently thrown doubt on some of the depth zonations determined by forams (Proc. Geol. Assoc., 1975)

Neale: Bearing in mind my previous remarks, I think that eventually it will be possible to give some general indication of depths in relative terms such as inner neritic, outer neritic, bathyal, etc.

CRETACEOUS OSTRACODA - SOUTH ATLANTIC

Alwine Bertels

Facultad de Ciencias Exactas y Naturales,
Universidad de Buenos Aires, República Argentina
Consejo Nacional de Investigaciones Científicas y Técnicas,
República Argentina

Abstract

En el presente trabajo se compendian los aportes científicos efectuados por diversos autores relacionados con las asociaciones de ostrácodos que prevalecieron durante el Cretácico en áreas del Atlántico Sur.

Se analizan las cuencas sedimentarias en donde tuvieron lugar depositaciones durante el Cretácico.

Hacia fines del Jurásico y comienzos del Cretácico tuvieron lugar los primeros movimientos de origen tectónico que produjeron diversos tipos de estructuras en los actuales márgenes de los continentes sudamericano y Africano y que controlaron las áreas de sedimentación.

Durante el Neocomiano, con exclusión de las regiones australes de ambos continentes, la sedimentación es netamente continental. Las faunas de ostrácodos son en la mayoría de los casos comunes a ambas áreas.

Durante el Aptiano-Albiano se producen las primeras ingresiones marinas que originan depósitos de tipo evaporítico.

A partir de este intervalo predomina la sedimentación marina que se registra en diversas cuencas aunque no simultáneamente en todas ellas.

El ciclo Cretácico culmina con la transgresión del Maastrichtiano que ocupa vastas áreas en ambos continentes. Las asociaciones que se conocen hasta el presente, si bien presentan similitudes, no exhiben especies idénticas, lo cual lleva a la conclusión que durante el Maastrichtiano el Atlántico Sur estaba netamente delineado.

Summary

In the present work the scientific contributions made by several authors are synthesized relative to the ostracodal assemblages which prevailed during the Cretaceous in South Atlantic areas.

At the end of the Jurassic and early Cretaceous the first tectonic events produced several structures which controlled the sedimentary areas.

During the Neocomian, excluding the southernmost regions in both the South American and African continents, non-marine sedimentation prevailed.

During Aptian-Albian times the first marine transgression took place which formed evaporitic beds.

From this time forward, marine sedimentation was predominant in the Cretaceous although it was not contemporaneous in all basins.

The Cretaceous cycle culminates with the Maastrichtian transgression which occupied vast areas in both the South American and African continents. The known ostracodal assemblages show similarities which would indicate that during the Maastrichtian the Atlantic Ocean was definitively delineated.

Introduction

The purpose of the present work is to synthesize the most diagnostic ostracodal assemblages which prevailed during Cretaceous time in the present African and South American continents.

The sedimentary basins in which Cretaceous deposits occur are analyzed in both continents; the faunal assemblages are listed following a latitudinal order according to the sedimentary basins from which they were recorded.

Acknowledgments

The author wishes to express gratitude to the Argentina National Council for Scientific and Technical Research, for the economical aid in the realization of the present work, and who generously provided the use of the Jeolco JSM-U3 scanning electron microscope with which some of the micrographs were taken.

The writer is also grateful to the Buenos Aires University, Department of Geology for providing facilities to achieve the completion of this work.

The author is indebted to Dr. Karl Krömmelbein, Kiel University, who kindly supplied the ostracode material and to Dr. John Neale, Hull University who generously photographed the material and sent the stereographic paired photographs of the non-marine ostracodes.

Particular thanks are also expressed to Dr. Carlos A. Rinaldi and to Mr. Patricio Ganduglia, who generously helped in the construction of the stratigraphic charts and range charts which illustrate this paper.

The Lower Cretaceous non-marine ostracodes are deposited at the collection of Dr. Karl Krömmelbein at the University of Kiel.

The illustrated marine material is deposited in the Laboratory of Micropaleontology, Facultad de Ciencias Exactas y Naturales, Buenos Aires, University.

South America

Lower Cretaceous

In the Sergipe/Alagoas Basin the Baizo Sao Francisco and the Sergipe groups are distinguished in which several subgroups and formations are recognized by Schaller (1969); some formations are only recognized in Alagoas or in Sergipe, although some formations extends over both areas Text Fig. 1.

Within the Coruripe Subgroup several lower Cretaceous sequences are recognized; the Coruripe Subgroup comprises the following Formations: Barra de Itiuba, Rio Pitanga and its equivalents in Alagoas: the Penedo and Morro Chavez Formations, Coqueiro Seco, subdivided in several members, and Ponta Verde.

The Coruripe Subgroup is correlated with the Bahia Supergroup of the Reconcavo/Tucano basin.

The Barra de Itiuba Formation is composed of olive and black claystones with fine intercalations of sandstones, siltstones and calcareous rocks.

In the Barra de Itiuba Formation Schaller (1969) recognized the zones of:

Cypridea (Morininoides) candeiensis Krommelbein   = (003)
Paracypridea brasiliensis Krommelbein   = (004)
Paracypridea obovata obovata (Swain)   = (005)
Cypridea (Morinina?) bibullata Wicher   = (006)
Petrobrasia marfinensis Krommelbein and
Cypridea (Sebastianites) fida Krommelbein   = (007)

Schaller (1969) correlates the Barra de Itiuba Formation with the Candeias, Marfin and Pojuca Formations of the Reconcavo/Tucano Basin.

The Penedo Formation is composed of poorly sorted sandstones with clayey and silty intercalations, mainly white coloured; it contains ostracodes of the zones (007), (006) and

(005); it is correlated by Schaller (op. cit.) with the Marfin, Pojuco and Sao Sebastiao Formations of the Reconcavo/Tucano Basin.

The Rio Pitanga Formation is conglomeratic with clayey and silty intercalations; it comprises the ostracode zones (007), (006) and (005) of Schaller (1969).

The Morro de Chaves Formation is composed of light gray calcareous strata and coquinoid marls with brown claystones intercalated. It is correlated with the Zones (007) and (008) and, in the Piacabucu area with the zones (009) and (010/011).

The Coqueiro Seco Formation is mainly composed of alternating of sandstones, claystones, and siltstones; it contains the (008) zone of Schaller (1969). The overlying Ponta Verde Formation did not yield ostracodes.

In the Sergipe/Alagoas Basin the non-marine sequences were succeeded by an Atlantic marine transgression which followed the principal tectonic structures of the basin; these sequences began with the transitional Muribeca Formation which contains evaporitic beds. The brackish-marine facies of this formation are of Aptian age and younger persisting the marine facies in a consecutive stratigraphic sequence into Lower Eocene times.

The Muribeca Formation overlies the Sao Francisco Group.

In the entire section brackish water-marine ostracodes of the genus Cytheridea? and those recorded by Schaller (1969) of Zones (010/011) occur. For the upper Member Schaller (op. cit.) inferred an Aptian age.

The transitional Muribeca Formation is overlaid by the mainly marine Sergipe Group. The Sergipe group is subdivided into several formations in which some marine ostracodes occur.

The Riachuelo Formation is composed of limestones, claystones and siltstones with benthonic and planktonic foraminifera. Krömmelbein and Wenger (1966) assigned an Upper Aptian to Albian age to this formation based on ostracoda. Krömmelbein (196) cited from the Riachuelo Formation Sergipella transatlantica Krömmelbein and Aracajuia benderi Krommelbein; both species were also found in Gabon, Africa.

Text Fig. 1. Stratigraphic units of South American Sedimentary Basins.

The Sergipe/Alagoas sequences might be correlated with the Reconcavo/Tucano Basin strata; there exists close correlation between Zones (001) to (007) of the Sergipe/Alagoas Basin with the Zones R-10 to R.2 of the Reconcavo/Tucano Basin although some species are lacking in both areas (Schaller, 1969).

In the tectonic Reconcavo/Tucano Basin the basement is composed of igneous, metamorphic and sedimentary rocks of Precambrian and Paleozoic age.

Viana (1971) subdivided the Bahía Supergroup into the Brotas, Santo Amaro, Ilhas and Massacará groups (Text Fig. 1).

The Dom Joao, Rio da Serra, Aratu, Buracica, Jiquiá and Alagoas Stages constitute the Reconcavo Series which, excluding the Alagoas Stage, correspond to the rock stratigraphic Bahía Supergroup. The Jurassic-Cretaceous boundary is not clearly defined in this basin.

Viana et al. (1971) recognized several ostracode zones and subzones which characterize the Reconcavo Series in this basin.

Many authors have worked out the area and contributed with various data about the ostracode fauna and its utility as zonal markers such as Swain (1946), Wicher (1959), Pinto and Sanguinetti (1962), Krommelbein (1962), Muller (1966), Krommelbein and Wenger (1966), Braun (1966), Viana (1967), Schaller (1969) and Viana et al. (1971). Other organic groups, such as pollen, were also intensively studied (Weber 1963, 1964).

Useful ostracode guide fossils have been recorded in this basin which, also are present in neighboring basins and in Africa.

The Reconcavo area of the basin is located near the present coast which presents fault structures whereas the Tucano area extends inland; consequently more facies changes are found in the Reconcavo basin.

The Upper Jurassic Brotas Group overlies the Precambrian crystalline complex and Paleozoic rocks. The Sergi Formation of the Brotas Group shows a gradational contact with the underlying Alianca Formation as well as with the overlying Itaparica

Formation of the Santo Amaro Group of a supposed Lower Cretaceous age. The age of the Itaparica Formation was dated by
means of ostracodes.

The Santo Amaro Group comprises the Itaparica and Candeias
formations.

The Itaparica Formation is composed of varied colored claystones and fossiliferous olive grey siltstones. The formation
occurs in Reconcavo and southern Tucano. Ostracodes of the
Cypridea kegeli Wicher subzone are the most significant.

The Candeias Formation is composed of claystones, siltstones intercalated by limestones, dolomites and thick beds of
sandstones; the member denominations corresponds to these facial changes. The most important ostracodes are those of the
zones of:

Theriosynoecum varietuberatum Grekoff and Krommelbein
Cypridea (Morininoides) candeiensis Krommelbein and
Paracypridea brasiliensis Krommelbein

The Candeias Formation is correlated by Viana et al. (1971)
with the lower part of the Barra de Itiuba Formation of the
Sergipe/Alagoas Basin.

The Salvador Formation is composed of fanglomerates of
crystalline, calcareous, metasandstones and metasiltstones. It
is laterally interdigitated with the Santo Amaro, Ilhas and
Massacará Groups and possesses the same ostracode content as
the formations with which it is interfingered. Viana et al.
(op. cit.) correlate the Salvador Formation with the Rio Pitanga Formation of the Sergipe/Alagoas Basin.

The Ilhas Group comprises the Marfim and Pojuca Formations.
The Marfim Formation is localized in the subsurface; it is composed of light grey to greenish fine grained to siltic sandstones. In the Reconcavo area the Marfim Formation is laterally interdigitated with the Salvador Formation. The contact
with the overlying Pojuca Formation is gradational.

The most important ostracodes are: Paracypri-
dea brasiliensis Krommelbein, Cypridea dromedarius Krommelbein,
Cypridea salvadoriensis Krommelbein and Candona? gregaria
Krommelbein.

The Marfim Formation is correlated by Viana et al. (1971)
with the middle part of the Barra de Itiuba Formation of the

Sergipe/ Alagoas Basin.

The main lithological features of the Pojuca Formation are the varied composition greenish gray sandstones, claystones, siltstones and brown oolitic limestones which sometimes grade into arenaceous limestones. The formation presents good stratification, with ripple marks. The contact with the overlying Sao Sebastiao Formation is gradational.

The most significant ostracodes species belong to the biozones (Viana et al., 1970):

> Paracypridea obovata obovata Swain and
>
> Cypridea (Morinina?) bibullata bibullata Wicher

The Massacara Group comprises the formerly designated Ilhas Formation and the Sao Sebastiao Formation. The Sao Sebastiao Formation is composed of reddish-yellow fine to coarse sandstones, friable, with intercalations of silty claystones. The upper part is in some areas conglomeratic. The significant ostracodes are: Petrobrasia marfinensis Krommelbein, Coriacina coriacea Krommelbein, Cypridea (Sebastianites?) sostensis sostensis Krommelbein and Cypridea (Sebastianites) fida fida Krommelbein. Viana et al. (1971) correlate the Sao Sebastiao Formation with part of the Penedo Formation in the Sergipe/Alagoas Basin.

The Marizal Formation is composed of coarse sandstones and conglomerates intercalated by siltstones, claystones and limestones. This rock-stratigraphic unit is also recognized in the Jatobá and Mirandiba basins. Viana et al., (1971) correlate this formation with the Carmopolis Member of the Muribeca Formation of the Sergipe/Alagoas Basin.

Viana et al., (1971) based on the ostracode content of the several non-marine stages of the Reconcavo Series proposed the following Cretaceous ostracode zones, which also are subdivided in several subzones in ascending order of age (Text. Fig. 1):

	RT-002 - Theriosynoecum varietuberatum
Rio da Serra Stage:	RT-003 - Cypridea (Morininoides) candeiensis
	RT-004 - Paracypridea brasiliensis
	RT-005 - Paracypridea obovata obovata
Aratu Stage:	RT-006 - Cypridea (Morinina?) bibullata bibullata
Buracica Stage:	RT-007 - Coriacina coriacea

RT-008 – <u>Cypridea</u> (<u>Sebastianites</u>) <u>fida</u>
<u>minor</u>

Jiquiá Stage: RT-009 – <u>Petrobrasia</u> <u>diversicostata</u>

   Moura (1972) described additional new species from the
Reconcavo/Tucano Basin.

   Some of the new species occur in Africa, such as <u>Cypridea</u>
cf. <u>C</u>. <u>primaria</u> Grekoff and Krömmelbein in the transitional
Lower Cocobeach Series; and in other areas of Brazil such as:
<u>Cypridea</u> (<u>Sebastianites</u>) <u>fida</u> <u>minor</u>, <u>Cypridea</u> (<u>Pseudocypridina</u>)
<u>fabeolata</u> and <u>Cypridea</u> <u>riojoanensis</u> which occur in the Sergipe/
Alagoas Basin.

   In the interior region of Brazil a northwestern extension
of the <u>Reconcavo/Tucano</u> Basin, the so-called Jatobá Basin, the
<u>Araripe</u> <u>Basin</u>, conspicuous in its gypsum deposits, and the
"Mirandiba" Basin were studied by Braun (1966).

   In the <u>Jatobá</u> <u>Basin</u> the Alianca (Upper Jurassic?-Lower
Cretaceous), Sergi, Candeias, Ilhas (=Marfim and Pojuca forma-
tions of the Reconcavo/Tucano Basin), Sao Sebastiao, Marizal,
Santana and Exu formations were recognized.

   The Candeias Formation overlies unconformably the Sergi
Formation; for the Candeias Formation the zones of
   <u>Cypridea</u> <u>opifera</u> Krömmelbein
   <u>Cypridea</u> (<u>Morininoides</u>) <u>candeiensis</u> Krömmelbein
   <u>Cypridea</u> <u>ambigua</u> Krömmelbein were proposed by Braun (<u>op</u>.
<u>cit</u>.).

   The Ilhas Formation contains the zones of:
   <u>Paracypridea</u> <u>obovata</u> <u>obovata</u> Swain
   <u>Cypridea</u> (<u>Morinina</u>) <u>bilbullata</u>
   <u>Paracypridea</u> <u>brasiliensis</u> Krömmelbein

   The Aptian-Albian? Sao Sebastiao Formation in their lower
section yields the zones of:
   <u>Cypridea</u> (<u>Sebastianites</u>) spp.
   <u>Coriacina</u> <u>coriacea</u> Krommelbein and
   <u>Paracypridea</u> <u>quadrirugosa</u>

   The upper Sao Sebastiao and Marizal Formation are non-
fossiliferous whereas the Santana Formation yielded species of
the genera <u>Candonopsis</u>, <u>Paraschuleridea</u>, <u>Heterocypris</u> and <u>Bi</u>-
<u>sulcocypris</u>.

In the _Araripe Basin_ or Chapada the Araripe region (Aptian -Albian) Beurlen (1963) presented a stratigraphic sequence for this area; the section was subdivided into the Crato, Coriri, Missao Velha, Santana and Exu formations.

Braun (op. cit.) compares the partly evaporitic Santana Formation with the marine Riachuelo Formation of the Sergipe/ Alagoas Basin; faunally it seems similar to the Codo Formation in the northern located Maranhao Basin.

The ostracode assemblage is composed of _Candonopsis_ spp., _Paraschuleridea_ spp., _Heterocypris_ sp., and _Bisulcocypris_ sp.

In addition Bate (1971) recorded from the Aptian-Albian Santana Formation of Serra do Araripe, Ceará, northern Brasil, phosphatized ostracods; they belong to the freshwater Subfamily Cypridinae, Family Cyprididae.

In northern Brazil Braun (1966) mentioned, from the _Maran-hao Basin_, the Codó Formation of Aptian-Albian age; it is cor- relative with the marine Riachuelo Formation of the Sergipe/ Alagoas Basin and with the Santana Formation of the Araripe re- gion. Although Braun noted the ostracode similarities between the Santana Formation and the Codó Formation, the present writ- er has no knowledge of the ostracode content of the Codó For- mation.

In the "_Rio de Peixe Basin_," Paraiba, Braun (1966) found _Darwinula_ sp. and _Cypridea_ _vulgaris_ Krommelbein; Braun assigned the strata to the Berriasian and correlated with the Lower Ilhas Group (=Marfim Formation) of the Reconcavo/Tucano Basin.

From the western Argentina _Neuquén Basin_ Musacchio (1971) recorded from La Amarga Group (Pichi Picún Leufú Formation) non-marine ostracodes. In the Cerro China Muerta area the non-marine La Amarga Group overlies concordantly and gradually the marine Agrio Formation; the Pichi Picún Leufú Formation is mainly composed of marls.

Although La Amarga Group was considered of Aptian-Albian age (Digregorio, 1972), Musacchio (1971) based on charophytes assigned the strata of the basal La Amarga Group to the Barre- mian.

In the same basin Musacchio (1975) describes from the Ray- oso Formation the ostracode _Rayosoana_ _quilimalensis_ Musacchio. Although the Rayoso Formation has been considered to be Cenomanian (Digregorio, 1972), Musacchio (1975), based on affinities of the new genus and its species, assigned the levels in which Rayosoana quilimalensis was found to the upper part of the Lower Cretaceous.

In the western part of the San Jorge Basin Musachio and Chebli (1975) recorded from the continental Chubut Group "Wealdian" type ostracode assemblage; it comes from the Chubut Group (Gorro Frigio Formation). Musacchio (1975) described these species. The Paso de Indios locality represents the southernmost locality in which "Wealdian type" ostracodes have been found.

Musacchio (op. cit.) found affinities with the Northern Hemisphere association, principally from those described from the Rocallosa Mountains; Musacchio (op. cit.) assigned to the assemblage an Aptian age.

In Argentina marine extra-Andean Lower Cretaceous sequences are mentioned from the Neuquén and Austral basins. In the Neuquén Basin the Agrio Formation was dated as Hauterivian-Barremian based on contained ammonites. The Agrio Formation contains marine ostracodes which are being studied. In the Austral Basin marine sedimentation took place since Valanginian times. The first marine ostracode, Novocythere santacruceana was recorded by Malumian, Masiuk and Rossi de García (1972) from subsurface strata of Aptian-Albian age.

Upper Cretaceous

In South America although Upper Cretaceous sediments are recorded in several basins, data about the ostracode content are scarce.

Upper Cretaceous sequences are found in the Brazilian Foz de Amazonas, Barreirinhas, Recife/Joao Pessoa, Sergipe/Alagoas, Araripe and in the Argentina Neuquén, Colorado and Austral basins.

In the Amazonas Basin marine sedimentation is represented by the Upper Cretaceous to Paleocene Amapá Formation (Shaller et al., 1971); this formation is mainly made up of carbonates. Petri (1954) recorded rare Cretaceous Foraminifera, but up to now ostracodes have not been studied.

The Barreirinhas Basin contains continuous sedimentary sequences which range in age from Aptian to the Tertiary (Lima,

1973); as yet ostracodes have not been recorded.

In the Araripe Basin the Exu Formation does not contain fossils (Braun, 1966).

In the Sergipe/Alagoas Basin in discordance over the marine Riachuelo Formation sediments of Middle Albian? to Lower Santonian are represented by the Cotinguiba Formation.

The Sapucari Member, made up of carbonates, yielded the ostracode Brachycythere (Brachycythere) sapucariensis Krömmelbein, of Turonian age. Krömmelbein (1964) correlates the South American Sapucari Formation with the African Sibang Formation and the Milango Formation, of probably Coniacian age, based on the presence of Brachycythere (Brachycythere) sapucariensis in these formations.

The Cotinguiba Formation is overlain by the marine Piacabucu Formation, which range in age from the Campanian to the Lower Eocene. Although planktonic Foraminifera were studied in detail (Schaller, 1969) no ostracodes have been recorded from this formation.

For the Calumbi Member Krömmelbein and Wenger (1966) assigned an Upper Campanian-Maastrichtian age. For the Marituba Member (Schaller 1969) assigned an age which range from Campanian to Lower Eocene.

In the Recife/Joao Pessoa Basin the Upper Cretaceous is represented by the fluvial Beberide and marine Gramame formations of the Paraiba Group; the former is unfossiliferous.

The marine Gramame Formation represents a transgressive calcareous facies (Mabesone, Tinoco and Couitinho, 1968). The Gramame Formation yielded microfossils such as foraminifera and ostracodes. The Cretaceous foraminiferal assemblages were studied by Tinoco (1967); the only known ostracodes are species of the genre Cytherella, Cytheropteron and Cythereis (Tinoco, 1967). The ostracode assemblage is being studied.

In Argentina Upper Cretaceous non-marine sediments extends over a large area of the Neuquén Basin, and form the basal beds of the Jaguel Formation and equivalents such as the Lower member of the Huantrai-co Formation (Bertels, 1972). These have yielded fresh-water ostracodes such as Candona? huantraicoensis Bertels, Ilyocypris triebeli Bertels and Wolburgia? neocretacea Bertels.

After this episode the sea encroached upon the Colorado and Neuquén basins. The fresh water and polyhaline deposits belong to the Lower Jagüelian Substage whereas the nearly marine to the Upper Jagüelian Substage.

From the Jagüel Formation Bertels (1974, 1975) described from the Neuquén Basin Upper Cretaceous (Lower Maastrichtian?) and Middle Maastrichtian ostracode assemblages. The assigned age was based upon planktonic Foraminifera (Bertels, 1972).

The described species are only known from Argentina, but they show close similarities with western African species, especially with those belonging to Veenia (Nigeria) and Anticythereis described by Reyment (1960) and Apostolescu (1961, 1963).

Africa

Neocomian

Neocomian non-marine deposits are found in western Africa in the Gabon, Congo and Angola sedimentary basins and in South Africa.

The continental sediments are of lagoonal, lacustrine and estuarine origin; these sequences reach large thickness; varying from 3000 to 6000 meters. They mostly belong to the Cocobeach Series of Gabon and were denominated by Grekoff and Krömmelbein (1967) as the West African Wealdian Series.

In the Gabon Basin the Cocobeach System (de Klasz and Micholet, 1970; Le Calvez, de Klasz and Brun, 1974) is of Early Cretaceous age comprising the Neocomian (Berrisian to Barremian) and Aptian Stages; their strata extend to the south into the Congo Basin and Angola.

The Cocobeach System overlies the Precambrian basement and crops out in a northwest-southeast trend.

The Lower Cocobeach Subsystem is separated from the Upper Cocobeach Subsystem by an inportant discordance connected with major tectonic events that marked a phase of the separation of Africa and South America.

The Lower Cocobeach Subsystem comprises several stratigraphic units (Text Fig. 2). These units yielded fresh-water and brackish-water ostracodes.

The ostracodes assemblages from the Lower Cocobeach Subsystem were studied by Krömmelbein (1966), Grekoff and Krommelbein (1967), Grosdidier (1967), Viana (1966), and de Klasz and Micholet (1970).

De Klasz and Micholet (1970) proposed the Series of Kango, Remboue, N'Toum, Gamba and Salifere D'Ezanga for the nomenclature used previously by Grekoff and Krömmelbein (1967).

Krömmelbein (1966) compares the Kélélé (lower Kango Series) beds ostracode association with that of Itaparica and basal Candeias of the Reconcavo/Tucano basin; in addition Viana (1966) found common components in the assemblages of the overlying Bokoué and Bikoumé beds (upper Kango Series).

De Klasz and Micholet (1970) note that, excluding the new species, ostracodes of the Bikoumé Member are common to those of the Candeias Formation in Brazil.

Krömmelbein (1966), Viana (1966) and Grekoff and Krommelbein (1967) correlated the Fourou Plage Member (lower Remboue Series) with the lower Ilhas. The faunal assemblages of the Lobé Member and Bifoune Member (middle Remboue Series) are similar; Krömmelbein (1966) correlated both members, as well as the overlying Moundunga Member, with the lower part of the Upper Ilhas Group of the Reconcavo/Tucano Basin in Brazil, whereas Grekoff and Krömmelbein (1967) compared the Moundounga and the underlying Bifoune members with the upper part of the upper Ilhas Group. In the Moundunga Member, (upper Remboue Series), Grekoff and Krömmelbein (1967) in addition found a different assemblage from that of the underlying member and the appearance of Cypridea nanorostrata Grekoff and Krömmelbein and the genus Petrobrasia which announce the appearance of the overlying Benguié assemblage.

The ostracode assemblage of the Benguié and N'Gwanzé Members (N'Toum Series) contain several species of Petrobrasia and Cypridea (Sebastianites).

The assemblages recorded by Krömmelbein (1966) belong to part of the Moundunga Member and the Benguié Member; the species cited by Viana (1966) and Grosdidier (1967) came from the lower Part of the Benguié Member.

The N'Gwanzé Member fauna does not show great differences from that of the underlying Benguié Member.

Text Fig. 2. Stratigraphic units of African sedimentary basins.

Grekoff and Krömmelbein (1967) recorded species common to both the Brazil and Gabon areas; these authors made precise correlations with the Brazilian Wealdian type sediments, especially with those of the Reconcavo/Tucano and Sergipe/Alagoas Basins.

In the Congo Basin Grekoff (1957) studied outcrops along the Lualaba River and borehole samples from Samba; both areas are located in the northern part of the Congo Basin. The sequences yielded rich faunal assemblages which were considered to be upper Jurassic and lower Cretaceous age. The same author (Grekoff, 1960) analyzed additionally the upper Jurassic-lower Cretaceous ostracode associations of samples from several regions of the Congo Basin such as: N'Deke, Elisabetha, Ubangi, Yakoko and Dekese. Grekoff (1957) made an exhaustive analysis of the ostracode assemblage of the Congo Basin Lualaba Series. The Lualaba Series is subdivided into the Stanleyville Stage and the Loia Stage. The Lualaba Series is composed of several beds numbered from 1 to 16 (base to top). The beds 1 to 14 constitute the Stanleyville Stage whereas the bed 15 is assigned to the Loia Stage. The Horizon 16 or Bokungu Beds were considered by Grekoff (1957) as an independent entity; it is composed of red claystones and sandstones. The assigned age was late Jurassic for the Stanleyville "Stage" and lower Cretaceous for the Loia "Stage" of the Lualaba Series. The facies which yielded ostracodes are of fresh-water or brackish-water environment and have been described by Marlière (1948, 1950) and Grekoff (1957, 1960).

The Lualaba Series assemblage was found to contain zonal markers.

The Samba (Congo) borehole association (Grekoff, 1957) is composed of 27 species of fresh- and brackish?-water ostracodes

Grekoff (1957) found several species common to both the Lualaba Series and the Samba Series (1 to 5), and correlated the horizons 3 to 14 of Lualaba with the Samba Series 3 (Stanleyville Stage) and horizons 15 and 15' of Lualaba with Series 3 and 4 of Samba (Loia Stage). In addition Grekoff (1957) found six species common to Gabon. These are: Paracypridea obovata, Cypridea diminuta, Paracypria longaensis liloensis,

Bradycypris rotunda, Metacypris sp. 390, Darwinula leguminella and Darwinula oblonga.

The boreholes sequences at N'Deke, Elisabetha, Ubangi, Yakoko and Dekese were considered by Grekoff (1960) correlative with horizons 15 and 16 of the Lualaba Series.

Based on the above mentioned ostracode content Grekoff (1957) correlated levels 2 to 15' of the Lualaba Series and series 3 to 5 from the borehole at Samba with the Middle Coco-beach subsystem. The Cocobeach System is nowadays almost always included in the Lower Cretaceous; therefore the sequences of the Lualaba Series would correspond to the Gabonese Bokoué Member of the Mayanga Formation (Kango Series) up to the Ben-guié Member of the Bingone Formation (N'Toum Series). The Stanleyville Stage could be approximately equivalent to the Kango Series whereas the Loia Stage might be approximately equivalent to the Remboue Series, based on Grekoff's data (1957).

Pinto and Sanguinetti (1961) based on the presence of Metacypris consobrina Jones, and vertebrates, believed that horizon 15' of the Lualaba Series (upper part of the Loia Stage) and series 3 of Samba are of Albian age.

In the Congo Basin Grekoff (1960) distinguished two Weal-dian (Cretaceous) assemblages: a lower Wealdian (=A) "complex" and an upper Wealdian (=B) "complex." The boundary between A and B was not clearly distinguishable and some species are found in both associations.

Grosdidier (1967) studied ostracodes assemblages from Ga-bon and Congo. In reference to the Congo assemblage Grosdidier pointed out that they do not show close similarities with the Brazilian fauna; on the other hand Grosdidier (op. cit.) be-lieved that they are stratigraphically higher than those of the middle Sao Sebastiao Formation of the Brazilian Reconcavo/Tu-cano Basin.

The new species described by Grosdidier (1967) from the Congo Basin are Cypridea loango Grosdidier, Cypridea bapounou Grosdidier, Cypridea hollensis Grosdidier, Damonella? tink-oussouensis Grosdidier, "Reconcavona" batike Grosdidier, Rhinocypris kroemmelbeini Grosdidier and Orthonotacythere mvili

Grosdidier. This association is equivalent to the N'Gwanzé Member of the Bingone Formation of Gabon.

In the Angola Cuanza Basin "Wealdian" type sediments were deposited (Cuvo Formation) (Brognon and Verrier, 1966), but the sequence does not contain non-marine ostracodes as far as the writer knows.

In South Africa Dingle (1969, 1971) recorded for the first time Lower Cretaceous marine ostracodes from the Uitenhage region and from the continental margin Agulhas Bank.

The Neocomian marine strata of the Uitenhage Group are confined to the coastal areas in the east and southeast of the country.

The classic subdivision of the Cretaceous section of the Uitenhage Basin (Text Fig. 2) is composed of the Enon Conglomerate, followed by Variegated Marls and Wood Beds (Wealdian facies but without ostracodes) and then by the Sundays River Formation.

This Cretaceous section overlies the Cape Sequence and is covered by Tertiary strata (Dingle, 1969).

Dingle (1969) found the age of the Sundays River Formation to be at least partly Upper Valanginian. From the continental margin Agulhas Bank (Dingle 1971) recorded marine ostracodes. The samples studied by Dingle were dated by means of Foraminifera and ammonite fragments as Barremian?.

Aptian

In western Africa, Aptian sequences are recognized in Gambia, Nigeria, Gabon, Congo, Angola and South Africa. Apostolescu (1963) studied subsurface samples from Gambia and Senegal. In both regions the samples yielded marine ostracodes.

In Gambia from the borehole at Sara-Kunda 1, Apostolescu (1963) recorded ostracodes from several levels which range in age from Aptian to Maastrichtian.

Based on the ostracodal content of the Sara-Kunda 1 sequence Apostolescu proposed five faunizones.

Faunizone 1 comprises the Aptian-Albian interval; the Aptian Stage was dated by means on Choffatella decipiens Schl. It only yielded the marine ostracode genus Schuleridea.

In the Gabon Basin the fluvio-lacustrine Gamba Series is
separated from the Lower Cocobeach Subsystem by a discordance
which marks an important phase in the separation of Africa and
South America (de Klasz and Micholet, 1970); the first marine
influences are manifested in the Gamba/Ezanga boundary evi-
denced by laminated claystones, with salt crystals, of the
Schistes de Cocobeach Member which, otherwise, constitutes the
base of the Ezanga Series.

In the Gamba Series ostracodes are rare; they were recorded by Krömmelbein (1966) and by Grosdidier (1966).

The Ezanga Series is transitional between the fluvio-
lacustrine beds and those dominantly marine. This series has
not yielded ostracodes.

In the Congo Basin the Exanga? Stage, which is laterally
equivalent to the salt-bearing strata in Gabon did not yield
ostracodes as far as the writer knows.

In the Angola Cuanza Basin during Aptian times, subsidence
of the central part of the basin controlled the cyclical deposition of carbonate-evaporitic beds (Brognon and Verrier, 1966).
No ostracodes have been recorded from these facies.

From the continental margin of South Africa Dingle (1971)
recorded marine Aptian ostracodes. The samples came from the
Agulhas Bank and yielded ostracodes.

The age of the sample was dated by the foraminifer Episto-
mina (Brotzenia) alveata. Although the age ranges from Barremian to Albian, an Aptian age is preferred by Dingle.

Albian

Albian deposits are widely distributed in West Africa;
they are encountered in Gambia, Nigeria, Cameroon, Gabon, Congo,
Angola and South Africa.

In West Africa, Albian and younger Cretaceous sequences
are mostly of marine origin; these strata are dated by means of
ammonities and planktonic Foraminifera.

From Gambia Apostolescu (1963) recorded, from subsurface
samples of the Sara-Kunda well, the following ostracodes: Orthonotacythere sp. A., sp. B., Cytheropteron sp. A., sp. B.,
Asciocythere sp. A., and Cythereis sp. A.

This assemblage belong to the Faunizone 1 of Aptian-Albian
age.

In the Nigeria Basin and in Cameroon, the first datable
marine sequence begins in Albian times but ostracodes have not
been studied from these areas.

In the Gabon Basin marine Albian deposits are represented
by the Madiéla Series of the N'Komi System. According to de
Klasz and Micholet (1970) the Madiéla Series represent a period
of gradual regression. This Series, particularly its upper
part, is securely dated by ammonites (Reyment, 1966); the ostracodes are being studied (de Klasz and Micholet, 1970).

In the Congo Basin Upper Cretaceous strata are mentioned
by Grekoff (1960) from the southwestern part of the Basin; they
are represented by the Kwango Series. This Series is subdivided into two stages: the lower Inzia Stage and the upper Nsele
Stage. According to Grekoff in some areas there is a discontinuity between the Lualaba Series and the overlying Kwango
Series; in the type area at Kwango the Lualaba Series is unknown. The Inzia Stage was subdivided into five horizons according to their litholotical features. The fossiliferous horizon is Kwango 5 from horizon 5 of the Inzia Stage; this level
yielded fresh-water ostracodes such as : Ilyocypris compressa Grekoff, Ilyocypris luzubiensis Grekoff,
Cypridea (Cypridea) hilariensis Grekoff, Darwinula kwangoensis
Grekoff, Afrocythere? K 536 Grekoff and "Metacypris" K 3099
Grekoff. Taking into account that the overlying Nsele Stage
contains the horizon in which P. de Saint Seine found a bird
belonging to the Family Dercetidae, known only from the Cenomanian, it is to suspect that the underlying Inzia Stage could
be of Albian age.

In the Angola Cuanza Basin the Albian Stage sequences are
represented by the Tuenza, Catumbela and Quissonde Formations
(Brognon and Verrier, 1966), but ostracodes have not been recorded from them.

Cenomanian

Cenomanian strata are found in Gambia, Nigeria, Cameroon,
Gabon, Congo Basin and Angola.

The West African Cenomanian was characterized by a period
of general regression in all coastal basins (Hourq, 1964).

In Gambia Apostolescu (1963) studied the subsurface sequences; a complete section of marine strata ranging in age

from Aptian to Maastrichtian are found. The Cenomanian to Turonian sequences yielded an assemblage of ostracodes which belong to the Faunizone 2 of Apostolescu (1963)

In the Nigeria Basin well dated marine Cenomanian is only known from Calabar; it yielded ammonites and microfossils; the last ones have not yet been studied (Reyment, 1966). From Cameroon Reyment (1965) mentioned the Mundeck Sandstone Formation of Albian-Cenomanian age. No microfossils have been recorded from this formation. In the Gabon Basin the Cenomanian is a period of regression in which red facies have a large geographic extend. The strata of the Cap Lopez Series contain a characteristic microfauna. Up to date no ostracodes have been studied.

In the Congo Basin the lithologic sequence of the Nsele Stage is composed of very fine sandstone with claystones at their base and locally with silicified beds. The ostracode fossiliferous horizon is at the base of the Nsele Stage. This horizon provided the following: Dolerocypris kinkoensis Grekoff and Paracypria makawensis Grekoff. The ostracodes came from horizons in which P. de Saint Seine found birds belonging to the Family Dercetidae known from the Cenomanian in other areas (Grekoff, 1960). It is important to note that almost all Cretaceous ostracode species of the Congo Basin are different to those found in adjacent basins. In the Angola Cuanza Basin during Cenomanian times the Cabo Ledo Formation was deposited. The lithologic composition of this formation are silty shales with intercalated limestones (Brognon and Verrier, 1966). The strata contain ostracodes and foraminifers (Reyment, 1966), but the ostracodes have not been studied.

Turonian

The Turonian was a period of extensive transgression in West Africa; strata of this age are recorded from Gambia, Nigeria, Cameroon, Gabon and Angola. Reyment (1966) mentioned sequences of this age from Niger Republic, Chad and Ivory Coast.

In Gambia subsurface borehole samples at Sara Kunda 1 yielded Turonian ostracodes; they were recorded by Apostolescu (1963).

In Nigeria the Turonian sea spread over a wide area; during this time sequences of the Eze-Aku Shale and Makurdi Formation in the Benue Valley in southeastern Nigeria, and from the Pindiga, Gongila, Dukul and Jessu formations in Northeastern Nigeria were deposited (Reyment, 1966); datable ammonite faunas are common in these strata.

The Lower Turonian Eze-Aku Shale of Nigeria yielded ostracodes and benthonic Foraminifera which have not yet been given detailed study (Reyment, 1965).

In Cameroon the Turonian, Coniacian, Santonian, and Campanian strata of the Mungo River Formation are well dated with ammonites and planktonic Foraminifera (Belmonte, 1966). Ostracodes have not been studied up to the present, to the writer's knowledge.

In the Gabon Basin the marine Turonian Azile Series yielded guide planktonic Foraminifera and ostracodes (Reyment, 1966), but the ostracodes have not yet been studied.

In Angola the sediments of the Lower Itombe Formation were deposited during Turonian times, conditions being much the same as during Cenomanian time. No ostracods have been recorded from the Lower Itombe Formation.

Senonian, sensu lato

In the Senonian, s. l., are included the Coniacian, Santonian and Campanian Stages.

In Gambia subsurface sequences yielded Coniacian to Campanian ostracode assemblages. The Coniacian assemblage constitutes Apostolescu's Faunizone 3, whereas his Faunizone 4 comprises Santonian to Campanian strata

An important contribution was made by Apostolescu (1961); this author studied Upper Cretaceous ostracode faunas from Sénegal, Ivory Coast, Togo, Dahomey and Mali. The studied Cretaceous sequences range in age from the Senonian to the Maastrichtian, the Maastrichtian being only found in the Togo-Dahomey basin.

In the Senegal Basin the Senonian sequences are found in subsurface; the strata are mainly composed of clays and intercalated sandstones.

In the <u>Ivory Coast Basin</u> the Senonian ostracodes come from bore holes at Yokobové. The known species are characteristic from pre-Maastrichtian strata of Senegal, Cameroon and Nigeria (i.e. Santonian to Campanian).

The Dzodze area of <u>Ghana</u>, at the borehole at Quiba yeilded Campanian Maastrichtian ostracodes.

In the <u>Nigeria Basin</u> Senonian, s.l., strata are widespread.

Reyment (1960) studied the Nigerian Upper Cretaceous Senonian and Maastrichtian ostracoda; the studied material comes from outcrops and boreholes from northern, western, and eastern Nigeria and from Cameroon.

. The Santonian Stage is present in the Nigeria Basin. As regards to the micropaleontology of the Nigeria Basin, which was a time of regression, several ostracode species have been described from the Lamja Shale of northeastern Nigeria (Reyment, 1960) and possible Santonian limestones in Benue Province of central Nigeria; from the Benua Province, Reyment (1966) cited <u>Cytherella austinensis</u> Alexander, <u>Brachycythere ekpo</u> Reyment, <u>Buntonia vanmorkhoveni</u> Reyment. From the Lamja Formation Reyment (1960) recorded species of <u>Ovocytheridea</u>, <u>Cytherella austinensis</u> Alexander, <u>Metacytheropteron paganum</u> (Reyment) and <u>Buntonia</u> spp. Reyment stated that these associations contain many species in common with the Egyptian Upper Cretaceous. Undoubtedly, Campanian is known in Nigeria from the Duala Embayment (Reyment, 1965).

In <u>Cameroon</u> the only known Senonian strata came from the Mungo River Formation; to these strata a Campanian age was assigned (Reyment, 1965), but ostracodes were not recorded.

In the <u>Gabon Basin</u> the Senonian strata are represented by the Anguille Series and Pointe Clairette Series; ostracodes have not been cited from the deposits (de Klasz and Micholet, 1970).

From the <u>Congo Basin</u> Reyment (1965) recorded Santonian strata from the Vonso area (Leopoldville). No ostracodes have been mentioned.

In Angola the sedimentation continued during the Senonian. During this time the Itombe, Teba and Dio Dande formations were deposited. The sequences are dated by means of planktonic foraminifera; no ostracodes have been recorded.

In South Africa from the Agulhas Bank, located at the continental margin Dingle (1971) recorded two ostracode species: ?Amphicytherura sp. and Brachycythere agulhasensis Dingle. The assigned age was early-middle Senonian based on the planktonic foraminifera Globotruncana spp. ex. gr. G. marginata (Reuss).

Maastrichtian

In western Africa the Maastrichtian was marked by a wide transgression which, during its climax seems to have stretched across West Africa to North Africa (Reyment, 1966).

Maastrichtian sequences were recorded from Gambia, Senegal, Ghana, Dahomey, Togo, Nigeria, Cameroon, Gabon, Angola and South Africa.

From Gambia Apostolescu (1963) recorded from boreholes at Sara Kunda 1 an assemblage of Faunizone 5 of Apostolescu (op. cit.).

In Senegal the borehole at Sangalkam yielded the Maastrichtian ostracode Bradleya vesiculosa?.

In Ghana the section at Kulí Jabafi contains the following Maastrichtian ostracodes: Ovocytheridea nuda Grekoff, Veenia? ughelli Reyment, Veenia? varriensis Reyment and Brachycythere oguni Reyment (Apostolescu, 1961, 1963).

Several borehole sequences from Togo and Dahomey such as those at Attitogon, Bopa, Issoba, Lokossa and Sehoué were studied by Apostolescu (1961). The Maastrichtian sections were found to be equivalent to the planktonic foraminiferal Zone of Abathomphalus mayaroensis. They yielded the ostracode Brachycythere armata Reyment.

In western Nigeria the oldest sediments, dated by ammonites, are of upper Maastrichtian age; they lie directly on the Precambrian basement.

Reyment (1961) studied surface samples from the Aule River locality near Auchi and from subsurface at the localities of

Gbekevo and Araromi. Reyment (1960) observed that a large number of new species would reflect local developments.

In eastern Nigeria, during the Maastrichtian, coal-forming conditions developed in the Anambra Embayment. In this region the coal measure sequence begins with the marine Nkporo Shale, passing upward into the largely non-marine Mamu Formation (Reyment, 1966). The coal levels may extend into the Danian (Stolk, 1963).

In Gabon the Maastrichtian Ewongué Series contains only arenaceous Foraminifera

In Angola the Maastrichtian Stage is represented by the lower part of the Rio Dande Formation. No ostracodes have been mentioned from these strata.

From South Africa Dingle (1971) recorded middle to late Maastrichtian assemblages from the Agulhas Bank.

Summary

During the upper Jurassic and Cretaceous Periods the continental sedimentary sequences which accumulated on both sides of the borderlands of the present Atlantic Ocean are intimately related to tectonic events. These tectonic events began in late Jurassic times and continued during the Cretaceous and Cenozoic, producing several tectonic structures such as faults, grabens and rift valleys which controlled the sedimentary depositional areas.

During the Neocomian fresh, brackish, and marine deposits are registered in both the South American and African continents.

In the South American continent the fresh and brackish water sedimentary sequences are confined to the northeastern area of Brazil and Argentina. The sedimentary basins in which non-marine Neocomian strata are recorded in South America are the Sergipe/Alagoas, Reconcavo (Tucano, Jatobé and Almada basins in Brazil; in Argentina non-marine Neocomian is found in the Neuquén and San Jorge sedimentary basins.

During the Neocomian marine influences are confined to the southernmost South American areas; marine sequences occur in the Neuquén and Austral basins.

In the African Continent non-marine strata are found in Cameroon, Gabon, Congo, Angola and South Africa. Marine Neocomian strata are also restricted to the southernmost area of Africa, being found in South Africa.

In summary, Neocomian marine transgressions took place in regions where the present South American and African continents were not united to form the Gondwana Supercontinent in Lower Cretaceous time.

The ostracodal assemblages of both continents show close relationships and a large part of the associations are formed by species common to both continents. Non-marine ostracodes are at this time useful zonal markers.

Text Fig. 3 summarizes the marine and non-marine sedimentary basins during Neocomian times.

During Aptian and Albian epochs, the first marine transgressions were registered; the transgression left its record in thick evaporitic beds which are common in Gabon, Congo, Angola in Africa and in the Sergipe/Alagoas and Reconcavo/Tucano basins in South America.

After this event the sea encroached over some areas although not contemporaneously in all basins; the first transgression recorded by mainly marine ostracodes is of Aptian and Albian times in South America (Sergipe/Alagoas Basin - Riachuelo Formation), and in western Africa in Ivory Coast, Nigeria, Cameroon, Gabon and Angola.

Common African and South American marine ostracodes are recorded during Aptian-Albian times.

Although Upper Cretaceous sedimentation is recorded in several sedimentary basins, there are few data about the ostracodal content.

During the Maastrichtian a wide transgression was registered in several South American and African basins. The ostracode assemblages from this Stage show endemic features. The known assemblages from western Africa and those from South America (Argentina) evidences a common origin as shown by the great similarities between the assemblages, especially by the presence of the genera Veenia (Nigeria), Anticythereis and Togoina in both continents.

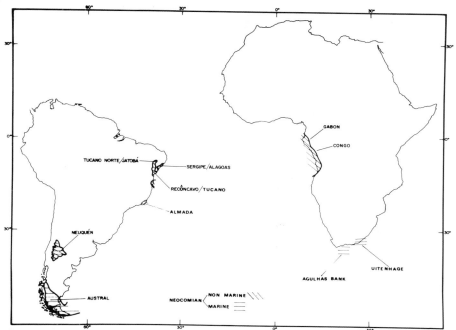

Text Fig. 3. Neocomian marine and non-marine sedimentary basins.

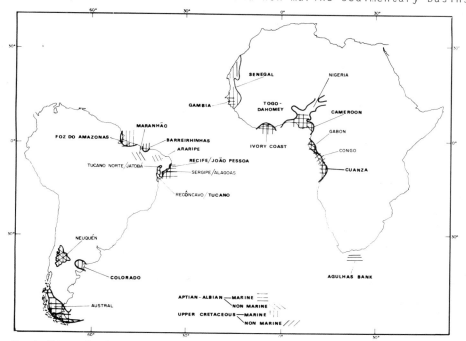

Text Fig. 4. Aptian, Albian and Upper Cretaceous marine and non-marine sedimentary basins.

Text Fig. 4 synthesizes the marine and non-marine sedimentary basins during the Aptian-Albian and Upper Cretaceous times.

References

Apostolescu, V., 1961. Contribution a l'étude paléontologique (Ostracodes) et Stratigraphique des Bassins Crétacés et Tertiaires de l'Afrique occidentale. Rev. Inst. Français du Pétr., v. XVI, no. 7-8, p. 779-867.

————, 1963. Essai de Zonation par les Ostracodes dans le Crétacé du Bassin du Sénégal. Rev. Inst. Français du Pétr., v. XVIII, no. 12, p. 1675-1694.

Bate, R. H., 1971. Phosphatized ostracods from the Cretaceous of Brazil, Nature, v. 230, no. 5293, p. 397-398.

Belmonte, Y. C., 1966. Stratigraphie du Bassin Sédimentaire du Cameroon. Proc. 2nd W. African Micropal. Coll. Ibadan, 1966, p.

Bertels, A., 1972. Ostrácodos de agua dulce del Miembro Inferior de la Formación Huantrai-co (Maastrichtiano inferior) Provincia del Neuquén, República Argentina. Ameghiniana, T. IX, no. 2, p. 173-182.

————, 1974. Upper Cretaceous (lower Maastrichtian?) ostracodes from Argentina. Micropaleontology, v. 20, no. 4, p. 385-397.

————, 1975. Upper Cretaceous (middle Maastrichtian) ostracodes of Argentina. Micropaleontology, v. 21, no. 1, pp. 97-130.

Beurlen, K., 1963. Geologia e Estratigrafia da Chapada do Araripe. XVII Congreso Brasileiro de Geologia, Recife, p.

Braun, P. G. O., 1966. Estragrafía dos sedimentos da parte interior da regiao nordeste do Brasil. Minist. Minas e Energia. Divi. Geol. e Mineral. Bol. no. 236, p. 5-79.

Brognon, G. and Verrier, G., 1966. Tectonique et sédimentation dans le bassin du Cuanza (Angola). In: Bassins sédimentaires du littoral africain. Ière partie: Littoral Atlantique. Assoc. des Services Géol. Africains., p. 207-252.

de Klasz, I., and Micholet, J., 1970. Elements nouveaux concernant la Biostratigraphie du Bassin Gabonais. IV Colloque African Micropal., p. 109-141.

Digregorio, J. H., 1972. Neuquén. En: Geología Regional Argentina, Acad. Nac. Ciencias. Cordoba., p. 439-505.

Dingle, R. V., 1969. Marine Neocomian Ostracoda from South Africa. Trans. roy. Soc. S. Afr., v. 38, p. 2, p. 139-163.

————, 1971. Some Cretaceous Ostracodal assemblages from the Agulhas Bank (South African continental margin). Trans. roy. Soc. S. Afr., v. 39, p. IV, p. 393-418.

Grekoff, N., 1957. Ostracodes du Bassin du Congo. Annales du Musée Royal du Congo Belge, Ser. in-8, v. 19, p. 1-97.

————, 1960. Ostracodes du Bassin du Congo. II. Crétacé. Ann. Musée Royal Congo Belge. Ser. in-8, v. 35, p. 1-70.

Grekoff, N. and Krömmelbein, K., 1967. Etude comparée des Ostracodes Mésozoïques continentaux des bassins Atlantiques: Serie de Cocobeach, Gabon et Série de Bahía, Brésil. Rev. Inst. Francais Pétr., v. XXI, no. 9, p. 1307-1353.

Grosdidier, E., 1967. Quelques ostracodes nouveaux de la Série Anté-Salifère ("Wealdienne") des bassins cottiers du Gabon et du Congo. Rev. Micropal., v. 10, no. 2, p. 107-118.

Hourq, V., 1964. Les grands traits de la géologie des bassins cotiers du groupe équatorial. Symp. W. Afr. Sed. Bas., 22 Intern. Congress Geol. New Delhi, p. 171-178.

Krömmelbein, K., 1964a. Ostracoden aus der marinen "Küsten-Kreide" Brasiliens. 1: Brachycythere (Brachycythere) sapucariensis n. sp. aus dem Turonium. Senck. Leth., v. 45, p. 489-495.

————, 1964b. Neue Arten der Ostracoden-Gattung Paracypridea Swain aus der Bahia-Serie des Reconcavo Bahiano (Oberjura?) Unterkreide, Wealden-Fazies, NE-Brasilien. Bol. Paranense de Geografia., nos. 10-15, p. 139-160.

————, 1964c. Über einige neue Arten der Ostracoden-Gattung Reconcavona Krömmelbein 1962 aus der NE-brasilianischen Bahia-Serie. Senck. Leth., v. 45, p. 29-41.

————, 1966a. On "Gondwana Wealden" Ostracoda from NE Brazil and West Africa. Proc. 2nd. West African Microp. Coll. Ibadan, p. 113-118.

————, 1966b. Preliminary remarks on some marine Cretaceous Ostracodes from northeastern Brazil and West Africa. Proceed. 2nd. West African Microp. Coll. Ibadan, p. 119-121.

————, 1967. Ostracoden aus den marinen "Küsten-Kreide" Brasilien. 2: Sergipella transatlantica n.g. n. sp. und Aracajuia benderi n.g. n. sp. aus dem Ober-Aptium/Albium. Senck. Leth., v. 48, p. 525-533.

Krömmelbein, K., and Wenger, R., 1966. Sur quelques analogies remarquables dans les microfaunes crétacées du Gabon et du Brésil oriental Bahia et Sergipe, p. 193-196.

Le Calvez, Y., de Klasz I., and Brun, L., 1974. Nouvelle contribution à la connaissance des microfaunes du Gabon. Rev. Española de Micropal., v. VI, no. 3, p. 381-400.

Lesta, P. J., and Ferello, R., 1972. Región extrandina de Chubut y norte de Santa Cruz. In: Geología Regional Argentina. Acad. Nac. Ciencias Cordoba., p. 601-653.

Lima, E. C., 1973. Bioestratigrafía da Bacía de Barreirinhas. Anais do XXVI Congreso Soc. Bras. Geol., v. 3, p. 81-91.

Mabesone, J. M., Tinoco, I. M., and Coutinho, P. N., 1968. The Mesozoic-Tertiary Boundary in Northeastern Brazil. Paleogeography, Palaeoclimatology, Palaeoecology, v. 4, p. 161-185.

Marliere, R., 1948. Ostracodes et Phyllopodes du Système du Karroo au Congo Belge. An. Mus. Roy. Congo Belge. Tervuren. Sc. Geol., v. 2, p. 1-15.

————, 1950. Ostracodes et Phyllopodes du Système du Karroo au Congo Belge et les régions avoisinantes. An. Mus. Roy. Congo Belge. Tervuren. Ser. in-8. Sc. Geol., v. 6, p. 1-43.

Moura, J. A., 1972. Algumas espécies e subespécies novas de ostracodes da bacia Reconcavo/Tucano. Beol. Tecn. Petrobrás. Rio de Janeiro, v. 15, no. 3, p. 245-263.

Musacchio, E. A., 1971. Hallazgo del género Cypridea (Ostracoda) en Argentina y consideraciones estratigráficas sobre la Formación La Amarga (Cretácico Inf.) en la Prov. de Neuquén. Ameghiniana, T. VIII, no. 2, p. 105-125.

Musacchio, E. A., and Chebli, G., 1975. Ostrácodos no marinos y carofitas del Cretácico Inferior en las provincias de Chubut y Neuquén, Argentina. 1. Ostrácodos y carofitas del Grupo Chubut. 2. Rayosoana quilimalensis nov. gen. nov. sp. de la Formación Rayoso, Neuquén. Ameghiniana, T. XII, no. 1, p. 70-96.

300

Petri, S., 1954. Foraminíferos fosseis da Bacia do Marajó. Univ. Sao Paulo. Facult. Filosofia, Ciencias e Letras. Bol. 176, Geol., no. 11, p. 1-173.

Pinto, I. D., and Sanguinetti, Y. T., 1961. Observaciones on Metacypris (Ostracoda) from the Mesozoic of North America and Africa. Esc. Geol. P. Alegre. Bol. 9, p. 3-14.

———, 1962. A complete revision of the Genera Bisulcocypris and Theriosynoecum (Ostracoda) with the world geographical and stratigraphical distribution. Esc. Geol. P. Alegre, Pub. Esp: no. 4, p. 1-165.

Reyment, R. A., 1960. Studies on Nigerian Upper Cretaceous and Lower Tertiary Ostracoda. Part 1: Senonian and Maestrichtian Ostracoda. Acta Univ. Stockholm. v. VII, p. 1-238.

———, 1965. Aspects of the Geology of Nigeria. The stratigraphy of the Cretaceous and Cenozoic deposits. Ibadan Univ. Press., p. 1-145.

———, 1966. Brief Review of the stratigraphic sequences of West Africa (Angola to Senegal). Proceed. 2nd West Afr. Microp. Coll. Ibadan, p. 162-175.

Schaller, H., 1969. Revisao estratigráfica da bacía de Sergipe-Alagoas. Bol. Tec. Petrobras, Rio de Janeiro, v. 12, no. 1, p. 21-86.

Schaller, H., Vasconcelos, D. N., and Castro, J. C., 1971. Estratigrafía preliminar da Bacía sedimentar da Foz do Rio Amazonas. Soc. Bras. Geol. Anais do XXV Congreso, v. 3, p. 189-202.

Swain, F. M., 1946. Middle Mesozoic non-marine ostracodes. Jour. Paleontology, v. 20, no. 6, p. 543-555.

Tinoco, I. de M., 1967. Micropaleontología da faixa sedimentar costeira Recife-Joao Pessoa - Bol. Soc. Bras. Geol. 16, no. 1, p. 81-85.

Viana, C. G., 1966. Stratigraphic distribution of ostracoda in the Bahia Supergroup (Brasil), Proc. 2nd. W. Afr. Micropal. Coll. Ibadan. p. 240, 257.

Viana, G. F., da Gama, E. G., Jr., Araujo Simoes, J., Joura, J. A., dos Reis Fonseca, J., Alves, R. J., 1971. Revisao estratigrafica da bacia Reconcavo/Tucano. Bol. Tec. Petrobras, v. 14, no. 3/4, p. 157-192.

Plate I. (Stereographic paired photographs) Fig. 1 - Cypridea (Morininoides) candeiensis Krömmelbein, x53, a) lateral view; b) dorsal view Santo Amaro - Zone (003). Occurs in Sergipe/Alagoas - Jatoba and Africa. Fig. 2. - Paracypridea brasiliensis Krömmelbein, x45, a) lateral view; b) dorsal view Zones (003)-(004). Occurs in Sergipe/Alagoas - Reconcavo/Tucano - Almada and Africa. Fig. 3 - Ilhasina torosa Krömmelbein Zones (004)-(005) - Upper Santo Amaro - Lower Ilhas, x68. Occurs in Reconcavo/Tucano and Africa, a) lateral view; b) dorsal view. Fig. 4 - Paracypridea obovata obovata (Swain), x38, 1946 Zones (005) Lower Ilhas. Occurs in Sergipe/Alagoas - Reconcavo/Tucano - Jatoba and Africa. a) lateral view; b) dorsal view.

Plate II. (Stereographic paired photographs) Fig. 1 - Salvadoriella redunca redunca Krömmelbein, x52, 1963 Zone (006) - Upper Ilhas. Occurs in Reconcavo/Tucano and Africa. a) lateral view; b) dorsal view. Fig. 2 - Paracypridea quadrirugosa weberi Krömmelbein, x31, a) lateral view; b) dorsal view. Zones (006)-(007). Upper Ilhas - Lower Sao Sebastiao. Occurs in Sergipe/Alagoas and in Africa. Fig. 3 - Petrobrasia

marfinensis Krömmelbein, 1962, x63, a) lateral view; b) dor-
sal view, Zone (007) - Sao Sebastiao. Occurs in Sergipe/
Alagoas and Reconcavo/Tucano. Fig. 4. Cypridea (Sebastian-
ites) sostensis sostensis Krommelbein, x53, a) lateral view;
b) dorsal view, Zones (007)-(008) - Sao Sebastiao - Jiquiá.
Occurs in Sergipe/Alagoas and Reconcavo/Tucano.
Plate III. Fig. 1 - Novocythere santacruceana Malumian, Musiuk
and Rossi de Garcia, x100, Austral Basin - Upper Aptian -
Albian. Fig. 2 - Alatacythere? rocana Bertels, x84, Neuquén
Basin - (Lower? Maastrichtian) Jagüeliano Stage. Fig. 3 -
Platycythereis? velata Bertels, x84, Neuquén Basin - (Lower?
Maastrichtian) Jagüeliano Stage. Fig. 4 - Veenia (Nigeria)
inornata Bertels, x100, Neuquén Basin - (Middle Maastrichtian)
Jagüeliano Stage. Fig. 5 - Anticythereis venusta Bertels,
x84, Neuquén Basin - (Middle Maastrichtian) Jagüeliano Stage.
Fig. 6 - Tumidoleberis australis Bertels, x100, Neuquén
Basin - (Middle Maastrichtian) Jagüeliano Stage. Fig. 7 -
Wichmannella magna Bertels, x100, Neuquén Basin - (Middle
Maastrichtian) Jagüeliano Stage. Fig. 8 - Veenia (Nigeria)
tumida Bertels, x100, Neuquén Basin - (Middle Maastrichtian)
Jagüeliano Stage. Fig. 9 - Veenia (Nigeria) jaguelensis
Bertels, x100, Neuquén Basin - (Middle Maastrichtian)
jaguelensis Bertels, x100, Neuquen Basin - (Middle Maastrich-
tian) Jagüeliano Stage. Fig. 10 - Veenia (Nigeria) punctata
Bertels, x100, Neuquén Basin - (Middle Maastrichtian Jagüeli-
ano Stage.

PLATE I

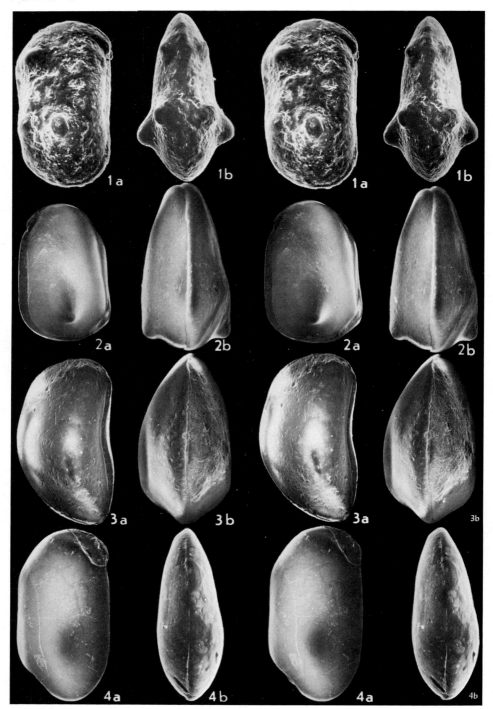

PLATE II

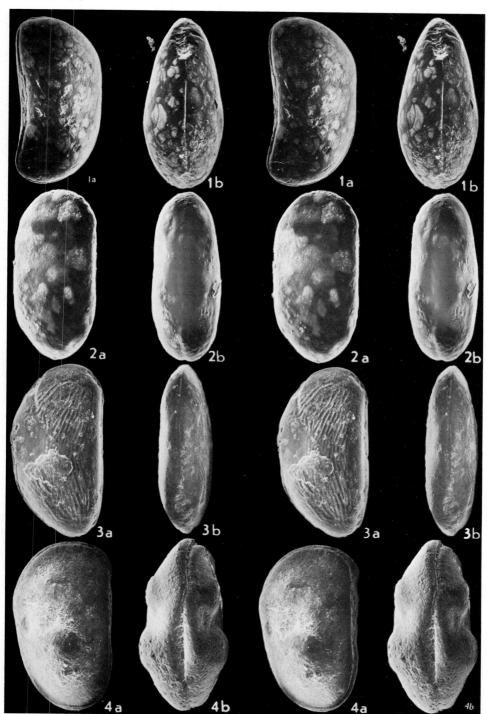

PLATE III

MESOZOIC RADIOLARIA FROM THE ATLANTIC BASIN AND ITS BORDERLANDS

Helen P. Foreman

Department of Geology, Oberlin College, Oberlin, Ohio 44074

Abstract

Mesozoic Radiolaria, primarily from Deep Sea Drilling cores, are surveyed. Maps indicate localities, ages, and usefulness of the occurrence. The new *Amphipyndax tylotus* Zone is defined and the next lower *Amphipyndax enesseffi* Zone is emended.

Introduction

It has long been recognized that Radiolaria in the Atlantic Basin are not as abundant or as well preserved as those from the Pacific (Murray, J., 1876). This has again been pointed out by Johnson (1976) who suggested that the poor preservation, at least, may be due to the greater amount of silica depleted detritus being deposited in the Atlantic Basin. Examination of Deep Sea Drilling Project Atlantic Basin Mesozoic cores again confirms this diagnosis. With only a few exceptions, paucity and poor preservation seem to be the rule.

Localities

The following four maps of the Atlantic Basin and its borderlands indicate with circles the locations of Deep Sea Drilling Project Sites, together with supplementary localities from which Mesozoic radiolarian bearing sediments have been recovered. The dark circles represent holes or land-based supplementary localities which contain sediments with the occasionally well-preserved Radiolaria, or Radiolaria in useful sequences. DSDP Legs 1 through 4, 10 through 15, 40 and 41, are included in this survey. Although Legs 36 and 39 in the southwest Atlantic and Leg 44 along the east coast of the United States also recovered Mesozoic radiolarian bearing sediments, their results are not included because the Initial Reports have not yet been published.

The localities on the maps (Figures 1-4) and the chart of occurrences (Figure 5) from which Mesozoic radiolarian bearing sediments have been recovered are listed below:

Deep Sea Drilling Project Atlantic Basin localities:

Leg 1	Site 4	24°28.68'N, 73°47.52'W	at depth of 5319.5 meters
	Site 5	24°43.59'N, 73°38.46'W	at depth of 5361 meters
Leg 2	Site 9	32°46.4' N, 59°11.7' W	at depth of 4965 meters
Leg 3	Site 13	06°02.4' N, 18°13.71'W	at depth of 4588 meters
Leg 4	Site 24	06°16.58'S, 30°53.46'W	at depth of 5148 meters
	Site 28	20°35.19'N, 65°37.33'W	at depth of 5521 meters
Leg 10	Site 95	24°09.00'N, 86°23.85'W	at depth of 1633 meters
	Site 97	23°53.05'N, 84°26.74'W	at depth of 2930 meters
Leg 11	Site 100	24°41.28'N, 73°47.95'W	at depth of 3525 meters
	Site 101	25°11.93'N, 74°26.31'W	at depth of 4868 meters
	Site 105	34°53.72'N, 69°10.40'W	at depth of 5251 meters
Leg 12	Site 118	45°02.65'N, 09°00.63'W	at depth of 4901 meters
Leg 13	Site 120	36°41.39'N, 11°25.94'W	at depth of 1711 meters
Leg 14	Site 136	34°10.13'N, 16°18.19'W	at depth of 4169 meters
	Site 138	25°55.37'N, 25°33.79'W	at depth of 5288 meters
	Site 140	21°44.97'N, 21°47.52'W	at depth of 4483 meters
	Site 144	09°27.23'N, 54°20.52'W	at depth of 2957 meters
Leg 15	Site 146	15°06.99'N, 69°22.67'W	at depth of 3949 meters
	Site 150	14°30.69'N, 69°21.35'W	at depth of 4545 meters
	Site 152	15°52.72'N, 74°36.47'W	at depth of 3899 meters
	Site 153	13°58.33'N, 72°26.08'W	at depth of 3932 meters
Leg 40	Site 361	35°03.97'S, 15°26.91'E	at depth of 4549 meters
	Site 363	19°38.75'S, 09°02.80'E	at depth of 2248 meters
	Site 364	11°34.32'S, 11°58.30'E	at depth of 2448 meters
Leg 41	Site 367	12°29.2' N, 20°02.8' W	at depth of 4748 meters
	Site 368	17°30.4' N, 21°21.2' W	at depth of 3366 meters
	Site 369	26°35.5' N, 14°59.9' W	at depth of 1752 meters

Non DSDP Atlantic Basin localities:

| Lamont V-18-129 | 41°42'S, 56°35'W | at depth of 2039 meters |
| Lamont V-17-144 | 40°34'S, 55°10'W | at depth of 2503 meters |

Marginal Atlantic Basin localities:

B191, Pre-Habana formation Cuba (Foreman, 1968)

RM312, Val Dorbia, Umbria, Italy (Foreman, 1968)

Marac Well # 1, Trinidad (Foreman, 1968)

BR1025, Via Blanca formation Cuba (Brönnimann & Rigassi, 1963)

BR1028, Via Blanca (?) formation Cuba (Brönnimann & Rigassi, 1963)

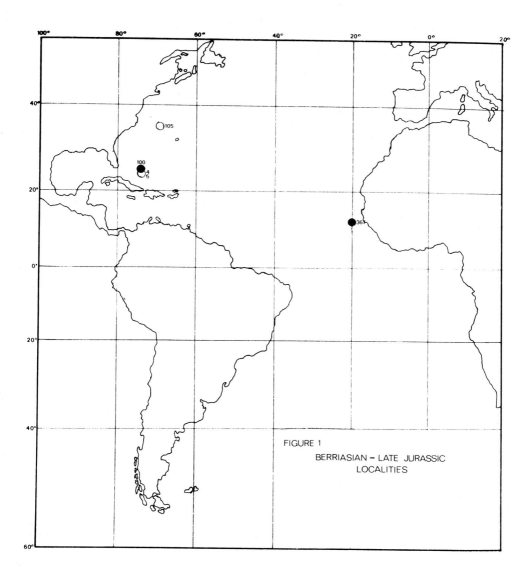

Figure 1. **Berriasian - Late Jurassic localities**. These are, as would be expected, few, dependent as they are on location, and drilling expertise. All the material so far examined was poorly preserved with abundances generally rare, occasionally common. However, Holes 367 and 100 are shaded to indicate useful sequences. Legs 1, 11, and 41 sampled this sequence.

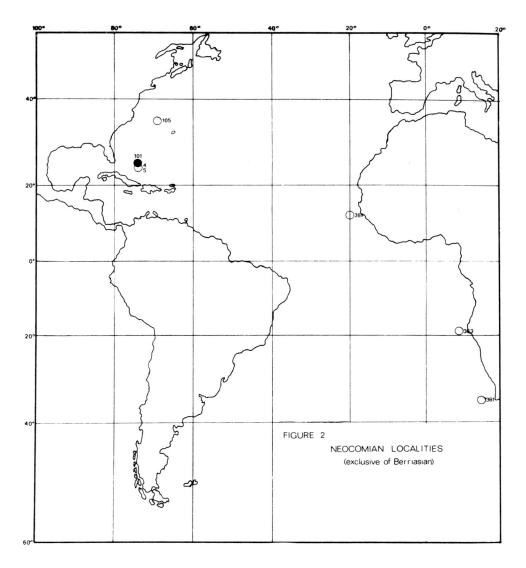

Figure 2. **Neocomian localities, exclusive of Berriasian**. Only at Site 101 of Leg 11 have Radiolaria in significant quantities in a useful sequence been recovered. At Site 105 there is a long sequence, but the Radiolaria are generally rare and poorly preserved.

The *Sethocapsa trachyostraca* Zone could be identified in Core 18 of Hole 105. This Zone could also be recognized in Core 29 of Hole 367. Holes 363 and 361 had, in general, only very rare, very poor Radiolaria, and no Neocomian Zones could be recognized. For the Neocomian, as for the Berriasian - Late Jurassic, only Legs 1, 11, and 41 are represented, together with the very minor occurrences of Leg 40.

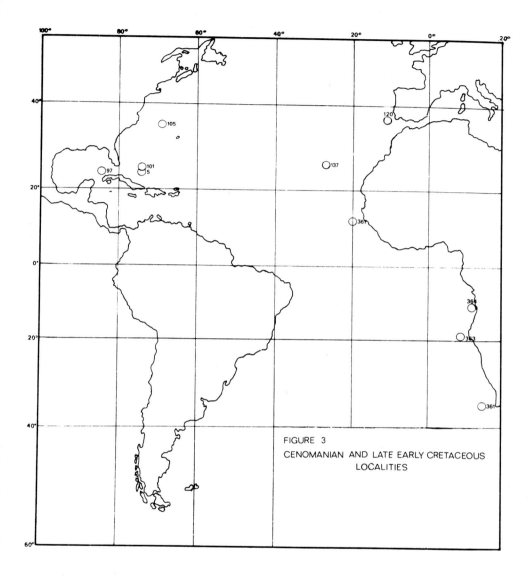

FIGURE 3

CENOMANIAN AND LATE EARLY CRETACEOUS
LOCALITIES

Figure 3. **Cenomanian and late Early Cretaceous localities.** With a few additions the map for the Cenomanian and late Early Cretaceous is very similar to that for the Neocomian with faunas generally poor and sparse. An exception might be Site 5 where Radiolaria from an Albian core show moderate preservation. Localities from Legs 1, 10, 11, 13, 14, 40, and 41 are identified.

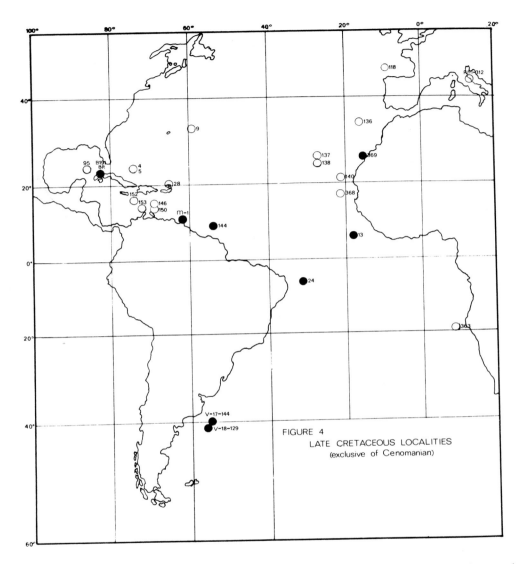

Figure 4. **Late Cretaceous localities**. These are by far the most abundant. The material is in general much better preserved and the shaded localities indicate in almost all instances moderate to good preservation in either a single core or in a sequence of a few cores. No long sequences are known except those from the Carribean Leg 15 which are poorly preserved and were considered by the shipboard scientists to be redeposited. For this time, some excellent land based localities with very good preservation are known. A single Campanian/Santonian sample (B191) collected by Bermudez from Habana, Cuba (Foreman, 1966) and some samples (BR), Campanian/Maestrichtian from Brönnimann and Rigassi (1963) also from Cuba, and early Santonian *Globotruncana concavata* Zone, sample (M-1) from Trinidad (Foreman, 1968).

What emerges is a pattern which finds Radiolaria present rather consistently in sediments of late Late Cretaceous age, minor occurrences in the Cenomanian-Albian and the late Neocomian, and again Radiolaria consistently present in the few holes that sampled the Early Neocomian/Late Jurassic (Tithonian-Kimmeridgian and possibly Oxfordian). This is in rather great contrast to the Pacific where Radiolaria in the late Late Cretaceous are generally lacking or when present poorly preserved with Maestrichtian Radiolaria completely missing. In the Pacific fairly common but relatively poor early Late Cretaceous and late Early Cretaceous Radiolaria are recognized and rich sequences of Aptian and Late Neocomian Radiolaria are known. No Radiolaria older than Tithonian have been recovered from the Pacific.

Radiolarian zonation

For some time now, three major groups of Late Cretaceous radiolarian fossils have been recognized: A Maestrichtian/Late Campanian fauna typically represented by the forms described by Foreman (1968) and recognized by Kozlova (1972) from Leg 14 of the Deep Sea Drilling Project; an Early Campanian fauna partly described by Pessagno (1963) from Puerto Rico and recognized in samples from Cuba collected by Bermudez, Brönnimann and Rigassi as well as samples from numerous Deep Sea Drilling localities; and a Cenomanian/Albian fauna, elements of which have been variously described by Dumitrica (1970), Aliev (1965), and Foreman (1975).

In the same way three major groups of Early Cretaceous Radiolaria have in the last few years been recognized: the Albian/Cenomanian fauna mentioned above, a Barremian to Hauterivian or Valanginian fauna, and an Early Neocomian to Late Jurassic one.

With the impetus provided by the long sequences recovered by the Deep Sea Drilling Project, a number of zonal schemes have recently been introduced in an attempt to break down this very coarse, very inadequate division. However, the radiolarian zonation of the Cretaceous (Figure 5) is still very much in a fledgling state. The first coarse zonation by Moore (1973) based on the Radiolaria recovered from Leg 17 divided the whole of the Cretaceous and late Late Jurassic into seven zones. Poor preservation, the lack of comparative material and inadequate calcareous control made the resulting zones difficult to apply. Riedel and Sanfilippo (1974) using a broader base introduced a coarse zonation, again of seven zones for the same period of time. Some of these zones were widely applicable but they again had very imprecise calcareous control. This zonation was modified by Foreman (1975) and is here modified further to make it applicable to the sequences studied from the Atlantic basin.

While these studies based largely on Deep Sea Drilling Project material were being made, Pessagno (in press) and Dumitrica (1975) produced, respectively, detailed zonations, based on well-preserved material with good calcareous control for the Late Cretaceous of California and the Cenomanian of Roumania. Dumitrica's zones were open-ended and his Late Cenomanian *Holocryptocanium nanum-Excentropyloma cenomana* Zone may be partly comparable to the lower part of Pessagno's Turonian *Alievium superbum* Zone, and his *Holocryptocanium barbui-H. tuberculatum* Zone probably extends into the Albian.

Figure 5 presents the various zonations in chronological order from left to right with the oldest zonation scheme recorded here at the left. In the figure the zones of Pessagno which could, for the most part, not be recognized in the Atlantic are correlated with the later zonations on the basis of age. The resulting uncertainty is indicated by dashed boundary lines. Also, the base of the *Theocapsomma comys* [group] is certainly lower than indicated by Riedel and Sanfilippo, the dashed line introduced here between it and the next lower zone again indicates uncertainty.

PESSAGNO (in press)		RIEDEL & SANFILIPPO 1974		FOREMAN 1975 emend.		DUMITRICA 1975	
AGE	ZONE	AGE	ZONE	AGE	ZONE	AGE	ZONE
Maestrichtian	Orbiculiforma renillaeformis	Approximately Maestrichtian	Theocapsomma comys [group]	Maestrichtian / E.Maest./L.Camp.	Amphipyndax tylotus		
Campanian	*Crucella espartoensis*: Patulibrachium dickensoni / Phaseliforma carinata / Patulibrachium lawsoni / Protoxiphotractus perplexus	Approximately Campanian	Amphipyndax enesseffi	Campanian	Amphipyndax enesseffi		
Santonian	Alievium gallowayi	Approximately Campanian-Coniacian	Artostrobium urna	Santonian	Artostrobium urna		
Coniacian	*Alievium praegallowayi*: Orbiculiforma vacaensis / Archaeospongoprunum triplum			Coniacian			
Turonian	*R.hessi*: A. superbum / A. venadoensis / Halesium sexangulum / Q. spinosa / Cassideus riedeli			Turonian L. Albian	Dictyomitra somphedia	Late Cenomanian	Holocrypt. nanum / E. cenomana
Cenomanian						Cenomanian	H. barbui / tuberculatum
		Approximately Coniacian-Albian	Dictyomitra veneta	Late Albian			
				Albian	Acaenictyle umbilicata		
				E. Albian / L. Aptian / E. Aptian/ Barremian			
		Approximately Albian-Barremian	Eucyrtis tenuis	Barremian/ Hauterivian or Valanginian	Eucyrtis tenuis		
		Approximately Hauterivian-Valanginian	Staurosphaera septemporata		Sethocapsa trachyostraca		
				Valanginian			
		Approximately Valanginian-Tithonian	Sphaerostylus lanceola	Valanginian or Berriasian/ Kimmeridigian or older	Sphaerostylus lanceola		FIGURE 5

Figure 5. **Radiolarian zonations**

The best DSDP Atlantic sequence so far of Maestrichtian/Late Campanian Radiolaria was recovered from Hole 369A, Cores 35-39. While preservation is quite good, the Radiolaria are not common. This together with the incursion of a boreal fauna in Cores 35 and 37 leaves the accuracy of many first and last appearances open to question unless there is some outside back-up control. Nevertheless, this sequence, because it did have calcareous control, was used as a base together with other samples, with and without calcareous control, throughout the Late Cretaceous to contruct a chart of occurrences (Figure 6) in an attempt to pin down the ranges or at least the order of appearance and extinction for some of the more common members of Late Cretaceous assemblages.

This chart gives some detail for the Campanian/Maestrichtian section but is deficient in that the early Late Cretaceous samples are too few in number. Five Zones are recognized:

The Cenomanian *Holocryptocanium barbui-H. tuberculatum* Zone and the Late Cenomanian *Holocryptocanium nanum-Excentropyloma cenomana* Zone, both of Dumitrica; the *Artostrobium urna* and *Amphipyndax enesseffi* Zones of Riedel and Sanfilippo emended here and the new *Amphipyndax tylotus* Zone.

This new *Amphipyndax tylotus* Zone and the consequently emended *Amphipyndax enesseffi* Zone are defined and amplified as follows. All first appearances and last occurrences are morphological unless otherwise stated.

Amphipyndax tylotus Zone Foreman, new zone

The base is defined as the first evolutionary appearance of *Amphipyndax tylotus*. It includes near its

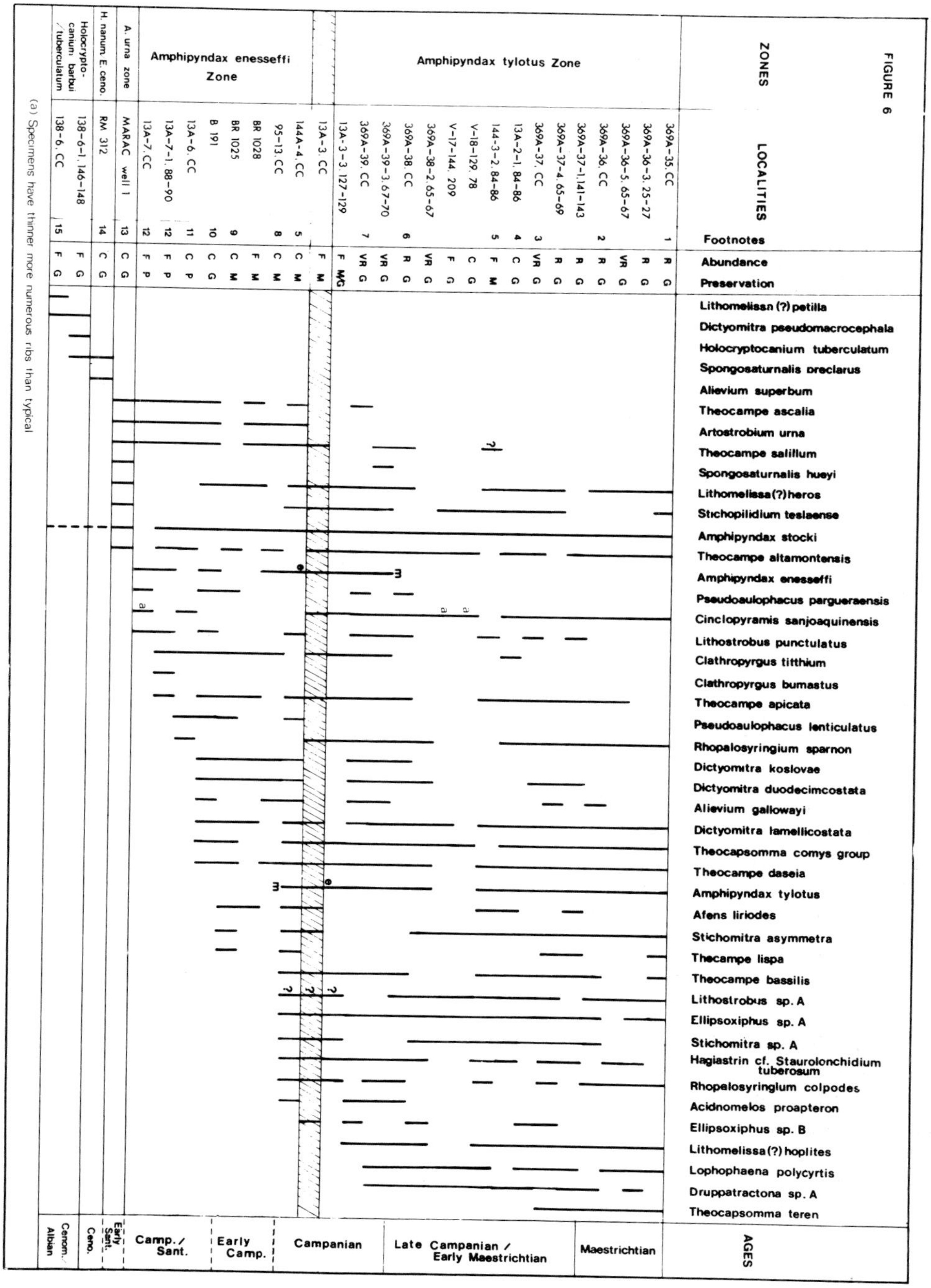

313

Figure 6. Chart of occurrences

base the first appearance of *Lithomelissa* (?) *hoplites* and *Lophophaena polycyrtis*. Also included are the last occurrences of *Clathropyrgus titthium* and *Afens liriodes*. The top is defined as the Maestrichtian/Danian boundary recognized by the last occurrence of the many species described from the California Maestrichtian (Foreman, 1968) which are recognized here, among them *Lophophaena* (?) *polycyrtis, Theocampe bassilis* and *Theocapsomma teren*.

Amphipyndax enesseffi Zone Riedel and Sanfilippo, emend. Foreman

The base is defined by the first morphological appearance of *Amphipyndax enesseffi* which may be approximately coincident with the first appearance of *Lithostrobus punctulatus*. It includes near its base the first appearance of *Clathropyrgus titthium* and *Theocampe apicata*. Also included are the first appearances of *Theocampe daseia, Afens liriodes* and *Theocampe bassilis* and the last occurrence of *Artostrobium urna*. The top is defined by the base of the next higher, *A. tylotus* Zone.

The *Amphipyndax tylotus* Zone is recognized by the presence of its nominal species alone or in greater abundance than its ancestor *A. enesseffi* in all the Atlantic Basin and Atlantic margin samples of comparable age, except the two high latitude samples V-17-144 and V-18-129. It also cannot be recognized by its nominal species in California. It can, however, be identified in all the California localities of appropriate age, i.e. Maestrichtian, or Lower Maestrichtian-?Campanian of Foreman (1968, Table 1), and the two high latitude Atlantic samples, by the presence of *Lophophaena(?)polycyrtis* and/or *Lithomelissa(?)hoplites* (Figure 7).

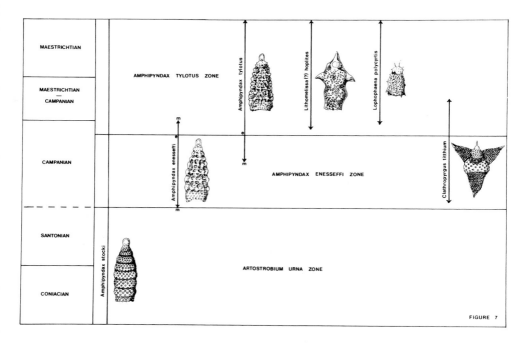

Figure 7. **Ranges for some diagnostic species of the** *Amphipyndax tylotus* **and** *A. enesseffi* **Zones.**

Taxonomic List

Species included in the chart of occurrences (Figure 6) are listed here in alphabetical order with one or two references to the literature where they are more completely described or a more complete synonymy is given.

Acidnomelos proapteron Foreman, in press. Illustrated here on Plate 1, Figure 12.

Afens liriodes Riedel and Sanfilippo, 1974, p. 775, pl. 11, fig. 11; pl. 13, figs. 14-16.

Alievium gallowayi (White) *in* Pessagno, 1972, p. 299, pl. 25, figs. 4-6; pl. 26, fig. 5; pl. 31, figs. 2, 3.

Alievium superbum (Squinabol) *in* Pessagno, 1972, p. 302, pl. 24, figs. 5-6; pl. 25, fig. 1; text-fig. 1.

Amphipyndax enesseffi Foreman, 1966, p. 356, text-figs. 7-11. Illustrated here on Plate 1, Figure 2.

Amphipyndax stocki (Campbell and Clark) *in* Foreman, 1968, p. 78, pl. 8, figs. 12a-c.

Amphipyndax tylotus Foreman, in press. Illustrated here on Plate 1, Figure 1.

Artostrobium urna Foreman, 1971, p. 1677, pl. 4, figs. 1-2.

Cinclopyramis sanjoaquinensis (Campbell and Clark) 1944, p. 22, pl. 7, fig. 2. Foreman, in press.

Clathropyrgus bumastus Riedel and Sanfilippo, 1974, p. 775, pl. 12, figs. 6-8.

Clathropyrgus titthium Riedel and Sanfilippo, 1974, p. 775, pl. 3, fig. 12; pl. 12, figs. 9-12.

Dictyomitra duodecimcostata (Squinabol) *in* Foreman, 1975, p. 614, pl. 1G, fig. 5; pl. 7, fig. 10.

Dictyomitra koslovae Foreman, 1975, p. 614, pl. 7, fig. 4.

Dictyomitra lamellicostata Foreman, 1968, p. 65, pl. 7, figs. 8a-b.

Dictyomitra pseudomacrocephala (Squinabol) *in* Foreman, 1975, p. 614, pl. 7, fig. 10.

Druppatractona sp. A Foreman, in press. Illustrated here on Plate 1, Figure 3.

Ellipsoxiphus sp. A Foreman, in press. Illustrated here on Plate 1, Figure 4.

Ellipsoxiphus sp. B Foreman, in press. Illustrated here on Plate 1, Figure 5.

Hagiastrin cf. *Staurolonchidium tuberosum* Rust *in* Riedel and Sanfilippo, 1974, p. 779, pl. 14, figs. 5-8.

Holocryptocanium tuberculatum Dumitrica, 1970, pl. 16, figs. 102, 103a-c.

Lithomelissa (?) *heros* Campbell and Clark *in* Foreman, 1968, p. 25, pl. 3, figs. 5a,b; text-fig. 1, fig. 7.

Lithomelissa (?) *hoplites* Foreman, 1968, p. 26, pl. 3, figs. 2a-c. Illustrated here on Plate 1, Figure 7.

Lithomelissa (?) *petilla* Foreman, 1975, p. 616, pl. 1G, figs. 2,3; pl. 6, fig. 3.

Lithostrobus sp. A Foreman, in press. Illustrated here on Plate 1, Figure 9.

Lithostrobus punctulatus Pessagno, 1963, p. 210 (partim) pl. 1, fig. 1. Foreman, in press. Illustrated here on Plate 1, Figure 6.

Lophophaena (?) *polycyrtis* Foreman, 1968, p. 23, pl. 3, figs. 3a-c. Illustrated here on Plate 1, Figure 8.

Pseudoaulophacus lenticulatus (White) *in* Pessagno, 1963, p. 202, pl. 2, figs. 8-9.

Pseudoaulophacus pargueraensis Pessagno, 1963, p. 204, pl. 2, figs. 4-7; pl. 6, figs. 4,5.

Rhopalosyringium colpodes Foreman, 1968, p. 57, pl. 6, fig. 6.

Rhopalosyringium sparnon Foreman, 1968, p. 56, pl. 6, fig. 5.

Spongosaturnalis hueyi (Pessagno), in press. Foreman, 1975, p. 641, pl. 1A, fig. 6; pl. 4, fig. 10.

Spongosaturnalis (?) *preclarus* Foreman, 1975, p. 611, pl. 1A, figs. 4,5; pl. 4, fig. 8.

Stichomitra asymmetra Foreman, in press. Illustrated here on Plate 1, Figure 10.

Stichomitra sp. A Foreman, in press. Illustrated here on Plate 1, Figure 11.

Stichopilidium teslaense Campbell and Clark, *in* Foreman, 1968, p. 70, pl. 8, fig. 13.

Theocampe altamontensis (Campbell and Clark) *in* Foreman, 1968, p. 53, pl. 6, figs. 14,a,b.

Theocampe apicata Foreman, 1971, p. 1679, pl. 4, fig. 6.

Theocampe ascalia Foreman, 1971, p. 1678, pl. 4, fig. 4.

Theocampe bassilis Foreman, 1968, p. 50, pl. 6, fig. 10.

Theocampe daseia Foreman, 1968, p. 48, pl. 6, figs. 9a,b.

Theocampe lispa Foreman, 1968, p. 49, pl. 6, fig. 11.

Theocampe salillum Foreman, 1971, p. 1678, pl. 4, fig. 5.

Theocapsomma comys Foreman, 1968, p. 29, pl. 4, figs. 2a-c; *Theocapsomma comys* group, *in* Foreman, in press.

Theocapsomma teren Foreman, 1968, p. 32, pl. 4, fig. 4.

Triactinosphaera sp. Although not tabulated because of its rare occurrence, this form is illustrated here on Plate 1, Figure 13. It represents one of the few boreal forms present in the Atlantic Basin material as well as in the mid-continent Late Campanian.

Footnotes to Figure 6

These footnotes give the age or ages assigned to various samples together with a reference to the source. Letters "N" and "F" indicate that the information is based respectively on nannofossil or foraminiferal data.

1. F.- Maestrichtian, *Abathomphalus mayaroensis* Z., Krasheninnikov (in press). N.- Maestrichtian, (?) *Micula mura* Z., Cepek (in press).
2. F.- Maestrichtian, *Globotruncana gansseri* Z., Krasheninnikov (in press). N.- Maestrichtian, *Lithraphidites quadratus* Z., Cepek (in press).
3. N.- Late Campanian/Early Maestrichtian, *Arkhangelskiella cymbiformis* Z., Cepek (in press).
4. N.- Campanian, Milow (1970). Maestrichtian/Campanian, *Tetralithus nitidus trifidus* Z., Bukry and Bramlette (1970).
5. F.- Campanian/Early Maestrichtian, *Globotruncana fornicata - stuartiformis* Z., Beckman (1972). N.- Late Campanian/Early Maestrichtian, *Tetralithus gothicus trifidus* Z., Roth and Thierstein (1972), Bukry (1972). Although samples 144-3-2, 84-86 and 144A-4,CC are considered to be similar in age on the basis of nannofossil data, the presence in 144A-4,CC of common, well-developed *Amphipyndax enesseffi* unaccompanied by *A. tylotus* suggests an earlier age and the latter sample is here considered to be Campanian.
6. N.- Late Campanian/Early Maestrichtian, *Tetralithus trifidus* Z., Cepek (in press).
7. F.- Campanian, *Globotruncana elevata* Z., Krasheninnikov (in press). N.- Campanian, *Eiffelithus eximus* Z., Cepek (in press).
8. F.- Early Campanian, *Globotruncana elevata* Z., McNeely (1973). N.- Early Campanian/Late Santonian, Bukry (1973).
9. Campanian/E. Maestrichtian (age of Via Blanca form.) Bronnimann & Rigassi (1963).
10. F.- Campanian/Santonian, Pre Habana form., Bermudez (Foreman, 1968).
11. N.- Senonian, Milo (1970).
12. F.- Senonian, Saito (1970). N.-Senonian, Milo (1970).
13. F.- Early Santonian, *Globotruncana concavata* Z., J.B. Saunders (Foreman, 1968).
14. F.- Cenomanian, M.A. Chierici (Foreman, 1968).
15. R.- Cenomanian, Kozlova (1972).

Acknowledgements

Partial financial support for this study was provided by the National Science Foundation, Grant no. DES75-19288.

References

Aliev, Kh.Sh., 1965. Radioliarii nizhnemelovykh otlozhenii severovostochnogo Azerbaidzhana i ikh Stratigraficheskoe znachenie. (Radiolarians of the Lower Cretaceous deposits of northeastern Azerbaidzhan and their stratigraphic significance.): Izd. Akad. Nauk Azerbaidz. SSR, Baku, p. 3-124.

Beckman, J.P., 1972. The foraminifera and some associated microfossils of Sites 135 to 144. *In* Hayes, D.E., Pimm, A.C., et al., Initial Reports of the Deep Sea Drilling Project, v. 14, Washington (U.S. Government Printing Office). p. 389-420.

Brönnimann, P. and Rigassi, D., 1963. Contribution to the geology and paleontology of the area of the city of La Habana, Cuba, and its surroundings: Eclog. Geol. Helv. v. 56, no. 1, p. 193-480, pl. 1-25.

Bukry, D., 1972. Coccolith stratigraphy—Leg 14, Deep Sea Drilling Project. *In* Hayes, D.E., Pimm, A.C., et al., Initial Reports of the Deep Sea Drilling Project, v. 14, Washington (U.S. Government Printing Office), p. 487-494.

————1973. Coccolith statigraphy, Leg 10, Deep Sea Drilling Project. *In* Worzel, J.L., Bryant, W., et al., Initial Reports of the Deep Sea Drilling Project, v. 10, Washington (U.S. Government Printing Office), p. 385-406.

Bukry, D. and Bramlette, M.N., 1970. Coccolith age determinations Leg 3, Deep Sea Drilling Project. *In* Maxwell, A. E et al., Initial Reports of the Deep Sea Drilling Project, v. 3, Washington (U.S. Government Printing Office), p. 589-611.

Cepek, Pavel, in press (See Lancelot, Y. and Seibold. E. et al., in press.)

Dumitrica, P., 1970. Cryptocephalic and cryptothoracic Nassellaria in some Mesozoic deposits of Romania: Rev. Roum. Geol., Geophys., Geogr., Ser. Geol., v. 14, p. 45-124.

———— 1975. Cenomanian Radiolaria at Podul Dimbovitei (excursion B). *In* Micropaleontological guide to the Mesozoic and Tertiary of the Romanian Carpathians, 14th Europ. Micropal. Colloq., Bucharest (Inst. Geol. and Geophys.)

Foreman, H. P., 1966. Two Cretaceous radiolarian genera: Micropaleontology, v. 12, p. 355-359.

———— 1968. Upper Maestrichtian Radiolaria of California: Palaeontol. Assoc., London, Spec. Paper 3, p. iv + 1-82.

———— 1971. Cretaceous Radiolaria. *In* Winterer, E. L., Riedel, W. R., et al., Initial Reports of the Deep Sea Drilling Project, v. 7, Washington (U.S. Government Printing Office), p. 1673-1693.

———— 1975. Radiolaria from the North Pacific. Deep Sea Drilling Project, leg 32. *In* Larson, R.L., Moberly, R., et al., 1975. Initial Reports of the Deep Sea Drilling Project, v. 32, Washington (U.S. Government Printing Office), p. 579-676.

———— in press. Mesozoic Radiolaria in the Atlantic Ocean, off the west coast of Africa, Deep Sea Drilling Project, Leg 41.

Johnson, T.C., 1976. Controls on the preservation of biogenic opal in sediments of the eastern tropical Pacific: Science, v. 192, p. 887-890.

Kozlova, G. E., 1972 (See Petrushevskaya, M. G. and Kozlova, G. E., 1972.)

Krasheninnikov, V., in press. (See Lancelot, Y., and Seibold, E. et al., in press).

Lancelot, Y., and Seibold, E. et al., in press. Initial Reports of the Deep Sea Drilling Project, v. 41, Washington (U.S. Government Printing Office).

Maxwell, A. E. et al., 1970. Initial Reports of the Deep Sea Drilling Project, v. 3, Washington (U.S. Government Printing Office) xx + 806 p.

318

McNeely, B. W., 1973. Biostratigraphy of the Mesozoic and Paleogene pelagic sediments of the Campeche embankment area. *In* Worzel, J. L., Bryant, W., et al., Initial Reports of the Deep Sea Drilling Project, v. 10, Washington (U.S. Government Printing Office) p. 679-695.

Milow, D., 1970. (See Maxwell, A. E. et al. 1970).

Moore, T. C., 1973. Radiolaria from Leg 17 of the Deep Sea Drilling Project. *In* Winterer, E. L., Ewing, J. L., et al., Initial Reports of the Deep Sea Drilling Project, v. 17, Washington (U.S. Government Printing Office), p. 797-869.

Murray, John, 1876. Preliminary reports to Professor Wyville Thomson, F.R.S., Director of the civilian staff, on work done on board the *'Challenger'*: Roy. Soc. London, Proc., v. 24, p. 471-544.

Pessagno, E. A., Jr. 1963. Upper Cretaceous Radiolaria from Puerto Rico: Micropaleontology, v. 9, no. 2, p. 197-214.

————— 1972. Pseudoaulophacidae Riedel from the Cretaceous of California and the Blake-Bahama Basin (JOIDES Leg 1). *In* Cretaceous Radiolaria: Bull. Am. Paleontol., v. 61, p. 281-328.

————— in press. Radiolarian zonation and stratigraphy of the Upper Cretaceous portion of the Great Valley sequence, California Coast Ranges.

Petrushevskaya, M. G. and Kozlova, G. E., 1972. Radiolaria: Leg 14, Deep Sea Drilling Project. *In* Hayes, D. E., Pimm, A.C. et al., Initial Reports of the Deep Sea Drilling Project, v. 14, Washington (U.S. Government Printing Office), p. 495-678.

Riedel, W. R. and Sanfilippo, A., 1974. Radiolaria from the southern Indian Ocean, DSDP Leg 26. *In* Davies, T. A., Luyendyk, B. P., et al., Initial Reports of the Deep Sea Drilling Project, v. 26, Washington (U.S. Government Printing Office), p. 771-814.

Roth, P. H. and Thierstein, H., 1972. Calcareous nannoplankton: Leg 14 of the Deep Sea Drilling Project. *In* Hayes, D. E., Pimm, A. C., et al., Initial Reports of the Deep Sea Drilling Project, v. 14, Washington (U.S. Government Printing Office), p. 421-485.

Saito, T., 1970. (See Maxwell, A. E., et al., 1970.)

Plate 1. All figues are magnified 214x.

Fig. 1. *Amphipyndax tylotus*: 13A-2-1, 84-88. Sl. 2, Q19/0.

Fig. 2. *Amphipyndax enesseffi*: 369A-39-3, 69-70. Sl. 1, E24/3.

Fig. 3. *Druppatractona* sp. A: 369A-37,CC. Sl. 2, L30/3.

Fig. 4. *Ellipsoxiphus* sp. A: 369A-35,CC. Sl. 4, D19/2.

Fig. 5. *Ellipsoxiphus* sp. B: 369A-37,CC. Sl. 2, T28/0.

Fig. 6. *Lithostrobus punctulatus*: 369A-38,CC. Sl. 1, O18/0.

Fig. 7. *Lithomelissa* (?) *hoplites*: 369A-37,CC. Sl. 2, V21/2.

Fig. 8. *Lophophaena polycyrtis* V-18-129, 78 cm. St. 1241, D39/0.

Fig. 9. *Lithostrobus* sp. A: 369A-36-5, 65-67. Sl. 1, J28/1.

Fig. 10. *Stichomitra asymmetra*: 13A-2-1, 84-86. Sl. 1, L24/4.

Fig. 11. *Stichomitra* sp. A: 369A-36,CC. Sl. 5, N37/1.

Fig. 12. *Acidnomelos proapteron*: 369A-38,CC. Sl. 1, R24/1.

Fig. 13. *Triactinosphaera* sp.: 369A-37,CC. Sl. 1, M21/4.

Plate 1

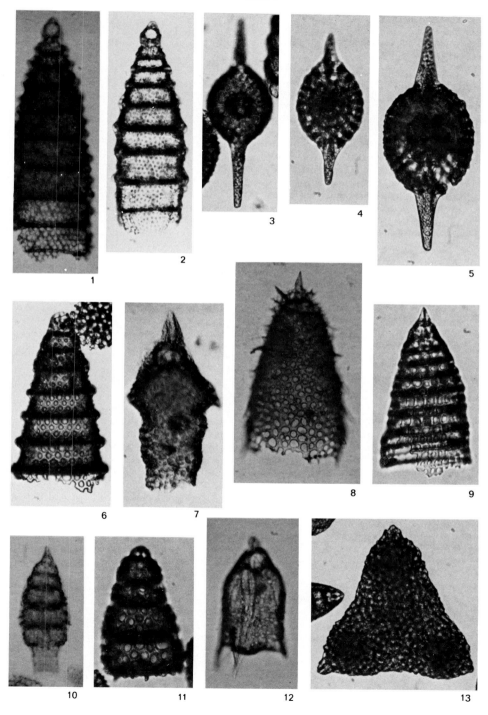

Discussion

Dr. W. W. Hay: Can the A. tylotus Zone be recognized in the
  mid-continent?

Foreman: No, the North and South Dakota Radiolaria I have examined of comparable age are almost completely different. The
  few species that are common to both sequences are only sporadically present in the Atlantic Basin sequence and have not
  been tabulated. One form is illustrated on Pl. 1, figure 13.

JURASSIC PALYNOSTRATIGRAPHY OF OFFSHORE EASTERN CANADA

by

JONATHAN P. BUJAK and GRAHAM L. WILLIAMS
Atlantic Geoscience Centre, Dartmouth, Nova Scotia

ABSTRACT

The Scotian Shelf and Grand Banks, offshore southeastern Canada, cover approximately 175,000 square miles and extend over 850 miles from northeast to southwest. More than 100 exploratory wells drilled to date have encountered Lower Paleozoic to Recent sediments, which frequently contain rich and age diagnostic palynomorph assemblages. Based upon the analysis of twenty wells, ten palynomorph zones have been defined within the Jurassic. These zones have been correlated with the standard European Jurassic stages, all of which are recognised, although the Hettangian-Sinemurian and Toarcian-Aalenian cannot, as yet, be separated. Both marine (dinoflagellates) and non-marine (spores) palynomorphs are utilized for zonation and for the interpretation of environments. On the Scotian Shelf the oldest encountered sediments, the Eurydice Formation, are Rhaetian-Hettangian, and are overlain by the Argo Salt, probably of Hettangian-Sinemurian age. The succeeding non-marine or marginally marine Iroquois Formation contains distinctive spore assemblages which have no known correlatives, but are tentatively dated Sinemurian-Pliensbachian. On the Grand Banks, marginally marine and evaporitic Hettangian-Sinemurian strata are overlain by the marine Whale Unit, whose dinoflagellates closely compare with Pliensbachian-Toarcian assemblages from Europe. Throughout the Middle and Late Jurassic, predominantly shallow marine deposition occurred in both areas, with some non-marine episodes. Dinoflagellate species diversity, low in the Early Jurassic and Aalenian-Bajocian, increased to a peak in the Kimmeridgian and subsequently declined in the Portlandian.

SOMMAIRE

La plateforme continentale de la Nouvelle Écosse et les Grands Bancs, situés au large de la partie sud-est du Canada, recouvrent environ

175,000 milles carrés et s'étendent, du nord-est au sud-ouest, sur plus de 850 milles.   Jusqu'ici, plus de 100 sondages d'exploration ont permis d'échantilloner des sédiments dont l'origine s'échelonne depuis le Paléozoïque inférieur jusqu'au Quaternaire, et qui renferment fréquemment des riches assemblages de palynomorphes permettant d'en établir l'âge.   L'analyse de vingt sondages a permis de distinguer dix zones palynomorphiques au sein du Jurassique.   On a établi la cor- relation entre ces zones et les étages jurassiques classiques de l'Europe, qui ont tous été identifiés, bien qu'on ne puisse pas encore séparer l'Hettangien-Sinémurien et le Toarcien-Aalénien.   On se sert à la fois des palynomorphes marins (dinoflagellés) et non marins (spores) pour établir les zones et pour interpréter les environnements.   Les sédiments les plus anciens de la plateforme continentale de la Nouvelle Écosse, ceux de la formation Eurydice, datent du Rhetien-Hettangien, et sont recouverts par la formation Argo ("Argo Salt"), qui remonte probablement à l'Hettangien-Sinémurien.   La formation Iroquois, non marine ou marginalement marine, qui y succède renferme des assemblages distinctifs de spores qui n'ont pas de correlatif connu, mais qu'on date expériment- alement du Sinémurien-Pliensbachien.   Sur les Grands Bancs, des strates hettangiennes-sinémuriennes, évaporitiques et marginalement marines sont recouvertes par le dépôt marin Whale ("Whale Unit"), dont les dino- flagellés correspondent de très près aux assemblages pliensbachiens- toarciens de l'Europe.   Durant tout le Jurassique moyen et supérieur, il s'est fait dans les deux secteurs des dépôts marins surtout en eau peu profonde, avec quelques épisodes non marins.   Le nombre d'espèces différentes de palynomorphes, faible au Jurassique inférieur et à l'Aalénien-Bajocien, s'est accru pour atteindre un maximum pendant le Kimmeridgien, et a diminué par la suite au Portlandien.

INTRODUCTION

The eastern Canadian continental shelf extends from the northern part of Georges Bank in the southwest to the Arctic Ocean in the north. Its southern part primarily consists of the Scotian Shelf and Grand Banks. These two areas cover more than 175,000 square miles and extend over 850 miles from northeast to southwest (text-figure 1).   Since the first two exploratory wells were drilled on the Grand Banks in 1966, over 100 wells have provided a unique opportunity for detailed geological study of the western North Atlantic margin.   Intensive investigations of many of

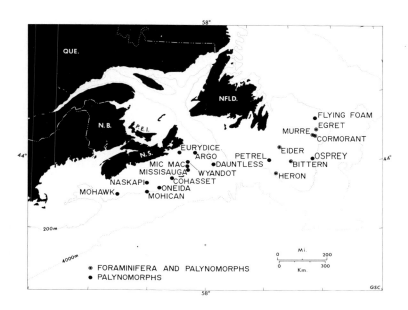

Text-figure 1. Scotian Shelf-Grand Banks wells in which
Jurassic sections have been studied.

these wells have provided lithostratigraphic, biostratigraphic and geochemical data which have been integrated with various geophysical parameters.

Biostratigraphic studies have been undertaken on foraminifera,
ostracods, nannoplankton, and various organic-walled microfossils, the
latter including chitinozoa and acritarchs in the Early Paleozoic,
acritarchs and spores in the Late Paleozoic, and dinoflagellates, spores,
and acritarchs in the Mesozoic and Cenozoic. The presence of dinoflagellates, chitinozoa, and acritarchs is usually indicative of marine
deposition. Spores, although of terrestrial origin, may also be
present in marine sediments and thus permit correlation of marine and
non-marine deposits.

This paper presents the palynological zonation, based on dino-
flagellates and spores, of the Jurassic studied in eleven Scotian Shelf
and nine Grand Banks wells (text-figure 1). The Scotian Shelf wells are
Mobil-Tetco Cohasset D-42, Mobil-Tetco Dauntless D-35, Shell Argo F-38,
Shell Eurydice P-36, Shell Mic Mac J-77, Shell Missisauga H-54, Shell
Mohawk B-93, Shell Mohican I-100, Shell Naskapi N-30, Shell Oneida O-25,

324

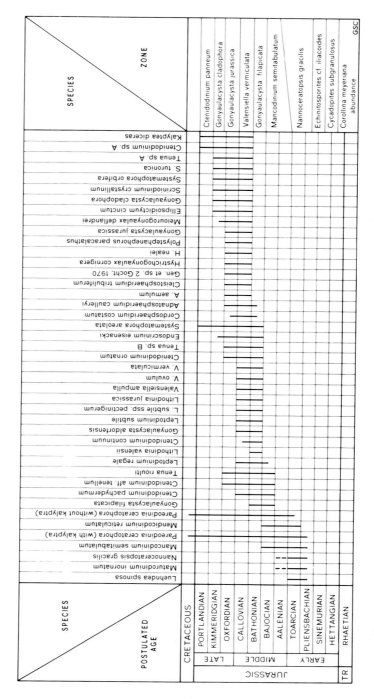

Text-figure 2A. Known stratigraphic ranges of Jurassic dinoflagellates, Scotian Shelf-Grand Banks.

The table lists species as columns and postulated ages as rows, with stratigraphic ranges shown as vertical bars.

Period	Epoch	Sub	Postulated Age	Occisucysta sp. A	Tenua hystrix	Stephanelytron caytonense	S. scarburghense	Hexagonifera jurassica	Stephanelytron redcliffense	Acanthaulax paliuros	Taeniophora iunctispina	Tenua villersense	Gonyaulacysta jurassica ssp. longicornis	Systematophora fasciculigera	Endoscrinium luridum	Gen. et sp. 1	Gonyaulacysta granulata	G. granuligera	Leptodinium egemenii	Parvocavatus tuberosus	Scriniodinium dictyotum	Leptodinium "arcuatum"	Gonyaulacysta ambigua	Herendeenia pisciformis	Prolixosphaeridium granulosum	Muderongia simplex	Epiplosphaera reticulospinosa	Gen. et sp. 2	Gonyaulacysta aculeata	Gonyaulacysta ehrenbergi	Imbatodinium antennatum	Wanaea spectabilis	Dingodinium jurassicum	Prolixosphaeridium deirense	Psaligonyaulax apatela	Imbatodinium kondratjevi	Polystephanephorus sarjeantii	Amphorula metaelliptica	Ctenidodinium culmulum	Ctenidodinium panneum	Lanterna sportula	Hystrichodinium pulchrum	Oligosphaeridium dividuum	Systematophora schindewolfi	Ctenidodinium schizoblatum	Pyxidiella sp.	Zone
CRETACEOUS																																															
JURASSIC	LATE		PORTLANDIAN																																											Ctenidodinium panneum
			KIMMERIDGIAN																																											Gonyaulacysta cladophora
			OXFORDIAN																																											Gonyaulacysta jurassica
	MIDDLE		CALLOVIAN																																											Valensiella vermiculata
			BATHONIAN																																											Gonyaulacysta filapicata
			BAJOCIAN																																											
			AALENIAN																																												Mancodinium semitabulatum
	EARLY		TOARCIAN																																											
			PLIENSBACHIAN																																											Nannoceratopsis gracilis
			SINEMURIAN																																											Echinitosporites cf. iliacoides
			HETTANGIAN																																											Cycadopites subgranulosus
TR.			RHAETIAN																																											Corollina meyeriana abundance

Text-figure 2B. Known stratigraphic ranges of Jurassic dinoflagellates, Scotian Shelf-Grand Banks.

GSC

326

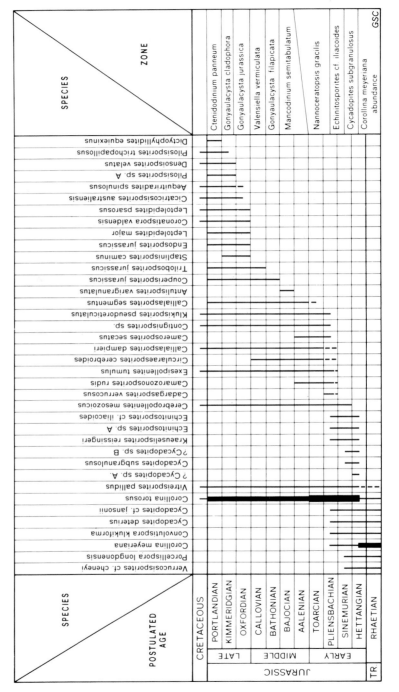

Text-figure 3. Known stratigraphic ranges of Jurassic spores, Scotian Shelf-Grand Banks.

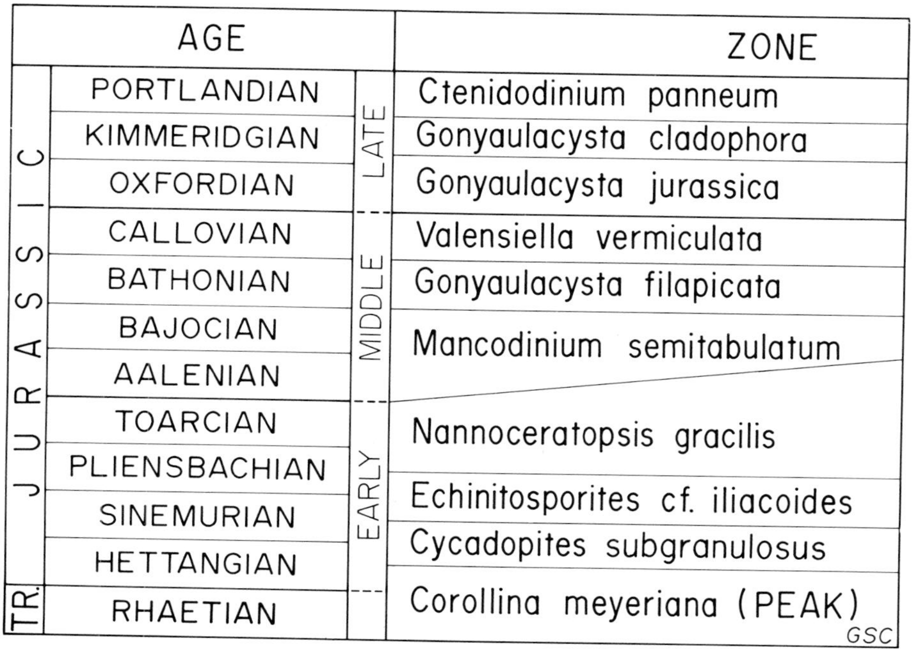

AGE			ZONE
JURASSIC	PORTLANDIAN	LATE	Ctenidodinium panneum
	KIMMERIDGIAN		Gonyaulacysta cladophora
	OXFORDIAN		Gonyaulacysta jurassica
	CALLOVIAN	MIDDLE	Valensiella vermiculata
	BATHONIAN		Gonyaulacysta filapicata
	BAJOCIAN		Mancodinium semitabulatum
	AALENIAN		
	TOARCIAN	EARLY	Nannoceratopsis gracilis
	PLIENSBACHIAN		
	SINEMURIAN		Echinitosporites cf. iliacoides
	HETTANGIAN		Cycadopites subgranulosus
TR.	RHAETIAN		Corollina meyeriana (PEAK)

GSC

Text-figure 4.   Jurassic palynomorph zonation, Scotian Shelf-Grand Banks.

and Shell Wyandot E-53.   The Grand Banks wells are Amoco-Imp Bittern M-62, Amoco-Imp Cormorant N-83, Amoco-Imp Heron H-73, Amoco-Imp Petrel A-62, Amoco-Imp-Skelly Egret K-36, Amoco-Imp-Skelly Osprey H-84, Amoco IOE Eider M-75, Amoco-IOE Murre G-67, and Mobil-Gulf Flying Foam I-13. The stratigraphic ranges of Jurassic dinoflagellates and spores, as determined from these wells, are shown in text-figures 2A, 2B and 3.

Ten formal zones are recognized in the Jurassic of the Scotian Shelf-Grand Banks, based on the stratigraphic distribution of dinoflagellates and spores (text-figure 4).   Nine of these are assemblage zones and one is a peak zone as defined in the American Commission on Stratigraphic Nomenclature, 1961, Articles 20g, 21.   The four highest Jurassic zones were erected by Williams (1975).   The remaining six zones are proposed formally herein for the first time, although H. Sabry verbally presented a paper, discussing the Early Jurassic Scotian Shelf palynostratigraphy, at the Eighth Annual Meeting of the American Association of Stratigraphic Palynologists at Houston in 1975.   The present study is based primarily on cuttings samples, so that it is necessary to use fossil "tops" (i.e., the latest or highest occurrences of a species) to define zones and correlate between wells.   Each assemblage zone takes its name from a fossil not found in sediments above that zone.   Where it can be established with some certainty, as when using sidewall or conventional cores, the "base", or oldest occurrence, of a species is also used to delineate zones and for correlation.   Peak zones are named after species which attain their maximum abundance within them.

The majority of Scotian Shelf-Grand Banks Jurassic species have been previously described from European surface sections.   The strati-graphic ranges of these species permit dating of the zones relative to European stage terminology.   It must be noted, however, that all age assignments are provisional and may be subject to modification as new data become available.

Correlation of the palynological and foraminiferal zonations is presented in Gradstein (this volume).

BIOSTRATIGRAPHY

Corollina meyeriana Peak Zone

Type section: Shell Eurydice P-36, 8920-9700 ft.

Other wells: Argo F-38, Heron H-73, Osprey H-84.

Characteristic species: Spores; *Convolutispora klukiforma* (Nilsson) Schulz, *Corollina meyeriana* (Klaus) Venkatachala and Góczán abundance (plate 1, figure 1), *C. torosus* (Reissinger) Klaus, *Verrucosisporites cheneyi* Cornet and Traverse, *Verrucosisporites* sp. A (plate 2, figure 3).

Species bases: None established.

Selected species tops: Spores; *C. meyeriana* abundance.

Discussion: Some horizons from this zone may be dominated by *C. torosus* with fewer *C. meyeriana*.   The assemblages containing abundant *C. meyeriana* are closely comparable to those documented by Cornet and Traverse (1975) from the Shuttle Meadow Formation of the Hartford Basin, eastern United States.   These authors give a detailed discussion of the age of the Shuttle Meadow Formation and assign a probable Rhaeto-Liassic (Rhaetian-Hettangian) age from macro- and microfossil evidence and isotope dating.   The *C. meyeriana* Peak Zone is recognised in only four wells (Argo F-38, Eurydice P-36, Heron H-73, Osprey H-84).   In Bittern M-62, Cormorant N-83, and Petrel A-62 the pre-late Pliensbachian Liassic cannot be subdivided.

Cycadopites subgranulosus Assemblage Zone

Type section: Shell Eurydice P-36, 2640-8814 ft.

Other wells: Argo F-38, Mohican I-100.

Selected species bases: Spores; *Cerebropollenites mesozoicus* (Couper) Nilsson, *Cycadopites subgranulosus* (Couper) comb. nov., *?Cycadopites* spp. A and B (plate 2, figures 4,7,8), *Echinitosporites* cf. *iliacoides* Schulz and Krutzsch, *Kraeuselisporites reissingeri* (Harris) Morbey (plate 1, figure 2; plate 2, figures 5,6).

Selected species tops: Spores; *Cycadopites subgranulosus*, *?Cycadopites* spp. A and B, *Verrucosisporites cheneyi*, *Verrucosisporites* sp. A.

Discussion: Laevigate, granulate, and verrucate species of *Cycadopites* and *?Cycadopites* are frequently common and characterise this zone. Assemblages are mostly dominated by *Corollina torosus* while *C. meyeriana* is uncommon.   The predominance of *C. torosus* over *C. meyeriana* throughout the *Cycadopites subgranulosus* Zone indicates that the zone is no older than Hettangian, following the reasoning of Cornet and Traverse (1975).   The presence of *Cerebropollenites mesozoicus* supports an age no older than Liassic and the occurrence of *Kraeuselisporites reissingeri* and *Cycadopites subgranulosus* in the *C. subgranulosus* Zone indicates that the zone is Hettangian-Sinemurian in age.   Dinoflagellates have not been recorded from this zone.

Echinitosporites cf. *iliacoides* Assemblage Zone

Type section: Shell Eurydice P-36, 1765-2575 ft.

Other wells: Argo F-38.

Selected species bases: None established.

Selected species tops: Spores; *Convolutispora klukiforma*, *Corollina meyeriana*, *Echinitosporites* cf. *iliacoides* (plate 1, figures 3,4; plate 2, figures 9,12), *Echinitosporites* sp. A (plate 2, figures 10,11), *Kraeuselisporites reissingeri*.

Discussion: Monosulcate pollen assigned to *Echinitosporites* cf. *iliacoides* are common in this zone.   *E. iliacoides* has only been described from Late Triassic sediments (see Scheuring, 1970, for discussion). *Convolutispora klukiforma* is characteristic of the European Liassic (Cornet and Traverse, 1975), and Geiger and Hopping (1968) noted that *Kraeuselisporites reissingeri* has a known range of Rhaetian-Pliensbachian. The zone is tentatively dated Sinemurian-early Pliensbachian.

Nannoceratopsis gracilis Assemblage Zone

Type section: Amoco-IOE Murre G-67, 6450-7870 ft.

Other wells: Argo F-38, Bittern M-62, Cormorant N-83, Eider M-75, Heron H-73, Mohican I-100.

Selected species bases: Dinoflagellates; *Luehndia spinosa* Morgenroth, *Mancodinium semitabulatum* Morgenroth, *Maturodinium inornatum* Morgenroth, *Mendicodinium reticulatum* Morgenroth, *Nannoceratopsis gracilis* Alberti, *Pareodinia ceratophora* Deflandre. Spores; *Callialasporites dampieri* (Balme) Dev, *Contignisporites* sp. (plate 2, figure 2), *Klukisporites pseudoreticulatus* Couper.

Selected species tops: Dinoflagellates; *Luehndia spinosa, Maturodinium inornatum, Nannoceratopsis gracilis.*

Discussion: Dinoflagellates first appear in the *Nannoceratopsis gracilis* Zone on the Scotian Shelf and indicate a late Pliensbachian-Toarcian/ Aalenian age, as they compare closely with assemblages of this age from northwest Europe (see Williams, in press, for discussion). On the Scotian Shelf, the coeval deposits are devoid of dinoflagellates, but can be dated using spores. The first appearance of *Callialasporites dampieri, Contignisporites* sp. and *Klukisporites pseudoreticulatus* is taken as the base of the *N. gracilis* Zone. Herngreen and de Boer (1974) noted that *Contignisporites problematicus* first appears in the Upper Pliensbachian. Pocock (1970) placed the base of *C. dampieri* in western Canada within the Pliensbachian-Toarcian.

Mancodinium semitabulatum Assemblage Zone

Type section: Amoco-IOE Eider M-75, 8545-10,240 ft.

Other wells: Argo F-38, Bittern M-62, Cormorant N-83, Mohican I-100, Murre G-67.

Selected species bases: Dinoflagellates; *Ctenidodinium pachydermum* (Deflandre) Gocht, *C.* aff. *tenellum* Deflandre, *Gonyaulacysta filapicata* Gocht, *Leptodinium regale* Gocht.

Selected species tops: Dinoflagellates; *Mancodinium semitabulatum, Mendicodinium reticulatum.*

Discussion: *Ctenidodinium pachydermum* and *Gonyaulacysta filapicata* first appear in the Bajocian of southern England (personal observation, G.L. Williams). *Mancodinium semitabulatum* and *Mendicodinium reticulatum* were originally described from the Late Pliensbachian of Germany by Morgenroth (1970). The association of all four species in this zone indicates a Bajocian age. The *M. semitabulatum* Zone is recognised in

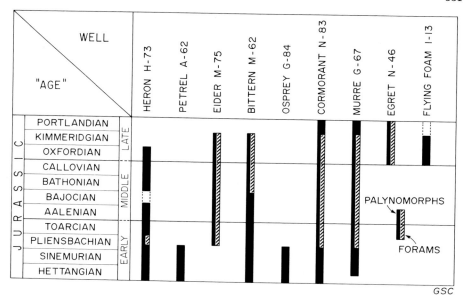

Text-figure 5. Stratigraphic distribution of Jurassic sediments in the examined Grand Banks wells.

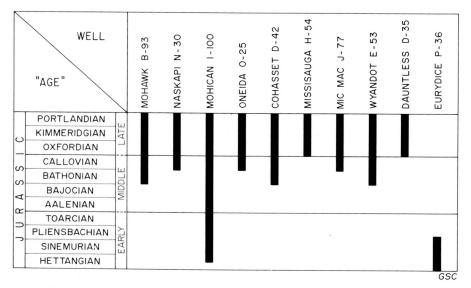

Text-figure 6. Stratigraphic distribution of Jurassic sediments in the examined Scotian Shelf wells.

six of the wells, four on the Grand Banks and two on the Scotian Shelf (text-figures 5,6).

Gonyaulacysta filapicata Assemblage Zone

Type section: Amoco-Imp Bittern M-62, 6600-8230 ft.

Other wells: Argo F-38, Cormorant N-83, Eider M-75, Heron H-73, Mohawk B-93, Mohican I-100, Murre G-67, Wyandot E-53.

Selected species bases: Dinoflagellates; *Cordosphaeridium costatum* (Davey and Williams) Gorka, *Ctenidodinium ornatum* (Eisenack) Deflandre, *Gonyaulacysta aldorfensis* Gocht, *Lithodinia jurassica* Eisenack, *Valensiella ovulum* (Deflandre) Eisenack, *V. vermiculata* Gocht.

Selected species tops: Dinoflagellates; *Gonyaulacysta filapicata*.

Discussion: Bathonian dinoflagellate assemblages have been described from Bulgaria by Dodekova (1975) and from Germany by Gocht (1970). Most of the species described by Gocht, including *Gonyaulacysta filapicata*, occur in the *G. filapicata* Zone. The zone is therefore dated Bathonian. A detailed discussion of the Bathonian index species documented by Gocht is given in Williams (in press). There is a marked increase in dinoflagellate species diversity from nine in the *Mancodinium semitabulatum* Zone (Bajocian) to 22 in the *G. filapicata* Zone (Bathonian), as shown in text-figure 7.

Valensiella vermiculata Assemblage Zone

Type section: Shell Naskapi N-30, 6590-6992 ft.

Other wells: Argo F-38, Bittern M-62, Cohasset D-42, Cormorant N-83, Eider M-75, Heron H-73, Mic Mac J-77, Mohawk B-93, Mohican I-100, Murre G-67, Oneida O-25, Wyandot E-53.

Selected species bases: Dinoflagellates; Gen. et sp. 2 Gocht, 1970, *Gonyaulacysta cladophora* (Deflandre) Dodekova, *G. jurassica* (Deflandre) Norris and Sarjeant. *Scriniodinium crystallinum* (Deflandre) Klement, *Systematophora orbifera* Klement. The dinoflagellates *Hexagonifera jurassica* Gitmez and Sarjeant, *Stephanelytron caytonense* Sarjeant and *S. scarburghense* Sarjeant appear in the upper part of this zone. Spores; *Trilobosporites jurassicus* Pocock.

Selected species tops: Dinoflagellates; *Gonyaulacysta aldorfensis*, *Leptodinium regale*, *L. subtile*, *Lithodinia jurassica*, *Stephanelytron caytonense*, *S. scarburghense*, *Valensiella ovulum*, *V. vermiculata*. Spores; *Circularaesporites cerebroides*.

Discussion: Williams (in press) discussed the known records of dinoflagellates from the type Callovian section and elsewhere. Comparison

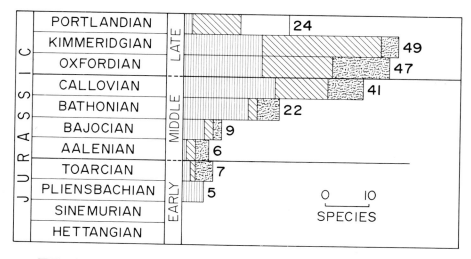

Text-figure 7. Species diversity of Jurassic dinocysts, Scotian Shelf-
Grand Banks.

of the *V. vermiculata* Zone assemblages with this data substantiates its
Callovian age assignment.   This zone is widespread in the Jurassic of
offshore eastern Canada and is apparently always marine with diverse
dinoflagellate assemblages, 41 species having been recorded.   The zone
is recognised in five Grand Banks and seven Scotian Shelf wells.

Gonyaulacysta jurassica Assemblage Zone

Type section: Shell Naskapi N-30, 6390-6590 ft.
Other wells: Argo F-38, Bittern M-62, Cohasset D-42, Dauntless D-35,
Egret K-36, Eider M-75, Flying Foam I-13, Heron H-73, Mic Mac J-77,
Missisauga H-54, Mohawk B-93, Mohican I-100, Murre G-67, Oneida O-25,
Wyandot E-53.
Selected species bases: Dinoflagellates; *Endoscrinium luridum* (Deflandre)
Gocht, *Gonyaulacysta granulata* (Klement) Sarjeant, *G. granuligera* (Klement)
Sarjeant, *Leptodinium egemenii* Gitmez, *Scriniodinium dictyotum* Cookson and
Eisenack.   The spores *Cicatricosisporites australiensis* (Cookson)
Potonié, *Densoisporites velatus* Weyland and Krieger, and *Pilosisporites*
sp. A (plate 2, figure 1) appear in the upper part of this zone.
Selected species tops: Dinoflagellates; *Adnatosphaeridium caulleryi*

(Deflandre) Williams and Downie, *Cordosphaeridium costatum, Ctenidodinium ornatum, Endoscrinium eisenacki,* Gen. et sp. 2 Gocht, 1970, *Gonyaulacysta jurassica.*

Discussion: This zone is dominated by *Gonyaulacysta jurassica* on the northeastern Grand Banks. Elsewhere species of *Ctenidodinium* replace *G. jurassica* as the dominant form. Marine conditions apparently prevailed throughout offshore eastern Canada during this time. The age of this zone is discussed by Williams (1975), who assigned an Oxfordian, possibly in part Early Kimmeridgian, age.

Gonyaulacysta cladophora Assemblage Zone

Type section: Shell Oneida O-25, 10,005-11,000 ft.

Other wells: Argo F-38, Bittern M-62, Cohasset D-42, Cormorant N-83, Dauntless D-35, Egret K-36, Eider M-75, Flying Foam I-13, Mic Mac J-77, Missisauga H-54, Mohawk B-93, Mohican I-100, Murre G-67, Naskapi N-30, Wyandot E-53.

Selected species bases: Dinoflagellates; *Dingodinium jurassicum* Cookson and Eisenack, *Epiplosphaera reticulospinosa* Klement, *Imbatodinium kondratjevi* Vozzhennikova, *Polystephanophorus sarjeantii* Gitmez, *Psaligonyaulax apatela* (Cookson and Eisenack) Sarjeant. The dinoflagellates *Amphorula metaelliptica* Dodekova, *Ctenidodinium culmulum* (Norris) Lentin and Williams, *C. panneum* (Norris) Lentin and Williams, and *Systematophora schindewolfi* (Alberti) Downie and Sarjeant appear in the upper part of this zone. Spores; *Pilosisporites trichopapillosus* (Thiergart) Delcourt and Sprumont.

Selected species tops: Dinoflagellates; *Endoscrinium luridum, Gonyaulacysta cladophora, G. granuligera, Hexagonifera jurassica, Leptodinium egemenii, Systematophora turonica* (Alberti) Downie and Sarjeant.

Discussion: Dinoflagellate species diversity reaches a maximum (49) for the Jurassic in the *G. cladophora* Zone. The zone is clearly delineated by the first appearance of 18 species and the last occurrence of 27 species. Dinoflagellates are consistently present, permitting recognition of the zone in 16 of the 17 wells having a Late Jurassic section. Kimmeridgian dinoflagellates have been well documented from Europe (Gitmez, 1970; Gitmez and Sarjeant, 1972) and show close affinities with assemblages from the *G. cladophora* Zone. Justification of the Kimmeridgian age assignment for this zone is given in Williams (1975).

Ctenidodinium panneum Assemblage Zone

Type section: Shell Mohawk B-93, 5315-5395 ft.

Other wells: Argo F-38, Cohasset D-42, Cormorant N-83, Dauntless D-35, Egret K-36, Flying Foam I-13, Mic Mac J-77, Missisauga H-54, ?Murre G-67, Naskapi N-30, Oneida O-25, Wyandot E-53.

Selected species bases: Dinoflagellates; *Pyxidiella* sp. Williams, 1975.

Selected species tops: Dinoflagellates; *Amphorula metaelliptica, Ctenidodinium culmulum, C. panneum, Systematophora areolata* Klement. Spores; *Couperisporites jurassicus* Pocock, *Dictyophyllidites equiexinus* (Couper) Dettmann, *Endosporites jurassicus* Pocock, *Pilosisporites* sp. A, *Trilobosporites jurassicus.*

Discussion: Dinoflagellate assemblages from this zone differ from those of the underlying Late Jurassic zones in containing fewer species. This, together with the paucity of specimens, possibly reflects local changing environmental conditions, apparently resulting from a marine regression. However, similar observations in other areas suggest a worldwide decline phase in dinoflagellate populations at this time. Only 12 Jurassic dinoflagellate species persist into the Cretaceous in offshore eastern Canada. The *Ctenidodinium panneum* Zone has been dated Portlandian by Williams (1975) on the basis of the dinoflagellates present. This zone, widespread on the Scotian Shelf and present on the eastern Grand Banks is generally absent on the western and central Grand Banks.

Several Scotian Shelf wells contain a distinctive spore assemblage between 200 and 300 ft. above the *C. panneum* Zone. This is typically characterised by numerous specimens of *Callialasporites dampieri, Contignisporites cooksonii* (Balme) Dettmann, *Densoisporites velatus, Klukisporites pseudoreticulatus* and sometimes an abundance of *Corollina torosus,* as in Argo F-38. The interval is provisionally included in the Berriasian because of lack of data regarding its true age, whereas the age of the *C. panneum* Zone can be justified by comparison with southern England.

PALEOECOLOGY

Lower Jurassic strata on the Scotian Shelf commonly contain abundant spores of low diversity and are devoid of dinoflagellates. During the Early Jurassic, the Eurydice, Argo, and Iroquois Formations were deposited. The Eurydice Formation is dated Rhaetian-Hettangian and, according to Jansa and Wade (1975) accumulated under predominantly arid conditions in a desert environment. The overlying Argo Formation, consisting mainly of salt deposits, is assigned a Hettangian-Sinemurian age and is possibly coeval with the Louann Salt of the Gulf of Mexico (H. Sabry, personal communication). Marginally marine conditions have been suggested for the succeeding Iroquois Formation (Jansa and Wade, *op. cit.*), which is herein dated Sinemurian-early Pliensbachian.

Marine late Pliensbachian-Aalenian sediments (*Nannoceratopsis gracilis* Zone), recognised in five Grand Banks wells, contain several diagnostic dinoflagellate species. These species are absent from the examined Scotian Shelf wells, so that it is not certain if late Pliensbachian and Aalenian sediments were deposited in this area. Bajocian dinoflagellates (*Mancodinium semitabulatum* Zone) have been observed in four Grand Banks wells, but in only two Scotian Shelf wells, Argo F-38 and Mohican I-100. Species diversity is low, making correlation with the European type sections difficult. From Bathonian to Kimmeridgian time, marine conditions increasingly prevailed on both the Scotian Shelf and Grand Banks. This is reflected in the steady increase in dinoflagellate species diversity, which attained a peak in the Kimmeridgian (text-figure 7). The depositional environment is interpreted as predominantly inner neritic. More restricted marine to non-marine environments characterised the Portlandian and prevailed into the Early Cretaceous on the Scotian Shelf. The Grand Banks succession at this time is complicated by a major unconformity.

ACKNOWLEDGEMENTS

The authors are grateful to Hassan Sabry for fruitful discussions on Early Jurassic palynostratigraphy, to M. S. Barss, D. Umpleby and J. Wade for their constructive criticism of this paper, and to G. L. Cook and G. M. Grant for drafting the text-figures.

PLATE 1   Magnification shown on figures.

1.   *Corollina meyeriana* (Klaus) Venkatachala and Góczán. Shell Eurydice P-36, cuttings sample 2840-2880 ft: Hettangian-Sinemurian. GSC No. 47929.

2.   *Kraeuselisporites reissingeri* (Harris) Morbey. Shell Eurydice P-36, cuttings sample 2840-2880 ft: Hettangian-Sinemurian. GSC No. 47930.

3,4.   *Echinitosporites* cf. *iliacoides* Schulz and Krutzsch. Shell Eurydice P-36, cuttings sample 2840-2880 ft: Hettangian-Sinemurian. GSC No. 47931 and 47932.

REFERENCES

American Commission on Stratigraphic Nomenclature, 1961. Code of
  stratigraphic nomenclature: Am. Assoc. Petrol. Geol. Bull.,
  v. 45, p. 645-660.
Cornet, B. and Traverse, A., 1975. Palynological contributions to the
  chronology and stratigraphy of the Hartford Basin in Connecticut and
  Massachusetts: Geoscience and Man, v. 11, p. 1-33.
Dodekova, L., 1975. New Upper Bathonian dinoflagellate cysts from north-
  eastern Bulgaria: Paleont. Stratigr. and Lithol., v. 2, p. 17-32.
Geiger, M.E. and Hopping, C.A., 1968. Triassic stratigraphy of the
  southern North Sea basin: Philos. Trans. R. Soc. London, ser B, v. 254,
  p. 1-39.
Gitmez, G.U., 1970. Dinoflagellate cysts and acritarchs from the basal
  Kimmeridgian (Upper Jurassic) of England, Scotland and France: Bull.
  Br. Mus. nat. Hist. (Geol.), v. 18, p. 231-331.
Gitmez, G.U. and Sarjeant, W.A.S., 1972. Dinoflagellate cysts and
  acritarchs from the Kimmeridgian (Upper Jurassic) of England, Scotland
  and France: Bull. Br. Mus. nat. Hist. (Geol.), v. 21, p. 171-257.
Gocht, H., 1970. Dinoflagellaten-Zysten aus dem Bathonium des Erdölfeldes
  Aldorf (NW-Deutschland): Palaeontographica, Abt. B, v. 129, p. 125-165.

PLATE 1

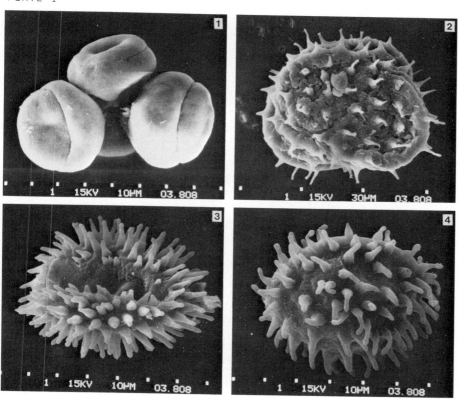

Herngreen, G.F.W. and de Boer, K.F., 1974. Palynology of Rhaetian, Liassic and Dogger strata in the Netherlands with emphasis on the Achterhoek area:Geologie en Mijnbouw, v. 53, no. 6, p. 343-368.

Jansa, L.F. and Wade, J.A., 1975. Geology of the continental margin off Nova Scotia and Newfoundland; *in* Offshore Geology of Eastern Canada, Vol. II: Geol. Surv. Can., Paper 74-30, v. 2, p. 51-105.

Morgenroth, P., 1970. Dinoflagellate cysts from the Lias Delta of Lühnde/ Germany: Neues Jahrb. Geol. Paläontol., Abh., v. 136, no. 3, p.345-359.

Pocock, S.A.J., 1970. Palynology of the Jurassic sediments of western Canada. Part 1: Terrestrial species: Palaeontographica, Abh. B, v. 130, p. 12-72.

Scheuring, B.W., 1970. Palynologische und palynostratigraphische Untersuchungen des Keupers im Bölchentunnel (Solothurner Jura): Schweiz. Paläont. Abh., v. 88, p. 1-119.

Williams, G.L., 1975. Dinoflagellate and spore stratigraphy of the Mesozoic-Cenozoic, offshore eastern Canada; *in* Offshore Geology of Eastern Canada, Vol. II: Geol. Surv. Can., Paper 74-30, v. 2, p. 107-161.

_____, In Press. Dinocysts: Their palaeontology, biostratigraphy and palaeoecology; *in* Oceanic Micropalaeontology, Academic Press, London, England.

PLATE 2   Figures 1-3 x 440; Figures 4-12 x 715.

1. *Pilosisporites* sp. A.   Mobil-Tetco Cohasset D-42, cuttings sample 10,570-10,600 ft: Kimmeridgian. GSC No. 47933.

2. *Contignisporites* sp.   Mobil-Tetco Cohasset D-42, cuttings sample 11,070-11,100 ft: Kimmeridgian. GSC No. 47934.

3. *Verrucosisporites* sp. A.   Shell Eurydice P-36, cuttings sample 6600-6640 ft: Hettangian-Sinemurian. GSC No. 47935.

4. *?Cycadopites* sp. A.   Shell Eurydice P-36, sidewall core 5048 ft: Hettangian-Sinemurian. GSC No. 47936.

5,6 *Kraeuselisporites reissingeri* (Harris) Morbey, Shell Eurydice P-36, sidewall core 5048 ft: Hettangian-Sinemurian. GSC No. 47937.

7,8 *?Cycadopites* sp. B.7, Shell Eurydice P-36, sidewall core 2911 ft: Hettangian-Sinemurian. 8, Shell Eurydice P-36, sidewall core 3229 ft: Hettangian-Sinemurian. GSC No. 47938 and 47939.

9,12 *Echinitosporites* cf. *iliacoides* Schulz and Krutzsch.   Shell Eurydice P-36, sidewall core 2575 ft: Sinemurian-early Pliens-bachian. GSC No. 47940 and 47941.

10,11 *Echinitosporites* sp. A.   Shell Eurydice P-36, cuttings sample 2840-2880 ft: Hettangian-Sinemurian. GSC No. 47942.

PLATE 2

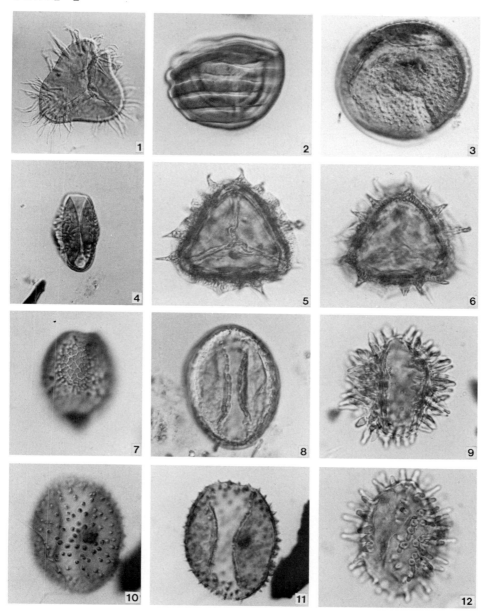

COMPARISON OF LOWER AND MIDDLE CRETACEOUS PALYNOSTRATIGRAPHIC ZONATIONS IN THE WESTERN NORTH ATLANTIC

by

Daniel Habib
Department of Earth and Environmental Sciences
Queens College, New York City

ABSTRACT

Cretaceous palynomorph zonations, based respectively on the stratigraphy of dinoflagellate cysts and sporomorphs, are presented for six Deep Sea Drilling Project sites in the western North Atlantic. Eight dinoflagellate zones and four sporomorph zones are described from the reference section at site 105, which ranges in age from Berriasian (core 30) to Cenomanian (core 9).

Published nannofossil and planktonic foraminiferal studies of site 105 permit the direct comparison of part of the palynomorphic zonations with a published geochronological scale. The sporomorph zonation is compared with that published for the largely nonmarine Lower Cretaceous facies of the adjacent U. S. Atlantic coastal plain.

INTRODUCTION

Zonations are proposed for two groups of palynomorphs, dinoflagellate cysts and sporomorphs, occurring in Lower and Middle Cretaceous sections recovered at six sites in the western North Atlantic by the Deep Sea Drilling Project (Figure 1). The dinoflagellate zonation published by Habib (1976) for Neocomian assemblages is revised in part and extended to the Cenomanian, and a sporomorph zonation ranging from Berriasian to Cenomanian is proposed. The zonations are based primarily on the stratigraphy at site 105, although two dinoflagellate subzones are included from species which are stratigraphically continuous at site 391 only. The dinoflagellate and sporomorph zonations established at site 105 are correlated for at least part of the section, through sites

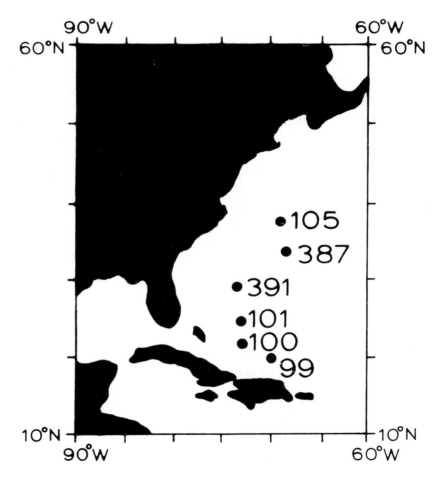

Figure 1.   Geographic distribution of investigated
sites, Deep Sea Drilling Project.

99, 100, 101, 387, and 391 (cf. Habib, 1976, Text-Figure 4).

A total of 161 samples were studied, collected from 21 cores representing the reference section at site 105.

This interval is approximately 250 meters thick, ranging from the stratigraphic level of core 30 (Berriasian) through that of core 9 (Cenomanian). The cores immediately below and above the investigated interval are essentially devoid of palynofossils. However, the investigated interval is highly palyniferous, with respect to dinoflagellates, sporomorphs, and small acritarchs. The latter group is currently under investigation.

BIOSTRATIGRAPHY

The ranges of 65 species of dinoflagellates and sporomorphs in the Lower and Middle Cretaceous section at DSDP site 105 are presented in Figure 2. These species are considered to be of chronostratigraphic value, based on their stratigraphic ranges in the section or their ages as derived from the literature. Separate dinoflagellate and sporomorph zonations are proposed from the ranges of selected stratigraphically continuous species at site 105 (Figure 3). Separate zonations are preferable to a single combined palynomorph zonation for the following reasons. First, a separate sporomorph zonation permits the potential correlation of North Atlantic sections with the largely nonmarine Lower Cretaceous of the adjacent U.S. Atlantic Coastal Plain, where dinoflagellates are rare or absent. Also, in the study of a number of sections within the North Atlantic the relative abundance of

344

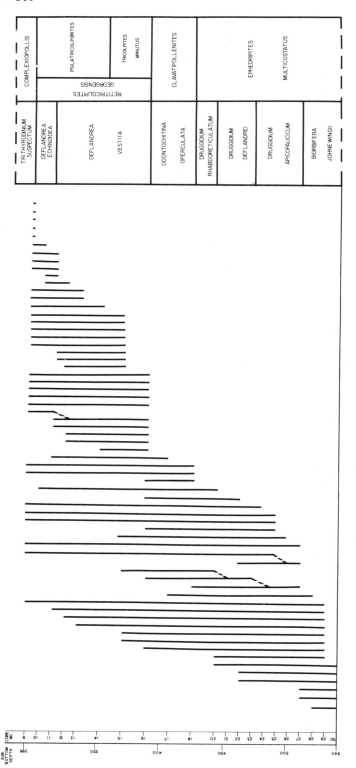

Figure 2. Stratigraphic range chart and zonations at site 105. Sporomorph taxa indicated by a • .

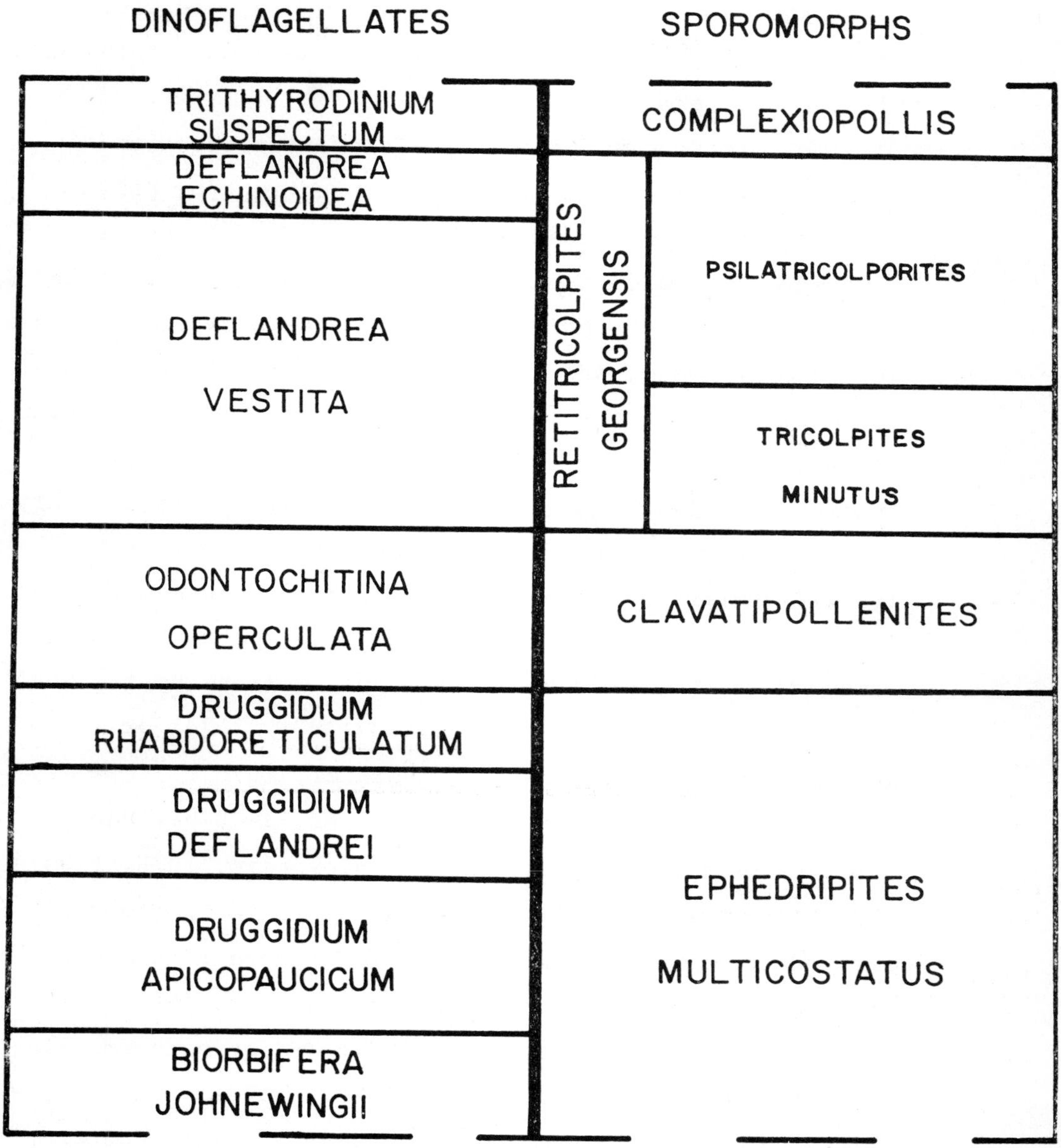

DINOFLAGELLATES		SPOROMORPHS	
TRITHYRODINIUM SUSPECTUM		COMPLEXIOPOLLIS	
DEFLANDREA ECHINOIDEA	RETITRICOLPITES GEORGENSIS	PSILATRICOLPORITES	
DEFLANDREA VESTITA			
		TRICOLPITES MINUTUS	
ODONTOCHITINA OPERCULATA		CLAVATIPOLLENITES	
DRUGGIDIUM RHABDORETICULATUM		EPHEDRIPITES MULTICOSTATUS	
DRUGGIDIUM DEFLANDREI			
DRUGGIDIUM APICOPAUCICUM			
BIORBIFERA JOHNEWINGII			

Figure 3. Dinoflagellate and sporomorph zonations at site 105.

dinoflagellates and sporomorphs varies, in some cases with
one palynomorph group represented to the almost total exclusion of the other.

The Cretaceous (Berriasian-Cenomanian) section at
site 105 is valuable as a reference section for palynomorph stratigraphy because of the number of cores recovered, the small stratigraphic intervals between collected samples, the paleontological ages provided by other
groups of fossils, and that by ranging from Berriasian to
Cenomanian, it is the most complete section. The section
is dated in large part on the basis of the nannofossil
studies of Thierstein (1976), Wilcoxon (1972), and Bukry
(1972); the planktonic foraminifers studied by Luterbacher
(1972); and the ammonite aptychi studied by Renz (1972).
It is dated otherwise by dinoflagellate cysts, from the
study of Lower Cretaceous stratotype sections by Millioud
(1967, 1969) and Davey and Verdier (1971, 1974) or independently dated sections (e.g. Verdier, 1975).

The interval represented by cores 33-22 has been
dated as Berriasian-Valanginian (Thierstein, 1976, figure
5). Thierstein (1976) dated core 20 within the ages of
early Hauterivian to early Barremian. Core 18 has been
dated by ammonite aptychi within the ages of late
Valanginian-Hauterivian (Renz, 1972) and by nannofossils
as Hauterivian (Wilcoxon, 1972; Bukry, 1972). It is considered in this paper to be not older than late Hauterivian based on the first appearance of the dinoflagellate
Odontochitina operculata (Wetzel), which has its lowest
occurrence in the uppermost Hauterivian in the Angles
section of France, according to Millioud (1969). Core 17
is considered to be of Barremian or Aptian age. Wilcoxon

(1972) dated the core as Barremian on the basis of nanno-fossils, and Luterbacher (1972) recovered an assemblage of small and primitive foraminifers which he stated are similar to faunules occurring in sections ranging from late Hauterivian to early Aptian.

The interval represented by cores 16-11 is considered to be Albian, based on the stratigraphic range of the dinoflagellate *Deflandrea vestita* (Brideaux). According to Sverdlove and Habib (1974) this species became phyletically extinct in the time represented by core 11, which has been dated on the basis of foraminifers as latest (Vraconian) Albian (Luterbacher, 1972). This age assignment is supported by the first occurrence of tricolpate pollen grains, e.g. *Retitricolpites georgensis* Brenner, in core 16. Core 9 is considered to be of Cenomanian age, based on the first appearance of the dinoflagellate *Trithyrodinium suspectum* (Manum and Cookson) and on the occurrence of the first Normapolles pollen genera, *Complexiopollis* and *Atlantopollis*.

DINOFLAGELLATE ZONATION

Eight dinoflagellate zones are distinguished in the reference section at site 105, based on the stratigraphic first appearance or phylogenentic appearance of a selected dinoflagellate species defining the lower boundary of the corresponding zone (Figure 2). The Neocomian zonation proposed by Habib (1976) in the western North Atlantic is extended to the Cenomanian Stage, with the description of the following new zones: DEFLANDREA VESTITA, DEFLAN-DREA ECHINOIDEA, TRITHYRODINIUM SUSPECTUM Zones.

The OLIGOSPHAERIDIUM COMPLEX Zone proposed by Habib (1976) is modified, based on the dinoflagellate stratigraphy occurring elsewhere in the North Atlantic. Its two original subzones, the lower DRUGGIDIUM RHABDORETICULATUM and higher ODONTOCHITINA OPERCULATA, are proposed as zones in its place because they are correlatable at sites for which there is adequate core recovery, e.g. site 391. Also at site 391 (Habib, *in press*), *Oligosphaeridium complex* (White) appears stratigraphically below the first occurrence of *Druggidium rhabdoreticulatum*.

1. BIORBIFERA JOHNEWINGII Zone.

Defined as the interval from the lowest occurrence of *Biorbifera johnewingii* up to (but not including) the first appearance of *Druggidium apicopaucicum* (Habib, 1976). This interval ranges from the stratigraphic levels represented by samples DSDP-105-30-2, 90-92 cm. through 105-27-2, 64-66 cm. The zone is dated as Berriasian to early Valanginian, based on the nannofossil zonation published by Thierstein (1976) for this site.

A number of species which are restricted to the Neocomian in the western North Atlantic have their first stratigraphic appearance in the BIORBIFERA JOHNEWINGII zone. In addition to the nominative species, these include *Diacanthum hollisterii* Habib, and *Pyxidinopsis challengerensis* Habib. The Upper Jurassic species *Amphorula metaelliptica* Dodekova has its highest stratigraphic occurrence in this zone. Other species which appear in the zone include *Polysphaeridium warrenii* Habib, *Gonyaulacysta helicoidea* (Eisenack and Cookson) and

Wallodinium krutzschii (Alberti). Stratigraphically persistent species which appear in the BIORBIFERA JOHNEWINGII Zone at site 391, but which were not found at sites 99, 100, 101, and 105, include *Phoberocysta neocomica* (Gocht) and *Pseudoceratium pelliferum* Gocht. *Muderongia* sp. cf. *M. simplex* Alberti was found restricted to this zone at site 391 only, as well.

2. DRUGGIDIUM APICOPAUCICUM Zone.

Defined by Habib (1976) as the interval from the lowest occurrence of the nominative species up to the phylogenetic appearance of *Druggidium deflandrei* (Millioud). At site 105, this zone is represented as the interval from samples DSDP-105-27-2, 13-15 cm. through 105-24-1, 40-42 cm. It is dated as early Valanginian to late Valanginian based on the nannofossil zonation of site 105 (Thierstein, 1976).

Scriniodinium attadalense Cookson and Eisenack, *S. campanula* Gocht, *Dingodinium cerviculum* Cookson and Eisenack, *Cyclonephelium distinctum* Deflandre and Cookson, and *Hystrichodinium voigtii* (Alberti) appear in the DRUGGIDIUM APICOPAUCICUM Zone at site 105.

3. DRUGGIDIUM DEFLANDREI Zone.

Phylozone characterized as the interval from the phylogenetic appearance of *Druggidium deflandrei* up to the phylogenetic appearance of *D. rhabdoreticulatum* Habib. At site 105, *Oligosphaeridium complex* appears at the same stratigraphic horizon as *D. rhabdoreticulatum;* elsewhere

it appears at a lower level (Habib, *in press*). In the
western North Atlantic, the stratigraphic ranges of
Biorbifera johnewingii, Pyxidinopsis challengerensis, and
Diacanthum hollisterii terminate within this zone.
Meiourogonyasulax stoveri Millioud first occurs in this
zone at site 105.

The DRUGGIDIUM DEFLANDREI Zone is represented as the
interval from samples DSDP-105-23 core catcher through
105-20-1, 101-103 cm. It is considered to range in age
from late Valanginian at its lower boundary (cf.
Thierstein, 1976) to Hauterivian (or early Barremian).

4. DRUGGIDIUM RHABDORETICULATUM Zone.

New zone defined as the interval from the phylogenetic appearance of the nominative species up to the
stratigraphic appearance of *Odontochitina operculata*. At
site 105, this zone is represented as the interval from
samples DSDP-105-20-1, 63-65 cm. to core 18. The zone is
dated as Hauterivian or Barremian.

This zone was originally described as a subzone by
Habib (1976). It is elevated to the status of a zone
based on its geographic range in the western North
Atlantic (Habib, *in press*).

5. ODONTOCHITINA OPERCULATA Zone.

New zone defined by the interval from the stratigraphic appearance of the nominative species up to the
appearance of *Deflandrea vestita* (Brideaux). This interval is represented at site 105 from the stratigraphic

level of sample DSDP-105-18 core catcher through that of
sample 105-16 core catcher. It is dated as late Hauterivian or Barremian to early Albian. The zone was originally described as a subzone (Habib, 1976). *Subtilisphaera
perlucida* (Alberti) is restricted to the ODONTOCHITINA
OPERCULATA Zone in the western North Atlantic appearing
at its base and disappearing at or very near the upper
boundary. At site 391, *Phoberocysta neocomica* and
Pseudoceratium pelliferum disappear within the zone. The
stratigraphic disappearances of *P. neocomica* and *S.
perlucida* within the ODONTOCHITINA OPERCULATA Zone at site
391 serve to divide it into two subzones there (Figure 4);
a lower PHOBEROCYSTA NEOCOMICA Subzone which on the basis
of dinoflagellate evidence is dated as late Hauterivian
or Barremian to early Aptian, and a higher SUBTILISPHAERA
PERLUCIDA Subzone which is dated as Aptian to early
Albian (Habib, *in press*).

In the western North Atlantic, *Callaiosphaeridium
asymmetricum* (Deflandre and Courteville) appears within
the ODONTOCHITINA OPERCULATA Zone (and within the PHOBERO-
CYSTA NEOCOMICA Subzone at site 391). A number of species
either appear or disappear stratigraphically in the upper
half of the zone in the western North Atlantic. Those
which disappear include *Wallodinium krutzschii, Polysphaeridium warrenii, Druggidium apicopaucicum, D. deflandrei, Dingodinium cerviculum, Meiourogonyaulax stoveri*
and *Gonyaulacysta cassidata* (Eisenack and Cookson).
Species which appear stratigraphically in the upper half
are *Cribroperidinium muderongensis* (Cookson and Eisenack),
Hystrichokolpoma ferox (Deflandre), *Chlamydophorella nyei*
(Cookson and Eisenack), *Cleistosphaeridium ancoriferum*
(Cookson and Eisenack), *Spiniferites cingulatus* (Wetzel),

S. ramosus (Ehrenberg), *Hystrichosphaeridium arundum*
Eisenack and Cookson, and *H. cooksonii* Singh. This phenomenon apparently represents a large evolutionary change of
dinoflagellate species near the boundary between the
ODONTOCHITINA OPERCULATA and overlying DEFLANDREA VESTITA
Zones. Alternatively, it may be related to the change of
lithofacies in the western North Atlantic, from nannoplankton ooze below to carbonaceous clay above (Lancelot,
Hathaway, and Hollister, 1972). This change in species
and lithofacies occurs within the ODONTOCHITINA OPERCULATA
one elsewhere in the North Atlantic as well, e.g. sites
99 and 101 (OLIGOSPHAERIDIUM COMPLEX Zone of Habib, 1976).

6. DEFLANDREA VESTITA Zone.

New zone defined as the interval from the appearance
of the nominative species below up to the phylogenetic appearance of *Deflandrea echinoidea* Cookson and Eisenack
above. It is represented at site 105 from the stratigraphic level of sample DSDP-105-16-2, 110-112 cm. through
that of 105-11-3, 20-22cm. Sample DSDP-105-11-2, 75-77
cm. contains the first morphotype assigned to *D.*
echinoidea; sample 105-11-1, 130-132 cm. contains the last
morphotype of *D. vestita*. This zone is dated as Albian to
latest (Vraconian) Albian.

The following species appear in the DEFLANDREA VESTI-
TA Zone: *Gonyaulacysta exilicristata* Davey, *Hexagonifera*
chlamydata Cookson and Eisenack, *Palaeohystrichophora*
infusorioides Deflandre, *Litosphaeridium siphoniphorum*
(Cookson and Eisenack), *Achomosphaera ramulifera*
(Deflandre), *Epelidosphaeridia spinosa* Davey, *Hystricho-*

sphaeropsis ovum Deflandre, and *Protoellipsodinium* sp. *Phoberocysta ceratioides* (Deflandre) was observed in this zone at sites 101 and 391.

7. DEFLANDREA ECHINOIDEA Zone.

New zone defined as the interval from the phylogenetic appearance of the nominative species below up to the stratigraphic appearance of *Trithyrodinium suspectum*. At site 105, this interval is represented from the level of sample DSDP-105-11-2, 75-77 cm. through that of 105-9-6, 80-82 cm. It is dated as latest Albian-Cenomanian.

Species which appear in this zone include *Trigonopyxidia ginella* (Cookson and Eisenack), *Diplofusa gearlensis* Cookson and Eisenack, *Dinogymnium acuminatum* Evitt, Clarke, and Verdier, and *Subtilisphaera pontis-mariae* (Deflandre).

8. TRITHYRODINIUM SUSPECTUM Zone.

New zone defined from the appearance of the nominative species in core 9 (DSDP-105-9-6, 38-40 cm.). The interval above core 9 is not known, as the cores immediately above are essentially devoid of dinoflagellate fossils. It is dated as Cenomanian. *Microdinium veligerum* Davey occurs in this zone.

The dinoflagellate zonation presented above is correlatable, at least in part, through sites 99, 100, 101, 387 and 391, as well as site 105. Figure 4 shows the zonation in the western North Atlantic within the framework of a chronostratigraphic scale.

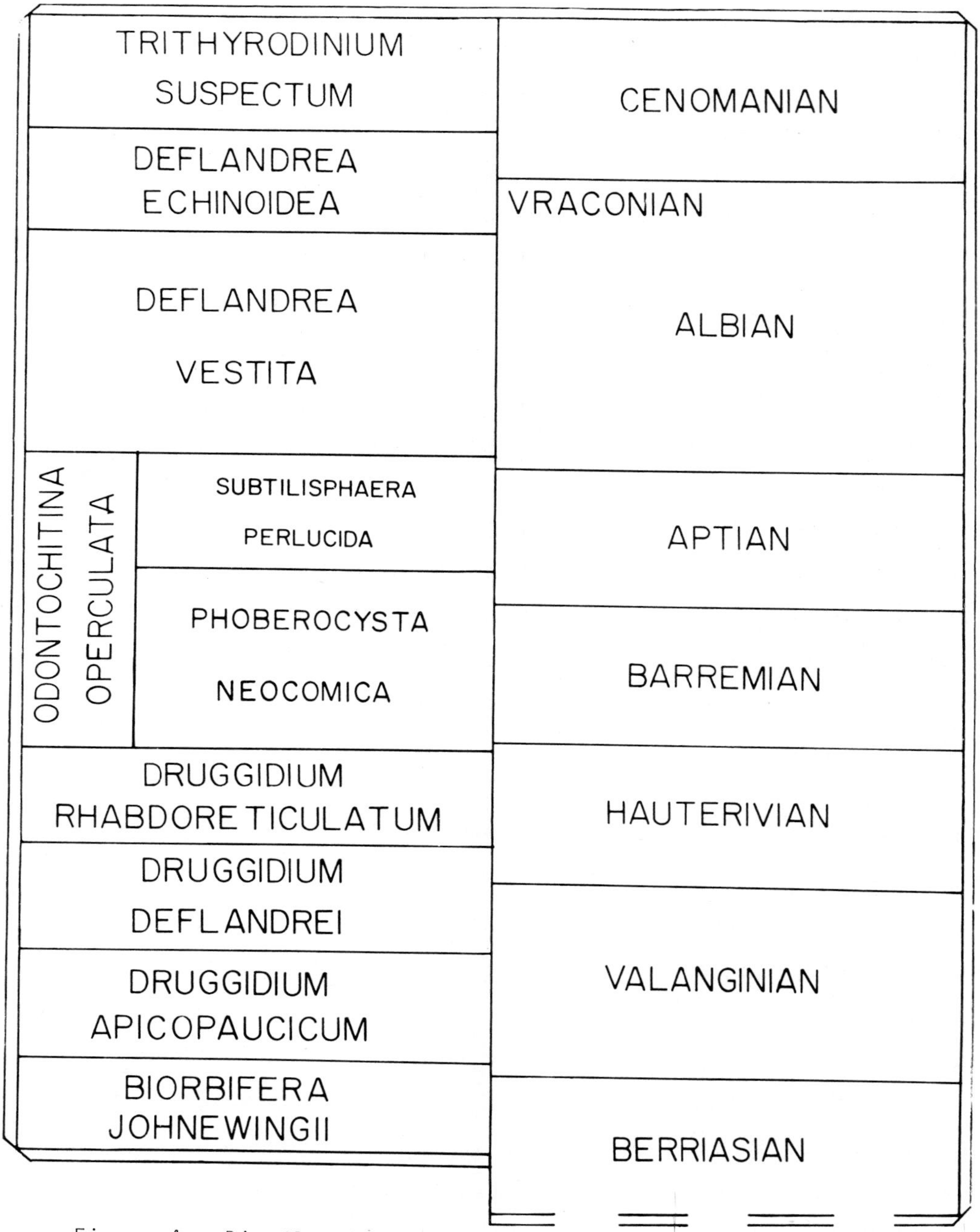

TRITHYRODINIUM SUSPECTUM		CENOMANIAN
DEFLANDREA ECHINOIDEA		VRACONIAN
DEFLANDREA VESTITA		ALBIAN
ODONTOCHITINA OPERCULATA	SUBTILISPHAERA PERLUCIDA	APTIAN
	PHOBEROCYSTA NEOCOMICA	BARREMIAN
DRUGGIDIUM RHABDORETICULATUM		HAUTERIVIAN
DRUGGIDIUM DEFLANDREI		VALANGINIAN
DRUGGIDIUM APICOPAUCICUM		
BIORBIFERA JOHNEWINGII		BERRIASIAN

Figure 4. Dinoflagellate zonation compared against a chrono-stratigraphic scale.

SPOROMORPH ZONATION

Four sporomorph zones are distinguished based on the overlapping first occurrences of pollen species defining the lower boundaries of their respective zones. In ascending stratigraphic order, these are the EPHEDRIPITES MULTICOSTATUS, CLAVATIPOLLENITES, RETITRICOLPITES GEORGENSIS, and COMPLEXIOPOLLIS Zones. The RETITRICOLPITES GEORGENSIS Zone is divided into a lower TRICOLPITES MINUTUS Subzone and a higher PSILATRICOLPORITES Subzone.

Pollen in the sporomorph genus *Ephedripites* are reported from early Neocomian (Berriasian-Valanginian) sediments in the North Atlantic. *Ephedripites multicostatus* Brenner is stratigraphically persistent in these sections and, hence, is considered to be a valuable zonal species.

Ephedripites has long been considered a useful stratigraphic indicator for dating sediments as Barremian or younger. Couper (1964) indicated that "specimens of *Ephedra*-like pollen" enter the chronostratigraphic column in the upper Hauterivian or lower Barremian. Wolfe and Pakiser (1971) indicated a maximal age around the Hauterivian-Barremian boundary for assemblages in the Atlantic Coastal Plain containing taxa such as *Ephedripites* (and *Clavatipollenites hughesii* Couper). In the Salisbury Embayment of the Atlantic Coastal Plain, *Ephedripites* occurs stratigraphically well below the first specimens of *Clavatipollenites*. In their study of the Bethards well in the Salisbury Embayment, Robbins, Perry, and Doyle (1975) report *Ephedripites* in their lowest sample studied, up to approximately 730 meters below the lowest occurrence of *Clavatipollenites* and just 40 meters approximately above the base of the sedimentary section.

The stratigraphically lowest specimens assigned to *Clavatipollenites* (as *Clavatipollenites* sp.) in the North Atlantic are characterized by monosulcate grains with a subcircular to elliptical outline in polar view. The exine is clearly tectate-columellate with a continuous tectum, as revealed in optical section. These grains differ from *Clavatipollenites hughesii* Couper in the continuity of the tectum layer, and from *Clavatipollenites* cf. *C. hughesii* sensu Doyle, van Campo, and Lugardon (1975) which possesses an exine structure intermediate between semitectate and tectate-perforate. *Clavatipollenites* sp. first appears in the western North Atlantic in assemblages of Barremian or late Hauterivian age (very near the base of the ODONTOCHITINA OPERCULATA zone at site 105) *C. hughesii* appears higher, in the SUBTILISPHAERA PERLUCIDA Subzone at site 391.

Reticulate tricolpate and tricolporoidate pollen grains first appear in Albian sediments (DEFLANDREA VESTITA zone) in the North Atlantic, in species of the genera *Retitricolpites* and *Tricolpites*. These species range into the Cenomanian. They occur in successive horizons with tricolporate species (*Psilatricolporites*) and triporate species of the Normapolles group (*Complexiopollis*).

1. EPHEDRIPITES MULTICOSTATUS Zone

New zone defined by the interval from the first appearance of the nominative species up to (but not including) the first occurrence of *Clavatipollenites*. It is represented at site 105 as the interval from samples DSDP-105-29 core catcher through 105-20-1, 63-65 cm.

A number of sporomorph species occur in the EPHEDRI-
PITES MULTICOSTATUS Zone, which date it as not older than
Early Cretaceous. These include *Leptolepidites epacror-
natus* Norris, *Cicatricosisporites potomacensis* Brenner,
C. brevilaesuratus Couper, *C. hughesii* Dettmann, *Monosul-
cites glottus* Brenner, *Appendicisporites problematicus*
(Burger), *A. potomacensis* Brenner, and *Parvisaccites ra-
diatus* Couper. However, except for *M. glottus* and *P. ra-
diatus,* these species are rare or have a discontinuous
stratigraphic range in the western North Atlantic, and
hence, are of limited value for zonation. Norris (1969)
reported several of these species from the approximate
position of the Jurassic/Cretaceous boundary in the non-
marine Purbeck facies of southern England.

The most abundant sporomorphs in this zone are species of bisaccate pollen (*Pinuspollenites, Alisporites*)
and *Classopollis torosus* (Reissinger), which together
comprise up to 84 percent of the sporomorph assimblages.
Exesipollenites tumulus Balme, *Vitreisporites pallidus*
(Reissinger), *Eucommiidites minor* Groot and Penny, and
Ginkgocycadophytus nitidus (Balme) are common as well.

2. CLAVATIPOLLENITES Zone.

New zone defined by the interval from the lowest occurrence of species in *Clavatipollenites* up to the stratigraphic appearance of *Retitricolpites georgensis*
Brenner. It is represented at site 105 as the interval
from sample DSDP-105-18 core catcher through sample 105-
16-2, 110-112 cm.

Clavatipollenites sp., *C. hughesii* Couper, and *C.
tenellis* Paden-Phillips and Felix all occur in this zone,

although only the first species was found in the lowest
sample. *Liliacidites dividuus* (Pierce), *L. peroreticula-
tus* (Brenner), and *L. trichotomosulcatus* Singh appear in
the upper half of the zone.

Classopollis torosus, Pinuspollenites spp., and *Alisporites bilateralis* Rouse continue to dominate the sporomorph flora.

3.   RETITRICOLPITES GEORGENSIS Zone.

New zone defined by the interval from the first appearance of *Retitricolpites georgensis* up to the first
appearance of *Complexiopollis* sp. At site 105, the zone
is represented from the level of sample DSDP-105-16-2,
70-73 cm. to that of sample 105-9 core catcher. Two subzones are distinguished. The lower subzone is defined
by the appearance of *Tricolpites minutus* (Brenner), which
appears in the lowest sample. The higher subzone is defined by the appearance of *Psilatricolporites* sp. (cf.
"*Tricolporopollenites*"*triangulus* Groot, Penny, and Groot)
in sample DSDP-105-13-2, 100-102 cm.

In addition to *R. georgensis* and *T. minutus,* a number of tricolpate and tricolporoidate species appear at
or very near the base of this zone. These include *Retitricolpites sphaeroides* Pierce, *R. magnificus* Habib, *Tricolpites auritus* (Bolkhovitina), *T. micromunus* (Groot and
Penny), and *Psilatricolpites psilatus* Pierce. These species range through the RETITRICOLPITES GEORGENSIS Zone
into the next higher zone. *Rugubivesiculites reductus*
Pierce (and *R. rugosus* Pierce) first occur in the PSILAT-
RICOLPORITES Subzone, along with *Striatopollis paraneus*

(Norris). Occurring near the top of the PSILATRICOLPORITES Subzone are the first triangular oblate triporate grains with simple, thickened pores located at the radial corners. These pollen are referred to the genus *Triporopollenites*.

Classopollis torosus continues to be the single most abundant sporomorph species near the base of the RETITRICOLPITES GEORGENSIS Zone. However, within the lower half of the zone, within the TRICOLPITES MINUTUS Subzone, it declines markedly in abundance and is rare or absent in samples higher in this zone and in the overlying zone. Species of tricolpate dicotyledonous angiosperm pollen become abundant in its place and remain so through the RETITRICOLPITES GEORGENSIS Zone.

4. COMPLEXIOPOLLIS Zone.

New zone defined by the appearance of the Normapolles genus *Complexiopollis* sp. at its base. It is represented at site 105 in core 9, through the interval of samples DSDP-105-9-6, 114-116 cm. through 105-9-1, 60-62 cm. The top of the zone is unknown, as the cores immediately above core 9 are essentially devoid of sporomorphs.

The COMPLEXIOPOLLIS Zone reflects the continuing increase in number of species and morphological diversification of angiosperm pollen in the Middle Cretaceous of t the western North Atlantic. In addition to *Complexiopollis* sp. a second genus of the Normapolles group, *Atlantopollis* sp., appears in this zone, as well as prolate tricolporate grains with well-defined circular pores (*Tricolporites* sp.) and spherical tricolporate grains occurring in obligate tetrads (cf. *Dicotetradites* sp.).

CONCLUSIONS

The individual zonations distinguished by dinoflagellates and sporomorphs respectively are valuable for relative age-dating in the western North Atlantic, as one zonation can complement the other and thereby afford greater precision. Thus, dinoflagellates provide more precise evidence for dating in the Neocomian, where four zones correspond to a single sporomorph zone (EPHEDRIPITES MULTICOSTATUS Zone). On the other hand, in the late Hauterivian-Cenomanian interval, both palynomorph groups appear to be of equal value for age-determination and correlation. It is in this interval expecially that a multifossil zonation incorporating the zonal taxa from both palynomorph groups might be useful. To this end, a third group of palynomorphs currently under investigation, small acritarchs of the *Micrystridium-Veryhachium* Group may contribute to such a zonation.

Based on the biochronostratigraphic determinations proposed in this article the correlation of the palynomorph zonations against a geochronological scale is attempted (Figure 5), according to the scale proposed recently by van Hinte (1976). Such an attempt is useful for the interval represented by the BIORBIFERA JOHNEWINGII-DRUGGIDIUM RHABDORETICULATUM Zones, as it is dated at site 105 by the same nannofossil zonation (Thierstein, 1976) which is presented in the geochronological scale (van Hinte, 1976). Of equal value is the boundary between the DEFLANDREA VESTITA and DEFLANDREA ECHINOIDEA zones, which is dated as Vraconian Albian on the basis of planktonic

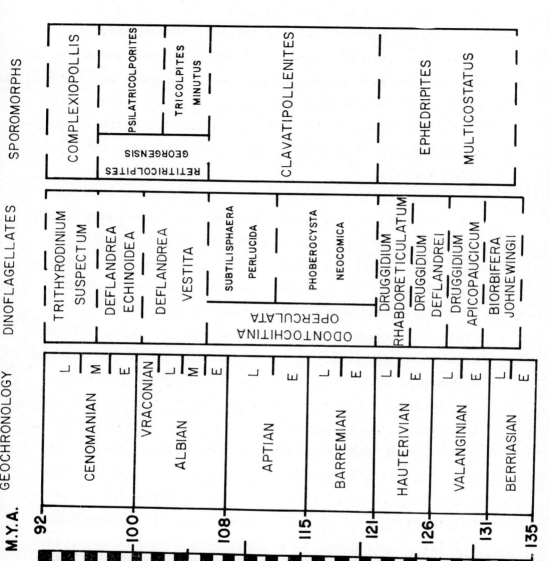

Figure 5. Palynomorph zonations correlated with the geochronological scale (part) published by van Hinte (1976).

foraminifers (*Rotalipora apenninica apenninica* (Renz), *Planomalina buxtorfi* (Gandolfi), cf. Luterbacher, 1972, p. 561) at site 105, and which occur also in van Hinte's scale. The ages of the remaining palynomorph zones are not as well founded, as the evidence is provided by species of dinoflagellates, nannofossils, and ammonite aptychi which are not directly comparable to the geochronological scale.

There is the potential for correlating the paleontologically dated sporomorph zonation in the western North Atlantic with the zonation established for the largely nonmarine coeval Cretaceous of the adjacent U.S. Atlantic Coastal Plain. A zonation has been established in outcrop and subsurface sections ranging from New Jersey to Virginia, based on sporomorph assemblages which indicate a range in age from Barremian (Zone I) to middle Turonian or younger (Zone V). Although the sequence is largely nonmarine, a major part of it has been compared with independently dated marine Aptian and Albian sporomorph assemblages of southern England (Kemp, 1970).

The zonation has developed mainly through the comprehensive studies of Brenner (1963), Doyle (1969, 1973), Wolfe and Pakiser (1971), and Robbins, Perry, and Doyle (1975). It is based to a large extent on the occurrence of 'restricted' species, most of which are rare in the Coastal Plain (Brenner, 1963; Wolfe and Pakiser, 1971, p. 337) and therefore of limited value for correlation. These species occur in the North Atlantic as well, although they are equally rare. However, the Coastal Plain sequence contains the same succession of pollen morphotypes which occurs regularly in the sections studied from the North Atlantic. Based mainly on the description of the Coastal Plain stratigraphy summarized by Robbins, Perry, and Doyle (1975), a comparison is proposed for the North Atlantic zonation and Zones I-IV of the Atlantic Coastal Plain (Figure 6). Zone I, defined by Brenner (1963), can be correlated with the CLAVATIPOLLENITES Zone in the western North Atlantic based on the occurrence of species of *Clavatipollenites*. The uppermost part of Zone I and lower part of Zone II (subzones IIa and IIb) contain the first tricolpate and tricolporoidate species of reticulate dicotyledonous pollen, including *Retitricolpites georgensis* and *Tricolpites minutus*, and can be correlated with the TRICOLPITES MINUTUS Subzone of this study. Zones IIc and III (cf. Robbins, Perry, and Doyle, 1975) contain *Rugubivesiculites rugosus* and *R. reductus*, as well as the first oblate triangular tricolporate grain grains, which suggests a correlation with the PSILATRICOLPORITES Subzone. Zone IV is characterized by the first Normapolles genera, specifically *Complexiopollis* and *Atlantopollis*, and prolate tricolporate and tetrad taxa, all of which occur also for the first time in the COMPLEXIOPOLLIS Zone.

The EPHEDRIPITES MULTICOSTATUS Zone is not represented in the zonation published for the Coastal Plain. However, it is believed to represent the interval containing *Ephedripites* in the Bethards well (Robbins, Perry, and Doyle, 1975) below the first specimens of *Clavatipollenites*. The close comparison of the two sporomorph floral successions permits the age-correlation of the North Atlantic sections with the Coastal Plain. Zone I is dated as late Hauterivian or Barremian to early Albian; IIa and IIb as middle to late Albian; IIc-III late Albian (including Vraconian) to Cenomanian; and Zone IV as Ceno-

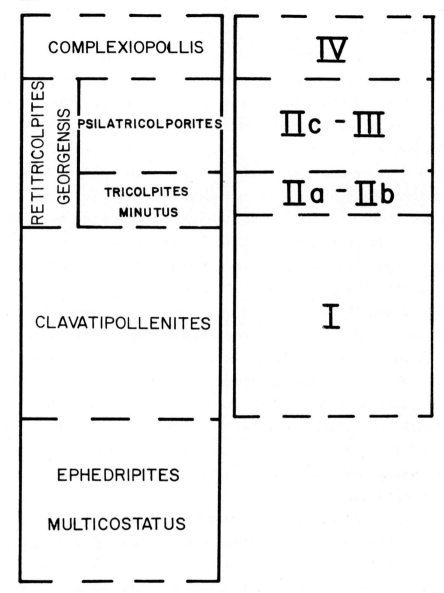

Figure 6. Comparison of North Atlantic and adjacent U.S. Atlantic Coastal Plain sporomorph zonations. Coastal Plain zonation after Robbins, Perry and Doyle (1975).

manian.

These age determinations correspond rather closely with that indicated for the Coastal Plain zonation dated by the sporomorph flora. The correlation of the COM-PLEXIOPOLLIS Zone with Zone IV is strengthened by the oc-currence of molluscs in the marine Woodbridge Clay member of the lower part of the Raritan Formation in New Jersey, which dates the upper part of Zone IV as middle Ceno-manian (Wolfe and Pakiser, 1971).

ACKNOWLEDGMENTS

This study was supported by a grant from the National Science Foundation, GA-39991.

E. I. Robbins reviewed the manuscript.

Discussion

Dr. R. Aurisano: The high cingulum which is characteristic of Druggidium is rather unique. Do you have any ideas concern-ing the relationship of Druggidium to other dinoflagellates?

Habib: Druggidium is similar in general morphology to Micro-dinium Cookson and Eisenack. However, it differs mainly by the position and type of archaepyle.

CITED REFERENCES

Brenner, G.J., 1963. The spores and pollen of the Potomac Group of Maryland: Maryland. Dept. Geol., Mines, Water Res., Bull. 27, p. 1-215.

Bukry, D., 1972. Coccolith stratigraphy, Leg XI, Deep Sea Drilling Project. In: Hollister, C.D., Ewing, J.I., and others, Initial Reports of the Deep Sea Drilling Project, Volume XI, p. 475-482.

Couper, R.A., 1964. Spore-pollen correlation of the Cretaceous rocks of the northern and southern Hemispheres: Soc. Econ. Paleontol. Mineral., Spec. Publ. 11, p. 131-142.

Davey, R.J., and Verdier, J.P., 1971. An investigation of microplankton assemblages from the Albian of the Paris Basin: K. Nederl. Akad. Wetensch., Afd. Natuurk, Verh., v. 26, p. 5-58.

_____, and _____, 1974. Dinoflagellate cysts from the Aptian type sections at Gargas and La Bedoule, France: Palaeontology, v. 17, pt. 3, p. 623-653.

Doyle, J.A., 1969. Cretaceous angiosperm pollen of the Atlantic Coastal Plain and its evolutionary significance: Jour. Arnold Arboretum. v. 50, no. 1, p. 1-35.

_____, 1973. The monocotyledons: their evolution and comparative biology. V. Fossil evidence on early evolution of the monocotyledons: Quart. Rev. Biol., v. 48, n. 3, p. 399-413.

_____, Campo, M. Van, and Lugardon, B., 1975. Observations on exine structure of *Eucommiidites* and Lower Cretaceous angiospern pollen: Pollen et Spores, v. 17, n. 3, p. 429-486.

Habib, D., 1976. Neocomian dinoflagellate zonation in the western North Atlantic: Micropaleontology, v. 21, n. 4, p. 373-392 (October, 1975).

_____, in press. Palynostratigraphy of the Lower Cretaceous section at Deep Sea Drilling Project Site 391, Blake-Bahama Basin, and its correlation in the North Atlantic.

Hinte, J.E. van, 1976. A Cretaceous time scale: Bull. Amer. Assoc. Petrol. Geol., v. 60, n. 4, p. 498-516.

Kemp, E.M., 1970. Aptian and Albian miospores from southern England: Palaeontographica, v. 131, pt. B, p. 73-143.

Lancelot, Y., Hathaway, J.C., and Hollister, C.D., 1972. Lithology of sediments from the western North Atlantic, Leg XI, Deep Sea Drilling Project, In: Hollister,C.D., Ewing, J.I., and others, Initial Reports of the Deep

Sea Drilling Project, Vol. XI, p. 659-666.

Luterbacher, H.P., 1972. Foraminifera from the Lower
Cretaceous and Upper Jurassic of the northwestern At-
lantic. In: Hollister, C.D., Ewing, J.I., and others,
Initial Reports of the Deep Sea Drilling Project,
Volume XI, p. 561-593.

Millioud, M.E., 1967. Palynological study of the type
localities at Valangin and Hauterive: Rev. Palaeobot.
Palynol., v. 5, p. 155-167.

_____, 1969. Dinoflagellates and acritarchs
from some western European Lower Cretaceous type localities: Internat. Conf. Planktonic Microfossils,
First Geneva, 1967, Proc., v. 2, p. 420-434.

Norris, G., 1969. Miospores from the Purbeck Beds and
marine Upper Jurassic of southern England: Palaeontology, v. 12, pt. 4, p. 575-630.

Renz, O., 1972. Aptychi (Ammonoidea) from the Upper
Jurassic and Lower Cretaceous of the western North At-
lantic (Site 105, Leg XI, Deep Sea Drilling Project).
In: Hollister, C.D., Ewing, J.I., and others, Initial
Reports of the Deep Sea Drilling Project, Vol. XI,
P. 607-630.

Robbins, E.I., Perry, W.J., and Doyle, J.A., 1975. Palynological and stratigraphic investigations of four
deep wells in the Salisbury Embayment of the Atlantic
Coastal Plain: U.S. Geol. Surv., Open File Report,
No. 75-307, p. 1-120.

Sverdlove, M.S., and Habib, D., 1974. Stratigraphy and
suggested phylogeny of *Deflandrea vestita* (Brideaux)
comb. nov. and *Deflandrea echinoidea* Cookson and
Eisenack: Geoscience and Man, v. 9, p. 53-62.

Thierstein, H.R., 1976. Calcareous nannoplankton biostratigraphy at the Jurassic-Cretaceous boundary: Bur.
Rech. Geol. Min., Mem., v.86,p. 85-94.

Verdier, J.P., 1975. Les Kystes de dinoflagellés de la
section de Wissant et leur distribution stratigraphique au Crétacé Moyen: Rev. Micropaléontologie, v. 17,
n. 4, p. 191-197.

Wilcoxon, J.A., 1972. Upper Jurassic-Lower Cretaceous
nannoplankton from the western North Atlantic Basin.
In: Hollister, C.D., Ewing, J.I., and others, Initial
Reports of the Deep Sea Drilling Project, Volume XI,
P. 427-458.

Wolfe, J.A., and Pakiser, H.M., 1971. Stratigraphic interpretations of some Cretaceous microfossil floras of
the Middle Atlantic States: U.S. Geol. Surv., Prof.
Paper 750-B, p. B35-B47.

Upper Cretaceous Dinoflagellate Zonation of the Subsurface

Toms River Section Near Toms River, New Jersey

Richard Aurisano and Daniel Habib

Rutgers University, New Brunswick, N. J.; Queens College, Flushing, New York

ABSTRACT

Thirteen peridinioid dinoflagellate species are used to zone the Upper
Cretaceous subsurface section at the Toms River Chemical Well No. 84 site
in the New Jersey Coastal Plain. Seven of the thirteen species are considered to be restricted entirely to the Upper Cretaceous from their occurrence in the Toms River samples and also partially from their distribution
in Upper Cretaceous sediments in North America, western Europe, the Soviet
Union and Australia. These are *Chatangiella vnigrii*, *Subtilisphaera pontis-
mariae* , *Chatangiella manumii* , *C.* **new species** , *C. victoriensis, Deflandrea minor,* and *Diconodinium firmum.*

Six zones are distinguished in this section based on the stratigraphic
ranges of selected dinoflagellates. In ascending stratigraphic order,
these are the CHATANGIELLA VNIGRII zone, CHATANGIELLA MANUMII zone, CHATAN-
GIELLA **NEW SPECIES** zone, DICONODINIUM FIRMUM zone, DEFLANDREA OEBISFEL-
DENSIS zone, and the DEFLANDREA BAKERII zone.

One new species, *Chatangiella* **new species** , is proposed. Relative
percentage distribution data reveal an inverse relationship between *Palaeohystrichophora infusorioides* and *P. paucisetosa.* The possibility is raised
that these species are morphotype variants of one species, dependent on
some environmental factor(s).

INTRODUCTION

The purpose of this study is to propose a biostratigraphic zonation
for Upper Cretaceous peridinioid dinoflagellate cysts in the New Jersey
Coastal Plain near Toms River. The Upper Cretaceous section was recovered

from the Toms River Chemical Well No. 84 (see Figure 1, Location Map).
The samples from this section were made available by the U.S. Geological
Survey.

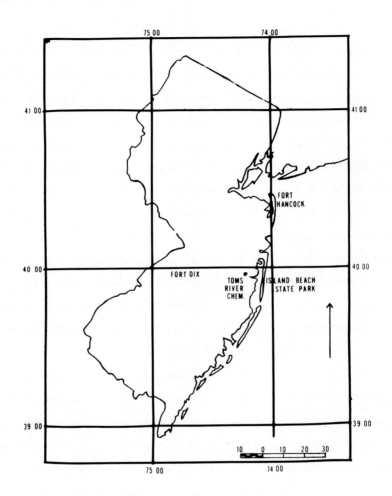

Figure 1. Sketch map of New Jersey showing
approximate position of Toms River Chemical
Well No. 84.

Investigations of Upper Cretaceous dinoflagellate cysts have been undertaken in Australia (Cookson, 1955, 1965; Cookson and Eisenack, 1958, 1960, 1961, 1962, 1968, and 1970), Europe (Alberti, 1959, 1961; Clarke and Verdier, 1967; Davey, 1970; Wilson, 1971), Africa (Davey, 1969), South America (Menendez, 1965), North America (Felix and Burbridge, 1973; Manum and Cookson, 1964; Harland, 1973; McIntyre, 1974; Drugg, 1967) and the USSR (Vozzhennikova, 1967). These investigations show that many dinoflagellate species have geographically wide ranging stratigraphic potential.

The Toms River section is approximately 2,000 feet in thickness and ranges in age from Late Albian (Doyle, 1973 - Pollen Zone IIC) to Tertiary. The part of the section studied is that interval from subsurface depths of 1031 feet to 508 feet and ranges in age from Mid- to Upper Campanian to Tertiary. The study was restricted to this interval because peridinioid cysts are most abundant in the Upper Cretaceous and because little biostratigraphic work in this interval has been published. The Lower and lower Upper Cretaceous part of the Toms River section was zoned by Doyle (1973) utilizing pollen.

Peridinoid dinoflagellate cysts, especially cavate species, were studied because of their abundance, ease of recognition and generally good preservation.

MATERIALS AND METHODS

Twenty samples of sediment, from the stratigraphic interval between 1031 feet and 508 feet depth, cored from the Toms River Chemical Well No. 84, were examined.

LITHOSTRATIGRAPHY

Based on data made available by W.J. Perry of the U.S. Geological Survey, the Toms River section (TR1031 through TR508) consists of the following lithostratigraphic units in ascending stratigraphic order: Woodbury Formation, Englishtown Formation, Marshalltown Formation, Wenonah Formation, Monmouth Group, Hornerstown Formation and the Vicentown Formation. These units have been identified on the basis of Gamma Ray and Self Potential Curves and lithological descriptions by the U.S. Geological Survey.

BIOSTRATIGRAPHY

R.K. Olsson (pers. comm.) provided ages based on foraminiferal data for the Toms River section studied. The ages assigned are Campanian, Maestrichtian, and Tertiary (Figure 3). The samples above the stratigraphic level of sample TR599 (e.g., TR591, TR577, TR554, TR531, and TR508) are of Tertiary age. The stratigraphic interval from level 1031 to somewhere between level 761 and 738 is Campanian. The interval from somewhere between levels 761 and 738 to level 599 is Maestrichtian.

Dinoflagellate data suggest that the entire studied section (samples TR1031 to TR508) ranges in age from no older than Turonian to Paleocene. The Turonian lower limit is based upon the appearance of *Chatangiella vnigrii* (Lentin and Williams, 1975) in the lowest sample studied, TR1031. Caution is exercised in setting this age limit because *C. vnigrii* has been cited in the literature only once at this writing (Vozzhennikova, 1967), from the Turonian of Kazakhstan, USSR. The geographic and time stratigraphic range of this species has not been firmly established. Other species occurring in sample TR1031 are *Deflandrea echinoidea* (Cookson and Eisenack, 1960), *Palaeohystrichophora infusorioides* (Deflandre, 1934) and *P. paucisetosa* (Deflandre, 1943). Williams *et al* (1974) reported the occurrence of *P. infusorioides* in Early to Middle Albian material dated with foraminifera from the Scotian Shelf of eastern Canada. *Deflandrea echinoidea* has been found in material as old as latest Albian (Vraconian) in the western North Atlantic (Sverdlove and Habib, 1974). *Palaeohystrichophora paucisetosa* however, has not been reported from material older than Coniacian (Sarjeant, 1967). Caution is exercised against assigning a Coniacian lower age limit for the section because *P. paucisetosa* has not been frequently cited in the literature, and there is evidence from this study that *P. infusorioides* and *P. paucisetosa* may be morphotype variants dependent upon some environmental factor(s). Doyle's work (1973) on angiosperm pollen types in the Atlantic Coastal Plain, including the lower section of the Toms River core (2000 feet to 1250 feet) places the Cenomanian/Turonian boundary at approximately 1250 feet, the approximate limit of Pollen Zone IV.

The Cretaceous/Tertiary boundary based upon dinoflagellates would be placed somewhere between the levels of samples TR599 and TR623. This is reasonably close to the foraminiferal boundary between samples TR591 and TR599. The dinoflagellate species *Deflandrea oebisfeldensis* (Alberti, 1959) makes its first appearance in the Upper Eocene of central Europe (Alberti, 1959), Paleocene through Lower Oligocene of the Volga region and the Eocene of the Baltic region (Vozzhennikova, 1967). In addition, the distinctive outline of its periblast and archeopyle resembles those of other noted Tertiary species such as *D. phosphoritica* (Eisenack, 1938) and *D. andromedensis* (Vozzhennikova, 1967). Since dating with foraminifera is better established and more reliable, the probability is greater that the Cretaceous/Tertiary boundary lies between samples TR599 and TR591. This suggests the extension of the stratigraphic range of *D. oebisfeldensis* to Late Maestrichtian.

A Campanian/Maestrichtian boundary is suggested between the levels of samples TR668 and TR646, based upon the stratigraphic disappearance of *Chatangiella manumii* (Vozzhennikova, 1967) and *Palaeohystrichophora infusorioides* in sample TR668, and the first appearance of *Deflandrea diebelii* (Alberti, 1959), in sample TR646. *C. manumii* has been found in Turonian to Campanian material in Kazakhstan and Santonian material in western Siberia (Vozzhennikova, 1967). *Palaeohystrichophora infusorioides* is found to

range from Early-Middle Albian to Campanian in material dated with foraminifera from the Scotian Shelf (Williams *et al*, 1974), and the Grand Banks of
New Foundland (Jenkins *et al*, 1974). *Deflandrea diebelii* ranges from Lower
Maestrichtian to lower Upper Maestrichtian in material dated with belem-
nites from Mons Klint, Denmark (Wilson, 1971), and is found in Danian stratotype material from Stevns Klint, Denmark (Wilson, 1971). Jenkins *et al*
(1974), found *D. diebelii* to range from Maestrichtian to Paleocene in material dated with foraminifera from the Grand Banks, New Foundland. Thus, it
appears that sample TR668 can be no younger than Campanian and sample TR646
no older than Maestrichtian. Stage boundaries cannot be distinguished in
the interval between samples TR668 and TR1031, based on dinoflagellates.
Dinoflagellate species found in this interval, which ranges in age from
Campanian to no older than Turonian, are *Deflandrea echinoidea, Chatangiella manumii, C. vnigrii, Subtilisphaera pontis-mariae, Chatangiella*

(new species), *Palaeohystrichophora infusorioides, P. paucisetosa, Chatangiella victoriensis,* and *Diconodinium firmum.*

The published stratigraphic ranges of these species are as follows:
Deflandrea echinoidea, latest Albian to Campanian (Davey, 1970; Harland,
1973; Sverdlove and Habib, 1974); *Chatangiella manumii,* Turonian to Campanian (Vozzhennikova, 1967); *Subtilisphaera pontis-mariae,*Albian (Jenkins *et
al,* 1974; Williams *et al,* 1974) and Coniacian to Campanian (Sarjeant, 1967);
Diconodinium firmum, Upper Campanian (Harland, 1973).

Palaeohystrichophora infusorioides and *P. paucisetosa*

The relative percentage distributions of selected peridinoid cysts,
other cavate cysts, proximate and chorate cysts, were computed for each
sample assemblage studied. (See Table 2).

The data show an inverse relationship between the relative percentage
distributions of *Palaeohystrichophora infusorioides* and *P. paucisetosa.*
For example, at the stratigraphic level of 1031 feet, the relative percen-

TABLE 2

Relative percentage distribution of selected peridinioid cysts, other cavate, proximate and chorate cysts.

SPECIES	TR1031	TR1009	TR987	TR964	TR875	TR852	TR829	TR806	TR761	TR738	TR715	TR668	TR646	TR623	TR599	TR591	TR577	TR554	TR531	TR508
Palaeohystrichophora infusorioides	15.49	4.03	56.20	60.00	32.61	55.14	20.49	76.92	12.56	40.95	35.58	1.19								
Palaeohystrichophora paucisetosa	24.34	55.24	13.57	1.79	3.26	3.42	35.69		2.24	3.02	14.90									
Deflandrea echinoidea	13.27	16.53	7.36	12.14	9.78		0.35	1.81	1.79			1.98	7.17	0.95	10.80	17.66				
Chatangiella vnigrii	1.33				1.09	5.82														
Subtilisphaera pontis-mariae						3.42														
Chatangiella manumii							9.19	2.71	13.45	9.48	9.62	6.72								
Chatangiella n. sp										0.43		2.77								
Chatangiella victoriensis											0.48		1.08	53.55						
Diconodinium firmum													1.19	3.94	13.74					
Deflandrea minor											0.86	0.40	14.34							
Deflandrea diebelii														2.15		0.28				
Deflandrea oebisfeldensis															1.41	4.28	0.31		2.70	1.05
Deflandrea bakerii																	0.31		1.20	0.35
OTHER CAVATE CYSTS	3.10	1.61	2.71	7.14	15.22	5.48	8.12	7.69	11.22	5.17	3.37	5.53	16.13	13.74	25.12	22.21	6.02	2.15	13.49	1.70
PROXIMATE CYSTS	12.39	6.45	6.98	9.64	16.30	11.64	12.37	8.14	29.15	19.40	21.63	44.27	27.96	16.11	21.13	25.36	44.58	1.23	1.20	5.96
CHORATE CYSTS	30.09	16.13	13.18	9.29	21.74	15.07	13.78	2.71	29.60	20.69	14.42	35.97	27.24	1.90	41.55	30.70	49.40	96.01	81.42	90.88

DEPTH OF SAMPLE

tage of *P. infusorioides* is 15.49 while for *P. paucisetosa* it is 24.34. At the level of 1009 feet, the relative percentage of *P. infusorioides* is 4.03 while for *P. paucisetosa* it is 55.24. Thus, from level 1031 to level 1009, the percentage distribution of *P. infusorioides* decreases while that of *P. paucisetosa* increases. At the level of 987 feet, the relative percentage of *P. infusorioides* is 56.20 while that of *P. paucisetosa* 13.57. From level 1009 to level 987, *P. infusorioides* increases while *P. paucisetosa* decreases in relative percentage distribution. This observation is made at each level containing these species except 875, 852 and 761, 738 where the percentage of *P. paucisetosa* remains fairly constant (less than one percent change).

It appears that when moving from one stratigraphic level to another, an increase in relative percentage distribution of one species is followed by a corresponding decrease in the other. The inverse relationship between *P. paucisetosa* and *P. infusorioides* apparently shows no corresponding relationship to lithological changes.

P. infusorioides and *P. paucisetosa* have similar morphologies. They are distinguished on the basis of periblast ornamentation, average size, and endoblast. The periblast of *P. paucisetosa* has hair-like processes (setae) which are longer than those of *P. infusorioides* and more sparsely distributed. In addition, the periblast of *P. paucisetosa,* on the average, is smaller than that of *P. infusorioides*. The endoblast of *P. paucisetosa* is rarely if ever appressed against the periblast.

The question is raised: Are *P. infusorioides* and *P. paucisetosa* two separate species which have different specific environmental needs or are they morphotype variants of the same species, i.e., a species population whose average characteristics vary with respect to some environmental factor(s) such as salinity, temperature, trace elements, etc.

This is an important consideration because *P. infusorioides* and *P. paucisetosa* have different stratigraphic ranges. If *P. infusorioides* and *P. paucisetosa* are morphotype variants of one species, any zonation utilizing the ranges of both species is invalid. Further investigation into this issue is warranted and until more information is available, it would be safer to consider the combined stratigraphic ranges of *P. infusorioides* and *P. paucisetosa* when making a zonation.

DINOFLAGELLATE ZONATION

Six zones are distinguished from the Upper Cretaceous/Tertiary section at Chemical Well No. 84 near Toms River, New Jersey, based on the stratigraphic ranges of selected peridinioid cysts. In ascending stratigraphic order, these are the CHATANGIELLA VNIGRII zone, CHATANGIELLA MANUMII zone, CHATANGIELLA **NEW SPECIES** zone, DICONODINIUM FIRMUM zone, DEFLANDREA OEBISFELDENSIS zone, and the DEFLANDREA BAKERII zone.

Of the thirteen species whose ranges are shown in Figure 3, seven are considered to be restricted entirely to the Upper Cretaceous, from their

occurrence in the Toms River (TR) samples and also partially from their
distribution in Upper Cretaceous sediments in North America, western
Europe, and the Soviet Union. These are *Chatangiella vnigrii, Subtilisphaera pontis-mariae, Chatangiella manumii, C.* **new species** *, C. victoriensis, Diconodinium firmum,* and *Deflandrea minor.* In addition, one species
is considered to be restricted to the Lower Tertiary. This species is *Deflandrea bakerii.*

CHATANGIELLA VNIGRII ZONE

Concurrent range zone defined by the stratigraphic interval from the
lowest occurrence of the nominative species up to but not including the
lowest occurrence of *Chatangiella manumii.*

Other cavate cysts which appear in the CHATANGIELLA VNIGRII zone include *Palaeohystrichophora infusorioides, P. paucisetosa, Deflandrea echinoidea,* and *Subtilisphaera pontis-mariae.* The most conspicuous feature of
this zone is the absence of *C. manumii.*

The CHANTANGIELLA VNIGRII zone is the lowest zone in the Toms River
samples studied. Its lower limit is not known since samples below 1031
feet were not investigated. Since its upper limit is approximately 829
feet, the CHATANGIELLA VNIGRII zone is at least 202 feet thick. The zone
is considered to range in age from upper Lower Campanian to lower Upper
Campanian on the basis of foraminiferal data. An age range of no older
than Turonian to Campanian is suggested by the published ranges of *C. vnigrii* and *C. manumii.*

CHATANGIELLA MANUMII ZONE

Concurrent range zone defined by the stratigraphic interval from the
lowest occurrence of the nominative species up to but not including the
lowest occurrence of *Chatangiella* **new species**

Other cavate cysts which appear in the CHATANGIELLA MANUMII zone include *Palaeohystrichophora infusorioides, P. paucisetosa,* and *Deflandrea*

echinoidea.

The CHATANGIELLA MANUMII zone ranges in age from lower Upper Campanian to Uppermost Campanian or Lowermost Maestrichtian on the basis of foraminiferal data. Dinoflagellates suggest an age range of no older than Turonian to Campanian. It occurs at the Toms River site from approximately 829 feet to 738 feet and includes samples TR829, TR806, and TR761.

CHATANGIELLA **NEW SPECIES** ZONE

Concurrent range zone defined by the stratigraphic interval from the lowest occurrence of the nominative species up to but not including the lowest occurrence of *Diconodinium firmum.*

Palaeohystrichophora infusorioides, P. *paucisetosa, Deflandrea echinoidea,* and *Chatangiella manumii* appear stratigraphically within the CHATANGIELLA **NEW SPECIES** zone. In addition, C. *victoriensis* and *Deflandrea minor* make their first appearance in this zone.

The CHATANGIELLA **NEW SPECIES** zone ranges in age from Upper Campanian or Lower Maestrichtian to Maestrichtian on the basis of foraminiferal data. Dinoflagellates suggest a Campanian to Maestrichtian age. It is found at the Toms River site from approximately 738 feet to 668 feet and is approximately 70 feet thick. Samples TR738 and TR715 are included in this zone.

DICONODINIUM FIRMUM ZONE

Concurrent range zone defined by the stratigraphic interval from the lowest occurrence of the nominative species up to but not including the lowest occurrence of *Deflandrea oebisfeldensis.*

Cavate cysts which appear stratigraphically in the DICONODINIUM FIRMUM zone include *Deflandrea echinoidea, Chatangiella victoriensis,* and *Deflandrea minor.* D. *diebelii* makes its first appearance within this zone, a species whose published range is Maestrichtian to Danian (Alberti, 1959; Wilson, 1971).

The DICONODINIUM FIRMUM zone ranges in age from lower Upper Maestrichtian to upper Upper Maestrichtian on the basis of foraminiferal data. Dinoflagellates suggest an age range of Campanian or Maestrichtian to Maestrichtian or Paleocene. It is found at the Toms River site from approximately 668 feet to 599 feet and has a thickness approximating 69 feet. Samples TR668, TR646, and TR623 are included in this zone.

DEFLANDREA OEBISFELDENSIS ZONE

Concurrent range zone defined by the stratigraphic interval from the lowest occurrence of the nominative species up to but not including the lowest occurrence of *Deflandrea bakerii*.

Deflandrea diebelii disappears stratigraphically within the DEFLANDREA OEBISFELDENSIS zone. *D. bakerii* makes its first appearance within this zone.

The DEFLANDREA OEBISFELDENSIS zone has a lower age limit of Upper Maestrichtian to Paleocene based on foraminiferal data. Dinoflagellate data disagrees and places the lower age limit in the Paleocene. The upper boundary of this zone lies within the Early Tertiary.

DEFLANDREA BAKERII ZONE

Concurrent range zone defined by the stratigraphic interval from the lowest occurrence of the nominative species. Because Toms River samples above 508 feet were not studied, the upper limit of this zone has not been determined.

D. oebisfeldensis occurs in the DEFLANDREA BAKERII zone.

The DEFLANDREA BAKERII zone is restricted to the Tertiary in age. The upper boundary of the zone is not known.

A COMPARISON OF LITHOSTRATIGRAPHY WITH BIOSTRATIGRAPHY

It is difficult to compare biostratigraphic boundaries with lithostratigraphic boundaries because the lithostratigraphic units are based on the

380

results of a continuous gamma log, whereas the biostratigraphic zones are

based upon the stratigraphic ranges of dinoflagellates as represented by 20

samples (See Figure 2).

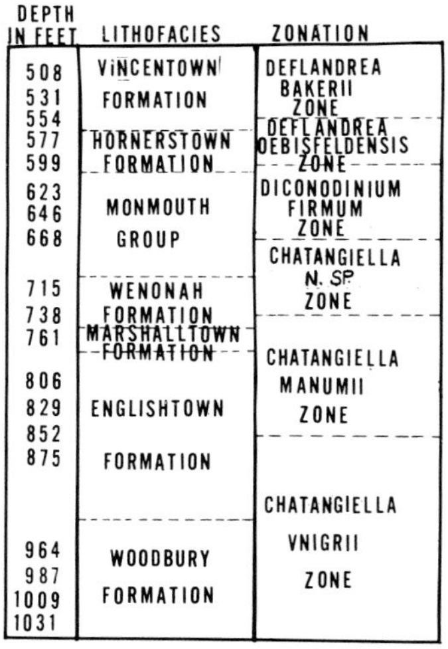

DEPTH IN FEET	LITHOFACIES	ZONATION
508 531 554	ViNCENTOWN FORMATION	DEFLANDREA BAKERII ZONE
577 599	HORNERSTOWN FORMATION	DEFLANDREA OEBISFELDENSIS ZONE
623 646 668	MONMOUTH GROUP	DICONODINIUM FIRMUM ZONE
715 738 761	WENONAH FORMATION MARSHALLTOWN FORMATION	CHATANGIELLA N. SP ZONE
806 829 852 875	ENGLISHTOWN FORMATION	CHATANGIELLA MANUMII ZONE
964 987 1009 1031	WOODBURY FORMATION	CHATANGIELLA VNIGRII ZONE

Figure 2. Chart comparing dinoflagellate
zonation with corresponding lithofacies.

Based upon the ages discussed in the beginning of the biostratigraphy

section, the Tertiary/Cretaceous boundary lies somewhere at the base of the

Hornerstown Formation and the Maestrichtian/Campanian boundary, based on

foraminiferal data, lies somewhere at the base of the Wenonah or top of the

Marshalltown. Dinoflagellate data place this boundary somewhere in the

middle of the Monmouth Group. The rest of the section ranges in age from

no older than Turonian to Campanian based on selected dinoflagellates and

Campanian, based on foraminifera.

ACKNOWLEDGMENTS

This paper was made possible through the generous gift of the B. P.
Alaska Exploration Company in support of R. Aurisano's thesis research
and by a grant from the National Science Foundation (NSF - GA - 39991)
at Queens College.

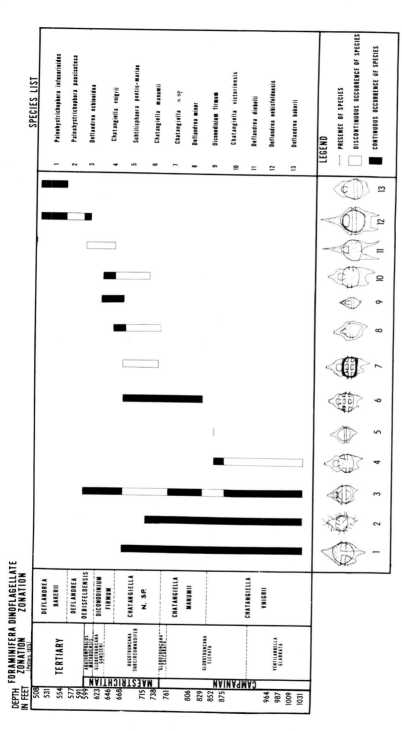

FIGURE 3

REFERENCES

Alberti, G., 1959. Zur Kenntnis der Gattung *Deflandrea* Eisenack (Dinoflag.) in der Kreide und im Alttertiär Nord--Mitte Deutschlands. Mitt. Geol. Staatsinst, v. 28, p. 93-105.

Clarke, R.F.A. and Verdier, J.P., 1967. An investigation of microplankton assemblages from the chalk of the Isle of Wight, England. Vorh. K. Ned. Akad. Net., v. 24, n. 3, 96p.

Cookson, I.C., 1956. Additional microplankton from Australian Late Mesozoic and Tertiary sediments. Austral. J. Mar. Freshw. Res., v. 7, p. 183-191.

_____ and Eisenack, A., 1958. Microplankton from Australian and New Guinea upper Mesozoic sediments. Proc. Roy. Soc. Victoria, v. 70, p. 19-79.

_____, 1960. Microplankton from Australian Cretaceous sediments. Micropaleontology, v. 6, n. 1, p. 1-18.

_____, 1961. Upper Cretaceous microplankton from the Belfast No. 4 Bore, southwestern Victoria. Proc. Roy. Soc. Victoria, v. 74, p. 69-76.

_____, 1962. Additional microplankton from Australian Cretaceous sediments. Micropaleontology, v. 8, n. 4, p. 495-507.

_____, 1968. Microplankton from two samples from Gengin Brook No. 4, Borehole, Western Australia. J. Roy. Soc. W. Austral., v. 51, p. 110-122.

_____, 1970. Cretaceous microplankton from the Eucla Basin, Western Australia. Proc. Roy. Soc. Victoria, v. 83, p. 137-157.

_____ and Manum, S., 1964. On *Deflandrea victoriensis* n. sp., *D. tripartita* Cookson & Eisenack, and related species. Proc. Roy. Soc. Victoria, v. 77, p. 521-524.

Davey, R.J., 1970. Non-calcareous microplankton from the Cenomanian of England, northern France, and North America, part II. Bull. Br. Mus. Nat. Hist. (Geol.) v. 18, n. 8, pp. 335-397.

_____, 1969. Some dinoflagellate cysts from the Upper Cretaceous of northern Natal, South Africa. Paleontol. Afr., V. 12, pp. 1-23.

Deflandre, G., 1934. Sur les microfossiles d'origine planctonique conservés à l'état de matière organique dans les silex de la craie. C.R. Acad. Sci. Paris, 199, 966-968.

_____, 1935. Considérations biologiques sur les microorganismes d'origine planctonique conservés dans les silex de la craie. Bull. Biol. Fr. Belg., v. 69, p. 213-244.

_____, 1936. Microfossiles des silex crétacés. Première partie. Generalities. Flagellés; Ann. Paléont., v. 25, p. 151-191.

_____, 1940. Microfossiles de quelques silex de la craie blanche de Vendôme. Bull. Soc. Hist. Nat. Toulouse, 75, 155-159.

_____, 1943. Sur quelques nouveaux Dinoflagellés des silex crétacés. Bull. Soc. Géol. Fr., v. 13, p. 499-509.

_____, 1966. Addendum à mon Mémoire: Microfossiles des silex crétacés. Cah. Micropaléontol. (Arch. orig. Cen. Docum. C:N.R.S. no. 419) ser. 1, no. 2, p. 1-9.

_____ and Cookson, I.C., 1955. Fossil microplankton from Australian Late Mesozoic and Tertiary sediments. Austral. J. Mar. Freshw. Res., v. 6, p. 242-313.

_____ and Courtville, H., 1939. Note préliminaire sur les microfossiles des silex crétacés du Cambresis. Bull. Soc. Fr. Microsc. 8, pp. 95-106.

_____ *et al*, 1970. Re-issue of Deflandre and Cookson, 1955 (in French), by Laboratoire de Micropaléontologie de L'Ecole Practique des

Hautes Etudes. Institute de Paléontologie du Museum, Paris. avec addendum et post pace, pp. 1-70, 1-54.

Douglas, J.G., 1960. Microplankton of the Deflandreidae group in western district sediments. Min. Geol. J., 6(4), pp. 17-32.

Downie, C. and Sarjeant, W.A.S., 1966. The morphology, terminology and classification of dinoflagellate cysts, in Davey, R.J., Downie, C., Sarjeant, W.A.S., and Williams, G.L., Studies on Mesozoic and Cainozoic dinoflagellate cysts; Bull. of the British Museum (Nat. Hist.) Geol. supplement 3, p. 10-17.

_____, Evitt, W.R., and Sarjeant, W.A.S., 1963. Dinoflagellates, hystrichospheres and the classification of the Acritarchs. Stanford Univ. Publ., Geol. Sciences, 7, 1-16.

Doyle, J.A., 1973. The monocotyledons: their evolution and comparative biology, Part V. Fossil evidence on early evolution of the monocotyledons. Quart. Review of Biology, v. 48, n. 3, p. 399-413.

Drugg, W.S., 1967. Palynology of the Upper Moreno Formation (Late Cretaceous-Paleocene) Escarpado Canyon, California. Palaeontographica, Abt. B, v. 120, p. 1-71.

Eisenack, A., 1938. Die Phosphoritknollen der Bernsteinformation als überlieferer Tertiären Planktons. Sehr. Phys. Ökon. Ges., v. 70, p.181-188.

_____ and Cookson, I.C., 1960. Microplankton from Australian Lower Cretaceous sediments. Proc. Roy. Soc. Victoria, v. 72, p. 1-11.

Evitt, W.R., 1961. Observations on the morphology of fossil dinoflagellates. Micropaleontology, v. 7, n. 4, pp. 385-420.

_____ 1967. Dinoflagellate studies II. The archeopyle. Stanford Univ. Publ., Geol. Sci., v. 10,n. 3, 84p.

_____ 1969. Dinoflagellates and other organisms in palynological preparation. Aspects of Palynology, R.H. Tschudy and R.A. Scott (Eds.) Wiley-Interscience, pp. 439-479.

_____ 1970. Dinoflagellates. A selective review. Geoscience and Man. v. 1,p. 29-45.

_____ and Davidson, S.E., 1964. Dinoflagellate studies I. Dinoflagellate cysts and thecae. Stanford Univ. Publ., Geol. Sci., v. x, n. 1, p. 3-12.

_____ and Wall, D., 1968. Dinoflagellate studies IV. Theca and cyst of recent freshwater *Peridinium limbatium* (Stokes) Lemmermann. Stanford Univ. Publ., Geol. Sci., v. xii, n. 2, 15p.

Felix, C.J. and Burbridge, P.P., 1973. A Maestrichtian age microflora from Arctic Canada. Geoscience and Man, v. vii, p. 1-29.

Habib, D., 1972. Dinoflagellate stratigraphy, Leg XI, Deep Sea Drilling Project, in Hollister, C. Ewing, J. Hathaway, J., Paulus, F., Lancelot, Y., Habib, D., Poag, C.W., Luterbacher, H.P., Worstell, P. and Wilcoxon, J.A.. Initial Reports of the Deep Sea Drilling Project, v. XI, Washington (U.S. Government Printing Office), xxii to 1077p.

_____ and Warren, J.S., 1973. Dinoflagellates near the Cretaceous/Jurassic boundary. Nature, v. 241, n. 5386, p. 217-218.

Harland, R., 1973. Dinoflagellate cysts and acritarchs from the Bearpaw Formation (Upper Campanian) of southern Alberta, Canada. Palaeontology, v. 16, part 4, pp. 665-706.

Jenkins, W.A.M. *et al*, 1974. Stratigraphy of the Amoco 10E 4-1 puffin B-90 Well, Grand Banks of Newfoundland. Can. Geol. Survey Paper, pp. 74-61.

Koch, R. and Olsson, R.K., 1974. Abt. Microfossil biostratigraphy of the uppermost Cretaceous beds of New Jersey. In Abstracts with Programs, v. 5, n. 1, GSA, p. 45.

Lejeune - Carpentier, 1942. L'etude microscopique des silex. Péridiniens nouveaux ou peu connus. (Dixième note). Ann. Soc. Géol. Belg. v. 65, p. B181-B192.

Lentin, J.K. and Williams, G.L., 1975. A monograph of fossil peridinioid dinoflagellate cysts. Bedford Inst. of Oceanography,Rept.Ser.BI-R-75-16.

Malloy, 1972. An Upper Cretaceous dinoflagellate cyst lineage from Gabon, West Africa. Geoscience and Man, v. 4, p. 57-65.

Manum, S., 1963. Some new species of *Deflandrea* and their probable affinity with Peridinium. Arbok Norst Polar Inst. 1962, p. 55-67.

_____and Cookson, I.C., 1964. Cretaceous microplankton in a sample from Graham Island, Arctic, Canada, collected during the second "Fram - expedition" (1898-1902). Norske Vid - Akad., Skrifter, I, Mat.-Naturv., 17, 1-36.

McIntyre, D.J., 1975. Morphologic Changes in *Deflandrea* from a Campanian section, District of Mackenzie, N.W.T., Canada. Geoscience and Man, v. xi, pp. 61-76.

Menendez, 1965. Microplancton fosil de sedimentos Terciarios y cretacecos del nort de Tierra del Fuego (Argentina). Ameghiniana, v. 9,p. 7-15.

Minard, J.P. *et al*, 1974. Preliminary report on geology along Atlantic continental margin of northeastern United States. AAPG Bull., v. 58, n. 6, part II of II, p. 1169-1178.

Owens, J.P. and Sohl, N.G., 1969. Shelf and deltaic paleoenvironments in the Cretaceous-Tertiary formations of the New Jersey Coastal Plain. Geology of Selected Areas in New Jersey and Eastern Pennsylvania and Guidebook of Excursions, Rutgers Univ. Press, pp. 235-278.

_____, 1973. Glauconites from New Jersey - Maryland Coastal Plain: Their K-Ar ages and applications in stratigraphic studies. GSA Bull., v. 84, p. 2811-2838.

Petters, Sunday, 1976. Upper Cretaceous subsurface stratigraphy of Atlantic Coastal Plain of New Jersey. AAPG Bull., v. 60, n. 1, p. 87-107.

Sarjeant, W.A.S., 1967. The stratigraphic distribution of fossil dinoflagellates. Rev. Palaeobotan. Palynol., 1, pp. 323-343.

_____, 1966. Further dinoflagellate cysts from the Speeton Clay: Studies on Mesozoic and Cainozoic dinoflagellate cysts. R.J. Davey *et al*, (eds.), Brit. Mus. (Nat. Hist.) Bull., Geol. Supp. 3, p. 199-214.

Snead, R.G., 1969. Microfloral diagnosis of the Cretaceous - Tertiary boundary, central Alberta. Research Council Alberta, Bull. 25, 148p.

Stover, L.E., 1973. Paleocene and Eocene species of *Deflandrea* (Dinophyceae) in Victorian coastal and offshore basins, Australia. Spec. Publ. Geol. Soc. Aust., 4: pp. 167-188.

Sverdlove, M. and Habib, D., 1974. Stratigraphy and suggested phylogeny of *Deflandrea vestita* (Brideaux) *comb. nov.* and *Deflandrea echinoidea* Cookson and Eisenack. Geoscience and Man, v. 9, p. 53-62.

Upshaw, C.F., 1964. Palynological zonation of the Upper Cretaceous Frontier Formation near Dubois, Wyoming: Palynology in oil exploration, A. T. Cross (ed.), SEPM Spec. Publ., n. 11, p. 153-168.

Valensi, L., 1955. Etude micropaléontologique des silex du Magdalenien de Saint-Amand (Cher). Bull. Soc. Préhist. Fr., 52, p. 584-596.

Vozzhennikova, T.F., 1965. Introduction to the study of fossil peridinean algae. Akad. Nauk. SSSR, Sib. Otd., Inst. Geol. Geofiz., ER., 156p.

_____, 1967. Fossil peridinians of the Jurassic, Cretaceous and Paleogene deposits of the USSR. Akad. Nauk. SSSR, Sib. Otd., Inst. Geol. Geofiz., Tr., 347p.

Wall, D., 1971. Biological problems concerning fossilizable dinoflagellates. Geoscience and Man, v. 3, p. 1-15.

_____ and Dale, B., 1967. The resting cysts of modern marine dinoflagellates and their paleontological significance. Rev. Paleobotan. Palynol., 2, pp. 349-354.

_____, 1968. Modern dinoflagellate cysts and evolution of the Peridiniales. Micropaleontology, v. 14, n. 3, p. 265-304.

_____,1973. Paleosalinity relationships of dinoflagellates

on the Late Quaternary of the Black Sea - A summary. Geoscience and Man,
v. 7, p. 95-102.

Williams, G.L. and Lentin, J.K., 1973. Fossil dinoflagellates: index to
genera and species. Geol. Sur. Can., Dept. of Energy, Mines and Resources, Paper 73-42, 176p.

_____ et al, 1974. Statigraphy of the Shell Naskapi N. 30 Well,
Scotian Shelf, eastern Canada. Canadian Geol. Survey Paper 74-50,p.1-12.

Wilson, G.J., 1971. Observations on European Late Cretaceous dinoflagellate
cysts. Proceedings of the II Planktonic Conference, Roma 1970.

_____ , 1967. Microplankton from the Garden Cove Formation, Campbell
Island, New Zealand, J. Bot., v. 5, p. 223-240.

EXPLANATION OF PLATE 1

1. Chatangiella sp., approx. 630x. Sample TR668, slide no. RA-1, 7.7x 105.5.
Ventral view. Note the broadly arched epitract, two-walled endoblast, and
distinctive cingulum. 2. Chatangiella sp.. Same specimen as in fig. 1
focusing on the endoblast. 3. Chatangiella sp.. Same specimen as in fig. 1
in dorsal view. Note the partite cingulum and precingular and postcingular
plates. 4. Chatangiella manumii (Vozzhennikova) Lentin and Williams,
1975. Approx. 740x. Sample TR829, slide no. RA-14, 0.7 x 96.3. Ventral view.
Note weakly arched epitract, cingulum outlined by conical projections, and
conical projections scattered about the periblast. 5. Chatangiella victoriensis (Cookson and Manum) Lentin and Williams, 1975. Approx. 630x. Sample
TR623, slide no. RA-15, 30.5 x 106.8. 6. Chatangiella vnigrii (Vozzhennikova) Lentin and Williams, 1975. Approx. 630x. Sample TR1031, slide no. RA-
10, 8.9 x 97.5. Ventral view. Note distinctively arched epitract and conical
apical horn. Also, note sulcus and conical projections on periblast.
7. Deflandrea bakerii Deflandre and Cookson. Approx. 630x. Sample TR554,
slide no. RA-9, 0.4 x 103.5. Note sulcus and weakly indicated cingulum.
Also, note the evenly scabrate ornamentation. 8. Deflandrea echinoidea
Cookson and Eisenack. Approx. 500x. Sample TR1009, slide no. RA-7, 6.4 x
105.3. 9. Deflandrea minor Alberti. Approx. 630x. Sample TR646, slide no.
RA-21, 9.3 x 84.2.

EXPLANATION OF PLATE 2

1. Deflandrea oebisfeldensis Alberti. Approx. 400x. Sample TR591, slide no.
RA-8, 9.5 x 91.0. Note the large trapezoidal archeopyle. 2. Diconodinium
firmum Harland. Approx. 1000x. Sample TR668, slide no. RA-6, 1.8 x 116.1.
Note bifurcating apical process. 3. Subtilisphaera pontis - mariae
(Deflandre) Lentin and Williams, 1975. Approx. 860x. Sample TR852, slide
no. RA-9, 0.4 x 87.5. 4. Palaeohystrichophora paucisetosa Deflandre. Approx.
1200x. Sample TR829, slide no. RA-14, 3.1 x 110.0. 5. P. paucisetosa.
Approx. 1200x. Sample TR1031, slide no. RA-8, 11.5 x 95.0. 6. Palaeohystrichophora infusorioides Deflandre. Approx. 860x. Sample TR806, slide no.
RA-1, 6.0 x 82.1. Ventral view. Note sulcus. 7. P. infusorioides. Approx.
860x. Same specimen as in Figure 1 with focus on endoblast. 8. P. infusorioides. Approx. 860x. Same specimen as in Figure 1 in dorsal view. Note
continuous cingulum. Also, contrast relative length of seta with those in
Figures 3 and 4 of P. paucisetosa. 9. P. infusorioides. Approx. 860x.
Same specimen as in Figure 1 with focus on endoblast in dorsal view.

386

PLATE 1

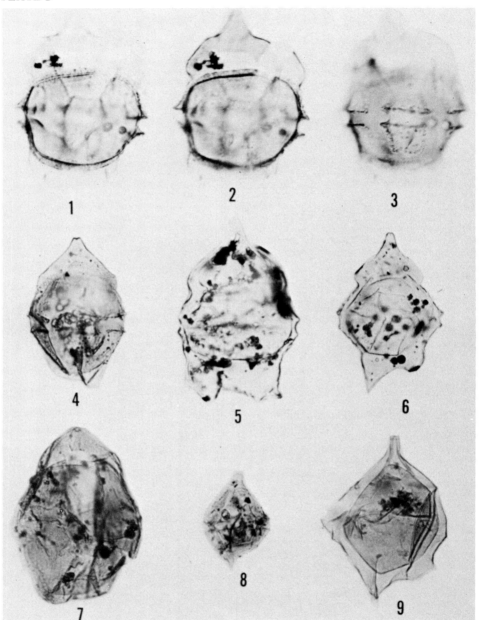

PLATE 2

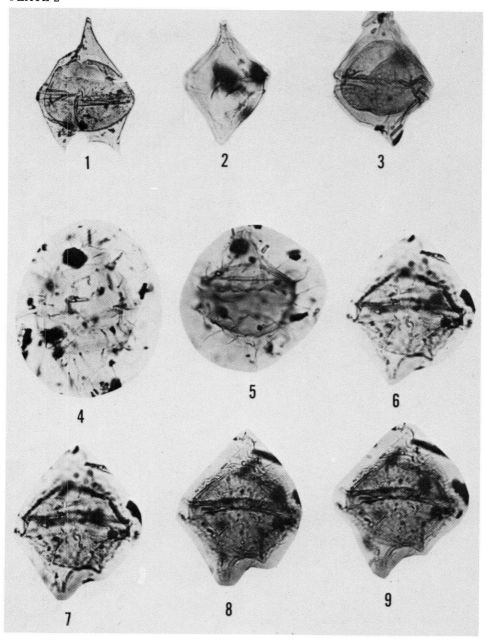

NORTH ATLANTIC CENOZOIC FORAMINIFERA

by

W. A. Berggren

Department of Geology and Geophysics, Woods Hole Oceanographic Institution
Woods Hole, Massachusetts 02543
and
Department of Geological Sciences
Brown University, Providence, Rhode Island   02912

Abstract

Cenozoic foraminiferal biostratigraphy, biogeography and ecology in the North Atlantic and its marginal borderland areas is reviewed. Planktonic foraminiferal faunal diversity and provincialization exhibit fluctuating patterns which indicate a complex historical relationship between changing paleogeography, paleooceanography and paleoclimatology. Late Neogene climatic deterioration is reflected in increased planktonic foraminiferal provincialization and the development of separate zonations for high and low latitudes as well as a reduction in amphi-Atlantic cosmopolitanism among shallow water benthonic Foraminifera. In a similar manner stratigraphic and geographic distribution patterns among benthonic Foraminifera have been dependent upon the changing geometry of the Atlantic plate.

The apparently slow rate of morphologic evolution in most shallow and deep-water benthonic foraminiferal phyletic lineages restricts their usefulness in fine-scaled biostratigraphic studies (the larger foraminifera are an obvious exception), but there has been a renewed interest in using them in paleobiogeographic, paleo-oceanographic and paleobathymetric studies.

The evolution of the modern bathyal-abyssal benthonic foraminiferal fauna about 15 Ma allows paleobathymetric and paleocirculation interpretations by analogy with living faunas valid over most, if not all, of the Neogene, and in a general way for the Paleogene as well.

Résumé

Il s'agit de la biostratigraphie cénozoique foraminiférale, la biogéographie et l'écologie dans l'Atlantique du Nord et les côtes qui l'entourent. La diversité faunale planctonique foraminiférale et la provincialisation montrent des motifs fluctuants qui indiquent un rapport historique compliqué parmi une paléogéographie, une paléoocéanographie et une paléoclimatologie changeantes. La déterioration climatique de la Late Néogène est réfléchie dans la provincialisation foraminiférale augmentée et le développement des zonations séparées des hautes et basses latitudes, ainsi qu'une réduction dans le cosmopolitanisme amphi-Atlantique parmi les foraminifères benthoniques des hauts-fonds. D'une façon semblable, les motifs de distribution stratigraphique et géographique parmi les foraminifères benthoniques dépendent de la géometrie changeante de la plate de l'Atlantique.

L'allure évidemment lente de l'évolution morphologique dans la plupart des lignages phylétiques foraminiférales benthoniques se trouvant dans les hauts-fonds et les eaux profondes limite leur utilité dans des études biostratigraphiques sur une petite échelle (les foraminifères qui sont plus grands évidemment font l'exception), mais il y a un renouvellement de l'intérêt dans leur utilisation dans des études paléobiogéographiques, paléoocéanographiques et paléobathymétriques.

L'évolution de la faune bathyal-abyssal-benthonique foraminiférale moderne d'environ 15 Ma permet a des interprétations paléobathymétriques et paléocirculations par analogie avec des faunes vivantes. Ces interprétations sont valables pour la plupart, sinon toute, le Néogène et dans un sens général, le Paléogène.

INTRODUCTION

Cenozoic planktonic foraminiferal biostratigraphy may be said to have matured, if not been born, in the Caribbean and Gulf Coast. Monographic investigations in the 1950's, primarily by oil company micropaleontologists, of the Paleogene and Neogene of the Caribbean-Gulf Coast region demonstrated the utility of planktonic Foraminifera in local and regional stratigraphic studies there and around the margins of the North Atlantic Basin.

The advent of the Deep Sea Drilling Project in the mid-1960's provided the material and impetus for detailed investigations on the geographic and stratigraphic distribution of taxa and the formulation of biostratigraphic zonations of increasing resolution. About 40-50 Cenozoic planktonic foraminiferal zones have been recognized in tropical regions. Correlation of these zones with the paleomagnetic (t = 25-0 Ma), magnetic anomaly (t = 65-0 Ma) and radiometric (t = 65-0 Ma) time-scales is providing a geochronologic framework in which the Cenozoic evolution of the Atlantic Ocean can be quantified.

A summary of recent advances in Cenozoic planktonic foraminiferal biostratigraphy and biogeography of the Atlantic Ocean has been recently completed (Berggren, in press), so that less emphasis is devoted to this group than the benthonic Foraminifera in this summary. North Atlantic Cenozoic planktonic foraminiferal data are contained in the Preliminary Reports of the Deep Sea Drilling Project (vols. 1-4, 10-15) and will be contained in the forthcoming volumes 37, 38, 44, and 45).

Studies on Cenozoic (exclusive of Holocene and living) deep-sea benthonic Foraminifera in the North Atlantic are still in their infancy. Prior to 1970 there were virtually no pre-Pleistocene deep sea cores available. Between 1968-1975 the Glomar Challenger has obtained numerous cores from several areas of the North Atlantic (legs 1, 2, 11, 12, 14, 38, 43, 45 and the Gulf Coast leg 10) and a general picture of the distribution of deep-sea Cenozoic faunas is beginning to emerge.

The use of benthonic Foraminifera in biostratigraphy has been hampered by lack of careful comparative taxonomic studies which represent the basic data base for biostratigraphic and biogeographic interpretations. The apparently slow rates of morphologic evolution in most shallow and deep-water phyletic lineages has restricted their usefulness in fine-scaled biostratigraphic studies (the larger Foraminifera are an obvious exception), but there has been a renewed interest, traceable to their increased recovery from the deep-sea floor by the Deep Sea Drilling Project, in using them in paleobiogeographic, paleooceanographic and paleobathymetric studies. This usefulness will depend, in turn, upon 1) the ability to extrapolate present distribution patterns into the past and 2) reliable data on (paleo)biogeography of benthonic Foraminifera. It has been found that 1) above is applicable over approximately the last 15 my; beyond that it is possible to extricate oneself from the horns of the dilemma by making geophysically-based sea-floor topographic reconstructions which then serve as a basis for evaluating paleobathymetric distribution patterns of pre-late Neogene taxa. In the case of 2) above, presently available data strongly suggest that the biogeographic distribution of benthonic foraminifera is strongly linked with past ocean-continent geometry and attendant paleoclimatic and paleocirculation history (Berggren and Hollister, 1974). This is particularly true for the North Atlantic Ocean and its circum-marginal areas.

The discussion of benthonic Foraminifera below represents a synthesis of a large amount of literature in addition to current work being conducted at our own laboratory. Most of the basic literature on Atlantic-Caribbean Cenozoic benthonic foraminiferal faunas up to 1969 was cited in Berggren and Phillips (1971). The interested reader is referred to this source for details. Only the more pertinent references cited there as well as those from the interval since 1970 will be specifically cited here.

A discussion of the tectonic-stratigraphic **setting** of the North Atlantic (including the Norwegian-Greenland Sea), and the marginal areas in North America and Europe provides the framework for a summary of the main distribution patterns of foraminiferal faunas in these three areas.

TECTONIC-STRATIGRAPHIC SETTING

North Atlantic

The tectonic-stratigraphic evolution of the North Atlantic Ocean has been reviewed in terms of sea-floor spreading and plate tectonics by Berggren and Phillips (1971), Pitman and Talwani (1972), Emery and Uchupi (1972), Berggren and Hollister (1974), Laughton (1975), and in Woodland (ed., 1975), among others. The main events in the Cenozoic evolution of the North Atlantic may be enumerated as follows:

1) In the Paleocene a change from a two to three plate geometry resulted from a shift in active spreading from the Labrador Sea (which continued opening until the Middle Eocene) to the Reykjanes Ridge which resulted in the separation of Greenland and Eurasia and the opening of the Norwegian Sea. At this time Rockall Bank (DSDP Site 116, 117) began subsiding to its present depth (1200 m). Concomitantly, but causally unrelated, Orphan Knoll (DSDP Site 111) subsided to its present depth (1800 m) during the latest Cretaceous-Paleocene time.

    The Mesozoic taphrogenic stage of rifting of the North Sea changed to an intracratonic drifting stage in the Early Cenozoic (mid-Paleocene). During the last 60 m.y. the North Sea Basin has been characterized by regional subsidence and the formation of an intracratonic basin in which terrigenous sediments up to 3.5 km in thickness have been deposited.

2) In the Middle Eocene spreading in the Labrador Sea ceased and Greenland became part of the North American plate. The Pyrenean orogeny resulted in the uplift of a central ridge in the floor of the Bay of Biscay (DSDP Sites 118 and 119).

3) During the Early to Middle Miocene Iceland appears to have formed as a subaerial feature, perhaps related to a mantle plume.

North America

The eastern part of North America is a coastal geosyncline containing great thicknesses of ?late Paleozoic, Mesozoic and Cenozoic sediments which slope gently away (seaward) from the central stable (cratonic) portion of the continent (Murray, 1961). Cenozoic sediments are relatively undisturbed and underlie a low-relief coastal plain which extends some 6000 km from north of Newfoundland to Honduras in Central America. The combined average width of the emerged and submerged portion of the province in the Atlantic region is 250 km; around the Gulf of Mexico the average width decreases from 600 km in the United States to 160 km in Mexico (Tampico). To the south it continues to narrow and then expands to nearly 900 km in width (N-S) through El Pitén (Guatemala) and Yucatan Peninsula.

The seaward boundary of the coastal province is placed at the seaward-

facing scarp or slope of the continental slope and rise (op. cit.; see also
Hedberg, 1976).

The classic picture that the regional Mesozoic and Cenozoic structural expression of the coastal province has been controlled by gravity
dominated deformation and that sedimentary alignments are accordant with
present day regional structure (Murray, 1961) has recently been questioned
(Brown *et al.*, 1972). In the view of the latter principal vertical
mobility has been caused by block faulting or flexing, accompanied by
rotational alignment of the axes of positive and negative regional
structures, and occurring concomitantly with differential rates of relative
subsidence for coastal segments that are juxtaposed along elements of an
intersecting hingebelt system, components of which have one of three
principal alignments - NE-SW, NW-SE or N-S. First and second order stratigraphic units are then delineated.

Cenozoic sediments of the Gulf and Atlantic Coastal provinces are
part of a large geosynclinal sedimentary complex of continental, estuarine,
deltaic and marine deposits. The integral relationship between bio- and
lithofacies has perhaps nowhere been better demonstrated than in this
complex region where the cyclical, sequential replacement of one lithotope
by another, has been delineated over two generations by micropaleontologists
engaged in commercial (petroleum) exploration and State and U.S. Geological
Survey explorations.

Some 12-15,000 meters of Mesozoic-Cenozoic sediments have been
estimated to occur in the Gulf Coast and 6000 meters for the Atlantic Coast
(roughly a 2:1 ratio) by Murray (1961), but recent estimates based on seaward extrapolations of onshore thickening trends for stratigraphic units in
North Carolina wells suggests that thicknesses in the two areas may be
comparable, on the order of 12-15 km (Brown *et al.*, 1972).

In the Atlantic Coastal Plain Cenozoic sequence carbonates predominate
in the southern half, whereas marls and continental clastics predominate in
the northern half (Maher, 1971). Outcrops of marine Paleogene and Neogene
have been dredged in submarine canyons along the east coast and on the
Blake Plateau - Bahama Banks region and Georges Bank. Subsurface stratigraphic correlation of the Atlantic Coastal Plain has recently been compiled
by Maher (1971), based on an investigation of numerous state survey water
wells and the JOIDES Phase I wells (J1-J6) off the Florida Coast.

Europe

The Cenozoic tectonic evolution of western Europe has been reviewed
by Rutten (1969) and its (bio)stratigraphic evolution by Brinkmann (1960)
and Pomerol (1972) on which whose publications the following brief summary
is based.

Because the present day coast of western Europe was evolved during
the Cenozoic, marine sediments are limited largely to the vicinity of
present coastal regions, except in the Alpine region (ex Tethys). Paleogene
sediments crop out in various basins extending some 1500 km from Scania
(southern Sweden) to the Aquitaine Basin (SW France) and NW Spain, whereas
Neogene sediments occur in scattered sequences from Denmark to SW Spain.

Central Europe may be divided into two primary zones of subsidence,
an E-W Baltic and a N-S Rhenish, a structural alignment which developed in
the Late Permian.

The North Sea and Anglo-Paris-Belgian basins were in existence since
the early Mesozoic (Jurassic). The latter formed a continuous, inter-
connected zone of subsidence on the margin of the Fenno-Scandian shield to
the north and separated from the North German Basin by the Ardennes-Rhenish
Massif and the Upper Rhine Ridge.

In Paleocene time the formerly unified Anglo-Gallic Basin was broken up into several units - the London, Hampshire, Paris and Belgian basins - and become, paleogeographically, a part of the same marine region as northern Germany. The North Sea Basin is the most prominent feature in the sedimentary history of western Europe north of the Alpine geosyncline. During its sedimentary history transverse ridges divided it into smaller basins which were intermittently linked with one another during the Mesozoic and Cenozoic. Significant subsidence began in the Paleocene in the North Sea Basin as a result of tensional forces related to the opening of the Norwegian Sea.

During the Cenozoic the depositional center of the North Sea Basin was gradually displaced in a northward or northwesterly direction by the gradual retreat of the sea from the marginal areas of northwestern Europe. The general regression of the sea and differential tectonics in the different basins resulted in the termination of marine deposition in the Anglo-Paris Basin at the end of the Paleogene. The North Sea Basin is the only one in which sedimentation has continued to the present time. Linked in the paleogeographic sense during the Paleocene with the North Sea Basin, the Anglo-Paris-Belgian basins have been essentially dormant for the past 25 million years, although periodic subsidence during the Neogene has resulted in sporadic, discontinuous sedimentary sequences in these basins, particularly during the Late Miocene, Pliocene and Pleistocene.

Thicknesses of Cenozoic sedimentary rocks range from about 250 m in the Paris Basin to 800 m in the Hampshire Basin to over 3000 m in the central part of the North Sea.

In SW Europe marine Cenozoic strata occur in the Aquitaine Basin of SW France, in the tectonically active coastal region of NW Spain and in the Late Neogene Guadalquivir Basin of SW Spain.

Stratigraphy was born in Europe with the investigations of William Smith in the early 19th century. The classic subdivision of Cenozoic marine sediments was originally done in western Europe. In particular, the classic subdivision of the Paleogene is based on marginal marine sequences developed in the basins of NW Europe (London, Hampshire, Paris, Belgian, North German), where individual stage units are generally identified with marine transgressions and concomitant changes in biota. Deeper water Paleogene sediments are exposed in the Aquitaine Basin (SW France) and along the NW coast of Spain. Neogene deposits are rare and poorly exposed in the marginal areas of western Europe except in the E-W striking linear reentrant of the Atlantic Ocean in the Guadalquivir Basin - in SW Spain. As a result, most of the Neogene stratotype sections are located in the more complete sequences developed in the Mediterranean region. For a relatively up-to-date summary of Tertiary stratigraphy and related boundary problems the interested reader is referred to Berggren (1971).

PALEOGENE

Cenozoic biogeography of the North Atlantic and marginal (i.e. adjacent) land areas is intimately related to the continuing process of the fragmentation of Laurasia. The opening of the Norwegian Sea during the Paleogene and attendant structural reorganization of the NE Atlantic leading to the eventual establishment of an Arctic Province with direct deep-water connection to the North Atlantic resulted in distinct faunal changes as a result of these modifications to land and sea configurations. The opening of the North Atlantic to the Arctic resulted in the formation of North Atlantic Deep Water, cooling of surface waters in the North Atlantic, enhanced latitudinal provincialization of planktonic foraminiferal faunas, shallowing of the CCD (and probable concomitant changes in distribution patterns of bathyal benthonic foraminiferal faunas) and widespread development of shallow, warm water *Nummulites* facies during the Eocene.

Superimposed upon these changes were those caused by a gradual lowering of temperature during the Cenozoic and the gradual southward displacement of paleolatitudes as the crust moved northwards over the mantle. There has been an approximately 15°-20° southward displacement of latitudes in the northern hemisphere during the Cenozoic (Berggren and Phillips, 1971).

One of the major events which affected Paleogene foraminiferal evolution and biogeography was the sudden and significant 5°C cooling of bottom waters (from *ca*. 10°-5°C) in the Southern Ocean (Kennett and Shackleton, 1976) near the Eocene/Oligocene boundary, *ca*. 38 Ma, resulting in the formation of the psychrosphere (see also Savin *et al*., 1975).

Planktonic Foraminifera

The Gulf and Atlantic Coastal Plain Paleogene succession has been placed within a planktonic foraminiferal biostratigraphic framework by the studies of Loeblich and Tappan (1957), Olsson (1960, 1970a,b), Eames *et al*. (1962), Blow (1969), and Barker and Blow (1976), among others. Similar studies have been conducted on Paleogene sediments of western Europe but the faunas tend to be poorer owing to the generally shallower-water environment of deposition.

Paleogene planktonic foraminiferal biostratigraphy and biogeography of the North Atlantic has recently been reviewed by Berggren (1972, in press) and Berggren and Hollister (1974) so that only a brief summary is presented here.

Early Paleocene (Danian) faunas exhibit morphologic conservatism, low diversity and geographic cosmopolitanism. Distinct diversification and provincialization occurred by mid-Paleocene time and by the end of the Paleocene three distinct, latitudinally-distributed assemblages are recognizable. Subbotinids and globular-chambered acarininids characterize the high latitude assemblages; middle latitudes are characterized by subbotinids, acarininids and morozovellids, whereas low latitudes are characterized by acarininids and morozovellids. Temperature-related fluctuations in faunal associations are seen throughout the remainder of the Paleogene. The most noticeable change occurs at the end of the Eocene and during the Early Oligocene when a marked faunal replacement occurred accompanied by lowered species diversity and the development of two groups of morphologically conservative globigerinids, the "small" and "large" globigerinids, which characterized high and low latitudes, respectively. Globigerinitids (*unicava*, *dissimilis*) and globoquadrinids are common elements in Upper Oligocene sediments in the North Atlantic. Because of the sparseness and/or irregular distribution of the fauna, a strict biostratigraphic zonation was not found possible in the high latitude North Atlantic Paleogene. More useful was a multiple zonation based on the association of various faunal elements. Only an approximate correlation with the tropical Atlantic zonation is possible.

Paleogene planktonic foraminiferal assemblages in the North Sea are characterized by acarininids-subbotinids in the Paleocene-Eocene and by small globigerinids and non-keeled globorotaliids in the Oligocene. In the Norwegian-Greenland Sea, planktonic Foraminifera are relatively sparse during the Paleogene; Eocene faunas consist almost solely of *subbotinids* (*S. patagonica*, *S. linaperta*) and a few, small acarininids.

Benthonic Foraminifera

Paleogene smaller benthonic foraminiferal faunas exhibit a marked cosmopolitanism which may be explained as a result of widespread equitable climatic conditions and low polar-equatorial thermal gradients.

In the Paleocene three main depth-controlled benthonic foraminiferal assemblages (i.e. facies) have been recognized: a continental shelf fauna, termed the "Midway-type fauna" (MF), a continental slope and abyssal plain fauna, the "Velasco-type fauna" (VF), and a shallow, warm-water carbonate shelf fauna, the "Tethyan carbonate fauna" (TCF) (Berggren and Aubert, 1975). The Midway fauna achieved essentially world wide distribution and occurs in Atlantic Paleocene sediments as far north as 70°N Lat. (Eastern Greenland). The Midway fauna is characterized by species of *Cibicidoides* [*alleni* (Plummer) = *proprius* Brotzen, *howelli* (Toulmin), *succedens* (Brotzen)], Anomalinoides [*acutus* Plummer), *midwayensis* (Plummer)], *Gavelinella* [*danica* (Brotzen), *neelyi* (Jennings)], and *Osangularia plummerae* Brotzen, as well as various lagenids (nodosariids, lenticuliids, vaginulinids), polymorphinids and textulariids.

The "Tethyan carbonate fauna" (TCF) is characterized by distinct assemblages of larger and smaller Foraminifera. Two aspects have been distinguished: a) *Lockhartia-Daviesina-Sakesaria*, and *Rotalia hensoni-Operculinoides bermudezi* assemblages confined to 5°-35°N latitude from Pakistan to French Guiana and b) *Rotalia trochidiformis-Thalmannita-Boldia-Laffiteina* assemblages of tropical and boreal distribution. Elements of both assemblages occur in the Caribbean and Mediterranean (Tethyan) regions. Elements of assemblage b) occur in Paleocene strata of western Europe (Netherlands).

Paleogene sedimentation on the margin of the North Atlantic occurred in a series of cycles of transgressions and regressions. These are nowhere better demonstrated than in the studies on the benthonic foraminiferal faunas of the Anglo-Paris Basins (Wright and Murray, 1972; Murray and Wright, 1974). These authors grouped similar assemblages into quantitatively defined faunules which were then used to compare depositional histories on the two sides of the English Channel: marginal marine environments (nearshore shelf, deltaic marine, lagoonal, fluviomarine, and marsh) in the Hampshire Basin and a region of enclosed shallow shelf, lagoonal, littoral environments of normal or hypersalinity in the Paris Basin and Western Approaches. Marked faunal differences occur between the two basins:

1. Textulariina - few species in common to the two areas.

2. Miliolina - costate, alveoline and arenaceous individuals and species far more abundant in the Paris Basin and Western Approaches.

3. Rotaliina - a) the London Clay has a distinct and restricted fauna of *Praeglobobulimina*, *Pullenia*, *Uvigerina*, *Alabamina*, *Gyroidina*, *Anomalinoides*, *Cibicides*, *Epistominella*; b) the Western Approaches is characterized by *Gypsina*; *Asterocyclina*, *Halkyardia*, *Asanpina*, *Rotalia*, and *Pararotalia* (rare in the Bartonian and Lutetian of the Paris Basin and present only in the Upper Bracklesham beds (Upper Lutetian of the Hampshire Basin); c) although the Eocene beds of the Paris Basin and Western Approaches have many species in common, the composition of the faunas is markedly different, e.g. Paris Basin: *Nonion*, *Elphidium*, *Cibicides*, polymorphinids (even in miliolid dominated assemblages); Western Approaches: *Pararotalia*, *Rotalia* (in miliolid dominated assemblages); d) some

species which occur typically in the Paris Basin occur briefly in the Hampshire Basin in Upper Bracklesham beds (Fisher beds XVII/ 21).

On the other hand, *Orbitolites* and *Fasciolites* were shown to have similar distributions to the two areas.

A distinct dependence of the assemblages on lithotope is postulated in the three areas. Thus, in the Hampshire Basin rivers drained Jurassic-Cretaceous sediments with large clay content. In the Paris Basin rivers drained massifs with little clay content. Sea-grass and shallow water attached forms flourished in the resulting clear water. In the Western Approaches and Contentin the low sediment input also resulted in clear, shallow water in which sea-grass, bryozoans, and calcareous algae flourished. The distribution of sea-grass was considered a major factor in controlling the distribution of benthonic Foraminifera.

The greater amounts of fresh water entering the Hampshire Basin are reflected in estimates of hyposalinity (32-33 °/oo) there vs. normal salinity for the Paris Basin. During middle Eocene time salinities in the Paris Basin-Contentin region may have reached 38 °/oo (*Orbitolites*, peneroplids). Summer temperatures in excess of 22°C are also suggested by these faunas. Additional information on the Paleogene biostratigraphy and biogeography of the Anglo-Paris Basin (including foraminiferal data) can be found in the thorough reviews by Curry (1965, 1966).

Additional data on the Paleogene benthonic foraminiferal biostratigraphy and paleoecology of NW Europe may be found in the studies on the Eocene (Kaaschieter, 1961) and Oligocene (Batjes, 1958) of Belgium and the Mid-(Ulleberg, 1974) and Late (Larsen and Dinsen, 1959) Oligocene of Denmark. In general the lagoonal-neritic nature of these faunal assemblages reflect the continuity through the Paleogene of fluctuating, marginal depositional conditions along the NW European coastal areas.

Paleocene deep water benthonic foraminiferal faunas are known from several tectonically active basins on the perimeter of the Atlantic in Eastern Mexico (Barker and Berggren, in press), and the Caribbean (Trinidad, Venezuela, etc.). Their taxonomy, stratigraphic and biogeographic distribution in the deep Atlantic Ocean are currently being studied by R.C. Tjalsma (W.H.O.I.; see also Tjalsma and Lohmann, 1975, in press; Lohmann and Tjalsma, 1975, in press).

Four major assemblages were distinguished by a principal component analysis. Three of them contain a diverse, calcareous fauna and were found to be restricted to a pelagic sedimentary environment. They are:

1. *Nuttallides truempyi* assemblage

2. *Gavelinella beccariiformis* assemblage

3. *Nuttallides crassaformis* assemblage

The fourth, *Saccorhiza ramosa* assemblage, is dominated by various primitive agglutinated species and was found to be restricted to a continental margin, terrigenous sedimentary environment.

Using the relative plate motion of Phillips and Forsyth, the paleomagnetic data of McElhinny and the "backtracking" method of Berger it was possible to study the paleolatitudinal and paleobathymetric distribution of these assemblages. Assemblage 2, occurring throughout the depth range studied (∿ 1000-4500 m) during the Early Paleocene, became progressively restricted to shallower sites during the Middle-Late Paleocene preceding its extinction during the latest Paleocene. Assemblage 1 consisting of long-range taxa and characterizing deep sites during the Middle Paleocene, follows this trend and occurred also at intermediate depths during the Late Paleocene. It is the dominant assemblage at all depths during the earliest Eocene and tends to be more important in the mid-high latitude (30°-50°) sites. Assemblage 3 is restricted during the Paleocene to low-mid latitude (0°-30°) sites.

A drastic taxonomic replacement occurred near the Paleocene/Eocene boundary (primarily at the species level). The significance of this is not yet clear.

Eocene deep water sediments are known from the Caribbean (Barbados, Cuba, Trinidad, Venezuela), and Central America (Mexico), but although their benthonic foraminiferal faunas have been described by various investigators (Berggren and Phillips, 1971), they have not been analyzed for comparative (geographic) taxonomy or quantitative biogeography. An examination of various bathyal and abyssal faunas from the Eocene and Oligocene suggests continued cosmopolitanism.

It is within the larger foraminifera that distinct provincialism becomes noticeable. Adams (1967) has admirably summarized the geographic distribution of Tertiary larger Foraminifera in the Tethyan, American and Indo-Pacific provinces. The Tethys (or Mediterranean) and/or Indo-Pacific areas were the centers of dispersal of *Nummulites*, alveolinids, assilinids, *Fasciolites* and *Orbitolites* during the Paleogene. These forms occur sporadically in the Anglo-Paris (where they first appeared in late Early Eocene [Ypresian] time) and North German Basins where they are used in local and inter-regional biostratigraphic correlations. These forms are absent in the western Atlantic and, in fact, there is no warm-water carbonate province on the western side of the Atlantic north of the Caribbean and Florida-Gulf Coastal Plain region (e.g. the *Heterostegina*-"reef" of Late Oligocene age in Texas).

On the other hand larger Foraminifera common to both sides of the Atlantic may have had their origin during the Paleogene in the Caribbean (e.g. lepidocyclinids, miogypsinids). Numerous studies have been made on these various groups, most notably by Drooger and his students at Utrecht, Cole at Cornell University, Clarke at British Petroleum in London, and Blondeau and his colleagues in Paris. The miogypsinids, in particular, have been used in biostratigraphic studies of the Late Paleogene-Early Neogene in western Europe, particularly between the Aquitaine and North German Basins.

Important recent studies on the biostratigraphy of Paleogene larger Foraminifera of the Caribbean include those by Kugler and Caudri (1975) and Caudri (1975) on the Paleocene-Eocene of Soldado Rock, Trinidad and Barker and Blow (1976) on the Middle Eocene-Lower Miocene of the Tampico-Misantla Embayment, Mexico. The phylogeny of various orbitoidal larger Foraminifera in the Caribbean Eocene was discussed by Caudri (1975). In the latter paper Barker and Blow (1976) have placed Middle Eocene through Lower Miocene formational units in the Tampico-Misantla Embayment of Mexico in a planktonic foraminiferal biostratigraphic framework and calibrated the stratigraphic ranges of the larger Foraminifera to this scheme. The occurrence of *Pliolepidina tobleri* in the type area of the Tantoyuca Formation (in levels referrable to latest Zone P16 and/or earliest Zone P17 = Upper Eocene) and even at levels referrable to Zone P15 (Barker and Blow, 1976, p. 42, fig. 2) support Stainforth's (1965) arguments a) that it is a strictly Upper Eocene form b) refuting the argument put forth by Eames *et al.* (1962, p. 41, 42, 70, 71-81) in support of a major hiatus within the San Fernando Formation of Trinidad and, by extension, of a major, virtually worldwide, intra-Oligocene hiatus. Circum-Atlantic and Caribbean stratigraphic units temporarily excluded from the Oligocene by Eames *et al.* (1962)

have now been rehabilitated, and the Oligocene appears to have been reinstated in the hierarchy of Lyellian Cenozoic chronostratigraphy although its upper and lower limits, in terms of foraminiferal biostratigraphy, remain a matter of animated discussion.

The sequential appearance of *Lepidocyclina* (*Eulepidina*) *favosa* in mid-Oligocene (P19) levels of the Horcones Formation, *Miogypsinoides complanatus* from N2 = P21 (and questionably as low as N1 = P20), *Miogypsina gunteri* from the Upper Palma Real Formation (N2 = P21) and *M. cushmani* from the Meson Formation (N3 = P22) (*sensu* Barker in Barker and Blow, 1976) is similar to that which occurs in Europe and facilitates Late Oligocene trans-correlation based upon larger Foraminifera.

The stratigraphic distribution of Paleogene benthonic Foraminifera in the North Atlantic has been discussed in recent studies by Berggren (1972, 1974a) and Berggren and Aubert (1976a,b). In these studies the distribution of benthonic faunas was used in delineating the subsidence history of two continental remnants - Orphan Knoll in the Labrador Sea and Rockall Bank in the NE Atlantic - which were formed during early stages of continental rifting and sea-floor spreading.

In the case of Rockall Bank three benthonic foraminiferal zonules span about 5 m.y. of Late Paleocene-Early Eocene time reflecting progressively deepening deposition from near shoreline to about 200 m. The northernmost occurrence of *Nummulites* was recorded from the basal sediments just above basement on Rockall Bank and the specimens were recently described as a new species, *Nummulites rockallensis* by van Hinte and Wong (1974). The lowermost zonule, characterized by, *i.al. Nummulites rockallensis, Pararotalia tuberculifera,* and *Alabamina midwayensis* was probably deposited in water depths of less than 10 meters. The middle zonule contained an essentially Midway fauna, with numerous elements typical of the Gulf Coast Midway, as well as the Paleocene faunas described from Greenland (Hansen, 1970) and NW Europe (Brotzen, 1948). The upper zonule exhibited similarities with the "lower Eocene 3" faunas of NW Europe (Staesche and Hiltermann, 1940; Bettenstaedt, 1949) and represented depths of about 200 m, near the shelf-slope margin.

By Late Eocene time Rockall Bank had sunk to middle bathyal depths ($\sim$ 700-900 m). *Nuttallides truempyi* and an indeterminate osangulariid characterize the uppermost Eocene of Rockall Bank. Their extinction here, synchronous with various other bathyal forms elsewhere coincides with the Eocene/Oligocene boundary, a relatively sudden reduction in bottom water temperature of $\sim$ 5°C (10°C to 5°C) and the establishment of the psychrosphere (Savin *et al.*, 1975; Kennett and Shackleton, 1976). The Oligocene at Rockall Bank is characterized by *Heterolepa mexicana* (Cushman), *Siphonina tenuicarinata* (Cushman) and *S. advena* (Cushman). Common accessory forms include *Planulina renzi* Cushman and Stainforth, *P. marialana* Hadley, *Gyroidinoides girardanus* (Reuss), *Oridorsalis ecuadorensis* (Galloway and Morrey) and *Cibicidoides trincherasensis* (Bermudez), forms common in both Mediterranean and Caribbean mid-Cenozoic bathyal sequences. Sporadic, but persistent, stilostomellids, pleurostomellids and uvigerinids suggest middle bathyal depositional depth in Late Paleogene time. Neritic faunal elements, characteristic of the Chattian stage of NW Europe, occur at certain levels together with bathyal elements, indicating displacement from marginal areas of Rockall Bank and providing an insight into the Late Oligocene shallow water faunas in the area.

Rockall Bank has remained essentially at its present depth ($\sim$ 1100 m) throughout the Neogene. A gradual taxonomic replacement can be observed in the early-middle Miocene with the Late Paleogene fauna evolving into the present day fauna. There is no evidence of any related tectonic evidence accompanying the faunal change at about 15 Ma.

Eocene benthonic foraminiferal faunas at Orphan Knoll (Labrador Sea) are dominated by *Cibicidoides* (8 species), *Anomalinoides* (6 species,) *Pleurostomella* and *Bulimina* (5 species each), *Stilostomella* (4 species) and *Chrysalogonium* and *Ellipsodimorphina* (3 species each). Predominant taxa include *Nuttallides truempyi* (Nuttall), *Oridorsalis ecuadorensis* (Galloway and Morrey), *Gyroidinoides girardanus* (Reuss), *Buliminella grata* Parker and Bermudez, *Stilostomella curvatura* Cushman, *S. aculeata* Cushman and Renz, *Cibicidoides hercegovinensis* (de Witt Puyt) and *Bulimina orphan-ensis* Berggren and Aubert. This *Stilostomella-Pleurostomella-Nuttallides* assemblage suggests deposition at lower bathyal to abyssal depths (1000-2000 m) and supports the interpretation made by DSDP Leg 12 geologists that Orphan Knoll sank to its present depth during Paleocene time.

Essentially contemporaneous Eocene assemblages from the central, deeper part of the Labrador Sea consist almost wholely of agglutinated species of, *i.al. Ammodiscus, Bathysiphon, Cribrostomoides* and probably indicate deposition near or below the local carbonate compensation depth in deep cold waters (Berggren, 1972, p. 474).

North Sea benthonic foraminiferal faunas generally exhibit the following sequence during the Paleogene:

1) Early Paleocene (Danian) - calcareous assemblages (*Gavelinella danica* and various cibicidids, anomalinids) - whose ancestry may be traced into the underlying Cretaceous carbonate sequences - indicating deposition in an extensive carbonate shelf province at depths of less than 200 m, followed abruptly by

2) Paleocene (Late Danian-Thanetian) - agglutinated assemblages (*Rhabdam-mina, Glomospira, Cribrostomoides, Recurvoides, Spiroplectammina, Bolivinopsis, i.al.*) reflecting rapid subsidence, turbiditic facies and depositional depths in excess of 1000 m, followed by

3) Eocene (Ypresian)-Oligocene (Chattian) - Gradual balance between cal-careous and agglutinated faunal components reflecting differential rates of basin filling.

Paleogene benthonic foraminiferal faunas of the Norwegian Greenland Sea are known from the Iceland-Faeroes Ridge (DSDP Site 336, North Flank; 352, South Flank), Iceland Plateau (DSDP Site 348) and the Jan Mayen Ridge (DSDP Sites 346, 347, 349) (Talwani, Udintsev *et al.*, 1975). Agglutinated assemblages characterize the Late Paleogene (Oligocene) of the Iceland Plateau (348) and Jan Mayen Ridge (346, 347, 349). Two different sequences were found on the north and south flanks of the Iceland-Faeroes Ridge. On the south flank (352) Middle and Upper Eocene calcareous (*Cibicides* cf. *tenellus, Turrilina, Alabamina, Gyroidina*) and agglutinated (*Bathysiphon, Spiroplectammina spectabilis*) assemblages (indicative of middle neritic depths and similar to correlative faunas in NW Europe) are succeeded upwards by Oligocene faunas (*Angulogerina gracilis, Siphonina tenuicarinata, Gyroidina, Pullenia, Uvigerina, Bolivina*) similar to those recorded from Rockall Bank (Sites 116, 117; Berggren, 1972; Berggren and Aubert, 1976b). Oligocene assemblages on the north flank of the ridge (336) are character-ized by *Angulogerina gracilis, Sphaeroidina bulloides*, dentalinids, lagen-ids, polymorphinids, fissurinids, indicative of outer neritic-upper bathyal depths of water. The tentative conclusion was drawn that the north flank of the Iceland-Faeroes Ridge had no connection with the North Atlantic to the south essentially during most of the Paleogene and that its eventual subsidence was a post-Middle Oligocene phenomenon (Talwani, Udintsev *et al.*, 1975).

The Paleogene benthonic foraminiferal faunas indicate significant vertical movement of the sea floor in the Norwegian-Greenland Sea and are currently undergoing investigation in an attempt to unravel the complex Cenozoic geotectonic history of this area.

NEOGENE

The major paleogeographic and/or paleoclimatic events which affected circulation (and, concomitantly foraminiferal biogeography) in the North Atlantic Ocean during the past 25 Ma include:

1) Separation of the eastern and western Tethys by the junction of Africa and Eurasia about 18 Ma (Burdigalian Age)

2) Growth and expansion of the Antarctic Ice Cap during the Middle-Late Miocene (12-6 Ma)

3) Junction of Europe and North Africa, and gradual closure and eventual separation of western Tethys and Atlantic Ocean basins during the Late Miocene (10-5 Ma)

4) Reconnection of Atlantic and Mediterranean in Early Pliocene (5 Ma)

5) Early Pliocene (3.5-4.0 Ma) elevation of the Isthmus of Panama severing the marine connection and faunal interchange between the Atlantic and Pacific Oceans

6) Mid-Pliocene (3 Ma) initiation of polar glaciation in the Northern Hemisphere, the (probable) formation of the Labrador current as a significant water mass and the development of a Polar (Arctic) faunal province.

Planktonic Foraminifera

The gradual cooling of the earth during the Neogene resulted in stronger latitudinal thermal gradients and, concomitantly, in a more pronounced provincialization of planktonic foraminiferal faunas in the North Atlantic. At the same time the continued presence of the Gulf Stream as far north as the Labrador Sea until mid-Pliocene time (*ca.* 3 Ma) brought tropical faunas as far north as 54°N Lat. in the western Atlantic, whereas contemporaneous faunas in the central and eastern North Atlantic were of cool temperate nature.

The increasing provincialization of Atlantic planktonic foraminiferal faunas in the Late Neogene - due to the combined effects of changing continent-ocean geometry and climatic deterioration renders the application of a standard mondial tropical zonation scheme inoperable in the Atlantic Ocean, particularly from the Late Miocene on. As a result, a separate Late Miocene-Pliocene low-latitude (Berggren, 1973; in press) and high latitude (Berggren, 1972; Poore and Berggren, 1975) zonations have been developed for the North Atlantic which reflect the increasing provinciality of Atlantic faunas during the Late Neogene.

During the Early Miocene the dominant faunal elements in the North Atlantic (exclusive of the low-latitude, tropical region and the western Atlantic-influenced Gulf Stream) are the large, robust globoquadrinids and globigerinitids. The *Globorotalia miozea* and *G. puncticulata-crassaformis*

lineages can be used in biostratigraphic subdivision during the Middle-Late
Miocene and Early Pliocene, respectively. The appearance of *Globorotalia
inflata* and *Neogloboquadrina pachyderma* coincides with the first ice-rafted
detritus (at 3 Ma) and signifies the development of a polar faunal realm in
the circum-Arctic region. This represents the completion of the gradual
process of faunal provincialization which had been taking place during the
Cenozoic. After this three main faunal provinces can be seen in the North
Atlantic: tropical, boreal, polar.

In the North Sea planktonic Foraminifera became sparser during the
later Neogene as a result of basin filling and concomitant shallowing of
facies. In the Norwegian-Greenland Sea planktonic Foraminifera are
virtually absent during the Neogene until the mid-Pliocene (3 Ma) when,
with the advent of polar glaciation, a polar fauna appears dominated by
sinistral *Neogloboquadrina pachyderma*.

Quantitative (factor regression) investigations of planktonic foraminiferal assemblages by the CLIMAP group have resulted in the delineation of
oceanic paleoclimatic trends over the past 1 Ma. The more important results
of this major program include the elucidation of:

1) Repeated (perhaps as many as 30 or more) glaciations at high and mid-
   latitudes in the northern hemisphere over the past 1.5 Ma.

2) Repeated and drastic latitudinal displacement of climatic zones by as
   much as 20°-30°.

3) Dramatic changes in the North Atlantic circulation and evidence that
   most of the Norwegian-Greenland Sea has been ice-covered for about
   100,000 of the past 127,000 years.

The integrated geophysical, geochemical and paleontologic studies on
the deep-sea Late Pleistocene stratigraphic record of the North Atlantic
are leading to a better understanding of global climate over the past
million years. The goal of this program is the formulation of quantitative
synoptic paleotemperature maps for selected specific "moments" in time
(McIntyre, Moore *et al.*, 1976) which can, in turn, serve as boundary
conditions for modeling general atmospheric circulation patterns in the
Pleistocene (Gates, 1976) and serve as a reliable means to predicting
future global climatic trends.

The Neogene stratigraphic succession in the Atlantic Coastal Plain
has been put in a planktonic biostratigraphic framework by the investigations of Akers (1972), and in the Caribbean by Bolli (1957), Blow (1969),
and Lamb and Beard (1972). Of particular significance for Atlantic
Coastal Plain stratigraphy has been the demonstration that the Chipola
Formation of Florida is of Burdigalian Age (Zone N7) and the Yorktown
Formation – long considered Late Miocene – is actually of Pliocene (predominantly Zanclean) age (Akers, 1972). Subsequent work has shown, in
fact, that the "Yorktown" as used in the SE Atlantic Coastal Province
essentially spans the entire Pliocene (Hazel, in press).

In western Europe the Miocene of northern Belgium (Hooyberghs and
De Meuter, 1972), and the Upper Miocene of SW Spain (Verdenius, 1970;
Tjalsma, 1971; Berggren and Haq, 1976) have been put into planktonic foraminiferal zonal perspective. In the case of the Belgian Miocene planktonic
foraminiferal faunas revealed significant temporal gaps between the various
lithic units which had previously been considered to have been brief or
non-existent.

Benthonic Foraminifera

Neogene shallow water benthonic foraminiferal faunas exhibit a gradual decline in amphi-Atlantic cosmopolitanism as a result of more pronounced climatic gradients over the greater trans-oceanic and intercontinental distances caused by the combined effects of continental drift and sea-floor spreading. Shallow water Neogene marine deposits are extremely sparse in the U.S. Atlantic and Gulf Coastal provinces, as well as in Western Europe. Those that are present contain faunas with predominantly cibicidids, nonionids, elphidiids, rotaliids.

Neogene benthonic foraminiferal assemblages of the Atlantic Coastal Plain are typified by those described by Cushman (1930, 1933) from the Choctawhatchee Stage of Florida, Puri (1953) from the Alum Bluff Stage of Florida and by Gibson (1962) from Maryland-Virginia (see also Berggren and Phillips, 1971, for additional references). Similar studies have been made in Western Europe by ten Dam and Reinhold (1941, 1942) and van Voorthuysen (1958) on the Neogene faunas of Holland and Belgium, respectively, and by Margerel (1968) on the Pliocene of Brittany.

The disruption of the tropical Tethyan seaway in the Early Miocene led to separate and distinct phyletic trends among the larger Foraminifera. Thus the miogypsinids and lepidocyclinids had essentially run their course by earliest Middle Miocene time in western Europe and the Caribbean, whereas they continued to evolve in the eastern Tethys and Indo-Pacific region until near the Middle/Late Miocene boundary. Recent (unpublished) evidence indicates that some forms may have survived even longer.

Neogene bathyal benthonic foraminiferal faunas are known in the Caribbean (Cuba, Puerto Rico, Venezuela, Dominican Republic, etc.) and from the E-W reentrant of the Atlantic Ocean in SW Spain, the Guadalquivir Basin, among other places. Recent studies on Late Miocene benthonic foraminiferal faunas in two sections of the western Guadalquivir Basin (Berggren and Haq, 1976; Berggren *et al.*, in press) have allowed the delineation of intra-basinal evolution from middle bathyal (1500 m) to inner-middle neritic < 50 m) depths over an interval of about 5 Ma. The older section, El Cuervo (T = ∿ 11-10 Ma) contains a fauna dominated by *Cibicidoides mediocris* (Finlay) (= C. *pseudoungerianus* (Cushman, 1931, *non* Cushman, 1922)), *Melonis soldanii* (d'Orbigny), *Planulina wuellerstorfi* (Schwager) and such accessory forms as *Epistominella exigua* (Brady), *Oridorsalis umbonatus* (Brady), and various stilostomellids and was interpreted as having been deposited at lower bathyal (*ca.* 1500 m) water depths. The other section, Carmona-Dos Hermanos, stratotype of the Andalusian Stage, contains a faunal sequence which indicates upward shallowing from middle bathyal (*ca.* 800 m) depths at the bottom to inner neritic (< 50 m) at the top over a 4-5 my interval. A particularly sharp change in water depth (*ca.* 50-70 m) within the upper part of the Andalusian (*ca.* 5.5 Ma) was limited to an apparently contemporaneous significant eustatic fall in sea-level (based on oxygen isotope data in SW Pacific deep-sea cores) related to a major expansion of the Antarctic Ice Cap. A simultaneous and similar faunal change occurs in the Mediterranean Basin. Thus the contemporaneous elimination of benthonic foraminiferal faunas reflect a sudden eustatic sea-level fall, the former in a basin which maintained open connection with the Atlantic, the latter in a basin which temporarily lost connection with the Atlantic and became a basin of interior drainage during the Messinian Age. The implications of this Atlantic fauna from SW Spain are of major importance in understanding the Late Miocene geological history of the Mediterranean Basin (Van Couvering *et al.*, in press).

Neogene bathyal and abyssal benthonic foraminiferal faunas have been recorded from Rockall Bank (DSDP Site 116) in the North Atlantic and the

Bay of Biscay (DSDP Sites 118, 119) (Berggren, 1972), respectively. Early
Miocene assemblages at Rockall and Biscay are essentially the same as those
which occur in the Oligocene. In the Middle Miocene a change is seen in
both areas. Preliminary analysis of this material (Berggren, 1972)
suggests that modern bathyal-abyssal benthonic foraminiferal faunas (Barker,
1960) developed in the Middle Miocene about 15 Ma. The essentially
conservative evolution in these faunas enhances the possibility of paleoecologic, paleobathymetric and paleohydrologic interpretations over the
past 15 Ma (see above).

In the North Sea benthonic Foraminifera reflect gradually shoaling
conditions. Pliocene-Pleistocene assemblages are dominated by an
Elphidium-Cibicides-miliolid assemblages, in many instances.

In the Norwegian-Greenland Sea agglutinated assemblages characterize
the pre-glacial Neogene sequences of the Iceland and Vøring Plateaus,
whereas calcareous assemblages characterize the glacial late Pliocene-
Pleistocene sequences there and on the Jan Mayen Ridge and Iceland-Faeroes
Ridge. Dominant Neogene faunal elements in the Norwegian-Greenland Sea
sites are shown in tabular form below (table 1).

SUMMARY

1. The stratigraphic and geographic distribution of Cenozoic planktonic
    and benthonic foraminiferal faunas of the Atlantic Ocean and adjacent
    areas are best understood within the framework of dynamically changing
    relationships between ocean and continents and the resulting changes in
    circulation and climate which they have caused.

2. Planktonic foraminiferal faunas exhibit an increasing degree of latit-
    udinal provincialization during the Cenozoic. During the Late Neogene
    (*ca.* last 15 my) this provincialization has become so pronounced that
    multiple zonations for low and high latitudes must be substituted for
    (*i.e.* replace) the "standard" mondial, low latitude zonation(s)
    applicable during pre-Late Neogene time. The process of faunal
    provincialization, which began in the Early Eocene with the development
    of a distinct boreal fauna, was completed in the Mid-Pliocene with the
    development of a distinct polar fauna which was related to the initia-
    tion of northern hemisphere polar glaciation 3 Ma.

3. Planktonic Foraminifera have been useful in placing various Paleogene
    and Neogene formations in the marginal areas of North America and
    Europe in a time-stratigraphic framework. In several instances the
    planktonic Foraminifera have resolved long standing differences in age
    interpretations as well as demonstrated the temporal extent of
    hiatus(es) in supposedly continuous or quasi-continuous sequences.

4. Paleogene benthonic foraminiferal assemblages in the marginal areas of
    the circum-North Atlantic exhibit a high degree of cosmopolitanism
    which is somewhat reduced during the Neogene as a result of the
    combined effects of changing paleogeography and lowered worldwide
    temperatures. However, intra-provincial migration over long distances
    is seen, such as the trans-Arctic migration into northern Europe of
    Arctic-Pacific faunal elements during late Pleistocene interglacial
    intervals.

5. Evolutionary conservatism among bathyal and abyssal benthonic Foram-
    inifera appear to make paleobathymetric and paleocirculation

404

interpretations by analogy with living faunas valid over the past 15 my at least. These studies are in their infancy and hold great promise in paleooceanographic studies. Bathymetric and environmental data on shallow water living benthonic Foraminifera are generally applicable to paleoecologic studies throughout the Cenozoic.

6. Major taxonomic replacement among bathyal benthonic Foraminifera has been observed near the Paleocene/Eocene, Eocene/Oligocene and Early/ Middle Miocene boundaries, *ca*. 54, 38 and 15 Ma ago. The change at 38 Ma coincides with the establishment of the psychrosphere.

ACKNOWLEDGEMENTS

This summary was prepared at the request of Professor F.M. Swain for presentation at a Symposium on the Stratigraphic Micropaleontology of Atlantic Basin and Borderlands held at the University of Delaware, 14-16 June, 1976. It forms part of a continuing research program being conducted jointly on Cenozoic benthonic foraminifera by the Woods Hole Oceanographic Institution and the Societé Nationale des Pétrole d'Aquitaine (Pau, France). Research at Woods Hole is supported by a joint grant from the Chevron, Marathon, Mobil and Shell Oil Companies. I am indebted to Drs. R.C. Tjalsma (W.H.O.I.) and C. Wylie Poag (U.S. Geological Survey, Woods Hole) for a critical review of an early draft of this paper and to Dr. J.E. van Hinte (EXXON, EPR-E, Bordeaux) for his comments and general information on (as yet unpublished) foraminiferal faunas from the Norwegian-Greenland Sea (DSDP Leg 38). This is Woods Hole Oceanographic Institution Contribution Number 3796.

Discussion

Dr. F. Gradstein: The distinction of Paleogene Midway versus Velasco faunas, as I understand it, would reflect shallower versus deeper water conditions. I wonder if such an ecologic distinction is in part related to a certain broad latitudinal effect on depth distribution of species. Is it possible that Velasco faunas in higher latitudes, than where most of your work is derived from, occur at a shallower depth?

Dr. W. A. Berggren: At the present time we cannot answer that with certainty. At high latitudes in the North Atlantic I am not familiar with Velasco-type benthonic foraminiferal faunas. In the circum-Atlantic area Paleocene deposits are essentially shallow water in nature with Midway type faunas in them. In the few instances of deeper water Paleocene faunas that I am familiar with, they are of an agglutinated nature (North Sea Basin, Norwegian Sea, etc.). I would, in principle, doubt that Velasco elements appear at shallower depths in higher latitudes because I have not found these elements in Midway faunas at high latitudes.

REFERENCES

Adams, C.G., 1967. Tertiary Foraminifera in the Tethyan, American, and
    Indo-Pacific provinces: In Adams, C.G. and Ager, D.B. (eds.), Aspects
    of Tethyan biogeography: System. Assoc., Publ. 7, p. 195-217.

Akers, W.H., 1972. Planktonic Foraminifera and biostratigraphy of some
    Neogene formations, northern Florida and Atlantic Coastal Plain:
    Tulane Studies Geol., Paleontology, v. 9, nos. 1-4, 139 p., 60 pls.

Barker, R.W. and Berggren, W.A., in press. Paleocene and Early Eocene of
    the Rio Grande and Tampico embayments: foraminiferal biostratigraphy
    and paleoecology: Jour. Foram. Res.

Barker, R.W. and Blow, W.H., 1976. Biostratigraphy of some Tertiary
    formations in the Tampico-Misantla embayment, Mexico: Jour. Foram.
    Res., v. 6, no. 1, p. 39-58.

Batjes, D.A.J., 1958. Foraminifera of the Oligocene of Belgium: Inst.
    Roy. Sci. Nat. Belg., Mem. No. 143, 188 p., 13 pls.

Berggren, W.A., 1971. Tertiary boundaries: In Micropaleontology of the
    Oceans, B.F. Funnell and W.R. Riedel (eds.), Cambridge Univ. Press,
    p. 693-809.

Berggren, W.A., 1972. Cenozoic biostratigraphy and paleobiogeography of
    the North Atlantic: In Laughton, A.S., Berggren, W.A. et al., 1972,
    Initial Reports of the Deep Sea Drilling Project, vol. 12, Washington,
    p. 965-1001, 13 pls.

Berggren, W.A., 1973. The Pliocene time-scale: calibration of planktonic
    foraminiferal and calcareous nannofossil zones: Nature, v. 243,
    no. 5407, p. 391-397.

Berggren, W.A., 1974. Late Paleocene-Early Eocene benthonic foraminiferal
    biostratigraphy and paleoecology of Rockall Bank: Micropaleontology,
    v. 20, no. 4, p. 426-448.

Berggren, W.A., in press. Recent advances in Cenozoic planktonic foram-
    iniferal biostratigraphy, biochronology and biogeography: Atlantic
    Ocean: In Saito, T. and Riedel, W.R. (eds.), Ocean Plankton and
    Sediments, Micropaleontology, Spec. Publ.

Berggren, W.A. and Aubert, J., 1975. Paleocene benthonic foraminiferal
    biostratigraphy, paleobiogeography and paleoecology of the Atlantic-
    Tethyan regions: Midway-type fauna: Palaeogeogr., Palaeoclimatol.,
    Palaeoecol., v. 18, p. 73-192, 19 pls.

Berggren, W.A. and Aubert, J., 1976a. Eocene benthonic foraminiferal bio-
    stratigraphy and paleobathymetry of Orphan Knoll (Labrador Sea):
    Micropaleontology, v. 22, no. 2.

Berggren, W.A. and Aubert, J., 1976b. Late Paleogene (Late Eocene and
    Oligocene) benthonic foraminiferal biostratigraphy and paleobathymetry
    of Rockall Bank and Hatton-Rockall Basin: Micropaleontology, v. 22,
    no. 2.

Berggren, W.A. and Haq, B., 1976. The Andalusian Stage (Late Miocene): Biostratigraphy, biochronology and palaeoecology: Palaeogeogr., Palaeoclimatol., Palaeoecol., v. 20.

Berggren, W.A. and Hollister, C.D., 1974. Paleogeography, paleobiogeography and the history of circulation of the Atlantic Ocean: In Hay, W.W. (ed.), Sympos. on Geologic History of the Oceans, Soc. Econ. Pal. Min., Spec. Mem. 20, p. 126-186.

Berggren, W.A. and Phillips, J.D., 1971. The influence of continental drift on the distribution of Cenozoic benthonic Foraminifera in the Caribbean and Mediterranean regions: In Gray, C. (ed.), The Geology of Libya, p. 265-299, Catholic Press, Beirut.

Berggren, W.A., Benson, R.H., Haq, B., Riedel, W.R., Sanfilippo, A., Schrader, H.J. and Tjalsma, R.C., in press. The El Cuervo section (Andalusia, Spain): Micropaleontologic anatomy of an early Late Miocene lower bathyal deposit. Marine Micropaleontology, v. 1, no. 3.

Bettenstaedt, F., 1949. Paläogeographik des nordwestdeutschen Tertiär mit besonderer Berücksichtigung der Mikropalaontologie: Erdöl u. Tektonik in Nordwestdeutschland, p. 143-172, 1 pl.

Blow, W.H., 1969. Late middle Eocene to Recent planktonic foraminiferal biostratigraphy: In Brönnimann, P. and Renz, H.H. (eds.), 1st Internat. Conf. Planktonic Microfossils, Proc., v. 1, p. 199-422, 54 pls.

Bolli, H.M., 1957. Planktonic Foraminifera from the Oligo--Miocene Cipero and Lengua formations of Trinidad, B.W.I.: U.S. Nat. Mus. Bull. 215, p. 97-113.

Brinkmann, R., 1960. Geologic evolution of Europe, Hafner Publ. Co., N.Y., 161 p., 46 figs., 19 pls. (transl. from German).

Brotzen, F., 1948. The Swedish Paleocene and its foraminiferal fauna: Sver. Geol. Unders., Ser. C., no. 1, 493 (årsb. 42, no. 2), 140 p., 19 pls.

Brown, P.M., Miller, J.A. and Swain, F.M., 1972. Structural and stratigraphic framework, and spatial distribution of permeability of the Atlantic Coastal Plain, North Carolina to New York. U.S. Geol. Surv. Prof. Pap. 796, p. iii-iv + 1-79, 59 plates, 13 figs. 5 tabs.

Caudri, C.M.B., 1975. Geology and paleontology of Soldado Rock, Trinidad (West Indies). Part 2: The Larger Foraminifera: Ecolog. Geol. Helv., v. 68(3), p. 533-589, 30 pls.

Couvering, J.A. Van, Berggren, W.A., Drake, R.E., Aguirre, E., and Curtis, G.H., in press. The terminal Miocene event. Mar. Micropal., v. 1, no. 3.

Curry, D., 1965. The Palaeogene beds of south-east England: Proc. Geol. Ass., v. 76, p. 151-174.

Curry, D., 1966. Problems of correlation in the Anglo-Paris-Belgian Basin: Proc. Geol. Ass., v. 77, p. 437-467.

Cushman, J.A., 1930. The Foraminifera of the Choctawhatchee Formation of Florida: Florida State Geol. Surv., Bull. No. 4.

Cushman, J.A., 1933. The Foraminifera of the Choctawhatchee Formation of Florida. Florida State Geol. Surv., Bull. No. 6.

Dam, A. ten and Reinhold, T., 1941. Die stratigraphische Gliederung des niederländischen Plio-Plistozäns nach Foraminiferen: Meded. Geol. Sticht., Ser. C, V. no. 1, 66 p., 6 pls.

Dam, A. ten and Reinhold, T., 1942. Die stratigraphische Gliederung des niederländischen Oligo-Miozäns nach Foraminiferen. Meded. Geol. Sticht., Ser. C, V, no. 2, 106 p., 10 pls.

Eames, F.E., Banner, F.T., Blow, W.H., and Clarke, W.J., 1962. Fundamentals of mid-Tertiary stratigraphical correlation, Cambridge Univ. Press, 163 p., 17 pls.

Emery, K.O. and Uchupi, E., 1972. Western North Atlantic Ocean: Topography, Rocks, Structure, Water, Life and Sediments. Am. Assoc. Petrol. Geol. Mem. 17: 532 p.

Gates, W.L., 1976. Modeling the Ice-Age climate: Science, v. 191, no. 4232, p. 1138-1144.

Hansen, H.J., 1970. Danian Foraminifera from Nugssuaq, West Greenland. Meddel. Grønland, v. 193, no. 2, 132 p., 33 pls.

Hedberg, H., 1976. Ocean boundaries and petroleum resources. Science, v. 191, p. 1009-1018.

Hazel, J.E., in press. Distribution of some biostratigraphically diagnostic ostracodes in the Pliocene and Pleistocene of Virginia and northern North Carolina. U.S. Geol. Survey, Journ. Res., v. 5.

Hinte, J.E. van and Wong, T., 1975. *Nummulites rockallensis* n.sp. from the Upper Paleocene of Rockall Plateau (North Atlantic). Jour. Foram. Res., v. 5, no. 2, p. 90-101, 3 pls.

Hooyberghs, H.J.F. and de Meuter, F.J.C., 1972. Biostratigraphy and interregional correlation of the "Miocene" deposits of northern Belgium based on planktonic foraminifera; the Oligocene-Miocene boundary on the southern edge of the North Sea Basin: Meded. Koninkl. Acad. Wetersch., Lett. Schone Kunsten Belg., Kl. Wetensch., v. 34, no. 3, 47 p., 11 pls.

Kaaschieter, J.P.H., 1961. Foraminifera of the Eocene of Belgium: Inst. Roy. Sci. Nat. Belg., Mem. No. 147, 271 p., 16 pls.

Kennett, J.P. and Shackleton, N.J., 1976. Oxygen isotope evidence for the development of the psychrosphere 38 Myr ago: Nature, v. 260, p. 513-515.

Kugler, H.G. and Caudri, C.M.B., 1975. Geology and paleontology of Soldado Rock, Trinidad (West Indies). Part 1: Geology and biostratigraphy: Eclog. Geol. Helv., v. 68(2), p. 365-430, 2 pls.

Larsen, G. and Dinesen, A., 1959. Vejle Fjord Formationen ved Brejning: sedimenterne og foraminiferfaunaen (Oligocaen-Miocaen): Danm. Geol. Unders., v. 2, no. 82, 114 p., 9 pls.

Laughton, A.S., 1975. Tectonic evolution of the Northeast Atlantic Ocean: a review: Norges. Geol. Unders., v. 316, p. 169-193.

Loeblich, A.R., Jr. and Tappan, H., 1957. Planktonic foraminifera of Paleocene and Early Eocene ages from the Gulf and Atlantic Coastal Plains: Bull. U.S. Nat. Mus. 215, p. 173-198, pls. 40-64.

Lohmann, G.P. and Tjalsma, R.C., 1975. Geographic and bathymetric distribution of Paleocene deep-water benthic foraminifera from the western Atlantic Ocean: Abstracts, Benthonics '75, p. 27, Halifax.

Lohmann, G.P. and Tjalsma, R.C., in press. Paleocene deep water benthonic foraminifera from the western Atlantic Ocean. Part II: Geographic and bathymetric distribution.

Maher, J.C., 1971. Geologic framework and petroleum potential of the Atlantic Coastal Plain and Continental Shelf. U.S. Geol. Surv. Prof. Pap. 659, p. iii-iv + 1-98, 17 pls., 8 figs. 4 tabs.

Margerel, J.P., 1968. Les Foraminifères du Redonien. Thèse doct., Univ. Nantes, 207 p., 44 pls.

McIntyre, A., Moore, T.C. et al., 1975. The surface of the Ice-Age earth: Science, v. 191, no. 4232, p. 1131-1137.

Murray, G.E., 1961. Geology of the Atlantic and Gulf Coastal Province of North America, Harper and Brothers, New York, p. v-xvii + 1-692.

Murray, J.W. and Wright, C.A., 1974. Palaeogene Foraminiferida and palaeoecology, Hampshire and Paris Basins and the English Channel: Spec. Pap. Palaeontol. No. 14, Palaeontol. Assoc. London, 171 p., 20 pls.

Olsson, R.K., 1960. Foraminifera of latest Cretaceous and earliest Tertiary age in the New Jersey Coastal Plain: Jour. Paleontology, v. 34, p. 1-59.

Olsson, R.K., 1970a. Paleocene planktonic foraminiferal biostratigraphy and paleozoogeography of New Jersey: Jour. Paleontology, v. 44, p. 589-597.

Olsson, R.K., 1970b. Planktonic Foraminifera from the base of the Tertiary, Miller's Ferry, Alabama: Jour. Paleontology, v. 44, p. 598-604.

Pitman, W.C. and Talwani, M., 1972. Sea-floor spreading in the North Atlantic: Geol. Soc. America Bull., v. 83, p. 619-646.

Pomerol, C., 1973. Stratigraphie et Paléogéographie: Ère Cénozoique (Tertiare et Quaternaire), Doin, Paris, 269 p.

Poore, R.Z. and Berggren, W.A., 1975. Late Cenozoic planktonic foraminiferal biostratigraphy and paleoclimatology of the northeastern Atlantic: DSDP Site 116: Jour. Foram. Res., v. 5, no. 4, p. 270-293.

Puri, H.S., 1953. Contribution to the study of the Miocene of the Florida
    Panhandle: Florida Geol. Surv. Bull. No. 36, Pt. 2: Foraminifera,
    p. 71-213, 30 pls.

Rutten, M.G., 1969. The geology of Western Europe, Elsevier Publ. Co.,
    Amsterdam, 520 p.

Savin, S.M., Douglas, R.G., and Stehli, F.G., 1975. Tertiary marine paleo-
    temperatures: Geol. Soc. America, v. 86, p. 1499-1510.

Staesche, K. and Hiltermann, H., 1940. Mikrofaunen aus dem Tertiär Nord-
    westdeutschlands: Abh. Reichsanst. Bodenforsch., N.F. Berlin,
    no. 201, 26 p., 51 pls.

Stainforth, R.M., 1965. Mid-Tertiary diastrophism in northern South
    America: Fourth Caribbean Geol. Conf, Trinidad, p. 159-174.

Talwani, M., Udintsev, G. et al., 1975. Leg 38 - Deep Sea Drilling Project:
    Geotimes, v. 20, no. 2, p. 24-26.

Tjalsma, R.C., 1971. Stratigraphy and Foraminifera of the Neogene of the
    Eastern Guadalquivir Basin (southern Spain): Utrecht Micropaleontol.
    Bull., v. 4, p. 1-161.

Tjalsma, R.C. and Lohmann, G.P., 1975. Taxonomy and diversity of Paleocene
    deep water benthonic Foraminifera from the W. Atlantic: Abstracts,
    Benthonics '75, p. 48, Halifax.

Tjalsma, R.C. and Lohmann, G.P., in press. Paleocene deep water benthonic
    Foraminifera from the western Atlantic Ocean. Part I: Taxonomy.

Ulleberg, K., 1974. Foraminifera and stratigraphy of the Viborg Formation
    in Sofienlund, Denmark: Bull. Geol. Soc. Denmark, v. 23, p. 269-282.

Verdenius, J.G., 1970. Neogene stratigraphy of the western Guadalquivir
    Basin (southern Spain): Utrecht Micropalaeontol. Bull., v. 3, p.
    1-109.

Voorthuysen, J.H. van, 1958. Les Foraminifères Mio-Pliocènes et Quatern-
    aires du Kruisschans: Inst. Roy. Sci. Nat. Belg., Mém. No. 142,
    34 p., 10 pls.

Woodland, A.W. (ed.), 1975. Petroleum and the continental shelf of north-
    west Europe, v. 1, Geology, Halsted Press, J. Wiley and Sons, Inc.,
    New York, 501 p.

Wright, C.A. and Murray, J.W., 1972. Comparisons of modern and Paleogene
    foraminiferal distributions and their environmental implications.
    Mém. Bur. Rech. Géol. Min., v. 79, p. 87-96.

	ICELAND-FAEROES RIDGE (336,352)	ICELAND PLATEAU (348)	JAN MAYEN RIDGE (346, 347, 349)	VØRING PLATEAU (344)
Q	Islandiella spp. Cibicides spp. Pullenia bulloides Pullenia quinqueloba Nonion zaandamae Elphidium incertum miliolids	Planulina wuellerstorfi	Planulina wuellerstorfi Bulimina aculeata Islandiella norcrossi	Planulina wuellerstorfi Islandiella norcrossi
PLIOCENE L / E		Spirosigmoilinella Martinotiella communis		?
MIOCENE L				BARREN
MIOCENE M		Bathysiphon		?
MIOCENE E		Cyclammina- Psammosphaera- Cribrostomoides		Spirosigmoil- inella Spirolocammina ? BARREN

Table 1. Dominant benthonic foraminiferal elements in Neogene assemblages of the Norwegian-Greenland Sea.

PALEOGENE FORAMINIFERA - SOUTH ATLANTIC

Alwine Bertels

Facultad de Ciencias Exactas y Naturales,
Universidad de Buenos Aires, República Argentina.
Consejo Nacional de Investigaciones Científicas y Técnicas,
República Argentina

Resumen

En el presente trabajo de efectúa el análisis de las asociaciones foraminiferológicas que prevalecieron durante el
Paleogeno en el Atlántico Sur. Se compendian los resultados,
hasta ahora obtenidos, sobre los conjuntos faunísticos hallados
en las distintas cuencas sedimentarias, tanto en el continente
sudamericano como en el africano.

Los conjuntos planctónicos y bentónicos del Paleoceno al-
canzan una gran distribución areal en el Atlántico Sur. Durante el Eoceno se observa una diferenciación faunística latitudinal mas restricta que durante el Paleoceno; por otra parte,
en latitudes mayores a los 30º Sur las influencias australo-
asiáticas son marcadas, las cuales persisten durante el
Oligoceno.

Abstract

The several recorded foraminiferal assemblages from the
South American and African continents and those of the South
Atlantic Ocean are summarized in the present work.

Paleocene planktonic and benthonic assemblages have a wide
Atlantic areal distribution. During Eocene times foraminiferal
assemblages show more restricted latitudinal differentiations
than in the Paleocene and, at latitudes higher than 30° S
Australasian influences are evident, which persist throughout
the Oligocene.

Introduction

The purpose of the present work is to show in an approximate integral form the foraminiferal assemblage which prevailed
in the South Atlantic Ocean during Paleogene times.

To accomplish this purpose all sedimentary basins adjacent
to the South Atlantic Ocean are analyzed for the Paleogene
Period; chronostratigraphic and geographic order are followed;
this analysis is made from northern to southern latitudes in
both the South American and African continents.

The available data obtained from several oceanographic expeditions, provided also a great amount of information concerned with the marine sedimentary sequences in the Atlantic Ocean; these data are synthesized and the principal planktonic Zones are mentioned.

The data for this work were taken from all the available literature and also from unpublished theses dealing with the same topic.

It is accepted in this work that the Paleogene began approximately 65 m.y. ago, and that at this time the Atlantic Ocean was definitively outlined. Throughout the Tertiary, sea floor spreading, accompanied by the drive of the South American and African tectonic plates continued to operate.

Acknowledgments

The author wishes to express gratitude to the Argentina National Council for Scientific and Technical Research, for economical aid in the realization of the present work, and for generously providing use of the Jeolco J.S.M.-U/3 scanning electron microscope with which the micrographs were taken.

The writer is also grateful to the Buenos Aires University, Department of Geology, for the provided facilities for achieving the completion of the work.

Particular thanks are also expressed to Dr. Carlos A. Rinaldi and to Mr. Patricio Ganduglia, who kindly and generously helped in the construction of the stratigraphic charts and range charts which illustrate the paper.

The illustrated material is deposited in the Laboratory of Micropaleontology, Facultad de Ciencias Exactas y Naturales, Universidad de Buenos Aires, Argentina.

Paleogene of South America

Paleocene

Along the present South Atlantic South American borderlands and their continental margins Paleocene strata are developed in the Foz de Amazonas, São Luiz/Barreirinhas, northern Brazil and in the Recife/João Pessoa, Segipe/Alagoas, Roconcavo and Espirito Santo basins, located in northeastern and eastern Brazil. In Argentina, Paleocene sequences have been recognized in the Colorado, Neuquén, San Jorge and Austral basins; it is necessary to remark that the Austral Basin, in a northwest-southeast trend, extends into the southern region of the Chilean Republic.

In the Brazilian literature there is controversy about the interpretation of the extent, limits and nomenclature of several of the sedimentary basins; this lack of agreement is the result of many difficulties in interpreting the results of the tectonism which has been active along the present Atlantic coast of Brazil since the late Jurassic.

For instance, Amapá State, Marajó Island and Pará State, which were once considered by several authors to lie in independent basins, are included now in the single Foz de Amazonas or Amazonas Basin.

From the sedimentary Foz de Amazonas Basin Schaller, Vasconcelos and Castro (1971) recognized the Limoeiro, Amapá and Marajó Formations of Paleocene age.

Based on palynological data the age of the Limoeiro Formation ranges from late Cretaceous to Paleocene. It rests upon the Precambrian crystalline basement or on Paleozoic remnants. The sequence is composed of sandstones and several conglomeratic levels; the sandstones contain intercalated claystones. Its origin is partly fluviatile and partly marine. No Foraminifera have been recorded from this formation.

The Limoeiro Formation is overlain in this basin by the Marajó Formation which on lithological features was subdivided by Schaller, Vasconcelos and Castro (op. cit.) into four members (Text Fig. 1). The formation lacks foraminiferal faunas.

The Amapá Formation crops out in the Pará and Amapá States and occurs in their continental margins; it is mainly composed of limestones with intercalated claystones and siltstones, often glauconitic and pyritic.

The age of the Amapá Formation ranges from Paleocene to the Miocene. Although some Foraminifera are present, they are poorly represented and not clearly indicative of Paleocene age. The São Luiz/Barreirinhas Basin contains marine sedimentary sequences which range in age from the Aptian to the Tertiary. Although a palynological zonation was made (Lima, 1973), there are no foraminiferal records from this part of the basin.

From the continental margin of the São Luiz/Barreirinhas Basin the Globorotalia pseudobulloides Zone was recognized by Noguti (1975).

Paleocene sections become more complete in northeastern Brazil whereas in southeastern regions the Paleocene is absent.

In the Recife/João Pessoa Basin Tinoco (1967) and Mabesone, Tinoco and Coutinho (1968) mentioned the Paleocene Maria Farinha Formation which transitionally overlies the Upper Cretaceous Gramame Formation. The exposed section of the Maria Farinha Formation may represent a regressive facies; it is composed of clastic limestones, finer at the base and coarser to the top. A disconformity separates the Maria Farinha Formation from the overlying continental sequence in the region of the Barreiras Group of Pliocene or younger age.

From the cited fauna and the absence of Globoconusa daubjergensis (Bronnimann) the authors (Mabesone et al., 1968) inferred that part of the Danian is not well represented.

Taking into account the cited planktonics, although Globoconusa daubjergensis is absent as well as Globigerina eogubina, a large part of the Danian and Montian stages may be recognized.

In northeastern Brazil Schaller (1969) studied sedimentary sequences of the Sergipe/Alagoas Basin and the Reconcavo/Tucano and Jatobá basins. These basins are separated by a horst over which undifferentiated Cretaceous exists.

The basement of these basins consist of Precambrian and Paleozoic rocks, represented by gneisses, migmatites, metasedimentites, intrusive, extrusive and hypabyssal rocks.

In the Sergipe/Alagoas Basin continuous marine sections of Aptian to Lower Eocene age are present.

The Paleocene Muribeca Formation was studied by Schaller (1969); this author recognized two zones: Z 117 and Z 118. Z 117 is based on Chiloguembelina morsei (Kline), Globoconusa daubjergensis (Bronnimann), Globorotalia compressa (Plummer) Globorotalia pseudobulloides (Plummer) and Globorotalia perclara Loeblich and Tappan; Zone 118 is proposed on the content

of <u>Globorotalia</u> <u>angulata</u> (White), <u>Globorotalia</u> <u>velascoensis</u>
(Cushman) and <u>Globorotalia</u> <u>acuta</u> (Toulmin).

EPOCH	SERIES / EUROPEAN STAGES	P in ZONESMRY	T in ZONESMRY	PLANKTONIC FORAMINIFERAL ZONES CARIBBEAN AREA	NIGERIA LITHOSTRATIGRAPHIC UNITS (ADEGOKE, 1966) WESTERN NIGERIA	EASTERN NIGERIA	GABON CHROMOSTRATIGRAPHIC UNITS (LE CALVEZ, KLASZ, BUN, 1974)	ANGOLA ROCK STRATIGRAPHIC UNITS CUANZA BASIN (BRONGKOM AND VERRIER, 1966)	BRAZIL ROCK STRATIGRAPHIC UNITS	ARGENTINA MARINE STAGES
PLEISTOCENE	CALABRIAN	N 22		GLOBOR. TRUNCATULINOIDES					CHUI Fm.	PLATIANO QUERANDINIANO BELGRANIANO
PLIOCENE	ASTIAN PIACENSIAN ZANCLEAN	N 21 / N 20-21 / N 19 / N 18		GLOBOR. ALTISPIRA ALTISPIRA / GLOBOR. PUNCTULATINOIDES	BENIN Fm.	BENIN Fm.	AKOSSO		PARÁ GROUP: TUCANARÉ Fm.	
MIOCENE	TORTONIAN MESSINIAN	N 17		GLOBOROTALIA MARGARITAE / GLOBOROTALIA DUTERTREI			N'TCHENGUÉ		AMAPÁ Fm. {TAMOATÁ SEQUENCE / TAMBAQUI SEQUENCE}	ENTRERRIANO AND/OR PARANIANO
	SERRAVALLIAN	N 16 / N 15 / N 14 / N 13 / N 12 / N 11 / N 10 / N 9 / N 8		GLOBOROTALIA FONSI S.L.			M'BÉGA		MARAJÓ Fm. {ARAGUARI MR / MEXIANA MR}	(SUBSURFACE)
	LANGHIAN			PRAEORBULINA GLOMEROSA				LUANDA CACUACO	PIRABAS Fm.	
	BURDIGALIAN	N 6		GLOBIGERINATELLA INSUETA			MANDOROVÉ	UPPER QUIFANGONDO Fm.	PARÁ GROUP: PIRARUCU Fm.	
	AQUITANIAN	N 5 / N 4 / N 3		CATAPSYDRAX STAINFORTHI / CATAPSYDRAX DISSIMILIS				LOWER QUIFANGONDO Fm.		
OLIGOCENE	CHATTIAN	N...(P22) / (P21) / (P20) / P 19		GLOBOROTALIA KUGLERI / GLOBIG. CIPEROENSIS CIPEROENSIS / GLOBOROTALIA OPIMA OPIMA	OGWASHI - ASABA Fm.		N'GOLA	CUNGA Fm.	AMAPÁ Fm. TAMBAQUI SEQUENCE	LEONIANO
	RUPELIAN	P 18 / P 17		GLOBIGERINA AMPLIAPERTURA			ANIMBA		MARAJÓ Fm. ARAGUARI MR	JULIANO
	LATTORFIAN			CASSIGERINELLA CHIPOLENSIS/ HASTIGERINA MICRA		AMEKI Fm.				
EOCENE	BARTONIAN	P 16 / P 15		GLOBOROTALIA CERRO-AZU-LENSIS	AMEKI Fm.			GRATIDÃO Fm.	MARAJÓ Fm. {CURURÚ MR, MEXIA-NA MR / ARAGUARI MR}	(SUBSURFACE)
	PRIABONIAN	P 14 / P 13 / P 12 / P 11 / P 10 / P 9 / P 8		GLOBIGERAPSIS SEMIINVOLUTA / TRUNCOROTALOIDES ROHRI / PORTICULASPHAERA MEXICANA / GLOBOROTALIA LEHNERI / HANTKENINA ARAGONENSIS / GLOBOROTALIA PALMERAE / GLOBOROTALIA ARAGONENSIS	OSHOSUN Fm.		OZOURI		AMAPÁ Fm. {JACUNDÁ SEQUENCE / CANDIRU SEQUENCE}	'VENERICARDIA BEDS'
	LUTETIAN									
	YPRESIAN	P 7 / P 6		GLOBOROTALIA FORMOSA FORMOSA / GLOBOROTALIA REX	IMO SHALE	IMO SHALE	IKANDO	RIO DANDE Fm.	MARAJÓ Fm. {ARAGUARI MR / AFUÁ MR}	
PALEOCENE	THANETIAN	P 5 / P 4 / P 3		GLOBOR. PSEUDOMENARDII / GLOBOR. ROSALLA PUSILLA / GLOBOROTALIA ANGULATA / GLOBOROTALIA UNCINATA	EWEKORO Fm.				AMAPÁ Fm. CANDIRU SEQUENCE	
	MONTIAN	P 2		GLOBOROTALIA TRINIDADENSIS					MARIA FARINHA Fm.	SALAMANQUIANO
	DANIAN	P 1		GLOBOR. PSEUDOBULLOIDES / GLOBOROTALIA CONICOTRUNCATA						ROCANIANO

Text Fig. 1. Tertiary stratigraphic units from South America
and Africa.

Noguti and Fontana dos Santos (1972) and Noguti (1975) recognized in the Sergipe/Alagoas Basin and in its prolongation on the continental margin the Paleocene zones of: Globoconusa daubjergensis and Globorotalia pseudobulloides in descending order of age, whereas in the extension of the basin into the continental margin the same authors proposed additionally: the Globorotalia pusilla pusilla, Globorotalia (G.) pseudomenardii and Globorotalia (G.) velascoensis zones in descending order.

In the continental margin of the Reconcavo Basin, Noguti (1975) recognized the Paleocene Globorotalia pusilla pusilla Zone.

The Espirito Santo Basin yielded diagnostic planktonics; Noguti and Fontana dos Santos (1972) and Noguti (1975) proposed the Globorotalia pseudomenardii and Globorotalia velascoensis zones in ascending order.

In Argentina the oldest Tertiary sediments were deposited in the Neuquén, Colorado, San Jorge and Austral basins.

Danian strata were recorded from subsurface of the Colorado Basin by Kaasschieter (1963) and Malumian (1970b).

After the deposition of a thick terrestrial sequence of the Upper Cretaceous Neuquén Group, the Danian sea invaded the basin and deposited the Pedro Luro Formation. This formation is only recognized in subsurface; it is mainly composed of clayey to marly sediments. The sediments yielded planktonic and benthonic microfauna.

Based on the presence of Globorotalia pseudobulloides (Plummer) and Globorotalia compressa (Plummer) a Danian age was assigned to these strata (Kaasschieter, 1963).

Bertels (1964; 1969) studied the foraminiferal content of the Roca Formation and recognized the Cretaceous-Tertiary boundary over a large area in northern Patagonia, i.e. the Neuquén and Colorado basins.

The Lower Tertiary strata are represented by the Roca Formation which in the type area (Neuquén Basin) is of early Danian age based on the occurrence of Globoconusa daubjergensis (Bronnimann), Globorotalia pseudobulloides (Plummer) and Subbotina triloculinoides (Plummer). Benthonic Foraminifera are also abundant. The Roca Formation is composed of clays at the base grading into marls and limestones to the top indicating a regressive facies (Bertels, 1970). Bertels (1975) based on the occurrence of Globorotalia pseudobulloides (Plummer) recognized that Zone in the Roca.

In the San Jorge Basin Danian sedimentary sequences extend over large areas which shows lateral facial changes; the strata are exposed from the coast to the central region of the Chubut Province. These strata constitute the Salamanquiano Stage. The known formations are the Salamanca and Bororó formations.

The first micropaleontological studies in this basin were carried out by Camacho (1949) with subsurface samples; subsequent studies of this formation were undertaken by Mendez (1966), Masiuk (1967), and Bertels (1973, 1976a) based on surfaced sections. Mendez (1966) studied outcrop samples from the Salamanca Formation at Punta Peligro locality (Chubut Province). The assigned age was Middle to Upper Danian. Masiuk (1967) analyzed outcrop samples located at Puesto Alvarez, on the lower course of the Rio Chico River; the assigned age was Middle to late Upper Danian for this sequence. Bertels (1973) studied the microfaunal assemblage of the Cerro Bororó Formation which crops out in the central area of the Chubut Province.

Based on the presence of Globorotalia compressa (Plummer) a late Upper Danian age was suggested for the Cerro Bororó Formation.

Bertels (1976) studied the microfaunal content of the sedimentary sections of the Salamanca Formation at the type section of Bajada de Hansen locality (Andreis, Mazzoni and Spalletti, 1975).

The micropaleontological content comprises planktonic and benthonic Foraminifera. The most conspicuous planktonic is Globorotalia (Turborotalia) compressa (Plummer), and Bertels (1975) proposed the zone of Globorotalia (Turborotalia) compressa for strata of the upper or Hansen Member of the Salamanca Formation.

Benthonic Foraminifera constitute rich and diagnostic assemblages.

The Austral Basin is the southernmost basin and extends in a northwest-southwest trend comprising the southern regions of

Chile and Argentina.

In subsurface samples of the Austral Basin Malumian, Masiuk and Riggi (1974) found a more or less continuous sequence of strata which range in age from Valanginian to Miocene. Danian assemblages were recorded by those authors. The species found in the Austral Basin are basically the same as those recorded from northern areas.

In the Chilean part of the Austral Basin Brunswick Peninsula, Paleocene strata were studied by Herm (1966) and Charrier and Lahsen (1969).

From the southern Chile Brunswick Peninsula area, Charrier and Lahsen (1969) mentioned Upper Cretaceous and Lower Tertiary sediments. These authors included in the Paleocene the upper part of the Chorrillo Chico Formation, the San Jorge Formation and the lower part of the Agua Fresca Formation.

The Chorrillo Chico Formation consists of very hard, light-brown, silty, partly glauconitic argillite of tuffaceous appearance. Large lenticular limestone concretions are characteristic of the formation which yielded species of the genera Cyclammina and Allomorphina, etc. Charrier and Lahsen concluded an Upper Cretaceous-Lower Tertiary age for this formation.

The San Jorge Formation consists of poorly fossiliferous, glauconitic, brown argillite containing many limestone concretions. Charrier and Lahsen (1969) suggested a Paleocene age for these strata.

Eocene

The Eocene is represented in the Foz de Amazonas, Sergipe/ Alagoas, Reconcave, Espirito Santo, Campos (Rio de Janeiro) and Paraná Basin in Brazil, and in the Argentina Colorado, San Jorge and Austral Basins.

The northernmost basin is the Foz de Amazonas Basin. This basin comprises an emerged area (=sedimentary Marajó Basin) and a second situated on the continental margin; the basin is limited by the Guayanas Shield at the northeast and the Central Brazilian Shield at the southwest. The Gurupá arc separates this basin from the Paleozoic basin in the "Lower" Amazonas and the Tocantins arc separates it from the Maranhao sedimentary basin (Schaller, Vasconcelos and Castro, 1971). Forming part of the Foz de Amazonas Basin are the Amapá and Pará regions.

In the Amapá Formation several sequences have been recognized in which the following planktonic foraminiferal zones were recorded: Globorotalia wilcoxensis Zone below and Truncorotaloides rohri Zone above from the lower Candiru sequence. The Jacundá sequence (Upper Eocene-Oligocene) yielded the zonal species: Globorotalia cerroazulensis and Globorotalia opima opima. Schaller (1969) in the Sergipe/Alagoas Basin found the Eocene planktonics Globanomalina aff. pseudoiota Hornibrook and Globorotalia aff. G. wilcoxensis Cushman and Parker. Noguti and Fontana dos Santos (1972) and Noguti (1975) recorded from this basin the planktonic zones of: Globorotalia wilcoxensis, Truncorotaloides rhori, and Globorotalia cerroazulensis, in descending order of age.

Noguti and Fontana dos Santos (1972) and Noguti (1975) based on studies of sedimentary sequences of the Sergipe/Alagoas Basin continental margin recognized in ascending order of age the Zones of: Globorotalia wilcoxensis, Globorotalia quetra, Globorotalia palmerae, Globigerinoides higginsi, Orbulinoides beckmanni, Truncorotaloides rohri, Globigerapsis semiinvoluta, and Globorotalia cerroazulensis.

From the Reconcavo Basin (South of Bahía) Noguti (1975) recorded the planktonic Zones of: Globorotalia wilcoxensis, Globorotalia quetra, Globigerinoides higginsi, Truncorotaloides rohri, and Globorotalia cerroazulensis, in ascending order.

In the Espirito Santo Basin Noguti and Fontana dos Santos (1972) and posteriorly Noguti (1975) recognized the following Zones: Globorotalia wilcoxensis, Globorotalia quetra, Globigerinoides higginsi, Orbulinoides beckmanni, Truncorotaloides rohri, and Globorotalia cerroazulensis, in ascending order.

In the Campos Basin (Rio de Janeiro) Noguti (1975) recognized in ascending order: Globorotalia wilcoxensis Zone, Globorotalia quetra Zone, Orbulinoides beckmanni Zone, Truncorotaloides rohri Zone.

In the continental margin of the Paraná Basin Noguti (1975) recognized the Globorotalia cerroazulensis Zone.

From subsurface sequences of the Austral Basin, Santa Cruz Province, Argentina, Malumian (1968) found an Eocene fauna with marked affinities with that of the Agua Fresca Formation in which foraminiferal content was previously studied from the

Chilean region of this basin by Todd and Kniker (1952).

Todd and Kniker (1952) recorded Eocene Foraminifera from the Agua Fresca Formation of Magallanes Province, Southern Chile. Nevertheless, the age is Paleocene to early Eocene based on the planktonics found in surface and subsurface samples such as: Globigerina triloculinoides Plummer, Globorotalia compressa (Plummer), Globigerina aquiensis Loeblich and Tappan, Glorotalia membranacea (Ehrenberg), Acarinina triplex Subbotina, Globanomalina pseudoiota Hornibrook and Pseudohastigerina wilcoxensis Cushman and Ponton.

Subsequent subsurface studies in the Austral Basin were carried out by Malumian, Masiuk, and Riggi (1974); these authors recorded Truncorotaloides collactea (Finlay) and assigned a late Middle Eocene to early late Eocene age to this interval.

Outcrops samples of the San Julián Formation (Austral Basin) yielded a foraminiferal assemblage in which the planktonic zones of: Globigerina eocaena (late Eocene) and Globigerina officinalis (early Oligocene) were proposed (Bertels, 1975). The San Julián Formation is of late Eocene to early Oligocene age (Bertels 1975; 1976c).

Oligocene

The Oligocene is mainly recorded from subsurface samples from the continental margin of Brazil.

It is represented in the Foz de Amazonas Basin and its continental margin. From the Amapá Formation, in its Tambaquí Sequence, Schaller, Vasconcelos and Castro (1971) recorded the Globigerina ciperoensis ciperoensis Zone. In the continental margin from the Foz de Amazonas Basin Noguti and Fontana dos Santos (1972) and Noguti (1975) recorded the: Globigerina ampliapertura Zone, Globigerina opima opima Zone, and the Globigerina ciperoensis ciperoensis Zone, in ascending order of age.

Noguti and Fontana dos Santos (1972) and Noguti (1975) recorded from the continental margin of the Sergipe/Alagoas Basin the Oligocene zones of: Globigerina ampliapertura Globorotalia opima opima, and Globigerina ciperoensis ciperoensis, in ascending order. Noguti and Fontana dos Santos

(1972) and Noguti (1975) from the continental margin of the Reconcavo and Espirito Santo basins recorded, in ascending order, the zones of: Globigerina ampliapertura, Globorotalia opima opima, Globigerina ciperoensis ciperoensis, and Globorotalia kugleri.

In the Campos Basin (Rio de Janeiro) the zones of Globigerina ampliapertura and Globorotalia kugleri were recognized by Noguti (1975).

The continental margin of the Paraná Basin (Noguti, 1975) yielded fossils of the Oligocene Globigerina ampliapertura Zone.

In Argentina, subsurface samples from the Colorado Basin yielded planktonic Foraminifera that are of Oligocene age (Malumian, 1972). Benthonic Foraminifera are also abundant in the Oligocene section.

From the Austral Basin Bertels (1975, 1976b) mentioned planktonic and benthonic assemblages from the Oligocene type Monte León Formation and the Leoniano Stage.

Lithologically the Monte León Formation is composed of siltstones with a large pyroclastic content; at the top arenaceous sediments intergrade gradually with the non-marine Santa Cruz Formation.

The Leonian Stage is also represented in subsurface in this basin and in the San Jorge (surface) and Colorado (subsurface) basins.

The microfaunal assemblage yielded planktonic and benthonic Foraminifera; for the Leonian Stage Bertels (1975) proposed three planktonic foraminiferal zones in ascending order: i) Zone of Globigerina anguliofficinalis, ii) Zone of Globigerina ciperoensis, iii) Zone of Globigerina angulisuturales. The benthonic Foraminifera are also well represented.

Subsurface Oligocene sequences of the Austral Basin and their foraminiferal assemblages were studied by Malumian, Masiuk and Riggi (1971).

Paleocene

The Togo/Dahomey sedimentary basin is the western prolongation of the large Nigerian sedimentary basin (Slansky, 1963).

During the Danian (Slansky, op. cit.) regressive facies are registered which persist throughout the Paleocene. The lithological sequence of the early Paleocene is composed of several sedimentary facies at the base and at the top zoogenic calcareous strata.

The Paleocene foraminiferal assemblage was recorded by Slansky (1963).

In several regions of Nigeria Paleogene strata have been deposited; these sequences are found in surface exposures and in subsurface samples. In the present work the lithostratigraphic scheme of Adegoke (1966) is followed and the principal references are made to the western and eastern Nigeria formations which mostly were deposited in southern Nigeria.

The Southern Nigeria Basin is limited to the north and the east by the crystalline complex of the African Bouclier basement; to the west and south it extends to the continental margin. The Lower Tertiary sedimentary outcrops form an arc bordering the ancient complex. The southern basin is subdivided in the western and eastern basins; they were partially separated by a submarine basement ridge; this structure may have been partially responsible for the minor lithofacies differences known to occur between eastern and western Nigeria (Adegoke, 1966).

During the Paleocene marine sediments were deposited in both regions; the principal lithostratigraphic units are the Ewekoro Formation, exposed in western Nigeria and the Imo Shale which crops out in western and eastern Nigeria (Text Fig. 1). Paleocene strata are also well represented in subsurface strata.

The basal Ewekoro Formation is composed of biogenic limestones. The Imo Shale disconformably overlies this formation; its lithologic composition are greenish-gray, finely laminated shales containing interbedded sandstones and glauconite, culminating in dull red siliceous and sandy shales. According to Reyment (1965) the Imo Shale has three arenaceous equivalents in eastern Nigeria; Igababu Sandstone, Ebenebe Sandstone and Omuma Sandstone. The Paleocene Ewekoro Formation at its type locality contain Foraminifera recorded by Reyment (1965). Typical Danian planktonic Foraminifera were recorded from southeastern Nigeria from the Nsukka Formation (Reyment, 1965) and from southern Nigeria by Stolk (1963).

The Danian Stage is also well documented in coastal western Nigeria; subsurface samples at Araromi and Ilaro I and Ilaro II boreholes are composed of diagnostic microfaunal associations (Berggren, 1960; Reyment, 1960, 1966). Younger Paleocene is widely distributed in western Nigeria and the planktonic foraminiferal content compares with that of Central America (Reyment, 1966). In addition, in northeastern Nigeria the Kalambaina Formation contains the large foraminifer: Operculina canalifera d'Archiac; on the other hand species of Ranikotalia and Discocyclina were found in western and northwestern Nigeria. According to Reyment (1966) the Paleocene zone of Globorotalia trinidadensis is well developed and several subsequent zones are clearly delineated. The diagnostic taxa have been recorded by several authors from the paleocene of Nigeria (Berggren, 1960; Reyment, 1960, 1965, 1966; Stolk, 1963).

The Paleogene of Cameroon (Belmonte, 1966) comprises Paleocene and early to middle Eocene deposits, represented in western Cameroon, which is the continuation of the eastern Nigeria Basin.

The marine Paleocene is known from boreholes in the Duala Basin characterized by planktonic and benthonic Foraminifera (de Klasz, Le Calvez and Rérat, 1964a; Belmonte, 1966) and large Foraminifera such as Discocyclina gr. aguerreversi (Caudi) and small Nummulites (Blondeau, 1966). To the east the facies in the Duala Basin become more littoral, only Benthonic Foraminifera being found.

Reyment (1965) recorded the Mpundo Formation and the Logbaba beds in Cameroon as being of Paleogene to Neogene age. They contain planktonic, benthonic and large Foraminifera.

In Gabon the sedimentary basin was subdivided by de Klasz and Gageonnet (1963) into a western and an eastern part. The Paleocene is represented in both basins; post-Paleocene sediments were deposited only in the western part of the basin.

The Paleocene comprises the Ikando and lower Ozouri Stages.
The Lower Paleocene is represented by the Ikando Formation and
the Ikandoian Stage. The Ikando Formation is largely clayey
with a reefal carbonate level; it contains planktonic and benthonic Foraminifera recorded by de Klasz, Marie
and Meijer (1960), de Klasz and Gageonnet (1963), de Klasz,
Le Calvez and Rérat, (1964b) and Le Calvez, de Klasz and Brun
(1974).

The Ozuri Formation is composed of silicified sediments
with chert intercalated by claystones. The microfaunal assemblage of the Ozurian Stage was listed by de Klasz and Gageonet
(1963), de Klasz, Le Calvez and Rérat (1964b, c) and Graham,
de Klasz and Rérat (1965).

In Angola four sedimentary basins are distinguished
(Antunez, 1964): the Congo Basin, de Cuanza Basin, the Benguela Basin and the Moçamedes Basin.

From these basins only in the Cuanza Basin did the sedimentary fossiliferous sequence which began during Early Cretaceous time, continue during late Senonian-Paleocene time. The
central part of the basin contains strata of more or less silty
shales whereas in other parts of the basin the Paleocene is a
thick unfossiliferous sequence of yellow to tan sandstone and
red to tan argillite which may represent a fluvio-deltaic environment (Brognon and Verrier, 1966); toward the north, at Cacuaco and also in Luanda the wells encounter black unfossiliferous shale, argillite and finely laminated interbeds of dark
marl, containing numerous Foraminifera, especially carinate
planktonics (Broghon and Verrier, 1966).

In the Cuanza basin the Paleogene (Antunez, 1964) overlies
concordantly the Maastrichtian, their boundary being difficult
to determine. According to Happener (1958, in Antunez, 1964)
there exists a rich microfauna of Globorotalia and Globigerina
dated as early Eocene with questionable Paleocene at the base.

De Klasz, Marie and Meijer (1960) made reference to the
presence of Gabonella gigantea de Klasz and Meijer in the Dano-
Montian of Angola, a species which was also found in Gabon in
strata of the same age .

Eocene

The Eocene sedimentary sequence of Togo and Dahomey

(Slansky, 1963) is marked by clayey sediment at their base becoming glauconitic and marly at the top. Ypresian strata are developed in clayey facies at the base with glauconitic marly clay at the top.

Sequences attributed to the Lutetian vary between 10 and 170 meters of limestones with a phosphatic layer at the base and becoming light gray marly clay with thin gypsum beds at the top. The Lutetian seems to represent a regressive facies (Slansky, 1963).

In Dahomey the Oshosum Formation, which is also develped in Nigeria near Ifo, yeilded many planktonic Foraminifera such as Globorotalia aequa Cushman and Renz, Globorotalia pseudotopilensis Bronnimann and Globigerina soldadoensis Bronnimann, which are correlated with the sequence of the borehole at Otta, Western Nigeria, by Reyment (1965). Other Eocene microfaunas are locally abundant in Togo and Dahomey. There are also the large Foraminifera such as Discocyclina, Nummulites, and Clavigerinella. The planktonic, benthonic and large Foraminifera were recorded by Slansky (1963) and Reyment (1965).

The Eocene of Nigeria and adjacent areas was a period of regression. The sediments are locally fossiliferous and predominantly fine grained (Adegoke, 1966).

Diagnostic species were recorded from boreholes at Iju and Otta (Oshosun Formation) by Berggren (1960), Reyment (1959) and those recorded by de Klasz, Le Calvez and Rérat (1964b, c) and Adegoke (1966).

Berggren (1960) pointed out the absence in this area of the genus Hantkenina and other genera typical of the Eocene in Trinidad such as Globigerapsis and Porticulasphaera. Reyment (1965) suggested an Ypresian to Lower Lutetian age for the Oshosum Formation whereas Adegoke (1966) accepts a middle Eocene (Lutetian) age. In western Nigeria the Oshosun Formation is overlain by the Ameki Formation; in eastern Nigeria the Ameki Formation is the only recognized lithostratigraphic unit.

The type area of the Ameki Formation is in eastern Nigeria, where it is composed predominantly of greenish-gray, clayey sandstone and sandy claystone, with rapid lateral facies changes. In western Nigeria the Ameki Formation form a thin veneer

above the phosphate-bearing beds of the Oshosun Formation
(Adegoke, 1966). The Ameki Formation in eastern Nigeria is
richly fossiliferous. Outcrops of the Ameki Formation in
western Nigeria are non-fossiliferous and only few species were
recovered from boreholes at Otta and Iju (Berggren, 1960;
Reyment, 1960). The microfaunal assemblage of the Ameki Formation were recorded by Berggren (1960), Reyment (1960; 1965),
Stolk (1963) and Adegoke (1966). They are listed in Text Fig.
3. In addition Reyment (1966) mentioned the large foraminifer
Discocyclina sp. at some localities, i.e. Lakolabo. The Ameki
Formation is considered to be of Lutetian to Lower Bartonian
age by Adegoke (1966); Reyment (1965) assigned an early and
middle Eocene (Ypresian-Lutetian) age for this formation.

Stolk (1963) stated that during the early, middle and late
Eocene it is not possible here to subdivide the microfaunal
succession into the zones proposed by Bolli for the Caribbean
area. Locally, Stolk (1963) found the following biostratigraphic units: Late Eocene - Chiloguembelina martini cubensis Zone,
Middle Eocene - Cassigerinelloita amekiensis Zone, Early Eocene
- Globorotalia formosa Zone.

According to Reyment (1966) the late Eocene was
not well defined in Nigeria and evidences of the presence of
the Oligocene Stage is not extensive. On the other hand Stolk
(1963) mentioned the following late Eocene or early Oligocene
planktonic Foraminifera: Globigerina angustiumbilicata Bolli,
Chiloguembelina cubensis (Palmer), Pseudohastigerina aff. P.
micra Cole and Globorotaloides suteri Bolli.

The Eocene of Cameroon is not well characterized, being
found in isolated outcrops (Belmonte, 1966). In Kwa-Kwa the
basal Eocene contains a fauna of Nummulites indicating a basal
(Ypresian) Eocene age (Belmonte, 1966). In Bomono Eocene
strata are found which provided a poor foraminiferal microfauna
such as (Belmonte, 1966): Cancris cf. cocoaensis Cushman,
Planulina sp., Loxostomum sp. and Sagrina sp.. De Klasz, Le
Calvez and Rérat (1964a) recorded Bolivina crassicostata from
the Paleogene of Cameroon and Blondeau (1965) mentioned the
large foraminifer Nummulites (Operculinoides) furoni Blondeau
which is also found in Senegal and Ivory Coast. The middle

late Eocene and Oligocene stages are unknown in Cameroon
(Belmonte, 1966).

In the Gabon Basin sequences of Eocene age are mentioned
from the western part of the basin. In this area the Eocene
Upper Ozouri, Animba and N'Gola formations were recognized
(Le Calvez, Klasz and Brun, 1964). The microfauna of the Eocene Upper Ozouri Formation is not well known, but includes
many species in common with the overlying Animba Formation (de
Klasz and Gageonnet, 1963). The Eocene Animba Formation is
separated from the Ozouri Formation by a disconformity and its
upper part is overlain disconformably by marine Miocene strata.
The Animba Formation is mainly composed of phosphatic claystones with some detritic intercalations. The microfaunal content of this formation was studied by de Klasz and Rérat (1961)
de Klasz and Gageonnet (1963), de Klasz, Le Calvez and Rérat
(1964b, c), Graham, de Klasz and Rérat (1965) and Adegoke
(1966).

The late Eocene N'Gola Formation is only found in subsur-
face samples in boreholes at the southeast of Port Gentil.
This formation yielded the following Foraminifera (de Klasz and
Gageonnet, 1963): Hantkenina primitiva Cushman and Jarvis,
Bolivina striatellata Bandy and Mandjina excavata de Klasz and
Rérat. The N'Gola association, although it possesses its own
features, shows some similarities with the with the underlying
Animba Formation (de Klasz and Gageonnet, 1963).

The Eocene of Angola is represented by the Gratidão and
Cunga formations (Brognon and Verrier, 1966). The Gratidão
Formation is made up of whitish chalk, rich chert occurrences
in some places, and the presence of silicified limestones; it
is interbedded black bituminous marl and argillite (Brognon and
Verrier, 1966). According to Happener (1958, in Antunez, 1964)
there exists a rich microfauna of Globorotalia and Globigerina
dated as early Eocene.

The Cunga Formation is composed of gray marls with calcareous intercalations, partly silicified; it contains characteristic dolomitic concretions with Hastigerinella and

Hantkenina.

In South Africa (Chapman (1928) from a foraminiferal limestone of late Eocene age of the Alexandria Formation, recorded
several benthonics associated with the large Foraminifera
Discocyclina pratti (Mich.) and Discocyclina varians (Kaufmann).

Atlantic Ocean

In the last two decades throughout several cruises over
the oceans and the intensive development of marine research,
several projects were carried out. Oceanographic vessels such
as Vema, Chain, Glomar Challenger, etc., obtained deep-sea
cores from several areas of the South Atlantic Ocean. The Atlantic Ocean is of importance because the obtained data allowed
information to be added about the age of the cored sediments
and about tectonic plates, sea floor spreading and continental
drift geological hypothesis and other events which occurred
during the last 135 million years.

Several authors contributed with micropaleontological studies to the best knowledge of the sedimentary sequences found
in the South Atlantic Ocean. Saito, Ewing and Burckle (1966)
from cores obtained by the vessel Vema, recorded Upper Cretaceous to Pliocene sediments from the South Atlantic. Paleogene
sediments were obtained from the Rio Grande Rise latitude 28º
36'S and longitude 29º 01'W; at this site. Eocene sediments
with Hantkenina alabamensis Cushman, Hantkenina primitiva
Cushman and Jarvis, Globigerapsis index Finlay, and Globorotalia
centralis Cushman and Bermudez were recorded. This, the oldest
sediment cored from Rio Grande Rise is from a fault scarp on
the north side. Cifelli, Blow and Nelson (1968), from a dredge
sample recovered by the Chain from a fracture zone at 01º
23.5'S, 29º12 W recorded Paleocene and Lower Eocene planktonics.

The Paleocene (Thanetian) was mainly represented by Globorotalia (Globorotalia) velascoensis and Globorotalia (G.)
pseudomenardii whereas the Eocene sediments contained typical
Ypresian planktonic Foraminifera, mainly Globorotalia formosa
formosa, Globorotalia (Turborotalia) collactea and Pseudohastigerina eocenica, all typical of the Globorotalia formosa
formosa Zone.

Maxwell et al. (1970) recorded the results of the cruise
of the Glomar Challenger from Leg 3 which covered the South

Atlantic between Dakar, Senegal and Rio de Janeiro, Brazil.
The rock samples were obtained from seven sites located on the
mid-Atlantic Ridge flanks and two sites on the Rio Grande Rise;
most of the sediments cored contained planktonic Foraminifera
ranging in age from Campanian to late Pleistocene.

Maxwell et al. (1970) designated nine submarine subsurface formations ranging from Upper Cretaceous to Holocene in
the mid-Atlantic Province. The authors used the names of his-
toric exploratory vessels in an alphabetical order. They are
in order of increasing depth and range of age of each formation:

   Albatros Ooze, Pliocene-Quaternary
   Blake Ooze, Late Miocene-Pliocene
   Challenger Ooze, Late Miocene-Early Pliocene
   Discovery Clay, Late Oligocene-Middle Miocene
   Endeavor Ooze, Late Oligocene-Middle Miocene
   Fram Ooze, Early Oligocene-Early Miocene
   Gazelle Ooze, Middle-Late Eocene
   Grampus Ooze, Middle Eocene-Early Miocene
   Hirondele Ooze, Late Cretaceous-early Middle Eocene

Maxwell et al. (1970) found that the Early to Middle
Cenozoic planktonic foraminiferal faunas are diverse and bear a
close resemblance to those reported from tropical regions of
the world. Paleogene sediments were recovered at: Site 14
with the Zone of Cribohantkenina inflata, and at Site 19 with
the Zone of Hantkenina aragonensis. At sites 21 and 22 on the
Rio Grande Rise Lower Miocene to Maastrichtian assemblages were
recognized.

Blow (1970) made an extensive study of cores recovered
during Leg 3 of the Glomar Challenger cruise. Paleogene sequences with typical Caribbean planktonic Foraminifera were
recognized at sites 14, 17, 18, 19, 20, 21 and 22; at these
holes Paleogene planktonic zones were identified by Blow (op.
cit.) which are summarized below.

Funnell (1971) in the index of pre-Quaternary occurrences
in the oceans, recorded from the South Atlantic several locations in which Paleogene and particularly Eocene sediments
occur. This information is summarized below.

Berggren and Amdurer (1973) recorded Paleogene sediments,

also recovered in the South Atlantic by the Deep Sea Drilling
Project during the cruise of Leg 3. The locations of the Paleogene sites and their corresponding ages are summarized below.

Berggren (1973) synthesized the Paleogene planktonic foraminiferal faunas on legs I-IV from the Deep Sea Drilling Project in which additional and detailed information is given.

The several cores obtained from the Atlantic Ocean in
which Paleogene sediments are recorded are summarized below, in
which **the location,** as well as the age and the planktonic zones
found in the cored sediments are indicated. Text Fig. 2 shows
the location as well as the Paleogene sequences encountered
during the cruises of several vessels.

00º 15'S - 14º 25'W   Paleocene and Lower Eocene,
                                         in order of increasing age
                                        Zones of G. velascoensis
                                               G. pseudomenardii
                                                    G. aquiensis
01º 23.5'S - 29º 12W   Paleocene-Early Eocene (Than.-Ypresian)
28º 58.72S - 08º 00.70W-Late Oligocene-Early Mioc.-Pleist. Z.N3
28º 02.74S - 06º36 W -   Late Oligocene-Early Miocene Z.P19-N3
28º 19.89S - 20º 54.46W   Late Eocene-Early Miocene Z.P17-19
28º 31.47S - 26º 50.75W - Oligocene   Z.P18,19
                         Late Eocene   Z.P16-18
                         Middle Eocene   Z.P9-14
                         Early Eocene   Z.P8
                Basal Eocene or late Upper Paleocene Z.P6, 7
                         Paleocene   Z.P6
28º 32.08S - 23º 40.13 W Oligocene   Z.P18-N2
                         Late Eocene   Z.P14-17?
                         Middle Eocene   Z.P11,12
28º 35.10S - 30º 35.85 W Middle Eocene   Z.P10,11
                         Early Eocene   Z.P 8
                Early Eocene to Late Upper Paleoc. Z.P4,6
28º 36 S - 29º 01W   Late Eocene with Hantkenina alabamensis
                                         Hantkenina primitiva
                                         Globigerapsis index
                                         Globorotalia centralis
30º 00.31S - 35º 15.00W - Late Oligocene   Z.N 1-3
                         Middle Eocene   Z.P11,12

42º 32'S – 56º 29'W	Upper Cretaceous-Lower Tertiary
47º 29'S – 59º 21 W	Eocene
48º 29'S – 59º 21 W	Eocene
65º 43'S – 16º 25 E	Eocene

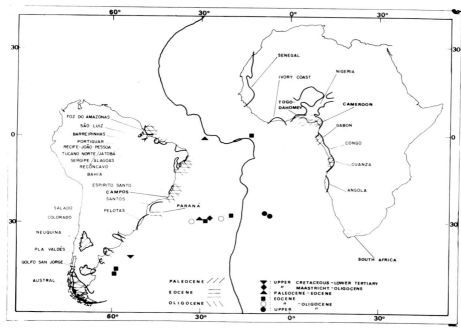

Text Fig. 2.   Locations at which Paleogene sequences in the
South Atlantic were found. The figure illustrates also the marine sedimentary basins in
which Paleogene transgressions took place. Some
of the Paleogene sequences were only recorded
from the continental margin.

Summary

In the South American continent Paleocene sequences are
more complete in northeastern Brazil than in southern regions,
such as in Argentina where only the Danian Stage is represented.

Paleocene planktonic Foraminifera are common to all South
Atlantic areas and are strongly related to those recorded by
Bolli (1966) from the Caribbean region.

Paleocene benthonic assemblages are spread over large
areas; they show many species common to several South American
basins. Although cosmopolitan and European species are found,
the prevailing assemblages are those recorded from the United

States Atlantic and Gulf Coast areas, particularly those of
the Midwayan Stage.

Eocene planktonic assemblages encountered north of Lat.
30°S are similar to those recorded from the Caribbean area,
whereas Australasian influences are observed in higher latitudes.

Eocene benthonic Foraminifera show similar features. In
addition, in the southernmost areas, such as the Austral Basin
some Pacific influences are registered.

Oligocene planktonics are only represented on the continental margin of northern South America, whereas in southern
latitudes, i.e. in the Colorado and Austral Basin Oligocene
transgression is recorded.

A large part of the Paleogene seems to be absent from
southeastern Brazilian basins.

No large Foraminifera are recorded from the South American
Paleogene sequences.

In the African continent Paleocene benthonics show some
common species to the Midwayan fauna, although some endemic
species do occur.

During Eocene times the endemism of benthonic foraminiferal species becomes stronger than during the Paleocene. Although in Gabon the microfaunal assemblages shows the strongest
endemic features, several new benthonic species are recorded
from Togo-Dahomey, Nigeria and Cameroon.

Large Foraminifera are recorded only from the Eocene of
the African continent such as Ranikotalia and Discocyclina in
South Africa and Nummulites from Cameroon.

Typical warm water planktonics, such as Hantkenina are
recorded only from Gabon and Angola, being absent in the
American South Atlantic.

Oligocene strata seems to be missing in Africa although
some long ranging species are recorded from Nigeria.

References

Adegoke, O. S. 1966. Eocene Stratigraphy of Southern Nigeria.
   Colloque sur l'Eocene. Paris. v. 8, pp. 23-49.
Andreis, R. R., Mazzoni, M. M. and Spalletti, L. A. 1973. Geo-
   logía y sedimentología del Cerro Bororó (Provincia de Chubut).
   Actas Quinto Congreso Geológico Argentino. T.III, p. 21-55.
Antuz, M. T., 1964. O Neocretaceo e o Cenozoico do litoral de
   Angola. Junta de Investigacoes de Ultramar, Lisboa. p. 1-259.

Belmonte, Y. C., 1966. Stratigraphie du bassin sedimentaire du Cameroun. Proc. 2nd. W. African Micropal. Coll. Ibadan, 1965.

Berggren, W. A., 1973. The Pliocene time-scale: calibration of planktonic foraminiferal and calcareous nannofossil zones: Nature, v. 243, no. 5407, p. 391-397.

─────, and Amdurer, M., 1973. Late Paleogene (Oligocene) and Neogene planktonic foraminiferal biostratigraphy of the Atlantic Ocean (Lat. 30° N to Lat. 30° S) Riv. Ital. Paleont. v. 79, no. 3, p. 337-392.

Bertels, A., 1964. Micropaleontología del Paleoceno de General Roca (Provincia de Rio Negro). Rev. Museo La Plata. Nueva Serie. Paleontología No. 23, T. IV, p. 125-184.

─────, 1969. Estratigrafía del límite Cretacico-Terciario en Patagonia Septentrional. Rev. Asoc. Geol. Arg. T. XXIV, No. 1, p. 41-54.

─────, 1970. Los Foraminíferos planctónicos de la cuenca Cretacico-Terciaria en Patagonia Septentrional (Argentina), con consideraciones sobre la estratigrafía de Fortín General Roca (Provincia de Rio Negro). Ameghiniana, T. VII, No. 1, p. 1-47.

─────, 1973. Bioestratigrafía del Cerro Bororó, Provincia del Chubut, República Argentina. Actas V. Congreso Geol. Arg. T. III, p. 71-90.

─────, 1975. Bioestratigrafía del Paleógeno en la República Argentina. Rev. Española de Micropaleontología, vol. VII, no. 3, p. 429-450.

─────, 1976a (in press). Bioestratigrafía del Paleoceno marino en la Provincia del Chubut, Republica Argentina.

Blondeau, A., 1966. Découverte de Nummulites au Cameroun. Proc. 2nd. W. African Micropal. Coll. Ibadan. p. 24-26.

Blow, W. H., 1970. A Deep Sea Drilling Project. Leg 3. Foraminifera from selected samples. In Maxwell, A. E. et al., 1970. Initial Reports of the Deep Sea Drilling Project, v. III. Washington (U.S. Government Printing Office) p. 629-661

Bolli, H. M., 1966. Zonation of Cretaceous to Pliocene marine sediments based on planktonic Foraminifera. Bol. Inf. Asoc. Benezolana de Geol. Min. y Petroleo., v. 9, no. 1, p. 1-32.

Brognon, G. and Vernier, G., 1966. Tectonique et sédimentation dans le bassin du Cuanza (Angola). In: Bassins sédimentaires du littoral africain. Iére partie: Littoral Atlantique. Assoc. des Services Géol. Africains. p. 207-252.

Chapman, F., 1928. On a Foraminiferal Limestone of Upper Eocene Age from the Alexandria Formation, South Africa. Annals of the South African Museum, vol. XXVIII, Pt. 2, p. 291-296.

Charrier, R. and Lahsen, A., 1969. Stratigraphy of Late Cretaceous-Early Eocene, Seno Skyring Strait of Magellan Area, Magallanes Province, Chile. American Association of Petroleum Geologists Bull. v. 53, no. 3, p. 568-590.

Cifelli, R., Blow, W. H., and Nelson, W. G., 1968. Paleogene Sediment fram a Fracture Zone of the mid-Atlantic Ridge. Journal of Marine Research. v. 26, p. 105-109.

de Klasz, I., and Gageonnet, R., 1963. Biostratigraphie du Bassin Gabonais. Colloque Internat. de Micropal. Dakar, p. 227-304.

434

de Klasz, I., Le Calvez, Y. and Rérat, D., 1964 a. Deux importantes espèces de Foraminifères du Miocène inferieur de l'
Afrique occidentale. C.R. Sommaire des Séanc. Soc. Géol.
France, Fasc. 5, p. 194-195.
————, 1964 b. Deux nouveaux Buliminidae (Foraminifères) du
Paléogène de l'Afrique occidentale. CR. Sommaire des Séanc.
Soc. Geol. France. Fasc. 5, p. 208-209.
————, 1964 c. Deux nouveaux genres de Foraminifères de Gabon (Afrique Equatoriale). C.R. Sommaire des Séanc. Soc.
Géol. France. Fasc. 6, p. 236-237.
de Klasz, I., Marie, P., and Meijer, M., 1960. Gabonella nov.
gen. un nouveau genre de Foraminifères du Crétacé Supérieur
et du Tertiaire basal de l'Afrique occidentale. Revue Micropal. v. 3, no. 3, p. 167-182.
Funnel, B. M., 1971. The occurrence of pre-Quaternary microfossils in the oceans. (In) Funnel, B. M. and Riedel, W.
(eds.) "Micropaleontology of Oceans," Cambridge, Cambridge
Univ. Press, 1967. p. 507-534.
Graham, J. J., de Klasz, I., and Rérat, D., 1965. Quelques importants Foraminifères du Tertiaire du Gabon (Afrique équatoriale). Rev . Micropal., v. 8, no. 2, p. 71-84.
Herm, D., 1966. Micropaleontological aspects of the Magellan-
ese Geosyncline, southernmost Child, South America. Proc.
2nd. W. African Micropal. Coll. Ibadan. p. 72-86.
Kaasschieter, J. P. H., 1963. Geology of the Colorado Basin.
Tulsa Geol. Soc. Digest, p. 177-187.
Le Calvez, Y., de Klasz, I. and Brun, L., 1974. Nouvelle con-
tribution à la connaissance des microfaunes du Gabon. Rev.
Española de Micropal. v. VI, no. 3, p. 381-400.
Lima, E. C., 1973. Bioestratigrafía da Bacia de Barreirinhas.
Anais do XXVI Congreso Soc. Brasilera de Geol. Belem., v. 3,
p. 81-91.
Mabesone, J. M., Tinoco, I. M., and Coutinho, P. N., 1968. The
Mesozoic-Tertiary Boundary in northeastern Brazil. Palaeo-
geography, Paleoclimatology, Palaecology, v. 4, p. 161-185.
Malumian, N., 1970b. Foraminíferos danianos de la Formación
Pedro Luro, Provincia de Buenos Aires, Argentina. Ameghini-
ana, T. VII, no. 4, p. 355-367.
Malumian, N., Masiuk, V., and Riggi, J. C., 1971. Micropaleon-
tología y sedimentología de la perforación SC-1 Provincia
Santa Cruz, República Argentina. Su importancia y correlaci-
ones. Rev. Assoc. Geol. Argentina. T. XXVI, no. 2, p.175-208.
Masiuk, V., 1967. Estratigrafía del Rocanense del Puesto P.
Alvarez. Curso inferior del Rio Chico, Prov. del Chubut. Rev.
Museo La Plata (Nueva Serie), T. V, p. 197-258.

Maxwell, A. E., et al., 1970. Initial Reports of the Deep
Sea Drilling Project, v. III., Washington, U.S. Government
Printing Office, xx+ 806 p.

Mendez, I. A., 1966. Foraminíferos, edad y correlación estratigráfica del Salamanquense de Punta Peligro (45° 30'S; 67°
11'W) Provincia del Chubut. Rev. Asoc. Geol. Arg., T. XXI,
no. 2, p. 127-159.
Noguti, I., 1975. Zonación Bioestratigráfica de los Foraminíferos planctónicos del Terciario de Grasil. Rev. Española de
Micropal., v. VII, no. 3, p. 391-401.
Noguti, I., and Fontana dos Santos, J., 1972. Zoneamiento preliminar por foraminíferos planctonicos do Aptiano ao Mioceno

na plataforma continental do Brasil. Bol. Tec. Petrobras. Rio de Janeiro, v. 13, no. 3, p. 265-283.

Reyment, R. A., 1959. Notes on some Globigerinidae, Globotruncanidae and Globorotaliidae from the Upper Cretaceous and Lower Tertiary of Western Nigeria. Rep. Geol. Survey Nigeria, p. 68-86.

————, 1960. Studies on Nigerian Upper Cretaceous and Lower Tertiary Ostracoda. Part 1: Senonian and Maestrichtian Ostracoda. Stockholm Contrib. Geol., 7:1-238, lám. 23.

————, 1965. Aspects of the Geology of Nigeria. The stratigrapyy of the Cretaceous and Cenozoic deposits. Ibadan Univ. Press, p. 1-145.

————, 1966. Brief review of the stratigraphic sequences of West Africa (Angola to Senegal). Proceed. 2nd. West African Micropaleontological Coll. Ibadan 1965, p. 162-176.

Saito, T., Ewing, M. and Burckle, L. H., 1966. Tertiary Sediment from the mid-Atlantic Ridge. Science, v. 151, no. 3714, p. 1075-1079.

Schaller, H., 1969. Revisao estratigráfica da bacía de Sergipe-Alagoas. Bol. Tecn. Petrobrás. Rio de Janeiro, v. 12, no. 1, p. 21-86.

Schaller, H., Vasconcelos, D. N., and Castro, J. C., 1971. Estratigrafía preliminar da Bacía sedimentar da Foz do Rio Amazonas. Soc. Brasileira de Geol. Anais XXV Congreso, v. 3, p. 189-202.

Slansky, M., 1963. Contribution à l'étude géologique du bassin sédimentaire cotier du Dahomey et du Togo. Mémoires Bureau de Recherches Géologiques et Min., no. 11, p. 1-270.

Stolk, J., 1963. Contribution à l'étude des corrélations microfauniques du Tertiaire inférieur de la Nigeria Méridionale. Coll. Intern. de Micropal., Dakar. p. 247-275.

Tinoco, I. de M., 1967. Micropaleontología da faixa sedimentar costera Recife-Joao Pessoa. Bol. Soc. Brasilera Geol., v. 16, no. 1, p. 81-85.

Todd, R., and Kniker, H. T., 1952. An Eocene Foraminiferal fauna from the Agua Fresca Shale of Magallanes Province, Southernmost Chile. Cushman Found. Foram. Res. Special Pub. no. 1, p. 1-28.

PLATE I

Danian assemblage

Fig. 1 - Globoconusa daubjergensis Brönnimann, x308, San Jorge Basin - Salamanquian Stage (Upper Danian)

Fig. 2 - Globorotalia (Turborotalia) pseudobulloides (Plummer), x342, Neuquén Basin - Rocanian Stage (Lower Danian)

Fig. 3 - Globorotalia (Turborotalia) compressa (Plummer, x308, San Jorge Basin - Salamanquian Stage (Upper Danian)

Fig. 4 - Pulsiphonina prima (Plummer), x205, Neuquén Basin - Rocanian Stage (Lower Danian)

Fig. 5 - Siphogenerinoides elegantus (Plummer), x171, Neuquén Basin - Rocanian Stage (Lower Danian)

Fig. 6 - Loxostomoides applinae (Plummer), x150, San Jorge Basin - Salamanquiano Stage (Upper Danian)

Fig. 7 - Alabamina midwayensis Brotzen, x171, Neuquén Basin - Rocanian Stage (Lower Danian)

Fig. 8 - Lenticulina klagshamnensis (Brotzen), x41, San Jorge Basin - Salamanquiano Stage (Upper Danian)

PLATE II
Eocene-Oligocene Assemblage

Fig. 1 - <u>Uvigerina</u> **tenuistriata** Reuss, x217, Austral Basin - Julian Stage (Eocene-Oligocene)

Fig. 2 - <u>Uvigerina</u> <u>abbreviata</u> Terquem, x181, Austral Basin - Julian Stage (Upper Eocene-L. Oligocene)

Fig. 3 - <u>Cribrorotalia</u> <u>lornensis</u> Hornibrook, x144, Austral Basin - Julian Stage (U. Eocene-L. Oligocene)

Fig. 4 - <u>Notorotalia</u> sp., x209, Austral Basin - Julian Stage (U. Eocene-L. Oligocene)

Fig. 5 - <u>Cribrorotalia</u> <u>hornibrooki</u> forma <u>typica</u> Malumian and Masiuk, x115, Austral Basin - Julian Stage (U. Eocene-L. Oligocene)

Fig. 6 - <u>Globigerina</u> <u>gortanii</u> Borsetti, x361, Austral Basin - Leonian Stage (Lower Oligocene)

Fig. 7 - <u>Globigerina</u> <u>officinalis</u> Subbotina, x361, Austral Basin - Julian and Leonian Stages (Oligocene)

PLATE III
Oligocene Assemblage

Fig. 1 - <u>Globigerina</u> <u>anguliofficinalis</u> Blow, x351, Austral Basin - Leonian Stage (Lower Oligocene)

Fig. 2 - <u>Globigerina</u> <u>ciperoensis</u> Bolli, x281, Austral Basin - Leonian Stage (Upper Oligocene)

Fig. 3 - <u>Globorotalia</u> <u>opima</u> Bolli, x351, Austral Basin - Leonian Stage (Upper Oligocene)

Fig. 4 - <u>Cribrorotalia</u> sp., x141, Austral Basin - Leonian Stage (Oligocene)

Fig. 5 - <u>Bolivina</u> <u>finlayi</u> Hornibrook, x211, Austral Basin - Leonian Stage (Oligocene)

Fig. 6 - <u>Nonionella</u> <u>magnalingua</u> Finlay, x281, Austral Basin - Leonian Stage (Oligocene)

Fig. 7 - <u>Uvigerina</u> <u>germanina</u> (Cushman and Edwards), x225, Austral Basin - Leonian Stage (Oligocene)

PLATE I

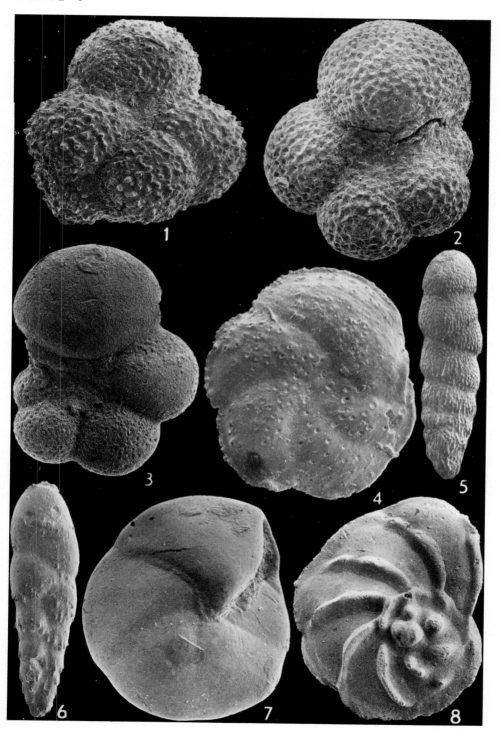

PLATE II

PLATE III

NEOGENE FORAMINIFERA - SOUTH ATLANTIC

Alwine Bertels and Marly Madeira-Falcetta

Facultad de Ciencias Exactas y Naturales

Universidad de Buenos Aires, República Argentina

Consejo Nacional de Investigaciones Científicas y Técnicas,
República Argentina

Escola de Geologia, Universidade Federal do Rio Grande do Sul

Porto Alegre, Brazil

Resumen

En el presente trabajo se sintetizan los aportes científicos efectuados por diversos autores, relacionados con las asociaciones foraminiferológicas que prevalecieron durante el Neógeno en el Atlantico Sur.

Se analizan diversas cuencas sedimentarias, tanto en el
continente sudamericano como en el africano en donde tuvieron
lugar depositaciones marinas durante el Neógeno.

Los conjuntos bentónicos del Mioceno de ambos continentes,
el sudamericano y africano, manifiestan escasas características
en común. Las asociaciones de sudamérica son aproximadamente
comparables a las de la Provincia Atlántica actual, en tanto
que las africanas muestran un marcado endemismo. Los conjuntos
planctónicos, en cambio, son similares en ambos continentes.

De acuerdo con los registros obtenidos hasta la actualidad,
los depósitos post Miocenos en el continente americano del sur
contienen faunas similares a las actuales de iguales latitudes;
las microfaunas africanas, por otra parte, no exhiben el marcado endemismo de épocas anteriores.

Abstract

In the present work the scientific contributions related
to the foraminiferal associations which prevailed during the
Neogene in the South Atlantic Ocean made by several authors are
synthesized.

Several South American and African sedimentary basins in
which Neogene deposits are registered are analyzed and the diagnostic taxa are listed.

The Miocene South American and West African benthonic
assemblages show little features in common; the South American
benthonic Foraminifera are approximately comparable to those of
the present Atlantic Province whereas the African assemblages

show a marked endemism. Miocene planktonic Foraminifera are comparable in both continents.

According to the records obtained up to the present, post Miocene South American deposits contain microfaunas more similar to each other than to those living today at the same latitudes; the West African microfaunas, on the other hand, do not exhibit the marked endemism of older epochs.

Introduction

The purpose of the present work is to show the most diagnostic fossil foraminiferal assemblages which prevailed in the South Atlantic Ocean during Miocene through Pleistocene. Holocene faunas will not be discussed.

The sedimentary basins in which Neogene deposits occur are analyzed in both continents; the faunal assemblages are listed following a latitudinal order according to the sedimentary basins from which they were recorded.

The available data obtained from several oceanographic expeditions are synthesized and the principal planktonic zones are mentioned.

Acknowledgments

The authors wish to express gratitude to the Argentina National Council for Scientific and Technical Research, for the economical aid in the accomplishment of the present work, and for generously providing the use of the Jeolco JSM - U-3 scanning electron microscope with which the micrographs were taken.

The writers are also grateful to Buenos Aires University, Department of Geology and to the Rio Grande do Sul University, Brazil for facilities provided in achieving the completion of this work.

Particular thanks are also expressed to Dr. Carlos A. Rinaldi and to Mr. Patricio Ganduglia, who kindly and generously helped in the construction of the stratigraphic charts and range charts which illustrate this paper.

The illustrated material is deposited in the Laboratory of Micropaleontology, Escola de Geología, Universidade do Rio Grande do Sul, Porto Alegre, Brazil.

Miocene

South America

The Amazonas Basin is located in northern Brazil. Miocene sediments were deposited into a graben with a southeast-northwest trend. The known Miocene lithostratigraphic units are the Marajó and Pirabas Formations and the Pará Group (Text Fig. 1 in Bertels, Paleogene, this volume).

The Marajó Formation is composed of clastic sediments which grade into the continental margin to a thick sequence of carbonates which constitute the Amapá Formation (Schaller et al., 1971).

The Marajó Formation extends over the whole area of Marajó Island and along the littoral areas of western Amapá and eastern Pará States. It overlies concordantly the Limoeiro Formation and is discordant with the overlying Pará Group. The Limoeiro Formation is considered to be Upper Cretaceous to Lower Tertiary (Schaller et al., 1971), based on palynological data.

Petri (1954) assigned a Miocene age to the sequence of the Marajó Formation and found reworked Cretaceous and Eocene Foraminifera. He considered that Globigerina cretacea d'Orbigny, Globigerina cf. triangularis White, Globorotalia compressa (Plummer), Globorotalia cf. crassata (Cushman) and Guembelina globulosa Ehrenberg were reworked species.

For the Miocene sequence Petri (1954) recognized four benthonic foraminiferal zones:

1) Amphistegina lessoni Zone
2) Bolivina plicatella Zone
3) Elphidium poeyanum elongata Zone
4) Quinqueloculina lamarckiana Zone

The ages assigned by Schaller et al. (1971), however, to the recognized members of the Marajó Formation are: Afuá Member: Paleocene; Cururú Member; Eocene; Mexiana Member: Eocene to Lower Miocene and Araguarí Member: Lower Miocene. Thus a problem exists as to age interpretation of the formation.

The reworked species mentioned by Petri (1954) indicate the existence in the area of Cretaceous and Paleocene strata such as was revealed by palynological studies.

The partly equivalent Amapá Formation occurs in the continental margin of the Amapá and Pará States. The sedimentary sequences are algal micrites with intercalations of greenish-gray claystones and siltstones, frequently glauconitic and pyritic.

The age of the Amapá Formation ranges from the Paleocene to the middle Miocene (Schaller et al., 1971):

a) Candiru Sequence, of Eocene age
b) Jacundá Sequence, to which a late Eocene to Oligocene age was assigned.
c) Tambaqui Sequence: planktonic zones of Globorotalia mayeri and Globorotalia fohsi peripheroacuta were identified; the age of the section is early to middle Miocene.

The Pará Group was deposited in the Amapá and Pará States (Foz de Amazonas Basin); their strata have been also bored in the continental margin of the Pará State.

The Pará Group has been subdivided into two formations by Schaller et al., (1971). These are the Pirarucu Formation, of clayey nature and the Tucanaré Formation, essentially arenaceous. The group has a fluviatile and paralic origin in the areas over the present continent, whereas it is of neritic facies on the continental margin. Throughout the whole section of the Tucanaré Formation Nummulites sp. have been found. The Zones of Orbulina universa and Globorotalia mayeri have been recognized in both formations. Schaller et al.(1971) attributed Miocene to Holocene age to the sedimentary sequence of the Group.

The upper part of the Amapá and Marajó formations grade in the southern border of the Pará state into sediments described previously by Petri (1957) as the Pirabas Formation.

The outcrop of the Pirabas Formation is composed of limestones and coquinas with arenaceous beds at the top. It is exposed along a large part of the littoral of Pará State. Petri (1957) assigned a Lower Miocene age to the formation based on the microfaunal assemblage, although he found many similarities with the middle Miocene associations from other regions of the world especially those of tropical America.

Based on the presence of Archaias angulata (Fichtel and
Moll) Elphidium poeyanum (d'Orbigny), Cancris sagra (d'Orb.),
Eponides repandus (Fichtel and Moll), Pyrgo subsphaerica (d'
Orbigny) and others, Petri (1957) infered shallow water deposi-
tion of the Pirabas Formation.

Noguti (1975) recorded from the continental margin of the
Amazonas Basin the Miocene Zones of Globigerina tripartita,
Praeorbulina glomerosa, Globorotalia fohsi peripheroronda,
Globorotalia fohsi peripherocauta and Globorotalia mayeri.

Ferreira (1973) recorded the occurrence of the Pirabas
Formation (early Miocene) in the Barreirinhas Basin, Maranhao
State.

In this Basin the Pirabas Formation is developed over the
whole basin discordantly overlying Cretaceous sediments and
covered by non-marine and littoral Quaternary deposits.

The Pirabas Formation in this basin is composed of calcareous marls and marly chalks at the base passing into blue
claystones, fossiliferous limestones, gray claystones and culminating into beige limestones. These limestones contain large
Foraminifera recorded as "Orbitoid foraminifera."

From the continental margin of the Sergipe/Alagoas Basin
Noguti (1975) recorded the Zone of Globigerina tripartita.

The continental margin of the Espirito Santo Basin yielded
a planktonic association in which the Planktonic Zones of Globigerina tripartita, Praeorbulina glomerosa and Globorotalia
mayeri were recognized by Noguti (1975).

In the continental margin of Rio de Janeiro (south of the
Campos Basin) Noguti (1975) found the Miocene Planktonic Zones
of Globigerina tripartita and Praeorbulina glomerosa.

The northeastern Brazilian Reconcavo Basin contains Miocene
sediments whose microfaunal content was studied by Petri (1972).

The absence of representatives of genera of the Miliolidae
and the genus Elphidium indicate an offshore environment.

South of Bahía (possibly Reconcavo Basin) Noguti (1975)
recorded the Zones of Globigerina tripartita and Praeorbulina
glomerosa of early Miocene age.

In the southeastern Brazilian Pelotas Basin (Closs, 1967,
1970; Fernandes, 1975 MS; Thiesen, 1975 MS; and Falcetta, 1976

MS) thick sequences of Miocene to Pleistocene age were deposited.

The Pelotas Basin is of tectonic nature; the sedimentary
sequences are found in subsurface and overly the granitic basement. The sediments are mainly composed of sandstones with
intercalated claystones; facies changes are observed near the
borders of the basin in which conglomeratic and arkosic levels
occur.

The predominant microfaunal assemblage studied by Closs
(1967, 1970) is composed of Globigerinoides bisphaericus, Orbulina suturalis, Globorotalia mayeri, Globorotalia obesa, Globigerinoides glomerosus, Globigerinoides transitorius, Globorotalia fohsi and Globorotalia menardii archaeomenardii. Closs
(op. cit) considered of importance the evolutionary trend of
Globigerina bisphericus - O. suturalis although he pointed out
the difficulties in establishing a definite correlation with
Miocene Stages. Fernandez (1975, MS) studied the uvigerinids
of this basin and found several species of Uvigerina. Bolivinids are also recorded from the pelotas Basin marine sequence
by Thiesen (1975, MS).

Recent work of one of the writers (M-M.F.) based on the
presence of Globorotalia acostaensis, Globigerina nepenthes and
other planktonic species, found that the
transgression in the area took place in late Miocene times,
i.e., approximately at Zone N 14 of Blow (1969). The Orbulina
series mentioned by Closs (1967, 1970) as well as other Lower
Miocene assemblages are probably reworked from adjacent areas.

In Argentina the northeastern and central Paraná Basin or
Chaco-Paranense Basin contains shallow water marine deposits of
the Entre Rios and Paraná formations (Entrerriano and/or Para-
niano stages) of late Miocene-early Pliocene? age.

The sediments of the Entre Rios and Paraná formations in
the Paraná basin extend inland over a large area. Surface exposures are found at the east and in subsurface westwards. The
sedimentary sequence is composed of claystones at the base and
sandstones and coquinas to the top.

The faunal assemblage as well as the lithological features

indicate shallow waters and an almost closed marine environment. This condition may be due to the Precambrian shield located at the eastern border of the basin. During Miocene-Pliocene times the Paraná Basin was probably connected with the Salado Basin which might have acted as a route for the large marine transgression.

The microfaunal assemblage described by Pisetta (1968, MS)
is composed of Buccella frigida (Cushman), Ammonia beccarii
(Linné) and Protelphidium tuberculatum (d'Oribigny).

In the Salado Basin Malumian (1970a) studied subsurface
sedimentary sequences. The Entrrerian Stage was found to be
represented by Globorotalia pachyderma forma sinistrorsa, Protelphidium tuberculatum, Cibicides berthelotti, Cibicides pseudoungerianus, and others.

The age of the Entrreriano Stage was considered to be of
late Miocene Age (Malumian 1970a). The microfaunal assemblage
indicates warmer sea water conditions than that of today at the
same latitude.

Sedimentary subsurface sequences were also studied by
Malumian (1972) in the Colorado Basin. The assemblage found by this author is listed in Text Fig. 1. The assigned age was Oligocene/Miocene based on the planktonics Globigerina woodi woodi, Cassigerinella chipolensis etc. Malumian
(1972) found that the planktonic assemblage is equivalent to
the Globigerina ciperoensis ciperoensis and Globorotalia opima
opima from tropical areas, and to Globigerina woodi woodi and
Globigerina euapertura from temperate regions.

The Austral Basin is the southernmost South American Basin. The sedimentary sequences range in age from the Valanginian to the Miocene.

Miocene Foraminifera were recorded by Malumian (1968) who
found an assemblage in subsurface samples.

Subsequent studies carried out by Malumian, Masiuk and
Riggi (1971) in the Austral Basin confirmed the marine Miocene
section in this basin. The following Miocene Foraminifera were
recorded: Buccella frigida Cushman, Nonion affine (Cushman),
Pullenia subcarinata d'Orbigny, Nonionella atlantica Cushman,
Lenticulina rotulata Lamark, Cibicides pseudoungerianus

(Cushman) and <u>Globigerina</u> ex. gr. <u>praebulloides</u> Blow.

Africa

In the African continent sequences referred to the Miocene are present in Nigeria, Cameroon, Gabon, Angola and South Africa.

In <u>Nigeria</u> (Reyment, 1965) Miocene strata are exposed in the western region; the recognized lithostratigraphic units are the non-marine Ogwashi-Asaba Formation and the marine Ijebu Formation. The last is located east of Lagos; after Reyment (op. cit.) it is partly marine and **may be partly** equivalent to the non-marine Ogwashi-Asaba Formation. The beds are poorly exposed and consists mainly of clays and sands, impregnated with bitumen. The marine facies contain benthonic Foraminifera.

Miocene sequences are also found in boreholes west of Lagos in western Nigeria; the strata contain <u>Eponidopsis</u> <u>eshira</u> de Klasz and Rérat, <u>Daucina</u> <u>ermaniana</u> Bornemann and <u>Daucinoides</u> <u>circumtegens</u> de Klasz and Rérat (Reyment, 1965).

Although the Ijebu marine beds lack planktonic Foraminifera, in Gabon <u>Eponidopsis</u> <u>eshira</u> de Klasz and Rérat is associated with <u>Globorotalia</u> of the <u>fohsi</u> group which would indicate a early-middle Miocene age.

Other typical species which occur in Gabon are present in southeastern Nigeria such as <u>Megastomella</u> <u>africana</u> <u>compressa</u> Faulkner, de Klasz and Rérat, <u>Nonion</u> <u>centrosulcata</u> de Klasz, Le Calvez and Rérat and <u>Gavelinella</u> <u>beninensis</u> Klasz, Le Calvez and Rérat.

In <u>Cameroon</u> Miocene sediments rest discordantly on older stratigraphic units.

In the Duala Basin the Miocene is represented by sandstones and claystones, with miogypsinids, intercalated by basalts (Belmonte, 1966).

In the central part of the Basin the Miocene reaches considerable thickness and is subdivided by Belmonte (1966) into:

a) basal beds consisting of claystones with intercalated sandstones.

b) a regressive sequence predominantly detritic which begins with nearly marine strata and ends with deltaic deposits. Belmonte (op. cit.) assigned to these Miocene strata a Burdigalian-Vindobonian age.

In Western Cameroon a limestone in the Missellele River (Reyment, 1965) has yielded diagnostic miogypsinids such as Miogypsina (Miolepidocyclina) burdigalensis Gümbel, M. (M.) negrii (Ferrero), M. (Miogypsinoides) complanata Schlumberger, M. (M.) bantamensis (Tan Sin Hok) and M. (M.) nigeriana Küpper.

Drooger (1966) recorded from deep wells in Cameroon approximately the same association of large Foraminifera as from outcrops in the Missellele River. Drooger (op. cit.) suggests a Burdigalian age for strata containing Miogypsinoides bantamensis Tam and M. burdigalensis Gümbel.

According to Reyment (1965) the beds of the Missellele River may be part of the Mpundu Formation.

Faulkner, de Klasz and Rérat (1963) have recorded Megastomella africana from a borehole at Uquo, 1.5 km southwest of Calabar, which indicates the presence of Miocene and permits correlation with the Mandorové Formation of Gabon of early? Miocene age.

In the Chang region are lignitic beds which in some levels contain Foraminifera.

Gazel et al. (1956, cited in Reyment, 1965) suggested a possible correlation of these beds with the non-marine Ogwashi-Asaba Formation of Nigeria.

In Gabon the marine Miocene sequences are exposed in the vicinity of Port Gentil, alongside the present coastline, where Miocene strata unconformably overlie Eocene sequences.

The sedimentary basin appears to be made of two separated units, divided by an intermediate high called the Lambarene horst (Belmonte, Hirtz and Wenger, in Reyment, 1965).

The Neogene sequence of Gabon seems to be similar to that of Nigeria and there are numerous Foraminifera in common.

In Western Gabon the recognized chronostratigraphic units are (Le Calvez, de Klasz and Brun, 1974): Mandorové, M'Bega and N'Tchengué which constitute the Alewana System. The Akosso Stage comprises strata belonging to the Miocene and younger deposits.

Many important contributions deal with the micropaleontology of the Gabon Basin (cf. de Klasz and Rérat, 1962; de Klasz and Gageonnet, 1963; de Klasz, Le Calvez and Rérat, 1964b;

Faulkner, de Klasz and Rérat, 1963; de Klasz, Le Calvez and Rérat, 1964a, c; Graham de Klasz and Rérat, 1965; de Klasz and Micholet, 1970).

The Mandorové Formation is detrital and clayey; based on the planktonic content, de Klasz and Micholet (1970) assigned a Lower Burdigalian age to this formation. These authors recorded from the Upper Mandorové Series the planktonics Globorotalia fohsi barisanensis, Globorotalia fohsi fohsi and Globorotalia fohsi lobata. Taking into account the planktonic zonation proposed by Blow (1969) the corresponding strata are of early-middle Miocene age comprising zones N12 to N13.

The M'Bega Formation, which overlies the Mandorové Formation is in general clayey with basal detrital beds. Arenaceous assemblages are found in the detrital beds. Most of the species found in the M'Bega Formation are common to the underlying Mandorové Formation. The following species appear in M'Bega: Nonion boueanum (d'Orbigny), Cassidulina laevigata carinata Cushman, Pulleniatina obliquiloculata Parker and Jones and Globorotalia menardii (d'Orbigny) miocenica Palmer.

De Klasz and Gageonet (1963) based on the large number of common Foraminifera to both Mandorové and M'Bega Formations assigned an early Miocene age to the M'Bega Formation.

The lower part of the N'Tchengué Formation is argillaceous, while the upper part is detrital with clayey intercalations.

The foraminiferal content is mainly like that of the underlying Mandorové and M'Bega Formations; only a few other species appear in the N'Tchengué Formation such as: Ptychomiliola separans (Brady) and Elphidium cf. crispum (Linné).

The most characteristic Foraminifera of N'Tchengué are species of Amphistegina which are found together with rare Operculinoides.

De Klasz and Gageonnet (1963) suggest a late Miocene age for at least part of the strata of the N'Tchengué Formation.

In Angola the known Miocene sequences are exposed in the Cuanza Basin and in the Benguela Basin, Brognon and Verrier

(1966). No Oligocene sediments are present in the Cuanza Basin. The Miocene is represented only by sediments of Aquitanian and Burdigalian age.

At the base of the Miocene section is recognized the Lower Quifangondo Formation which is black or variegated, highly gypsiferous and unfossiliferous. It rests unconformably on early, middle or late Eocene, or in some places, on Campanian strata (Brognon and Verrier, 1966). Above the unfossiliferous strata, black to brown shales with Globigerina dissimilis are recognized.

Following the strata of the G. dissimilis Zone, the Upper Quifangondo was deposited, consisting of silty shale, very rich in Foraminifera, and which grades upward into coquinoid limestone, with sandy limestone interbedded. This member corresponds with the Globigerinatella insueta Zone (Brognon and Verrier, 1966).

The Luanda Formation has brown shales at its base, containing Foraminifera of the Globorotalia fohsi Zone; upwards this formation is abruptly transitional to littoral and deltaic sand with sandstone interbeds that close the cycle. The marine sedimentary cycle began during early Aquitanian time and ended during Burdigalian time (Brognon and Verrier 1966).

The Miocene Cacuaco Formation was deposited in shallow water environment where limestone with algae, bivalves and echinoids was deposited.

In the Benguela Basin, Antunez (1964) recorded Neogene sediments consisting of sandstones, claystones, chalks and gypsiferous marls in which species of Globigerinoides were found.

Foraminiferal assemblages recorded by Rocha and Ferreira (1957) are composed of benthonics and the planktonics Globigerina bulloides d'Orbigny, Globigerinoides trilobus cf. irregularis Le Roy, Biorbulina bilobata (d'Orbigny), Orbulina suturalis Brönnimann, O. universa d'Orbigny and Globorotalia tumida (Brady); this assemblage represents a Miocene age (Rocha and Ferreira, op. cit.).

Only a few species of Angola such as Eponidopsis eshira Klasz and Rérat are common to Nigeria and Gabon. The rest of the recorded species are different to those of other areas.

452

Pliocene

As far as information is available, marine Pliocene is not represented on the present borderland of South America; in Africa, sediments younger than Miocene are referred to as "Post-Miocene."

Pleistocene

Pleistocene marine sediments are exposed in the Recife/João Pessoa Basin and in the Rio Grande Coastal Plain, Brazil, in Uruguay and in Argentina.

From the Brazilian Recife/João Pessoa Basin Tinoco (1958) described Foraminifera found in subsurface samples in Pernambuco State, northeastern Brazil.

The Quaternary sediments overlie the Upper Cretaceous Itamaracá and Gramame Formations and Paleocene sediments of the Maria Farinha Formation.

The Quaternary assemblage is mainly composed of: Textularia gramen d'Orbigny, abundant miliolids, Robulus rotulatus (Lamarck), Peneroplis bradyi Cushman, Peneroplis pertusus (Forskal), Archaias angulatus (Fichtel and Moll), Bolivina compacta Sidebottom, Bolivina tortuosa Brady, Bolivina pseudoplicata Heron-Allen and Earland, Bolivina variabilis (Williamson), Bolivina striatula Cushman, Discorbis mira Cushman, Discorbis obtusa (d'Orbigny), Ammonia beccarii parkinsoniana (d'Orbigny), Ammonia beccarii (Linné), Buccella peruviana campsi (Boltovskoy), Buccella frigida (Cushman), Reusella spinulosa (Reuss), Pattellina corrugata Williamson, Amphistegina lessoni d'Orbigny, Loxostomum limbatum (Brady), Nonion affine (Reuss), Nonionella atlantica Cushman, Elphidium discoidale (d'Orbigny), Elphidium advenum depressulum Cushman, Elphidium incertum (Williamson), Elphidium excavatum (Terquem), Elphidium galvestonense Kornfeld, Elphidium sagrum (d'Orbigny), Elphidium gunteri Cole, Cibicides bertheloti (d'Orbigny), Cibicides pseudoungerianus Cushman, Cibicides acknerianus (d'Orbigny), Dyocibicides biserialis Cushman and Valentine, Cassidulina laevigata d'Orbigny, Cassidulina sublobosa Brady, Planorbulina mediterranea d'Orbigny, Astrononion stelligerum (d'Orbigny) Globigerina bulloides d'Orbigny, Globigerina pachyderma (Ehrenberg) and Globigerinoides trilobus (Reuss).

The presence of abundant Miliolids and the genus Elphidium

indicates a shallow water environment whereas the total assemblage indicate warm waters, especially by the presence of representatives of Peneroplis, Archaias and Amphistegina (Tinoco,
1958).

Pleistocene and Holocene sequences in southern Brazil consists of marine, eolian and lagoonal deposits.

Subsurface samples from El Chui well were studied by Closs
and Madeira (1968). The Foraminiferal assemblage is composed
of: Textularia gramen d'Orbigny, Quinqueloculina seminulum
(Linné), Ammonia beccari parkinsoniana (d'Orbigny), Elphidium
discoidale (d'Orbigny), Elphidium excavatum (Terquem), Elphidium galvestonense Kornfeld, Elphidium gunteri Cole, Elphidium
selseyense (Heron-Allen and Earland), Amphistegina lessoni
d'Orbigny, Cibicides berthetoti (d'Orbigny), Cibicides pseudoungerianus (Cushman), Cassidulina laevigata d'Orbigny and
Nonionella atlantica Cushman.

In the Palmares do Sul locality, southern Brazil, a water
well yielded Pleistocene ostracode and foraminiferal assemblages (Bertels, Falcetta and Kotzian, 1976, in press).

The foraminiferal assemblage is mainly composed of the
same species, excluding Amphistegina lessoni, as those described from the El Chui well by Closs and Madeira (1968).

Throughout the vertical section the assemblage shows marine, mixohaline and fresh water fauna; these faunal changes
probably could be related to the Pleistocene fluctuations of
sea level in that area in response to possible glacial stages
and interstages.

The outcrops of the marine Pleistocene Chui Formation are
disposed parallel to the present coast line; lithologically it
is composed of semiconsolidated quartzitic sands with pelitic
intercalations.

The Holocene marine sediments so far have not yielded Foraminifera in southern Brazil.

In Uruguay the Pleistocene Fray Bentos, Camacho and El
Chui formations are recognized.

In Western Uruguay the Fray Bentos, of non-marine origin
and the marine Camacho Formation are distinguished.

The marine Camacho Formation is composed of white, well
sorted sands, greenish or yellowish clay and coquina. The

following Foraminifera are recorded by Delaney (1967): <u>Ammonia</u>
<u>beccarii</u> (Linné), <u>Ammonia</u> <u>beccarii</u> <u>parkinsonia</u> (d'Orbigny), <u>Bo-</u>
<u>livina</u> <u>striatula</u> (Cushman), <u>Buccella</u> <u>frigida</u> (Cushman), <u>Buccel-</u>
<u>la</u> <u>peruviana</u> <u>campsi</u> (Boltovskoy), <u>Elphidium</u> <u>discoidale</u>
d'Orbigny, <u>Elphidium</u> <u>excavatum</u> (Terquem) and <u>Quinqueloculina</u>
<u>seminulum</u> (Linné).

The marine Chui Formation has a limited extent and is confined to northeastern Uruguay (Delaney, 1967). Chui sediments
consit of well-sorted, predominantly quartz sand of medium
grain size.

No Foraminifera have been recorded from this formation,
but taking into account that it is the same stratigraphic unit
as that of southern Brazil, a similar microfaunal association
is to be expected.

In <u>Argentina</u> the principal manifestations of Quaternary
ingressions are represented by the Pampeano and Post-Pampeano
Series. The Pampeano Series comprises the Ensenadense and
Bonaerense Stages; the Post-Pampeano Series comprises the Bel-
graniano, Querandiniano and Platiano Stages (Fidalgo, de Francesco and Pascual, 1975).

The Ensenadense assemblage from the Quequén locality,
Buenos Aires Province, studied by Boltovskoy (1959), is composed of the same species as those living today at the same latitude.

The foraminiferal assemblage of the Platian Stage at Mar
Chiquita, Buenos Aires Province (Suarez Soruco, 1968 MS) is
composed of: <u>Cyclogyra</u> <u>involvens</u> (Reuss), <u>Quinqueloculina</u> <u>lam-</u>
<u>arckiana</u> d'Orbigny, <u>Quinqueloculina</u> <u>patagonica</u> d'Orbigny, <u>Quin-</u>
<u>queloculina</u> <u>seminulum</u> (Linné), <u>Pyrgo</u> <u>nasuta</u> Cushman, <u>Pyrgo</u> <u>per-</u>
<u>uviana</u> (d'Orbigny), <u>Pyrgo</u> <u>ringens</u> (Lamarck), <u>Miliolinella</u> <u>sub-</u>
<u>rotunda</u> (Montagu), <u>Oolina</u> <u>hexagona</u> (Williamson), <u>Oolina</u> <u>melo</u>
d'Orbigny, <u>Bolivina</u> <u>striatula</u> Cushman, <u>Bolivina</u> <u>variabilis</u>
(Williamson), <u>Bulimina</u> <u>patagonica</u> d'Orbigny, <u>Buliminella</u> <u>el-</u>
<u>egantissima</u> (d'Orbigny), <u>Buccella</u> <u>frigida</u> (Cushman), <u>Neoconor-</u>
<u>bina</u> <u>orbicularis</u> (Terquem), <u>Ammonia</u> <u>beccarii</u> <u>parkinsoniana</u>
(d'Orbigny), <u>Elphidium</u> <u>articulatum</u> (d'Orbigny), <u>Elphidium</u>
<u>discoidale</u> d'Orbigny and <u>Cibicides</u> <u>variabilis</u> (d'Orbigny).

The additional species recorded by Bertels (1976b, in
press) from the Platian Stage are: <u>Elphidium</u> <u>margaritaceoum</u>

Cushman, _Bulimina marginata_ d'Orbigny, _Bolivina pseudoplicata_
Heron-Allen and Earland, _Discorbis floridanus_ Cushman, _Discorbis peruvianus_ d'Orbigny, _Discorbinopsis aguayoi_ Bermudez and
Cibicides dispars (d'Orbigny).

Post Miocene of Africa

In the African continent Quaternary formations are found
normally in river valleys or littoral areas.

In _Dahomey_, Pliocene to Recent sequences have been mentioned by Slansky (1963). From subsurface samples the following assemblages have been found: _Globigerinoides ruber_ d'Orbigny, _Ammonia beccarii_ (d'Orbigny), _Globorotalia inflata_
d'Orbigny, _Nonion granosum_ d'Orbigny, _Nonion commune_ d'Orbigny,
Eponides procera Brady, and _Discopulvinulina bertheloti_ d'Orbigny.

After Slansky, (1963) this sequence may be attributed to
the Pliocene until the Recent.

In _Nigeria_ (Reyment, 1965) the Plio-Pleistocene Benin Formation is composed of clay and shales, deltaic to fully marine,
in part estuarine, lagoonal or fluviolacustrine in origin.
This formation yielded Foraminifera such as: _Candorbulina uni-
versa_ Jedlitschka, _Globoquadrina dehiscens_ (Chapman, Parr and
Collins), _Globigerinoides trilobus_ (Reuss), bolivinids and
Cibicides spp.

Post Miocene strata are represented also in _Gabon_ by the
Akkoso Series which overlie concordantly late Miocene sediments
in the western region whereas in the eastern area they discordantly overlie strata of Paleogene age.

The faunal assemblage is composed of (de Klasz and
Gageonnet, 1963): _Ptychomiliola separans_ (Brady), _Ammonia bec-
carii_ (d'Orbigny), _Nonion scaphum_, _Globigerina bulloides_ d'Orbigny, _Globigerinoides ruber_ d'Orbigny, _Globigerinoides saccu-
lifer_ (Brady), _Globigerinoides trilobus_ (Reuss), etc.

According to Bond (1965) the Pleistocene of Central and
South Africa is related to pluvial-interpluvial events; fossils
are generally lacking. The Quaternary sediments are fossil
dumbos, colluvial deposits and sands of eolian origin.

Atlantic Ocean

During several cruises Neogene South Atlantic cores were
obtained; from the recovered sections valuable data concerned

to the sedimentary sequences were published. Particular emphasis was given to the Plio-Pleistocene boundary and to the Pleistocene climatic changes inferred by means of planktonic Foraminifera.

Saito, Ewing and Burckle (1966) recorded Pliocene sediments from mid-Atlantic Ridge at the following coordinates: 17°39'S-15°06'W; 22°59'S - 06°46'W; 26°15'S - 03°01'W; 22°06'S - 00° 19'W.

From the Rio Grande Rise and Walvis Ridge these authors found: 25°27'S - 06°28'E Upper Miocene; 24°07'S - 05°45'E Upper Miocene; 29°02'S - 09°13'W Upper Miocene; 29°52'S - 36°48'W Miocene; 21°28'S - 40°05'W Pliocene.

Saito, Ewing and Burckle (op. cit.) found that the Miocene is characterized by Globoquadrina dehiscens, and the Pliocene by Globigerinoides fistulosus. These authors also suggested that the Rio Grande Rise and Walvis Ridge are recent features resulting from uplift in late Miocene or Lower Pliocene times.

Valuable information was provided by Blow (1970a, b) in the report of the results of Legs 3 and 4 of the Glomar Challenger Cruise. Leg 3 covered the South Atlantic between Dakar, Senegal and Rio de Janeiro, and some sites of Leg 4 lie in the South Atlantic.

The sites on Leg 3 in which Neogene sediments were obtained are sites 14-18 and 21-23. The locations of these sites and the planktonic Neogene Zones identified are the following:

Hole 14 - 28° 19.89S - 20° 56.46W	Early Miocene	N4-N6
Hole 15 - 30° 53.38S - 17° 58.89W	Pleistocene	N22
	Pliocene	N20,21
	Early Pliocene	N19
	Late Miocene	N18,19
	Middle Miocene	N15
	Early Miocene	N6,7
Hole 16 - 30° 30.15S - 15° 42.79W	Pleist. and Holoc.	N23
	Pleistocene	N22
	Pliocene	N19-21
	Late Miocene	N16-18
	Middle Miocene	N15
Hole 17 - 28° 02.74S - 6° 36.15W	Pleist. - Holocene	N22,23
	Late Miocene	N17,18

Hole 17 A 28° 02.74S - 6° 36.15W	Early Miocene	N4-7
Hole 18 - 27° 58.71S - 08° 00.70W	Pleistocene	N22
	Pliocene	N20,21
	Early Pliocene	N19
	Late Miocene	N18,19
	Middle Miocene	N15
	Early Miocene	N6,7
Hole 16 - 30° 30.15S - 15° 42.79W	Pleist. and Holoc.	N23
	Pleistocene	N22
	Pliocene	N19,21
	Late Miocene	N16-18
	Middle Miocene	N15
Hole 17 - 28° 02.74S - 6° 36.15W	Pleist.-Holocene	N22-23
	Late Miocene	N17,18
Hole 17A- 28° 02.74S - 6° 36.15W	Early Miocene	N4-7
Hole 18 - 27° 58.71S - 08°00.70W	Pleistocene	N22
	Early Miocene	N4-6
Hole 21 - 28° 35.10S - 30° 35.85W	Pliocene	N19,20
Hole 22 - 30° 00.31 S - 35.15.00W	Pleistocene	N22
	Lower Miocene	N4

On Leg 4 the sites lying in the South Atlantic and in which Miocene to Holocene sediments were recovered, are 23-25.

The location of the sites and the planktonic Zones identified are:

Hole 23 - 06° 08.75 S - 31° 02.60W	Pleistocene	N22
	Early Miocene	N4 with
	Miogypsina	
Hole 24 - 06° 16.30S - 30° 53.53W	Early Miocene	N4
Hole 25 - 00° 31.00S - 39° 14.40W	Pleistocene	N22
	Pliocene	N19
	Late Miocene-Early Plio.	N18
	Late Miocene	N16-18

Maxwell et al. (1970) recorded, from the Rio Grande Rise, stratigraphic sections at sites 21 and 22 of Leg 3 of the Glomar Challenger. Sequences range in age from the Campanian or older to the Pleistocene, the Oligocene and Miocene being missing at Site 21, and Pliocene directly overlies early Miocene at Site 22.

Several Neogene planktonic Zones were recognized by

Maxwell, et al. (op. cit.) at the following sites of Leg 3

  Site 15 Zone of Globigerinita dissimilis

  Site 16 Zone of Globorotalia acostaensis and
         Globorotalia merotumida

  Site 18 Zone of Globorotalia kugleri

  Funnel et al. (1971) recorded several Neogene sections found in the South Atlantic; Alberici, Barbieri, Iaccarino and Rossi (1973) mentioned Neogene localities at the western flank of the mid-Atlantic Ridge at Sites 15 and 16, Leg 3, of the Glomar Challenger.

  Berggren and Amdurer (1973) from studies of Leg 3 at sites 14, 17, and 18, which penetrated Neogene sediments, found close relationships with New Zealand and Atlantic Foraminifera during the Oligocene and Miocene and the correlations that can be made between New Zealand and South Atlantic Foraminiferal Zones. Berggren and Amdurer (op. cit.) also remarked that beginning with the Middle Miocene the South Atlantic faunas appear to exhibit a greater similarity with mid-low latitude Atlantic faunas than with New Zealand.

  Relative to the Atlantic Ocean the Brazilian continental margin was investigated in several areas; the foraminiferal results were obtained by Noguti and Fontana dos Santos (1972) and Noguti (1975).

  Miocene sequences were found in the prolongation into the continental margin of the Amapá, Pará, Sergipe, and Alagoas basins. The following planktonic zones were recognized: Globigerina tripartita, Praeorbulina glomerosa, Globorotalia fohsi peripheroronda, Globorotalia fohsi peripheroacuta, Globorotalia mayeri, and Globorotalia fohsi robusta, all of early Miocene age.

  Text Fig. 1 illustrates the location and the age of the obtained cores in the South Atlantic as well as around the sedimentary basins in which marine Miocene and younger sediments were recorded.

  Great emphasis has been placed on the Plio-Pleistocene boundary and the climatic changes that occurred during the Pleistocene; related to this topic several studies were undertaken in the South Atlantic, as well as in other oceans in search of data related to the temperature changes during the

Pleistocene. The data indicate that during the Pleistocene
there were periodic oscillations of temperature with an ampli-
tude of about 6°C with some differences between the different
oceans.

Ericson and Wollin (1956) studied cores in the equatorial
mid-Atlantic Ridge. Based on planktonic Foraminifera, these
authors found that during the Pleistocene the area was influenced by glacial climate, although the climate change was somewhat less drastic than in higher latitudes. Globorotalia
menardii menardii, Globorotalia menardii tumida and Globorotalia flexuosa were considered warm water species whereas Globorotalia punctulata, Globorotalia scitula and Globorotalia inflata are cool water species and related to glacial stages.

Rudiman (1971), from the piston cores of the equatorial
Atlantic, found that indicative warm water species are Globigerinita aequilateralis, Sphaeroidinella dehiscens, Globigerinoides conglobatus, Globigerinoides tenellus, Candeina nitida,
Globigerina rubescens, Globigerina digitata, Globigerina hexa-
gona and Hastigerina pelagica; cold species are: Globorotalia
hirsuta, Globorotalia punctulata, Globorotalia scitula and Globigerina quinqueloba. Taking into account these assemblages
Rudiman concluded that in the eastern equatorial Atlantic fracture zone, there are documented climatic variations over the
last 1.8 m.y. which can be used to correlate several cores.

Boltovskoy (1973) studied several South Atlantic submarine
cores between Lat 22 and 47 S and Long. 22 and 67 W. In this
area Boltovskoy recorded the Pliocene/Pleistocene boundary with
the evolutionary zonal markers Globorotalia tosaensis and Globorotalia truncatulinoides.

On the other hand, Boltovskoy (op. cit.) based on: 1) the
total relationships of warm and cool species; 2) the relation
Globorotalia menardii/Globorotalia inflata and 3) the coiling
direction of Globigerina pachyderma and Globigerina bulloides
found in the Pleistocene two cool and two warm periods which
were related to the Early Wurm-Principal Wurm and Riss-Wurm/
interstadial Wurm respectively.

Summary

South American Atlantic Miocene planktonic assemblages are
well represented in the Pelotas Basin (southern Brazil) and in

460

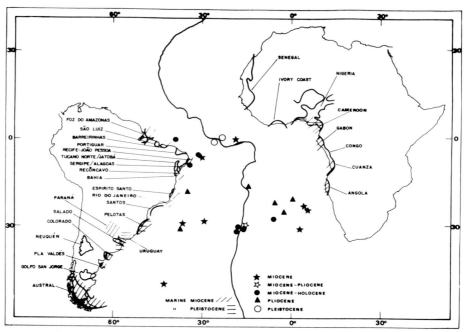

Text Fig. 1. Location of the cores recovered from the South
Atlantic and corresponding ages, and the sedi-
mentary basins in which Miocene and Pleistocene
Foraminifera were recorded.

the continental margin of the Amazonas and Sergipe/Alagoas basins.

The youngest Miocene marine sequences are recorded in the
Pelotas Basin in which middle-late Miocene planktonic Foraminifera are recorded by Falcetta (1976, MS).

Planktonic assemblages from low latitudes are comparable
to those of the Caribbean area, whereas those of the Pelotas
Basin and other basins from higher latitudes show more affinities with temperate assemblages.

At 30° S globorotaliids are scarce, being dominant globigerinids. The zonal scheme presented by Blow (1969) is a
useful one for latitudes near 30°S or higher, whereas at lower
latitudes that of Bolli (1966) seems to be more reliable.

Most of the uvigerinids and bolivinitids present in the
Pelotas Basin are restricted to the Miocene; although some are
of Australoasiatic origin, most of the species are related or
common to European ones.

Other benthonic Miocene assemblages show close relationships to those faunas living today in the western Atlantic
Ocean, and many cosmopolitan species are common components of
the associations.

African Miocene benthonic foraminiferal associations, although they contain some cosmopolitan and Caribbean species,
show strong endemic features; the planktonics are those known
from the Caribbean and equatorial areas. No Australoasiatic
influences are perceptible.

Large Foraminifera, although scarce are recorded from the
South American continent only in northern Brazil, that is, the
equatorial region. Below this latitude no large Foraminifera
were recorded.

In the African continent, large Foraminifera were recorded
from Gabon and Cameroon.

South American and African benthonic foraminiferal assemblages show clear latitudinal zonation.

In South America this is documented by the abundance of
warm water species down to latitudes of 20-25°S whereas south
of this latitude the faunas assemblage lack typical warm water
species or they become scarce.

Pleistocene South American assemblages are similar to
those of the present Atlantic Ocean.

In African basins, where post-Miocene marine sediments
were recorded, the strong endemic features, typical of older
faunal assemblages seem to have been replaced by more cosmopolitan faunal elements.

References

Alberici, A., Barbieri, F., Iaccarino, S. and Rossi., U. 1973.
    Considerazioni biostratigrafiche e paleoecologiche sul Neo-
    gene del Fianco Occidentale della dorsale Medio-Atlantica
    Meridionale (Foraminiferi dei "Sites" 15 e 16 del "Leg" III
    D.S.D.P. "L'Ateneo Parmense" - Acta Naturalia, Vol. IX Fasc.
    2, p. 137-151.
Antunez, M. T., 1964. O Neocretáceo e o Cenozoico do litoral
    de Angola. Junta de Investigacoes de Ultramar, Lisboa.
Belmonte, Y. C., 1966. Stratigraphie du bassin sédimentaire du
    Cameroun. Proc. 2nd W. African Micropal. Coll. Ibadan, 1965.
Berggren, W. A. and Amdurer, M., 1973. Late Paleogene (Oligo-
    cene) and Neogene planktonic foraminiferal biostratigraphy of
    the Atlantic Ocean (Lat. 30°N to Lat. 30°S) Riv. Ital. Pale-
    ont. v. 79, no. 3, p. 337-392.
Bertels, A., 1976b (in press). Micropaleontology and Paleogeo-
    graphy of the Upper Cretaceous and Cenozoic of Argentina.

Bertels, A., 1976. Paleogene Foraminifera - South Atlantic, this volume.

Bertels, A., Falcetta, M. M., and Kotzian, S. B., 1976. (in press). Micropaleontología (Foraminíferos y Ostrácodos) del Cuaternario de Palmares do Sul (Formación Chui), Brasil.

Blow, W. H., 1969. Late Middle Eocene to Recent planktonic foraminiferal biostratigraphy. Intern. Conference on Planktonic Microfossils. p. 199-422.

——————, 1970. A Deep Sea Drilling Project. Leg 3. Foraminifera from selected samples. In Maxwell, A. E., et al., 1970. Initial Reports of the Deep Sea Drilling Project, v. III. Washington (U.S. Government Printing Office) p. 629-661.

——————, 1970b. Deep Sea Drilling Project, Leg 4, Foraminifera from selected samples. In Maxwell, A. E., et al., 1970. Initial Reports of the Deep Sea Drilling Project, v. IV. Washington (U.S. Government Printing Office) p. 383-400.

Boltovskoy, E., 1959. Los foraminíferos de los sedimentos cuaternarios en los alrededores de Puerto Quequé (Provincia de Buenos Aires), Rev. Asoc. Geol. Argentina. T. XIV, nos. 3-4, p. 251-277.

——————, 1973. Estudio de testigos submarinos del Atlántico sudoccidental. Rev. Museo Arg. Ciencias Nat. "Bernardino Rivadavia" Geol. T. VII, no. 4, p. 213-340.

Bond, G., 1965. Quantitative approaches to rainfall and temperature changes in the Quaternary of southern Africa. International Studies on the Quaternary. VII Intern. Congress for Quaternary Res. Ed. Wright, H.E.Jr.,and Frey, D. G., p. 323-336.

Brognon, G. and Verrier, G., 1966. Tectonique et sédimentation dans le bassin du Cuanza (Angola). In: Bassins sédimentaires du littoral africain. Ière partie: Littoral Atlantique. Assoc. des Services Géol. Africains., p. 207-252.

Closs, D., 1967. Miocene planktonic Foraminifera from southern Brazil. Micropaleontology, v. 13, no. 3, p. 337-344.

——————, 1970. Estratigrafía da Bacia de Pelotas, Rio Grande do Sul. Iheringia, Geol., no. 3, p. 3-76.

Closs, D., and Madeira M. L., 1968. Cenozoic Foraminifera from the Chuy Drill Hole, northern Uruguay. Ameghiniana, T.V., no. 7, p. 229-236.

de Klasz, I., and Gageonnet, R., 1963. Biostratigraphie du Bassin Gabonais. Colloque Internat. de Micropal. Dakar. p. 277-304.

de Klasz, I., Le Calvez,Y.,and Rérat, D., 1964a. Deux importantes espèces de Foraminifères du Miocène inférieur de l'Afrique occidentale. C.R. Sommaire des Séanc. Soc. Géol. France, Fasc. 5, p. 194-195.

——————, 1964b. Deux nouveaux Buliminidae (Foraminifères) du Paléogene de l'Afrique occidentale. C.R. Sommaire des Séanc. Soc. Géol. France Fasc. 5, p. 208-209.

——————, 1964c. Deux nouveaux genres de Foraminifères de Gabon (Afrique Equatoriale). C.R. Sommaire des Séanc. Soc. Géol. France. Fasc. 6, p. 236-237.

de Klasz, I., and Micholet, J., 1970. Eléments nouveaux concernant la Biostratigraphie du Bassin Gabonais. IV Colloque African Micropal., p. 109-141.

de Klasz, I., and Rérat., 1962. Une nouvelle espèce d'Eponidopsis (Foraminifera) de l'Afrique occidentale. C.R. Sommaire Séanc. Soc. Géol. France. Fasc. 4, p. 112-113.

Delaney, P. J. V., 1967. Geomorphology and Quaternary coastal geology of Uruguay, p. 1-39.

Drooger, C. W., 1966. Notes on Miogypsina of Cameroon. Proceed. of the Second West African Micropl. Coll., Ibadan, p.

Ericson, D. B., and Wollin, G., 1956. Correlation of six cores from the equatorial Atlantic and the Caribbean, Deep-Sea Research, v. 3, no. 2, p. 104-125.

Falcetta, M. M., 1976 (MS). Foraminíferos Planctónicos e Estratigrafía do Cenozoico da Bacía de Pelotas, Brasil.

Faulkner, C. S., de Klasz, I., and Rérat, D., 1963. Megastomella nov. gen. nouveau Foraminifere de l'Afrique occidentale Rev. Micropal., v. 6, no. 1, p. 19-22.

Fernandes, J. M. G., 1975 (MS) O Género Uvigerina (Foraminiferida) do Cenozoico Superior na Bacia de Pelotas, Rio Grande do Sul. Brasil. Dissertacao de Mestrado.

Ferreira, C. S., Gonzalez, B. B., and Rodriguez, F. B. H., 1973. Occorrencia da Formacao Pirabas (Mioceno inf.) na Bacia de Barreirinhas, Maranhao. Rev. Brasileira de Geoc., v. 3.

Fidalgo, F., de Francesco, F. O., and Pascual, R., 1975. Geología Superficial de la Llanura Bonaerense. VI Congreso Geol. Arg. Relatorio Geol. Prov. Buenos Aires, p. 103-138.

Funnell, B.M., 1971. The occurrence of pre-Quaternary microfossils in the oceans. (In) Funnel, B. M. and Riedel W., (eds.) "Micropaleontology of Oceans," Cambridge, Cambridge Univ. Press, 1967. p. 507-534.

Graham, J. J., de Klasz, I., and Rérat, D., 1965. Quelques importants Foraminifères du Tertiaire du Gabon (Afrique équatoriale). Rev. Micropal.,v. 8, no. 2, p. 71-84.

Le Calvez, Y., de Klasz, I. and Brun, L., 1974. Nouvelle contribution à la connaissance des microfaunes du Gabon. Rev. Española de Micropal., v. VI, no. 3, p. 381-400.

Malumian, N., 1968. Foraminíferos del Cretacico Superior y Terciario del subsuelo de la Provincia Santa Cruz, Argentina. Ameghiniana, T.V., no. 6, p. 191-227.

————, 1970a. Biostratigrafia del Terciario marino del subsuelo de la Provincia de Buenos Aires (Argentina) Ameghiniana, T. VII, no. 2, p. 173-203.

————, 1972. Foraminíferos del Oligoceno y Mioceno del subsuelo de la Provincia de Buenos Aires. Ameghiniana, T. IX, no. 2, p. 97-137.

Malumian, N., Masiuk, V., and Riggi, J. C., 1971. Micropaleontología y sedimentología de la perforación SC-1 Provincia Santa Cruz, República Argentina. Su importancia y correlaciones. Rev. Asoc. Geol. Argentina. T. XXVI, no. 2, p. 175-208.

Maxwell, A, E. et al., 1970. Initial Reports of the Deep Sea Drilling Project, Volume III. Washington (U.S. Government Printing Office), xx + 806 pp.

Noguti, I., 1975. Zonación Bioestratigráfica de los Foraminíferos planctónicos del Terciario de Grasil. Rev. Española de Micropal. v. VII, no. 3, p. 391-401.

Noguti, I., and Fontana dos Santos, J., 1972. Zoneamiento preliminar por foraminíferos planctonicos do Aptiano ao Mioceno na

464

plataforma continental do Brasil. Bol. Tec. Petrobras. Rio de Janeiro. v. 13, no. 3, p. 265-283.

Petri, S., 1954. Foraminíferos fosseis da Bacia do Marajó. Univ. Sao Paulo. Facultade de Filosofia, Ciencias e Letras. Bol. no. 176, Geol. no. 11, p. 1-173.

————, 1957. Foraminiferos miocénicos da Formacao Pirabas. Univ. Sao Paulo, Facultade de Filosofia, Ciencias e Letras. Bol. no. 216, Geol. no. 16, p. 1-79.

————, 1972. Foraminíferos e o ambiente de deposicao dos sedimentos do Mioceno do Reconcavo Baiano. Rev. Bras. de Geociências. v. 2, no. 1, p. 51-67.

Pisetta, J. L., 1968 (MS). Descripción de una faúnula de Foraminíferos de la Provincia de Entre Rios. Trabajo Final de Licenciatura, Facultad de Ciencias Exactas y Naturales, Univ. Buenos Aires (unpublished).

Rocha, A. T., and Ferreira, J. M., 1957. Contribucao para o estudo dos Foraminiferos do Terciario de Luanda. García de Orta, v. 5, no. 2, p. 297-310.

Reyment, R. A., 1965. Aspects of the Geology of Nigeria. The stratigraphy of the Cretaceous and Cenozoic deposits. Ibadan Univ. Press. p. 1-145.

Rudiman, W. F., 1971. Pleistocene Sedimentation in the equatorial Atlantic: Stratigraphy and Faunal Paleoclimatology. Geol. Soc. Am. Bull. v. 82, no. 2, p. 283-301.

Saito, T., Ewing, M., and Burckle, L. H., 1966. Tertiary Sediment from the mid-Atlantic Ridge. Science, v. 151, no. 3714; p. 1075-1079.

Schaller, H., Vasconcelos, D. N., and Castro, J. C., 1971. Estratigrafia preliminar da Bacia sedimentar da Foz do Rio Amazonas. Soc. Brasileira de Geol. Anais XXV Congreso, v. 3, p. 189-202.

Slansky, M., 1963. Contribution à l'étude géologique du bassin sédimentaire côtier du Dahomey et du Togo. Mémoires Bureau de Recherches Géologiques et Min. no. 11, p. 1-270.

Suarez Soruco, J. R., 1968 (MS). Estudio micropaleontológico del cordón litoral de la localidad de Mar Chiquita, Provincia de Buenos Aires. Trabajo Final de Licenciatura, Facultad de Ciencias Exactas y Naturales, Universidad de Buenos Aires (unpublished).

Thiesen, Z. V., 1975 (MS). Bolivinitidae e Caucasinidae (Foraminiferida) do Cenozoico Superior da Bacia de Pelotas, Rio Grande do sul, Brasil. Dissertacao de Mestrado. Universidade Federal do Rio Grande do Sul, Brasil. (unpublished)

Tinoco, I. de M., 1958. Foraminiferos Quaternarios de Olinda, Estado de Pernambuco. Dep. Nac. Prod. Min. Geol. e Min. Monografia XIV, pp. 1-61.

Plate I

Fig. 1 - <u>Globorotalia conomiozea</u> Jenkins, x278,
         Pelotas Basin (Middle-Upper Miocene)
Fig. 2 - <u>Globorotalia crassula</u> viola Blow, x173,
         Pelotas Basin (Middle-Upper Miocene)
Fig. 3 - <u>Globigerina nephentes</u> Todd, x149,
         Pelotas Basin (Middle-Upper Miocene)
Fig. 4 - <u>Globoquadrina dehiscens</u> (Chapman,Parr and Collins),x186,
         Pelotas Basin (Middle-Upper Miocene)
Fig. 5 - <u>Globorotalia humerosa</u> (Takayanagi and Saito), x186
         Pelotas Basin (Middle-Upper Miocene)
Fig. 6 - <u>Globorotalia nympha</u> Jenkins, x186
         Pelotas Basin (Middle-Upper Miocene)
Fig. 7 - <u>Globigerina decoraperta</u> Takayanagi and Saito, x186
         Pelotas Basin (Middle-Upper Miocene)
Fig. 8 - <u>Globigerina druryi</u> Akers, x186
         Pelotas Basin (Middle-Upper Miocene)

PLATE I

CENOZOIC OSTRACODA - NORTH ATLANTIC

M.C.Keen, University of Glasgow, Scotland

ABSTRACT

The ostracod fauna of each Cenozoic epoch is described where possible from western Europe (Denmark, Belgium, England, France, N.Spain), the Atlantic seaboard of N.America, Greenland, Rockall, and from the N.Atlantic Basin. No species are known in common to the two sides of the N.Atlantic for the Palaeogene, although genera and species groups indicate some communication. Most Palaeogene ostracods can be related to Cretaceous forebears. The Neogene saw a renewal of the ostracod fauna, with wide and rapid dispersal of genera, with a few species present on both sides of the Atlantic. Fourteen ostracod zones are proposed for the Palaeogene of N.W.Europe, but these cannot be recognised at present in Aquitaine due to provincial differences.

RÉSUMÉ

On a décrit la faune d'ostracodes de tous les époques cénozoiques, si possible, de Danemark, Belgique, Angleterre, France, et Espagne; de la marge orientale de l'Amérique du Nord; de Rockall, et du bassin de l'Atlantique Nord. Au cours du Paléogène on ne trouve aucun des éspèces en tous les deux côtes de l'Atlantique Nord, bien que la présence du même genres suggèrent une liaison faible. Le Néogène voit l'apparition de plusieurs formes nouvelles avec un net renouvellement de la faune d'ostracodes. On trouve un certain nombre d'éspèces communes aux deux côtes de l'Atlantique. On a reconnu une succession de 14 biozones dans le Paléogène de l'Europe nord-ouest.

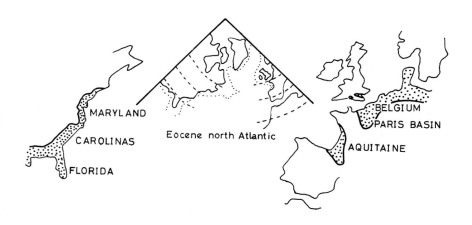

Text fig.1. Palaeogene areas discussed

468

Fig.2. Zonation of the Palaeogene, partly based on the work of Berggren and Curry.

SERIES	STAGE	PLANKTONIC FORAM ZONES	NANNOPLANKTON ZONES	DUCASSE OSTRACOD ZONES AQUITAINE	DUCASSE ZONES N. SPAIN	OSTRACOD ZONES FOR N.W. EUROPE		BRACKISH OSTRACOD ZONES Neocyprideis	FRESHWATER ZONES Virgatocypris	ENGLAND	PARIS BASIN	BELGIUM
OLIGOCENE	CHATTIAN	22	25			14	Flexus gutzwilleri	parallela			ETAMPES	BOOM / HENIS / NEERREPEN / GRIMMERTINGEN
	RUPELIAN	21 / 20	24 / 23	8		13 a	Hammatocythere hebertiana					
	LATTORFIAN	19	22 / 21	7		b	Quadracythere diversinodosa	Williamsoniana	tenuistriata			
EOCENE	BARTONIAN	18 / 17 / 16 / 15 / 14	20 / 19 / 18 / 17	6 / 5		a / 12 / 11	Haplocytheridea debilis / Cytheretta laticosta	colwellensis	edwardsi	SANNOIS / GYPSE / L.inornata / P. ludensis / MARINES / HEADON-OSBORNE / BARTON-BEDS	HAMSTEAD / BEMBRIDGE	ASSCHE / WEMMEL / LEDES / BRUXELLES
	AUVERSIAN	13 / 12	16 / 15		4	10 / 9	Cytheretta cellulosa / C. rigida	grisyensis	grisyensis	AUVERS-CRESNES	HUNTING BRIDGE	AALTER / MONS EN PREVELE
	BIARRITZIAN	11				8 b / a	C.eocaenica / C.nov	apostolescui		BIARRITZIAN XI.XVII / CALC GROSSIER	BRACKLE VII-X	
	LUTETIAN	10	14	4	3	7	Novocypris whitecliffensis	NO RECORD	NO RECORD	SHAM V / BEDS I-IV	CUISE	ARGILES / DYPRES
	CUISIAN	9 / 8	13		2	6 c / b / d	E.reticulatissum / C.scrobiculoplicata / C.nov			LONDON CLAY		
	YPRESIAN	7 / 6	12 / 11	3	1	5	Cytheretta nerva nerva	durocortor-iensis		WOOLWICH- BLACKHEATH	"SPARNACIAN" / BRACHEUX	DORMAAL / LINCENT
PALAEOCENE	THANETIAN	5	10 / 9	2		4	Paracytheretta reticosa	grandinatus		THANET BEDS		
	MONTIAN	4	8 / 7 / 6 / 5	1		3	Alatacythere cirrusa	NO RECORD	NO RECORD			
	DANIAN	3 / 2 / 1	4 / 3 / 2 / 1			2 / 1	Krithe montensis / Lornatoidella fissurata	murciensis			VIGNY ETC.	MONS / CIPLY

FAUNAL SUMMARY

A vast amount of work has been carried out on the Palaeogene ostracods of western Europe and North America, but as it varies from area to area it is impossible to treat all regions in similar detail. In Europe most workers have been concerned with the faunas of Belgium, the Paris Basin, southern England, Germany, and Aquitaine. The present state of knowledge makes it possible to attempt a zonation of the Palaeogene of these areas. In North America most work has been concentrated on the faunas of the Gulf states with relatively few studies along the Atlantic seaboard, so a zonal scheme cannot be justified at present. On the other hand, Neogene faunas are well known in the area from Florida to Maryland so that ostracods can be used for very accurate correlation. The only Neogene faunas described from the Atlantic coasts of Europe are from the classic localities of Aquitaine.

PALAEOCENE

Marlière (1958) and Deroo (1965) recognised three faunizones in the Upper Danian and Montian of Belgium and Holland, referred to as the "Couches à Cytherelloidea" (upper Pl), "Couches à Cytheretta" and "Couches à Triginglymus" (P2 = Montian). Mme Damotte (1964) was able to recognise the lower two in the Paris Basin, so they probably have a regional significance. They form the three lowest zones in the zonal scheme given below. The Seelandian fauna of Denmark is not very well known, although ostracods have been listed by Deroo and some described by Triebel. Particular importance is placed on the presence of Paracytheretta reticosa Triebel, which is also present in the Thanet Beds of Kent and the Palaeocene of Poland (Szcezechura, 1965). The ostracods of the Sables de Bracheux of the Paris Basin are the only Upper Palaeocene fauna to be described in detail from western Europe, so it is difficult to know how many of its species are truly indigenous. Of 26 described species (Apostolescu, 1956) only three are present in the Thanet Beds. In particular, the Hazelina of the Thanet Beds (undescribed species) is not the same as that of the Sables de Bracheux, "Puriana" sculpta APOSTOLESCU, while the absence of P.reticosa from the Sables de Bracheux is considered important, although bearing in mind that this may be due to geographical factors, i.e., it may be a northern "cold water" species. However, the evidence suggests that the Thanetian of the Paris Basin is rather younger than the type Thanet Beds. The distribution of important species are shows in Text Fig. **3.**

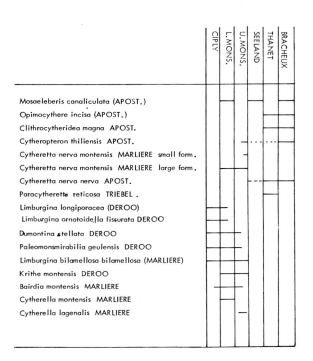

Fig.3. Distribution of ostracods in the Palaeocene of Europe.

The marine Thanetian is succeeded by continental deposits with a <u>Vetustocytheridea lignitarum</u> DOLLFUS (=<u>V.guitrancourtensis</u> APOST.), found throughout northern France, southern England, and possibly in northern Aquitaine. Other important species are <u>Neocyprideis durocortoriensis</u> APOST. and <u>Cytheromorpha aillyensis</u> BIGNOT.

Palaeocene faunas have been described from the north and south of the Pyrenees (Tambareau 1972, Ducasse 1972, Tambareau et Villatte 1974). The "Lower Thanetian" has a poorly preserved marine

ostracod fauna.   The "Upper Thanetian" has a brackish water fauna
in its lowest beds, with Vetustocytheridea sp., Cytheromorpha aff.
aillyensis, and Neocyprideis grandinatus TAMBAREAU.   The latter is
of great importance because it is intermediary between the
Maastrichtian N.murciensis DAMOTTE ET FOURCADE and N.durocortoriensis.
This is taken to represent zone 1 of Mme Ducasse (1972) described
from Campo in northern Spain (see Text Fig.2).   The upper part of
the Thanetian of the Pyrenees contains an undescribed marine fauna,
with two species related to Polish Palaeocene ostracods: Kingmaina
cf opima SZCEZECHURA and Trachyleberidea aff. semiplana SCZEZECHURA.
Mme Ducasse (1972) has also described Ilerdian faunas, discussed
below under the Eocene.

Mme Szcezechura (1971) has described a large Lower Palaeocene
fauna from Western Greenland.   Thirty nine species are illustrated,
many unassigned.   The fauna cannot be related to any in North
America or Europe.

Ostracods from the Danian Brightseat Formation of Maryland have
been described by Hazel (1968), where reference to other works may
be seen.   The fauna, consisting of thirty five species, is very
distinctive, having little in common with the overlying Aquia
Formation, but with thirteen species present in various Lower
Palaeocene Gulf Coast localities.   The ostracods indicate
correlation with the upper parts of the Kincaid Formation of Texas
and the Clayton Formation of Alabama.   The fauna is dominated
Vestustocytheridea and Opimocythere, indicating shallow marine inner-
sublittoral conditions.   The presence of Vetustocytheridea on
both sides of the Atlantic is interesting; Hazel points out that
the earliest species, V.fornicata fornicata (ALEXANDER) is very
similar to Haplocytheridea, and suggests it evolved from the latter.
Haplocytheridea is a very common genus in the Cretaceous and
Tertiary of N.America, but is rare in Europe.   So it is probable
that Vetustocytheridea, a shallow marine and brackish water ostracod
somehow migrated across the Atlantic to Europe, possibly via
Greenland where the ocean was narrow prior to ocean floor spreading?

The Aquia Formation is probably of Montian age in part, as
well as including younger members, and contains a large and
distinctive fauna.   One species of interest is that referred to
by Hazel as Eocytheropteron aff. thiliensis APOST., originally
described from the Sables de Bracheux of the Paris Basin.   Pooser

(1965) has described the fauna of the Black Mingo Formation of South Carolina which suggest that the formation straddles the Palaeocene–Eocene boundary, most of it being Eocene. He recognised two assemblage zones, Brachycythere interrasilis ALEXANDER and Cytherelloidea nanafaliensis HOWE, which could represent differences in age, environment, or both.

The distribution of some common Palaeocene and Eocene species from the Atlantic coast are shown in Text Fig.4.

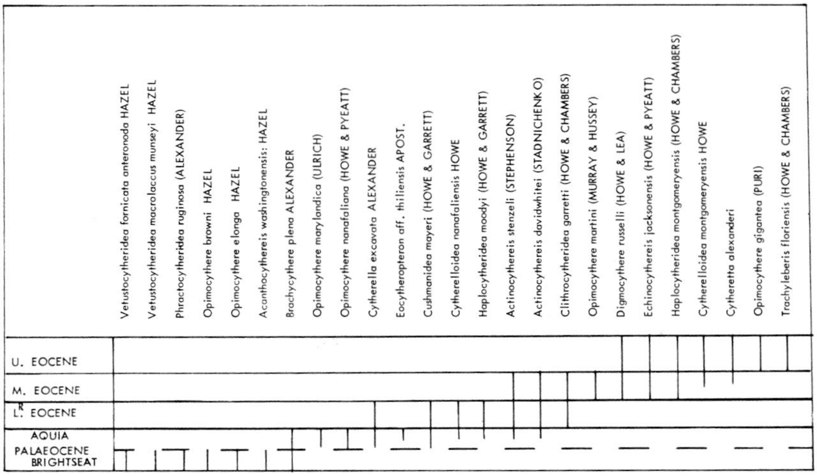

Distribution chart with the following species as column headers: Vetustocytheridea fornicata anteronoda HAZEL, Vetustocytheridea macroloccus muneyi HAZEL, Phractocytheridea ruginosa (ALEXANDER), Opimocythere browni HAZEL, Opimocythere elonga HAZEL, Acanthocythereis washingtonensis HAZEL, Brachycythere plena ALEXANDER, Opimocythere marylandica (ULRICH), Opimocythere nanafaliana (HOWE & PYEATT), Cytherella excavata ALEXANDER, Eocytheropteron aff. thiliensis APOST., Cushmanidea mayeri (HOWE & GARRETT), Cytherelloidea nanafaliensis HOWE, Haplocytheridea moodyi (HOWE & GARRETT), Actinocythereis stenzeli (STEPHENSON), Actinocythereis davidwhitei (STADNICHENKO), Clithrocytheridea garretti (HOWE & CHAMBERS), Opimocythere martini (MURRAY & HUSSEY), Digmocythere russelli (HOWE & LEA), Echinocythereis jacksonensis (HOWE & PYEATT), Haplocytheridea montgomeryensis (HOWE & CHAMBERS), Cytherelloidea montgomeryensis HOWE, Cytheretta alexanderi, Opimocythere gigantea (PURI), Trachyleberis floriensis (HOWE & CHAMBERS). Rows: U. EOCENE, M. EOCENE, L. EOCENE, AQUIA PALAEOCENE BRIGHTSEAT.

Fig.4. Distribution of ostracods in the Palaeogene of eastern U.S.A.

EOCENE

Eocene faunas have been described by Apostolescu (1955, 1956), Haskins (1968–72) and Keij (1957).

The base of the Ypresian stage is taken to define the base of the Eocene, which lies within zone P6. This stage is represented by the London Clay and Argiles d'Ypres. Three subzones can be recognised (Text Fig.2) in England and Belgium, although it is not clear to what extent these are facies controlled. The fauna is very characteristic, with a deeper water element represented by Brachycythere triangularis REUSS, Echinocythereis reticulatissum

EAGAR, Krithe londinensis JONES, and Trachyleberidea prestwichiana
(JONES AND SHERBORN), shallower water by Cytheretta scrobiculoplicata
(JONES), Hazelina aranea (JONES AND SHERBORN), Pterygocytheris
laminosa HASKINS, Schuleridea perforata insignis (JONES) and
Trachyleberis spiniferimma (JONES AND SHERBORN), and coastal waters
by Cytheridea primitia HASKINS and Cytheridea newburyensis GOKCEN.
The faunas have been described by Eagar (1965), Gokcen (1971),
Haskins (1968-72), and Willems (1975).

The Cuisian substage of the Paris Basin contains a rich fauna
(Apostolescu, 1956) including Novocypris whitecliffensis (HASKINS).
This species first appears in the Sables de Mons-en-Prevele of
Belgium with a fauna more characteristic of the London Clay, but
lacking Cytheretta scrobiculoplicata.   Novocypris is a genus more
typical of the Aquitaine Basin.

In Spain and Aquitaine the Ilerdian spans the Palaeocene-
Eocene boundary.   Mme Ducasse recognised three ostracod biozones
from Campo in northern Spain for the Ilerdian, numbered 2, 3, and 4
on Text Fig.2.   The lowest zone is characterised by a littoral
marine fauna with genera such as Bairdoppilata, Hermanites,
Pokornyella, and Quadracythere.   The middle zone is characterised
by a deep shelf or continental slope fauna, with Cytherella consueta
DELTEL, Echinocythereis, Krithe, and Pontocyprella aturica DELTEL.
The two species of Mlle Deltel are found in many Tertiary deep water
deposits of Aquitaine.   The upper zone marks a return to shallower
water.   Throughout the Pyrenean area the base of the Ilerdian is
marked by the appearance of Pokornyella citrea TAMBAREAU often
accompanied by Echinocytheris isabenana DERTLI (Tambareau, 1972).
Biozone 5 of Ducasse (1972) is equated with the Cuisian and the
first appearance of Novocypris eocaenica DUCASSE in northern Spain,
although this species occurred earlier in northern Aquitaine.

The Middle Eocene commences with the Lutetian stage.   The
oldest ostracod fauna recognised so far is found in Fisher Bed 1 of
Bracklesham (= lowest Fisher Bed VI of Whitecliff Bay; the
numerical subdivisions on Text Fig.2 are the Fisher Beds of
Whitecliff Bay), but the fauna is undescribed.   Above this, the
ostracod zone 8b contains the typical Lutetian fauna seen in the
Paris Basin (Apostolescu, 1955).   This subzone sees the first
appearance of Cytherelloidea damariacensis APOSTOLESCU,

474

Pterygocythereis cornuta (ROEMER), Schizocythere appendiculata
TRIEBEL, Schuleridea perforata perforata (ROEMER) and other species
indicated on Text Fig.5.   A major evolutionary radiation can be
recognised amongst the ostracods at this time (Keen 1972c).   The
upper part of the Lutetian, correlated with the Biarritzian, is
marked by the appearance of Cytheridea rigida rigida HASKINS (Text
Fig.5).   This species, together with Cytheretta forticosta KEEN and
Echinocythereis scabropapulosa JONES, differentiate ostracod Zone 9
from 8b.   All three are present in Zone 10, but accompanied by
descendants of Cytheretta eocaenica KEIJ: Cytheretta cellulosa KEEN
and Cytheretta carita KEEN; and a descendant of Cytheretta
costellata costellata BOSQUET, C.costellata grandipora KEEN.

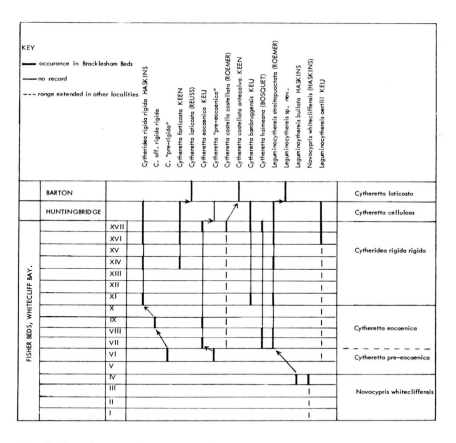

Fig.5. Distribution of ostracods in the Bracklesham Beds of England.

The Upper Eocene is equated with the Bartonian, where two ostracod zones can be distinguished. The lower is characterised by Cytheretta laticosta REUSS, Cytheridea intermedia REUSS, and Pterygocythereis bartonensis KEIJ, all restricted to this zone. The fauna can be recognised in the Barton Clay and in the Marnes à Pholadomya ludensis in the Paris Basin, but appears to be absent in Belgium. The upper zone contains freshwater, brackish, and marine horizons, and is characterised by an endemic fauna in southern England.

In northern Aquitaine Mme Ducasse (1969) has recognised eight biozones for the Eocene (Text Fig.2). These are really assemblage zones, and to a certain extent reflect environmental conditions. However, they have proved useful for correlation over a limited area. Many of the species are present in the Anglo–Paris–Belgium area, but have more extensive ranges, thus throwing some doubt on their identification. On the whole, these zones are not easy to correlate with northern Europe. Faunas from the Western Approaches and the English Channel will probably prove helpful in correlating Aquitaine and more northerly regions. A fauna recently described from the Loire valley near Nantes (Blondeau, 1971) probably belongs to the Cytheretta cellulosa zone, and may provide a link with Aquitaine.

Typical Atlantic coast Eocene ostracods are listed in Text Fig.4. The main Lower Eocene fauna is that described by Pooser (1965) from the Black Mingo Formation of South Carolina and from boreholes in North Carolina by Swain (1951). The former has already been discussed under the Palaeocene. Pooser also described a rich Middle Eocene fauna from the Warley Hill Formation and Santee Limestone. From the former he recorded Trachyleberis spinosissima (Jones and Sherborn) which he regards as a senior synonym of T.spiniferrima (JONES AND SHERBORN); it was suggested that the locality yielding this species was probably older than most Warley Hill localities. If the identification is correct, it suggests a lower Eocene age as this species is typical of the London Clay, belonging to ostracod zone 6b–c. Most of the Middle Eocene ostracods are also found in the Upper Eocene, and indeed are typical of the Jacksonion stage of the Gulf Coast. The Upper Eocene is difficult to recognise in the ostracod faunas of the Atlantic coast, but is probably represented by part of the Cooper Marl (see next section).

OLIGOCENE

There has been a great deal of debate in recent years over the position of the Eocene–Oligocene boundary. In the Paris Basin there is a marked faunal break between the Ludian Marnes à P.ludensis and Marnes à Lucina inornata on the one hand and the Sannoisian on the other: of 27 Sannoisian and some 30 Ludian species none are found in common (Keen, 1972a). In England a similar break is seen between the marine Middle Headon Beds and the Upper Hamstead Beds, the latter containing 13 species all of which are present in the Sannoisian or Rupelian of the Paris Basin. This Sannoisian fauna is also present in Belgium. The difficulty is in correlating these beds with the Lattorfian, which defines the base of the Oligocene. The presence of the typical Sannoisian ostracod Hemicyprideis montosa JONES AND SHERBORN just below the freshwater Bembridge Limestone of England and the Lattorfian Quadracythere diversinodosa (LIENENKLAUS) just above the limestone suggests that the Bembridge Beds might be correlated with the Lattorfian (Text Fig.2). Certainly the Middle Headon Beds have a strong affinity with the Barton Beds (Keen, 1968). The Rupelian has a very distinctive ostracod fauna (the Sannoisian is regarded as a facies of the Lower Rupelian), with little in common with older faunas or with the Lattorfian. Several of the species are present in Aquitaine, allowing accurate correlation. The Upper Oligocene is only really seen in Aquitaine if the study is restricted to the Atlantic seaboard. Here, the Couches à Phare St. Martin at Biarritz contain several species found in the Chattian of Germany and Switzerland, including Flexus gutzwilleri (OERTLI).

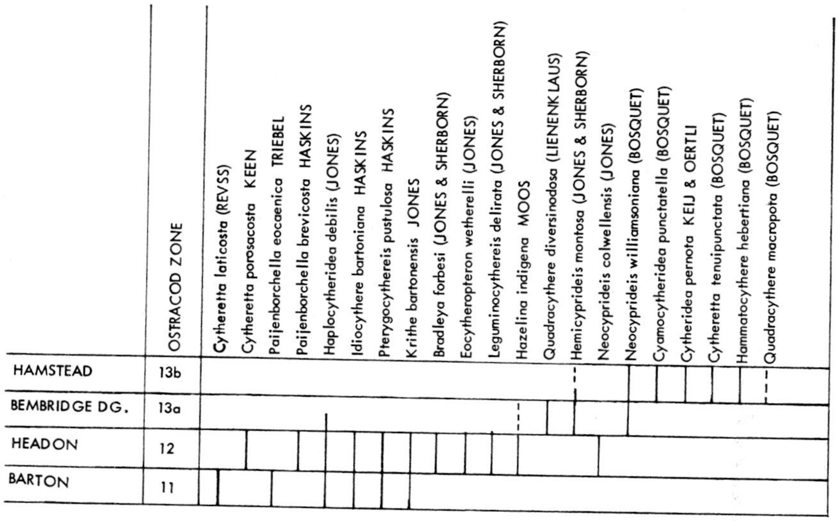

Fig.6. Distribution of ostracods in the Upper Eocene and Oligocene of England.

Along the eastern seaboard of N.America no Oligocene strata are definitely present. Swain has suggested that Cytheretta howei SWAIN might be typical for the Oligocene, but the age of the rocks containing it are not clear. Pooser placed the Cooper Marl in the Oligocene, but the fauna he records seems to indicate two ages, Upper Eocene and Miocene. The older group of samples contains Cytheretta alexander, HOWE & CHAMBERS, Haplocytheridea montgomeryensis HOWE & CHAMBERS and Trachyleberis floriensis (HOWE & CHAMBERS) which strongly suggest a Jacksonian age, i.e. Upper Eocene. Pooser suggests this might constitute a separate biostratigraphical unit. The same fauna is also recorded from a locality referred to the Miocene Duplin Marl by Pooser, and this unit must also be Upper Eocene. The younger assemblage, which is more typical of the Cooper Marl, was referred to the Henryhowella evax Assemblage Zone, and the species listed suggest a Miocene age. In fact, the main reason for an Oligocene age was the apparent overlap of Eocene and

Miocene species, which is obviously removed if two different faunas from two different horizons are present.

NEOGENE

Ostracod faunas of this age have been described from southern France. The controversial locality of Escornbéou in southern Aquitaine has yielded a rich fauna with some species in common with the Chattian of Biarritz, Germany, and Switzerland but also with many species usually regarded as typically Miocene: Callistocythere canaliculata (REUSS), Cnestocythere truncata (REUSS), Costa tricostata (REUSS), and Hemicythere deformis (REUSS). Many of the species are present in the classic localities of northern Aquitaine, but some typical Miocene genera such as Aurila and Puriana are absent. The Escornbéou horizon is undoubtedly older than the type Aquitainian, so if the latter defines the base of the Miocene, Escornbéou must be of Oligocene age.

The Aquitainian and younger ostracod faunas of the type area have been thoroughly described by Moyes (1965). He found that the distribution of many species was strongly influenced by the changes within a shallow coastal environment, so it was not easy to differentiate the classical stages. The Lower Miocene (Aquitanian + Burdigalian) formed one unit while the Upper Miocene could be divided into two. The Pliocene is mostly unfossiliferous. Carbonnel (1969) studied the ostracods of the Miocene of the Rhone Valley and was able to recognise four biozones which are discussed in the next section.

The Miocene strata of the eastern U.S.A. outcrop from Florida to Maryland and have yielded a very rich fauna, described by such authors as Ulrich & Bassler, Howe, McLean, and Puri. Many of the species have a long stratigraphical range with several still living. However, by using assemblages it is possible to use the ostracods in correlation. Puri (1952) has indicated that species of Cytheretta have very limited stratigraphical ranges and are ideal as zonal fossils. Their taxonomy does need sorting out though. Several of the American Cytheretta species resemble European ones and it is interesting to speculate whether these were derived from European immigrants. The Miocene saw the rapid dispersal of Neogene genera such as Aurila, Cyprideis, and Puriana with some species such as Hemicythere deformis (REUSS) recorded from both sides of the

Atlantic.   Man-made taxonomy probably hides many others.

CENOZOIC OSTRACODS FROM WITHIN THE ATLANTIC BASIN

The psychrospheric ostracods (i.e., those specially adapted to
abyssal environments) have been studied by Benson (1972a, 1972b).
In the North Atlantic the three genera Agrenocythere, Bradleya,
and Poseidonamicus have been recorded.   Agrenocythere was probably
derived from Oertliella during the Palaeocene.   The latter is
typical of shelf Cretaceous and Palaeogene deposits, O.aculeata
(BOSQUET) being very common during the Eocene.   O.ducassae BENSON
was described from the Bartonian of Biarritz and is also found in the
Oligocene of the Rockall Plateau; Benson suggests it lived on the
outer shelf and upper bathyal zone of western Europe and the N.E.
Atlantic.   Oertliella has prominent eye tubercles, which are lost
as the animal adapts to the psychrosphere where blind forms dominate.
Three species of Agrenocythere are recorded from the N.Atlantic:
A.antiquata BENSON (L-M Eocene of Europe, America, Carribean;
Oligocene of Rockall) regarded as upper bathyal; its descendant,
A.hazelae VAN DEN BOLD (Oligocene - Recent, N & S Atlantic,
Mediterranean, eastern Pacific, Carribean) truly psychrospheric;
and A.gosnoldi BENSON from the M.Eocene of submarine canyons along
the N.E. shelf of U.S.A., upper bathyal environment.   The latter
is part of a fauna mentioned by Gibson et.al. (1968) consisting
mostly of new species, but with the late Palaeocene Loxoconcha prava
ALEXANDER and the M.Eocene Trachyleberidea goochi SWAIN.   Benson
listed a rich fauna from the Rockall Plateau; a Palaeocene fauna
with such genera as Agrenocythere, Bairdia, Bythocypris, Cytherella,
Hazelina (cf.aranea), Hermanites, and Trachyleberidea (cf.
prestwichiana), probably indicating depths of 200-600 m; and an
Oligocene fauna with truly pschrospheric blind species,
Agrenocythere, Henryhowella, and Poseidonamicus with Bairdia,
Bythocypris, Cytherella and Krithe, indicating depths of 1000-1500 m
as today.   The Rockall story has been continued by Peypouquet
(1975) with drill samples from the Rockall trough.   He found a rich
Lower Miocene fauna with a generic composition similar to Benson's
Oligocene fauna, including the pschrospheric forms.   The fauna
became impoverished during the Miocene, although pschrospheric
species were always present.   The Pleistocene saw a great renewal
of the fauna, but without the pschrospheric forms.   Peypouquet

explains their absence then, as today, by a deficiency in P, N, & Si.
caused by a change in oceanic currents.

BIOSTRATIGRAPHIC ZONES

An attempt is made at a preliminary zonation for the Palaeogene
of the Anglo-Paris-Belgium area, with Germany, Denmark, and perhaps
the whole North Sea Basin.   Text Fig.2 indicates their relationship
to the stratigraphy of the area and to other zonal schemes.

Ostracod biozones.

1. Limburgina ornatoidella fissurata.   DEROO
    Base defined by first appearance of L.orn.fiss., top by first
    appearance of Krithe montensis.

2. Krithe montensis.   DEROO
    Base defined by first appearance of K.montensis, top by first
    appearance of Alatacythere cirrusa.

3. Alatacythere cirrusa.   DEROO
    Base defined by first appearance of A.cirrusa, top by first
    appearance of C.nerva nerva.

4. Paracytheretta reticosa.   TRIEBEL
                                        P.reticosa?
    Base defined by first appearance of C.nerva nerva, top by
    disappearance of P.reticosa.   Probably equivalent to range of
    P.reticosa.

5. Cytheretta nerva nerva.   APOST
    Base defined by disappearance of P.reticosa, top by disappearance
    of C. nerva nerva.

6a. Cytheretta sp.nov. cf. laticosta.
    Base defined by disappearance of C.nerva nerva, top by
    appearance of C.scrobiculoplicata.

6b. Cytheretta scrobiculoplicata (JONES)
    Base defined by appearance of C.scrobiculoplicata, top by
    appearance of E.reticulatissum.

6c. Echinocythereis reticulatissum (JONES)
    Base defined by appearance of E.reticulatissum, top by
    disappearance of C.scrobiculoplicata.

7. Novocypris whitecliffensis (HASKINS)
    Base defined by disappearance of C.scrobiculoplicata, top by
    disappearance of N.whitecliffensis.

8a. Cytheretta sp. nov aff.eoceanica.

Base defined by disappearance of N.whitecliffensis, top defined by appearance of Cytheretta eocaenica.

8b. Cytheretta eocaenica.   KEIJ

Base defined by appearance of C.eocaenica, top by appearance of C.rigida rigida.

9. Cytheridea rigida rigida.   HASKINS

Base defined by appearance of C.rig.rigida, top by appearance of C.cellulosa.

10. Cytheretta cellulosa.   KEFN

Base defined by appearance of C.cellulosa, top by appearance of C.laticosta.

11. Cytheretta laticosta (REUSS)

Base defined by appearance of C.laticosta, top by disappearance of Paijenborchella eocaenica.   Range zone of C.laticosta.

12. Haplocytheridea debilis (JONES)

Base defined by disappearance of Paijenborchella eocaenica TRIEBEL, top by appearance of Q.diversinodosa.

13a. Quadracythere diversinodosa (LIENENKLAUS)

Base defined by appearance of Q.diversinodosa, top by appearance of H.hebertiana.

13b. Hammotocythere hebertiana (BOSQUET)

Base defined by appearance of H.hebertiana, top by appearance of Flexus gutzwilleri.

14. Flexus gutzwilleri (OERTLI)

Base defined by appearance of F.gutzwilleri, top by appearance of H.deformis.

15. Hemicythere deformis (REUSS)

Base defined by appearance of H.deformis, top by appearance of Aurila.   This zone may belong to the Miocene.

Carbonnel (1969) has produced a zonal scheme for the Miocene of the Rhone Basin; this may not be applicable over a wide area, however.

Biozone A.   Loxoconcha linearis linearis CARBONNEL

This coincides with the Aquitanainian + Burdigalian.

Biozone B.   Neomonoceratina helvetica OERTLI

Lower Helvetian.

Biozone C.   Rhodanicites tripartita CARBONNEL

Upper Helvetian.

Biozone D.   Elofsonella amberii CARBONNEL

Equivalent to the Tortonian.

The Pliocene was not zoned, although characteristic species are listed.

Many lineages of ostracods can be traced through the Palaeogene and these offer great scope for correlation.   Genera such as Cytheretta, Cytheridea, and Leguminocythereis offer great scope, with Neocyprideis in the brackish water sediments.

No zonal schemes are available for the Atlantic Seaboard of North America, although it ought to be possible to create one. Certain genera appear to be ideal.   Opimocythere, as indicated by Hazel (1968) is an important genus in the Palaeogene with several widely dispersed species groups.   Vetustocytheridea also has potential, as does Haplocytheridea, which seems to occupy the same niche as Cytheridea in Europe.   Many species of Cytherilloidea seem to have short ranges.   In the Neogene a zonation based on Cytheretta seems possible.

Brackish water and freshwater ostracod zones are also indicated on Text Fig.2.

FACIES RELATIONSHIPS

These have mostly been dealt with in discussion of the faunas. Freshwater, brackish, and marine environments are readily recognised. Within the marine environment shallow coastal, shelf, outer shelf, bathyal, and abyssal have been mentioned.   The zonal scheme given above is mostly for the shallow and shelf seas; not enough information is available for deeper waters, but as there is quite a lot of mixing of these faunas there should be no difficulty in correlating shallow and deep water faunas.

REFERENCES

Apostolescu, V., 1955.   Déscription de quelques Ostracodes du
   Lutétien du Bassin de Paris: Cah. géol., no. 28/29, p. 241-279.

_____., 1956.   Contribution à l'étude des ostracodes de
   l'Eocène inférieur (s.l.) du Bassin de Paris: Rev.Inst.franç.,
   Pétrole, V.11, p.1327-1352.

Benson, R.H., 1972a.   Preliminary Report on the Ostracodes of
   DSDP Leg 12, sites 117 and 117A: Initial Reports of the Deep
   Sea Drilling Project, V.XII, p.427-432.

_____.,1972b.   The Bradleya Problem: Smithsonian contr.
   Paleobiol., no.12., 138p.

Blondeau, M.A., 1971.   Contribution à l'étude des Ostracodes
   éocènes des Bassins de Campbon et de Saffre (Loire-Atlantique):
   Thesis, Univ.Nantes, 150p.

Carbonnel, G., 1969.   Les ostracodes du Miocène Rhodanien: Docum.
   Lab.géol.Fac.Sci.Lyon, no.32, 469p.

Damotte, R., 1964.   Contribution à l'étude des "calcaires montiens"
   du Bassin de Paris: la faune d'Ostracodes: Bull.soc.géol.France,
   (7) V.VI, p.349-356.

Deroo, G., 1965.   Cytheracea (Ostracodes) du Maastrichtien de
   Maastricht (Pays-Bas): Meded.geol.Sticht., ser.C, no.2, 197p.

Ducasse, O., 1969.   Biozonation de l'Eocène nord-aquitain: Bull.
   Soc.geol.France, (7) V. XI, p.491-501.

_____., 1972.   Les ostracodes de la coupe de Campo: Revista
   Espan. Micropaleont., V.   , p.273-289.

Fagar, S.H., 1965.   Ostracoda of the London Clay (Ypresian) in the
   London Basin: I.   Reading District: Revue Micropaléont, V.8,
   p.15-32.

Gibson, T.G., Hazel, J.E., and Mello, J.F., 1968.   Fossiliferous
   rocks from submarine canyons off the northeastern United States:
   Prof.pap.U.S.Geol.Surv., 600-D.

Gökçen, N., 1971.   Les Ostracodes de l'Yprésien de l'ouest du
   Bassin de Londres: Bull.min.expl.Inst.Turkey, no. 75, p.69-86.

Haskins, C.W., 1968 - 1971.   Tertiary ostracoda from the Isle of
   Wight and Barton, Hampshire, England: Revue Micropaleont,
   V.10, p.250-260; V.11, p.3-12, p.161-175; V.12, p.149-170;
   V.13, p.13-29, p.207-221; V.14, p.147-156.

Hazel, J.E., 1968.   Ostracodes from the Brightseat Formation
   (Danian) of Maryland: J.Paleont., V.42, p.100-142.

Keen, M.C., 1968.   Ostracodes de l'Eocène supérieur et l'Oligocène inférieur dans les Bassins de Paris, du Hampshire et de la Belgique, et leur contribution à l'échelle stratigraphique: Colloque sur l'Eocène, Paris 1968, Mem.Bur.Rech.géol.min., no.58, p.137-145.

_____., 1972a.   The Sannoisian and some other Upper Palaeogene Ostracoda from north-west Europe: Palaeontology, V.15, p.267-325.

_____., 1972b.   Evolutionary patterns of Tertiary ostracods: Proc.Int.Geol.Congr., 24(7), p.190-197.

Keij, A.J., 1957   Eocene and Oligocene Ostracoda of Belgium: Mem. Inst.Roy.Sci.Nat.Belgique, V.136.

Marliere, R., 1958.   Ostracodes du Montien de Mons et resultats de leur étude: Mém.Soc.Belg.géol., V.5, p.1-53.

Moyes, J., 1965.   Les Ostracodes du Miocène Aquitain: Thesis, Univ.Bordeaux, 338 p.

Peypouquet, J.P., 1975.   La renouvellement de la faune d'Ostracodes du bassin de Rockall entre le Miocène et le Pléistocène supérieur: Bull. Soc. Géol. France, (7) V. XVII, p. 886-895.

Pooser, W.K., 1965.   Biostratigraphy of Cenozoic Ostracoda from South Carolina: Paleont.Contr.Univ.Kansas, Arthropoda, Art.8., 80p.

Puri H.S., 1952.   Ostracode genera Cytheretta and Paracytheretta in America: J.Paleont., V.26, p.199-212.

Swain, F.M., 1951.   Ostracoda from wells in North Carolina, Part 1, Cenozoic Ostracoda: Prof.pap.U.S.Geol.Surv., 234-A, 58p.

Szcezechura, J., 1965.   Cytheracea (Ostracoda) from the uppermost Cretaceous and lowermost Tertiary of Poland: Acta paleont.pol. V.10, p.451-564.

_____., 1971.   Paleocene Ostracoda from Nûgssuaq, West Greenland: Bull.Gronlands geol.under., no.94, 42p.

Tambareau, Y., 1972.   Thanetien supérieur et Ilerdien inférieur des Petites Pyrénees, du Plantaurel et des chainons audois: Thesis, Univ.Toulouse, 377p.

Tambareau, Y., and Villatte, J., 1974.   Le Passage Thanetien- Ilerdien dans la region de Campo: Bull.Soc.hist.nat.Toulouse, V.110, p.340-361.

Willems, W., 1975.   Ostracoda from the Ieper Formation of the Kallo well (Belgium): Bull.Soc.géol.Belg., V.82, p.511-522.

PLATE 1.

Figs. 1 and 3, Cytheridea primitia HASKINS; London Clay, Alum Bay;
 Fig.1, L = 0.55, Fig.2, L = 0.55.

Fig. 2, Cytheretta eocaenica KEIJ; Lutetian, Damery; L = 0.80.

Figs. 4 and 6, Cytheridea rigida rigida HASKINS; Bracklesham Beds,
 Whitecliff Bay; Fig.1, L = 0.70, Fig.2, L = 0.69.

Fig. 5, Cytheretta cellulosa KEEN; Sables de Beauchamp, Moiselles;
 L = 0.78.

Figs. 7 and 9, Cytheridea intermedia (REUSS); Barton Clay, Barton;
 Fig.7, L = 0.88, Fig.8, L = 0.86.

Fig. 8, Cytheretta haimeana (BOSQUET); Lutetian, Damery; L = 0.60.

Figs. 10 and 12, Cytheridea pernota OERTLI & KEIJ; Couches de
 Sannois, Cormeilles; Fig.10, L = 0.76, Fig.12, L = 0.75.

Fig. 11, Cytheretta scrobiculoplicata (JONES); London Clay,
 Kingsclere; L = 0.92.

Fig. 13, Cytheretta laticosta (REUSS); Barton Clay, Barton;
 L = 0.77,

Fig. 14, Cytheretta forticosta KEEN; Bracklesham Beds, Whitecliff
 Bay; L = 0.79.

Fig. 15, Cytheretta porosacosta KEEN; Middle Headon Beds, Colwell
 Bay; L = 0.75.

Fig. 16, Oertliella aculeata (BOSQUET); Marnes a P.ludensis,
 Verzy; L = 0.73.

Fig. 17, Cytheretta costellata cratis KEEN; Marnes a P.ludensis,
 Verzy; L = 0.73.

Fig. 18, Trachyleberidea prestwichiana (JONES AND SHERBORN); London
 Clay, Whitecliff Bay; L = 0.65.

Fig. 19, Paracytheretta reticosa TRIEBEL; Thanet Beds, Herne Bay;
 L = 0.78.

Fig. 20, Cytheretta costellata costellata (ROEMER); Bracklesham
 Beds, Selsey; L = 0.64.

PLATE 1

PLATE 2.

Figs. 1 and 4, Haplocytheridea montgomeryensis (HOWE and CHAMBERS);
Jackson, Louisiana; Fig.1, L = 0.83, Fig.4, L = 0.83.

Fig. 2, Echinocythereis jacksonensis (HOWE and PYEATT); Jackson
Creole Bluff, Louisiana; L = 0.98.

Fig. 3, Echinocythereis reticulatissima EAGAR; London Clay,
Sheppey; L = 0.93.

Fig. 5, Trachyleberis floriensis (HOWE and CHAMBERS); Jackson,
Creole Bluff, Louisiana; L = 0.82.

Fig. 6, Quadracythere diversinodosa (LIENENKLAUS); Bembridge Beds,
Whitecliff Bay; L = 0.66.

Fig. 7, Cytherelloidea montgomeryensis HOWE; Jackson Creole Bluff,
Louisiana; L = 0.66,

Fig. 8, Paijenborchella eocaenica TRIEBEL; Marnes a P.ludensis,
Verzy; L = 0.60.

Fig. 9, Digmocythere russelli (HOWE and LEA); Jackson Creole Bluff,
Louisiana; L = 1.20.

Fig.10, Cytherelloidea dameriacensis APOST. Bracklesham Beds,
Whitecliff Bay; L = 0.72.

Fig.11, Haplocytheridea debilis (JONES); Middle Headon Beds,
Headon Hill; L = 0.62.

Fig.12, Cytheretta alexander; HOWE and CHAMBERS; Jackson Creole
Bluff, Louisiana; L = 0.90.

Fig. 13, Pterygocythereis pustulosa HASKINS; Marnes a P.ludensis,
Verzy; L = 0.79.

Fig. 14, Leguminocythereis striatopunctata (ROEMER); Bracklesham
Beds, Whitecliff Bay; L = 0.96.

Fig. 15, Eocytheropteron wetherelli (JONES and SHERBORN), Middle
Headon Beds, Headon Hill; L = 0.70.

Fig. 16, Hammatocythere oertlii (DUCASSE); Upper Eocene, Blaye;
L = 0.85.

Fig. 17, Leguminocythereis oertlii KEIJ; Sables de Lede, Bambrugge;
L = 0.84.

Fig. 18, Paracypris contracta (JONES); Marnes a P.ludensis,
Verzy; L = 0.96.

Fig. 19, Leguminocythereis delirata
Middle Headon Beds, Headon Hill; L = 0.82.

488

PLATE 2

PLATE 3.

Fig. 1, <u>Cladarocythere apostolescui</u> (MARGERIE); Bembridge Limestone,
  Bouldnor Cliff; L = 0.73.

Fig. 2, <u>Echinocythereis scabra</u> (VON MUSNSTER); Escornbeou;
  L = 0.85.

Fig. 3, <u>Muellerina latimaryinata</u> (SPEYER); Chattion, Bunde;
  L = 0.69.

Fig. 4, <u>Cytheretta geoursensis</u> KEEN; Escornbeou; L = 107.

Fig. 5, <u>Crestocythere truncata</u> (REUSS); Escornbeou; L = 0.62.

Fig. 6, <u>Hammatocythere hebertiana</u> (BOSQUET); Rupelian, Auvers-St-
  George; L = 0.86.

Fig. 7, <u>Hemicythere deformis minor</u> MOYES; Escornbeou; L = 0.64.

Fig. 8, <u>Cytheretta tenuipunctata absoluta</u> KEEN; Marnes a Huitres,
  Cormeilles; L = 0.84.

Fig. 9, <u>Callistocythere canaliculata</u> (REUSS); Escornbeou; L = 0.55.

Fig.10, <u>Aurila conradi conradi</u> (HOWE and McGUIRT); Choctawhatchee
  Marl, Florida, L = 0.60.

Fig.11, <u>Protocytheretta Karlana</u> (HOWE and PYEATT); Chipola Marl,
  Florida; L = 1.04.

Fig.12, <u>Cytheretta orthozensis</u> MOYES; Helvetian, Orthez; L = 0.86.

Fig.13, <u>Henryhowella evax</u> (ULRICH and BASSLER); Chipola Marl,
  Florida; L = 0.82.

Figs.14 and 18, <u>Cytheretta burnsi</u> (ULRICH and BASSLER); Fig.14
  Calvert Formation, Calvert, L = 1.05; Fig.18 Area Zone, Red
  Bay, Florida, L = 1.10.

Fig.15, <u>Cytheretta sahni</u> PURI; Duplin Marl, Florida; L = 0.98.

Fig.16, <u>Cytheretta inaequivalvis</u> (ULRICH and BASSLER); Shoal River
  Facies, Alum Bluff, Florida; L = 0.90.

Fig.17, <u>Actinocythereis exanthemata</u> (ULRICH and BASSLER); Chipola
  Marl, Florida; L = 0.86.

490

PLATE 3

PLATE 4.

Fig. 1, <u>Virgatocypris grisyensis</u> (MARGERIE);   Ludian, Verzy;
L = 0.86.

Fig. 2, <u>Virgatocypris edwardsi</u> KEEN M.S.;   Upper Headon Beds,
Colwell Bay;   L = 1.13.

Fig. 3, <u>Virgatocypris tenuistriata</u> (DOLLFUS);   Sannoisian,
Cormeilles;   L = 1.02.

Figs. 4 and 5, <u>Hemicyprideis montosa</u> (JONES and SHERBORN);
Hamstead Beds, Bouldnor Cliff;   Fig.4, L = 0.78;   Fig.5, L = 0.80.

Fig. 6, <u>Cytheromorpha bulla</u> HASKINS;   Headon Beds, Headon Hill:
L = 0.58.

Fig. 7, <u>Hemicyprideis helvetica</u> (LIENENKLAUS);   Blaue Tor, Delemont,
Switzerland;   L = 0.70.

Fig. 8, <u>Neocyprideis colwellensis</u> (JONES);   Headon Beds, Headon Hill;
L = 0.63.

Fig. 9, <u>Neocyprideis durocortoriensis</u> APOST.;   Sparnacian, Epernay;
L = 0.81.

Fig.10, <u>Neocyprideis williamsoniana</u> (BOSQUET);   Sannoisian,
Cormeilles;   L = 0.83.

Fig.11, <u>Neocyprideis apostolescui</u> (KEIJ);   Sables de Beauchamp,
Moiselles;   L = 0.79.

Fig.12, <u>Agrenocythere antiquata</u> BENSON;   Upper Eocene, Trinidad;
L = 1.17.   Photograph kindly donated by Dr. R.H. Benso n.

Fig.13, <u>Agrenocythere gosnoldia</u> BENSON;   Middle Eocene, Atlantic
Slope off Cape Cod;   L = 1.14.   Photograph kindly donated by
Dr. R.H. Benson.

Fig.14, <u>Agrenocythere hazelae</u> (VAN DEN BOLD);   Miocene, Cipero
Formation, Trinidad;   L = 1.10.   Photograph kindly donated by
Dr. R.H. Benson.

Fig.15, <u>Candona daleyi</u> KEEN M.S.;   Lower Headon Beds, Headon Hill;
L = 0.87.

492

PLATE 4

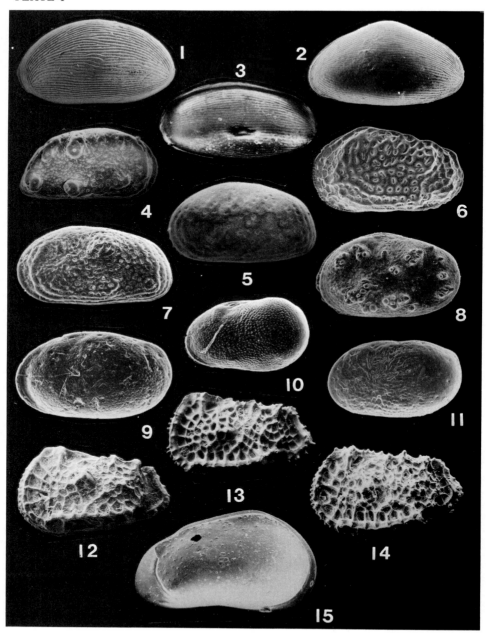

Discussion

Dr. F. M. Swain: In Pooser's paper the species lists for the
   so-called Cooper Marl, Oligocene, and for the late Miocene
   Duplin Marl were reversed in printing-he later sent a corr-
   ection sheet. This might account for your comment that the
   Oligocene forms seem to be of Miocene type.

Keen: That might explain it.

Swain: The other point is with reference to your remark
   about the late Eocene aspect of the described forms from
   the Atlantic coastal plain. Most of the collections from
   the Eocene that have been described from that area are
   early and middle Eocene, probably mostly middle Eocene.
   There is an excellent late Eocene ostracode fauna in South
   Carolina similar to that described by Huff from Mississippi. The Eocene forms from North Carolina are all, or
   nearly all early and middle Eocene types, not late Eocene.

Keen: I mainly got these from Pooser as the main reference.
   I certainly thought, looking at most of the species recorded
   for the middle Eocene in particular were ones that are found
   in the late Eocene of the Gulf Coast. I also thought that
   the so-called Oligocene formation, the Cooper Marl, seemed
   to me to be definitely late Eocene on the species he recorded.

Swain: The reason for that is the liklihood that some of the
   forms recorded for the Oligocene came from the underlying
   late Eocene.

CENOZOIC MARINE OSTRACODA OF THE SOUTH ATLANTIC

W. A. van den Bold

Louisiana State University, Baton Rouge

Abstract

Tertiary Ostracoda of the Atlantic Ocean south of the Tropic of Cancer are known from widely scattered clusters of localities on the continent and from a number of deep-sea cores. The latter have only been studied in detail for part of their fauna. Paleocene ostracodes have been studied in four areas: Central America (Guatemala and British Honduras), northern South America (Venezuela, Trinidad, Guiana), Argentina and West Africa. The distribution of some genera (e.g. *Buntonia*, *Soudanella*, *Togoina*) indicates similarities between faunas from widely different areas and suggests that during or shortly prior to the early Tertiary some shallow-water ostracodes made their way across the Atlantic either through shallow passages or attached to seaweed drifting with prevailing currents. Eocene ostracodes are known from West Africa and Trinidad, but from very different environments, which makes comparison hazardous. Oligocene ostracodes have only been reported from the Caribbean and from some deep-sea cores. Miocene ones have been described from West Africa (Gabon), the Caribbean, northern South America and Argentina (including southern Brazil). Here the presence of morphologically similar groups (*Soudanella*) in West Africa and southern South America and the report of the *Munseyella? punctata* group from southern Brazil and northern South America suggest connections or parallel faunal developments which will have to be investigated in more detail. The West African fauna bears a strong similarity to the Mediterranean one. For the Pliocene-Holocene, ostracodes are known from the same areas as in the Miocene and the distribution of the austral form *Patagonocythere* as well as the possible provinciality of such forms as *Cyprideis*, *Perissocytheridea*, *Sulcostocythere* and the distribution of *Callistocythere* and *Neomonoceratina* appear to be of great potential interest.

Zusammenfassung

Tertiär-Ostracoden des Atlantischen Ozeans südlich der Krebstwende sind von räumlich weitverbreiteten Lokalitäten und einer Anzahl Tiefsee-Bohrkerne bekannt. Nur ein kleines Teil der Fauna dieser letzten ist einer grundlichen Detail-Forschung ausgesetzt worden. Palaozäne Ostracoden sind in vier Gebiete studiert worden, nämlich Zentral Amerika, nordliches Südamerika, Argentinien und West Afrika. Faunistische Übereinstimming zwischen einzelnen Gebiete ist angedeutet durch karakteristische Verbreitung einiger Geschlechte und lässt frühtertiäre oder ein wenig ältere seichte transatlantische Meeresverbindungen, oder aber Beforderung von Flachwasser-ostracoden mittels von Meeres-Stromungen verfrachtetes Pflanzenmaterials, vermuten. Eozäne Arten sind aus West Afrika und Trinidad beschrieben, aber fazielle Unterschiede verschweren ein Vergleich der Faunen. Nur aus den Kariben und einigen Tiefsee-Bohrungen sind Oligozäne Ostracoden bekannt. Miozäne Arten sind von West Afrika (Gabon), dem Karaibischen Gebiet, Venezuela, Kolombien, Argentinien und südlichem Brazil gemeldet und das Vorkommen morphologisch ähnlicher Formen (wie *Soudanella* in Afrika und Südamerika) und die *Munseyella? punctata* Gruppe in nordlichem und südlichem Südamerika) zeigen Verbindungen oder parallele Entwicklungsgeschichten an. Die

West Afrikanische Fauna hat viele mediterrane Elemente. Im Jung-Caenozoi-
cum sind die Faunen untersucht in fast denselben Gebieten und weitere Stu-
dien der Verbreitung und Provincialität der Gattungen *Patagonocythere*, *Cyprideis*, *Perissocytheridea*, *Callistocythere*, *Sulcostocythere*, *Neomonoceratina* u.s.w. könnte wichtigen weiteren Daten hervorbringen.

PREVIOUS WORK

Distribution of Recent Ostracoda was summarized by Puri (1967), who recognized a North American Boreal Province (N. of the Carolinas), a Caribbean
Province, subdivided into Bahaman, South Floridan and Venezuelan subprovinces and a Gulf of Mexico Province. The South Atlantic was not divided into
provinces because of insufficient data and the Antarctic was not discussed.
Hartmann (1975) summarized the distribution of Recent Ostracodes from West
Africa and made comparisons to the Pacific Coast of South America. But
apart from Hartmann's (1955, 1956) studies of the fauna of southern
Brazil and some information on the Argentine coast by Whatley and Moguilevsky (1975) not much is known about the Recent ostracodes of the western
side of the Atlantic. The summary of the Argentine fossil ostracode-faunas
by Bertels (1975) greatly facilitates the task of making comparisons with
other areas. However, nothing is known of the fossil ostracodes on this
side of the Atlantic from about 5°N to 30°S. Data on the location of Neogene Caribbean subprovinces can be found in van den Bold (1970a, 1971c,
1974a, 1975).

Comparatively little is known about Tertiary faunas from the West
coast of Africa (Paleogene: Apostolescu, 1961 and Reyment, 1963-1966; Neogene: van den Bold, 1966h), and no information at all exists south of the
5°S parallel except a recent study by Dingle on the Paleocene of the east
coast.

LIMITATIONS OF THE DATA

Although in the Atlantic Ocean itself we have some very detailed information on some groups of deep-water ostracodes (*Agrenocythere*, *Bradleya*,
Poseidonamicus, *Abyssocythere*) thanks to Benson (1971, 1972) and the Deep-
sea Drilling Program, little is known of some of the other fossil groups,
and even the distribution of Recent forms is still very insufficiently
known. Also in the borderlands it is difficult to obtain an overall picture of the distribution because the information is generally from widely
separated localities and moreover severely restricted stratigraphically.
From West Africa the information is limited to the Danian-Lower Eocene
(Apostolescu, 1961, Reyment, 1963-1966) and the Lower Miocene to Plio-
Pleistocene of one small area only (van den Bold, 1966h). In Argentina

Tertiary faunas are reported from the Danian, Miocene (not always certain
from what levels) and Pleistocene (Bertels, Rossi de Garcia). In Guiana
only Paleocene faunas are known. In Venezuela and Trinidad there is fairly
continuous coverage from Paleocene to Pleistocene, but only deep-water
faunas are known from Middle and Lower Eocene and only shallow-water faunas
from the Upper Eocene. In the Caribbean there is only scattered (and
largely unpublished) information on anything older than Oligocene, although
the Neogene is fairly well known. In Central America Paleocene and some
Lower Eocene are known from Guatemala and British Honduras, no Middle and
Upper Eocene faunas have been reported and the Neogene is only known relatively well from Panama, Costa Rica and Guatemala.

   Therefore I have summarized the more important data for each region in
tabular form, using only the species that were fairly common or had a relatively wide distribution; in places where no detailed stratigraphic information was available the species are merely listed.

   Myodocopida and freshwater ostracodes have been omitted from the dis-
cussion because of the scarcity of references to them in the literature on
Tertiary ostracodes. The last were summarized by McKenzie (1973), while
Kornicker devoted many important papers to the distribution of Recent Myodocopida. A number of other forms have been left out of the discussion:
The species determined by Apostolescu (1961) as *Orionina* and *Ambocythere*
are certainly different from these genera. Several species referred to
Anticythereis have been transferred to other genera, but I have retained
those that resemble the type species and/or species referred to that genus
by Bertels, in order to arrive at some consistency. Some of the species
described as *Phacorhabdotus* (see also Omatsola, 1972) do not belong to that
genus. In these and some other cases it becomes virtually impossible to
make correct generic assignments without study of the original material and
therefore the species cannot be used for comparison with possibly similar
groups in other areas.

   Also many of the unornamented groups have been left out as it is very
difficult to draw conclusions from the dispersal of such genera as *Cytherella*, *Bythocypris*, *Bairdia*, *Propontocypris* and others, and it would be outside the scope of this paper to include detailed information on these. The
same is true of the genera *Haplocytheridea*, *Cyamocytheridea*, *Ovocytheridea*,
Cyprideis, *Cushmanidea* and others without study of the original material.

   One further problem is the identity of the genus *Rocaleberis* Bertels,
1969. The distinction from *Henryhowella* lies, according to Bertels, in the
presence of a vestibule and occasional branching of the marginal porecanals.

However, in *Henryhowella* ex gr. *asperrima*, there is a vestibule and branching porecanals both in the European material (e.g. Keij, 1957, Ruggieri,
1962) and in the Caribbean (van den Bold, 1960). Therefore, at the present
time, I cannot separate the two forms, and I have kept the name *Rocaleberis*
for the Danian species and indicated the distribution of *Henryhowella* in
other areas under the same heading with question marks. It should be noted
that Rossi de Garcia mentions a *Henryhowella* cf. *evax* from the Enterriense
in northern Argentina, and Sanguinetti one from the Miocene of the Pelotas
basin in southern Brazil.

The genus *Masiukcythere* Rossi de Garcia (1968) is described from the
Danian of Argentina. Bertels has ignored this genus in her summary of
Argentinian ostracodes (1975) and does not refer to it in other publications. I have therefore ignored it in this paper. A number of other
uncertainties will be mentioned in connection with the tables, the information on which is selective and therefore obviously biased.

ACKNOWLEDGEMENTS

Noordermeer and Wagner contributed a slide of the more prolific species of
the Paleocene of Guiana, which offered an opportunity for comparison with
species from the Paleocene of Venezuela and Trinidad. Bertels, by sending
representative specimens of a number of Argentine forms, helped me to develop a better understanding of the fauna of that region.

OSTRACODE DISTRIBUTION

There is a number of genera, which have been reported from single areas
only and these are considered here to be more or less provincial. Some
others are distributed over much larger regions and therefore may be considered to indicate possibilities of communication between faunas of those
regions. Genera which are of interest in these two ways have been summarized in Table 7 and suggestions for directions of dispersal are found in
Text-fig. 6. It must be pointed out, that this migration-pattern is based
entirely on what has been observed up to the present. We have to accept
the oldest reported occurrence as the first real presence of a genus, unless there is good reason to doubt this. In some cases extensive studies
have been published on the fauna of the underlying strata. If a particular
genus is not reported from these, it is assumed that the genus did not exist prior to its first reported occurrence. As both in Argentina and West
Africa the Cretaceous beds are reasonably well explored; I think we can
accept that a first occurrence reported in the Danian of these regions is
indeed the first presence of the genus. In northern South America the
Upper Cretaceous is much less known and hardly anything is known at present

on the ostracodes of the Cretaceous of the Caribbean and Central America.
On the other hand the Cretaceous of North America is much more extensively
studied and may be used in considerations of migration of ostracode-genera.
To the contrary in South Africa there is a fair amount of information of
Cretaceous species, but nothing at all is known from the Tertiary. In the
following tables and lists (in the appendix) our knowledge of the Cenozoic
ostracodes of the South Atlantic is summarized by region and stratigraphic
level.

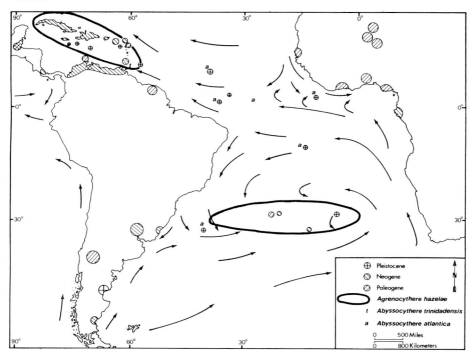

Text-fig. 1: Distribution of fossil deposits from which ostracodes
have been described in and around the South Atlantic, with the distribution of some deep-water species.

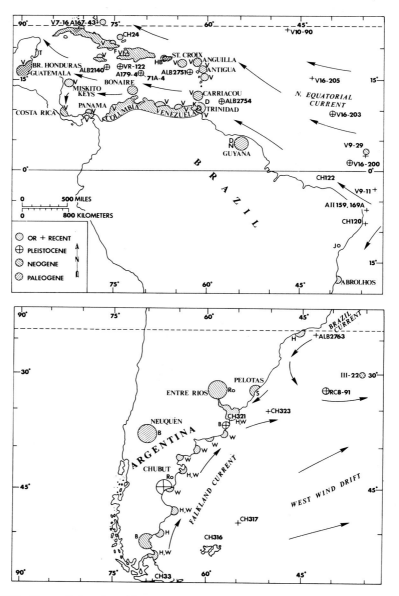

Text-figs. 2-3: Location map of the western South Atlantic. Letter symbols indicate author or sample: B: Bertels; D: Drooger; H: Hartmann; HB: Hulings & Baker; J: Jones; K: Keij; N: Noordermeer & Wagner; Ro: Rossi de Garcia; S: Sanguinetti; T. Teeter; V: van den Bold; W: Whatley & Moguilevsky.

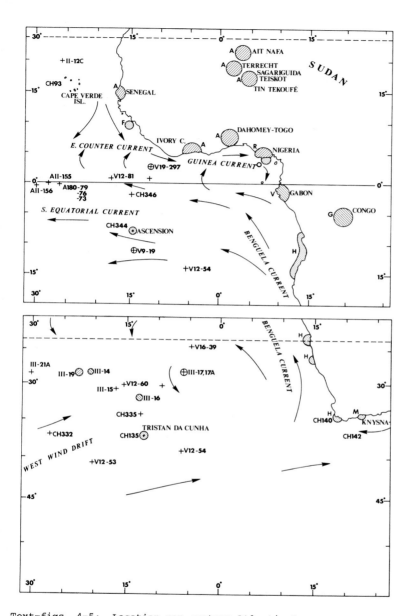

Text-figs. 4-5: Location map eastern Atlantic Ocean.
A: Apostolescu; F: Brady (Fonds de la Mer); G: Grekoff; M: Mad-
docks or Benson & Maddocks; O: Omatsola; R: Reyment; II-11: DSDP
(Benson); V: Vema; A: Atlantic; ALB: Albatros; RC: Chain (all
Benson); CH: Challenger.

502

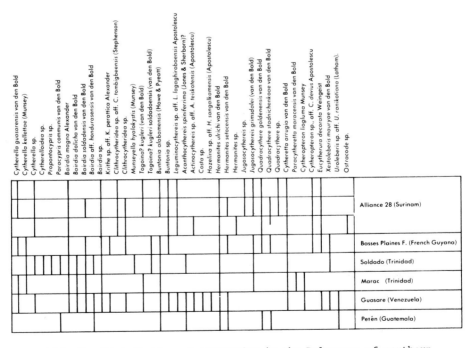

Table 1: Range of selected ostracode species in the Paleogene of
North Central Africa. For another arrangement of species of
the Paleocene see Kogbe, et al (1975, Table 5).

Table 2: Distribution of ostracodes in the Paleocene of northern
South and Central America.

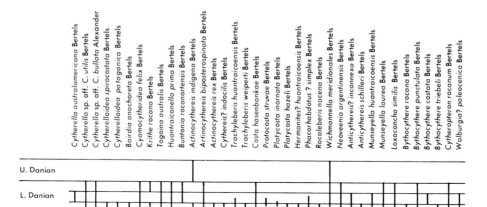

Table 3: Range of selected ostracode species in the Paleogene of
Argentina.

DISCUSSION

The following genera show a pattern in their distribution, which makes
them of special interest.

1: *Buntonia* is reported from the Central African Cretaceous (Apostolescu, Reyment), and shows in this region a continuous development throughout the known Tertiary. In Brazil Krommelbein (1975) reports this genus in
the uppermost Cretaceous. In Venezuela its earliest known presence is in
the Mito Juan Formation (Danian), and it continues into the Miocene. In
Central America its earliest known occurrence is in the Paleogene (van den
Bold, 1946), and in North America it has not been found until the early Eocene (Wilcox, Howe and Garrett, 1934). In southern South America it has so
far not been reported before the Miocene. Therefore, a dispersal from
Central Africa to northern South America and from there in northerly and
southerly directions seems indicated. The genus disappears in the Gulf
Coast in the Oligocene, but at present is found in the Gulf of Mexico at
depth of 60-500 m, whereas its earlier forms all seem to have been very
shallow-water species.

2: Similarly the genus *Soudanella* develops in the West African Cretaceous (Apostolescu, 1961) and has been reported from the African Miocene
(questionably) (van den Bold, 1966h), and Recent (Omatsola, 1972) and from
the Mediterranean Paleocene - Recent. In the Caribbean it appears for the
first time in the Lower Eocene (Chiptown of Jamaica) and in Argentina
it has not been reported earlier than Miocene (Becker, 1964). So far it

is not known from Central America and North America. Here dispersal from Central Africa, through South America, seems probable. These two examples of dispersal may be explained by dominating east-west currents in the equatorial Atlantic in Cretaceous-Early Tertiary times (compare Berggren and Hollister, 1975, fig. 16).

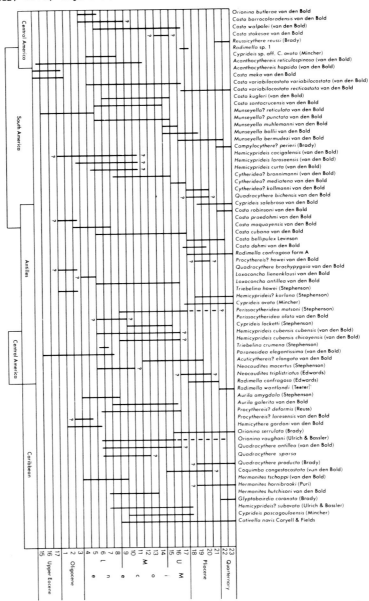

Table 4: Range of selected ostracode species in the late Paleogene and Neogene of the Caribbean. Provincialism of species indicated.

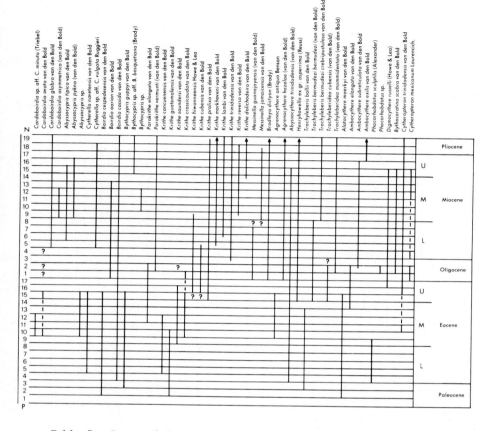

Table 5:   Range of deep-water ostracodes in the Tertiary of the Carib-
            bean.

3:   On the other hand, the genus *Togoina* has been identified by Bertels in the Cretaceous of Argentina.   In northern South America species,
formerly referred to *Brachycythere* (van den Bold, 1957), belonging to this
genus occur in the Paleocene, possibly extending into the Upper Eocene (van
den Bold, 1960).   In West Africa it was originally described from the Upper
Paleocene and Lower Eocene (Apostolescu, 1961).   Therefore dispersal seems
to have taken place in the opposite direction from that of *Buntonia* and
Soudanella.

4:   The genus *Basslerites* appears to have taken the same path as *Buntonia* but at a later time.   Its first reported occurrence is in the Eocene
of Nigeria (Reyment, 1966).   In the whole American area it first appears
in the Miocene, probably earlier in northern South America than in the

Caribbean and North America. So far it has not been reported from southern
South America.

    5: *Protobuntonia* is reported from the Cretaceous and Paleocene of
West Africa (Reyment, 1966). On the other side of the Atlantic it has been
reported from the Coniacian of Brazil (Krommelbein, 1975) and the Eocene
of Trinidad (van den Bold, 1960).

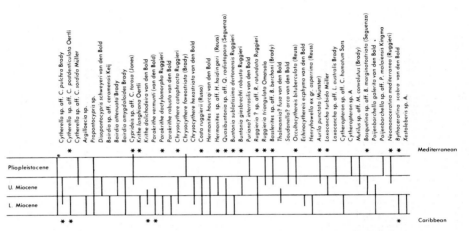

Table 6: Range of ostracode species in the Neogene of Gabon.

    6: *Procythereis* . Here we are in the peculiar position that the ear-
liest species attributed to this genus (in the Oligocene and Miocene of
Europe and the Caribbean) are questionable. It has been suggested that
they may belong to the genus *Pokornyella* (but have numerous anterior radial
porecanals) and could belong to the recently established genus *Pseudoaurila*
of Ishizaki and Kato (1976). It is certainly peculiar that these early
occurrences are in (sub) tropical regions, whereas the Recent occurrences
of *Procythereis* are from the antiboreal provinces. If the fossil and
recent forms really belong to the same genus, then this must have been established on both sides of the Atlantic prior to Miocene times and southward migration towards colder climates must have occurred since the Pliocene.

    7: *Neomonoceratina* is a common enough genus in the Mediterranean and
African Miocene-Recent tropical marine environment. So far it (*N. medi-
terranea*) has only shown up in Recent deposits on the other side of the
Atlantic, where it appears restricted to very shallow water (e.g., van
Morkhoven, 1972).

    8: Another case worth mentioning here is the distribution of *Callis-*

tocythere. Whereas this genus is common in the late Cenozoic of Europe (especially Mediterranean) and has been reported from both the Neogene and Recent deposits of South America, it is conspicuously absent in West Africa, the Caribbean and the United States Atlantic coast, except for some recent introductions in Florida and the Caribbean. There is only one occurrence reported in the late Oligocene (? early Miocene) of the Gulf Coast (Poag, 1972).

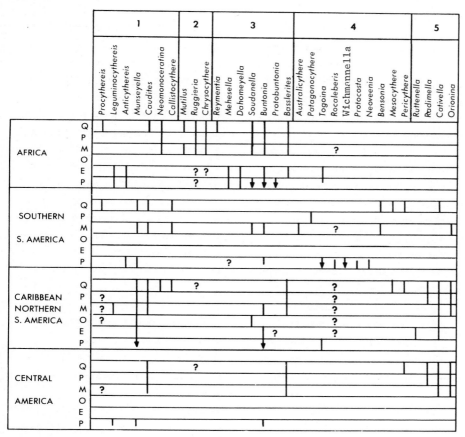

Table 7: South Atlantic distribution of selected ostracode genera.
1: worldwide distribution; 2: Mediterranean forms;
3: African forms; 4: Southern South American forms;
5: Caribbean forms.

Some groups with American distribution.

9. The Genus *Orionina* appears to be an American form. Its oldest reported occurrence is in the Chickasawhay of the Gulf Coast. Its present distribution is from southern Brazil to off the Atlantic coast of North America, and it also occurs on the Pacific side.

10: *Munseyella* is apparently of worldwide distribution, but the questionably assigned *Munseyella? punctata* group is only known from Eocene to Miocene in northern South America (van den Bold, 1958, 1960) and from the Miocene of southern Brazil, which may point to an extension of the Caribbean zoogeographic province to that region in the Miocene.

11: The genus *Cativella* has also not crossed the Atlantic. Omatsola (1972) reports *Cativella* and *Neocaudites* from Recent deposits of the Niger delta. However, this species of *Cativella* (*C. ixemojai*) seems different and is very similar to species of *Chrysocythere*. I also doubt the correct assignment of species of *Puriana* from West Africa. The earliest occurrence of *Cativella* is in the late Eocene of Trinidad (van den Bold, 1960). In the Caribbean it appears in later stages of the Miocene as one progresses farther north (van den Bold, 1974a), and in North America it only makes its appearance near the Pliocene boundary. It has been reported from the Quaternary of southern Brazil and the Antarctic. Here a northward and southward dispersal away from tropical America seems indicated. Recent distribution of the genus is scattered; whereas it is common between 10 and 20 m depth in the Gulf of Paria, it occurs scattered in the Gulf of Mexico between 30 and 100 m.

Brackish-water genera.

12: *Perissocytheridea* is a genus that was supposed to be restricted to tropical and subtropical America till Benson and Maddocks (1964) found it in the Knysna estuary of South Africa. According to Hartmann (1975), however, this species appears to be a somewhat abnormal representative. Diversification of the genus appears strongest in the Caribbean.

13: The distribution of *Cyprideis* has been discussed by several authors both on the Atlantic and African side of the Atlantic ocean. The eastern Atlantic species appear less diversified. Except for *Cyprideis similis* none of the species appears to occur on both sides, although some range over a considerable number of latitudes, possibly connected with dispersal by migrating waterfowl.

14: *Sulcostocythere* from South Africa appears to replace *Cyprideis* locally (Hartmann, 1975) in brackish-water environment; Masoli (1975), however, mentions it from the Ivory Coast.

Mediterranean forms in Africa.

15: Among these are *Chrysocythere*, *Ruggieria*, *Carinocythereis* and possibly *Mutilus*. They do not extend across the Atlantic although a species, tentatively identified as *Ruggieria* has been reported from Recent Caribbean waters (van den Bold, 1966c, Teeter, 1975). This species, however, is rather different from the Mediterranean forms (much stronger reticulation). *Carinocythereis* may be related to *Cativella* through common ancestors. *Paijenborchellina* (L. Eocene and Recent of Nigeria) has also not been found so far on this side of the Atlantic.

Endemic or provincial forms.

16: Africa: *Reymentia*, *Dahomeyella*, *Sulcostocythere*, also *Isohabrocythere*, *Isobuntonia*, *Evisceratocythere*, *Ovocytheridea*, *Nigeroloxoconcha*. Some of these have been reported from North Africa (Libya, Algeria, Egypt).

17: Southern South America: *Wichmannella*, *Huantraiconella*, *Protocosta*, *Neoveenia* have not been reported from other areas. However, forms resembling *Protocosta* occur in northern South America and Africa. *Mesocythere* and *Pericythere* also occur in the Caribbean area. A new species of *Australicythere* has been described from the Miocene of southern Brazil. Although indicated in text-fig. 6, it is still too early to draw any conclusions. Similar is the case of *Patagonocythere*.

18: Also of interest is the restriction of the genus *Ruttenella* to the Eocene of the southern Caribbean (van den Bold, 1960).

19: Puzzling is the distribution of *Cobanocythere* (originally described from the Pacific coast of Central America, Hartmann, 1959) in West Africa (Hartmann, 1975), but also mentioned by Reys from the Mediterranean and by Shornikov (1975) from Japan, while Siddiqui (1971) questionably assigned a species from the early Tertiary of Pakistan to this genus.

Deep-water Ostracoda.

20: Distribution of deep-water ostracodes in the Caribbean area (Table 5) affords some comparison with that in deep-sea cores. Unfortunately the study of the latter has so far been confined to the genera *Bradleya*, *Agrenocythere*, *Poseidonamicus* and *Abyssocythere* (Benson, 1971, 1972). In the Caribbean, *Agrenocythere* is known from Eocene to late Miocene, and Recent specimens have been found in deep-water cores in the Gulf of Mexico. In the DSDP cores *Agrenocythere hazelae* is known from Oligocene to early Miocene and from Pleistocene from around 30°S latitude (text-fig. 1). Recent specimens have been found in the North Atlantic, the Gulf of Mexico and the Pacific (Benson, 1972). Its predecessor *A. antiquata* is found in the Eo-

510

cene of the Caribbean, the eastern North Atlantic and Italy.

Abyssocythere trinidadensis is so far only known in the Caribbean, but a related species (*A. atlantica*) occurs in Quaternary cores in the Central Atlantic (Text-fig. 1).

The majority of specimens grouped under *Bradleya* ex gr. *dictyon* (Table 5) really belongs to that species, but they are accompanied over their entire stratigraphic range by smaller specimens, which might represent a shallow-water variant. For distribution of these genera the reader is referred to Benson (1971, 1972).

So far nothing has been published on Recent representatives of *Abyssocypris* and *Cardobairdia*. The latter has been found in water in excess of 900 m depth in the Gulf of Mexico. Although deep-water species of *Krithe*, *Cytherella*, *Argilloecia*, *Bythocypris* and *Bairdia* are well known, both from Recent and fossil material, a study of their distribution would require a more detailed investigation than can be undertaken here. It may be mentioned, however, that two species of *Bairdia*, common to deep-water deposits of the Tertiary of the Caribbean, are represented by identical or at least strongly related forms in the Gulf of Mexico, typically at depths between 900 and 1200 meters. It is interesting to note the similarity of Maddock's

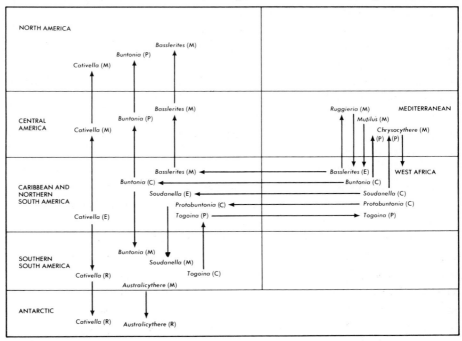

Text-fig. 6: Apparent direction of dispersal of some selected Ostracode genera in Africa and the Americas during the Cenozoic. (R) Recent, resp. Plio-Pleistocene; (M) Miocene; (E) Eocene; (P) Paleocene.

species *Bairdoppilata* sp. 3, from the Mozambique channel (1860 - 2980 m).
In the Caribbean Pleistocene (Atlantic 71A-4) *Ambocythere caudata*, (Recent,
North Atlantic), *Trachyleberis bermudezi* aff. *crebripustulosa* (identical to
Recent form from Gulf of Mexico), *Bythoceratina scabra* and *Zabythocypris
heterodoxa* were found.

Benson (1975) has argued on the basis of the deep-sea ostracodes, that
in late Eocene-early Oligocene times a change in faunal composition took
place, which may be correlated with lowering of temperature (and other
changes) in the deep-sea watermass (origin of the psychrosphere). As there
is a notable lack of data on shelf faunas of that time, no effect on
shallow-water faunas can be observed. In Trinidad, the only place where
upper Paleogene deep-water faunas have been studied, no significant change
can be detected, although minor changes - mainly on the species level -
take place, beginning in the late Eocene (Table 5).

Conclusion

Through the Paleocene and Eocene we see a fairly close relationship between
South American and West Central African ostracode faunas. This connection
could have been through fairly shallow water, but more likely was affected
by dispersal of species attached to plant material drifting mostly in east-
west direction. Of interest here is the almost complete lack of relationship between East and West African faunas (Dingle, 1976). After the Oligocene, most dispersal took place along the Atlantic coasts in north and
south directions.

BIBLIOGRAPHY

In order to keep the bibliography within reasonable dimensions, I have omitted all references to Recent fresh-water ostracodes and Recent Myodocopida, and placed the only two on Tertiary fresh-water ostracodes in brackets in the bibliographic index. This index shows that, whereas there are eight publications on the Paleogene of Central Africa, there are only two on the Neogene and four on Recent ostracodes. For southern South America these figures are 9, 7 and 4; in Central South America 0, 0, 8 (a paper by Sanguinetti on the Miocene of the Pelotas basin, southern Brazil, should be in press); in the Caribbean and northern South America the numbers are 13, 31 and 11 respectively. In the Atlantic basin itself 4 papers deal with Paleogene, Neogene and Recent ostracodes, and seven deal only with Recent ones. Finally, from the Antarctic part of the Atlantic 4 papers describe Recent ostracodes, none are on Tertiary forms. In the following index the abbreviations P: Palecgene, N: Neogene and R: Recent have been used.

Bibliographic index

AFRICA   Central: Apostolescu, 1961 (P); Bold, 1966 (N), 1968c (N) (Grekoff, 1958) (P,N); Keen, 1975 (N,R); Kogbe et al., 1975 (P); Masoli, 1975 (R); Omatsola, 1970a (R), 1970b (R), 1970c (R), 1971 (R), 1972 (R); Reyment, 1959 (P), 1960 (P), 1963 (P), 1964 (P), 1965 (P), 1966 (P); Reyment and Reyment, 1959 (P).

    South: Benson and Maddocks, 1964 (R); Hartmann and Hartmann-Schröder, 1975 (R); Klie, 1940 (R), Müller, 1908 (R).

SOUTH AMERICA
    South: Becker, 1964 (R); Bertels, 1966a (P), 1968b (P), 1969a (P), 1969b (P), 1969c (P), 1973 (P), 1975 (P,N); Malumian, 1970 (N); Ramírez, 1967 (R); Ramírez and Moguilevsky, 1971 (R); Rossi de Garcia, 1966a (P), 1966b (N), 1967a (P), 1967b (N), 1967c (N), 1971 (N), 1972 (?); Whatley and Cholich, 1974 (R), Whatley and Moguilevsky, 1975 (R).

    Central: Brady, 1880 (R); Hartmann, 1955a (R), 1955b (R), 1956 (R), 1957 (R), 1959 (R); (Jones, 1968) (N); Macedo, 1965 (R); Pinto and Ornellas, 1965 (R), 1970 (R).

CARIBBEAN AND NORTHERN
SOUTH AMERICA   Baker and Hulings, 1966 (R); Bold, 1946 (P,N), 1950a (P,N), 1950b (P,N), 1957a (P), 1957b (N), 1958a (R), 1958b (N), 1958c (N), 1960a (P), 1960b (P), 1961 (P,N), 1963a (N), 1963b (R), 1963c (P,N), 1963d (N), 1965a (R), 1965b (P,N), 1966a (N), 1966b (N), 1966c (R), 1966d (N), 1966e (N), 1966f (N), 1966g (N), 1966i (P, N), 1967a (N), 1967b (N), 1968a (N), 1968b (N), 1969a (N), 1969b (P,N), 1970a (N), 1970b (N), 1971a (P,N),

1971b (N), 1971c (N), 1972a (P,N), 1972b (N), 1972c
(N), 1973 (N), 1971c (N), 1972a (P,N), 1973b (N),
1972c (N), 1973 (P,N), 1974a (N), 1974b (N), 1974c (N),
1975a (N); Brady, 1870 (R); Brönnimann and Rigassi,
1963 (P), Drooger, 1960 (P), Drooger and Kaasschieter,
1958 (R); Keij, 1954 (R); Klie, 1939a (R), 1939b (R);
Lubimova and Sanchez, 1974 (P,N); Noordermeer and Wagner, 1969 (P); Pokorný, 1968 (R); Teeter, 1975 (R).

ATLANTIC   Benson, 1969, 1971, 1972; Benson and Sylvester-Bradley,
                    1971; Brady, 1880 (R), 1887 (R), 1911 (R); Hartmann,
                    1959a (R); Maddocks, 1969 (R); Puri, 1967 (R), 1971
                    (R).

ANTARCTIC   Brady, 1907 (R), Neale, 1967 (R), Scott, 1899 (R),
                    1912 (R).

Bibliography

Apostolescu, V., 1961. Contribution à l'étude paléontologique (ostracodes)
    et stratigraphique des bassins crétacés et tertiaires de l'Afrique.
    Rev. Inst. franç. pétrole et ann. combust. liq., v. 16, no. 7-8, p.
    779-867.
Baker, J. H. and Hulings, N. C., 1966. Recent Marine Ostracod Assemblages
    of Puerto Rico. Publ. Inst. Mar. Sci., Texas, v. 11, p. 108-125.
Becker, D., 1964. Micropaleontología del superpatagoniense de las locali-
    dades Las Cuevas y Monte Entrance. Ameghiana, v. 3, no. 10, p. 319-
    340.
Benson, R. H., 1969. Preliminary Report on the study of Abyssal Ostracoda.
    In: Neale, J. W. Tax. Morph. and Ecol. Rec. Ostr., Hull, Oliver &
    Boyd Edinburgh, p. 475-478.
_____, 1971. A New Cenozoic Deep-Sea Genus Abyssocythere (Crustacea, Os-
    tracoda: Trachyleberididae), with Descriptions of five new species.
    Smithsonian Contr. Paleobiol., no. 7, p. 1-20.
_____, 1972. The Bradleya Problem, with descriptions of two new Psychro-
    spheric Ostracode Genera, Agrenocythere and Poseidonamicus (Ostracoda:
    Crustacea). Smithsonian Contr. Paleobiol. no. 12, p. 1-138.
_____, 1975. The origin of the psychrosphere as recorded in changes of
    deep-sea ostracode assemblages. Lethaia, vol. 8, p. 69-83.
Benson, R. H. and Maddocks, R. F., 1964. Recent Ostracodes of Knysna estu-
    ary, Cape Province, Union of South Africa. Univ. Kansas Pal. Contr.,
    Art. 5, p. 1-39.
Benson, R. H. and Sylvester-Bradley, P. C., 1971. Deep-sea ostracodes and
    the transformation of ocean to sea in the Tethys. Bull. Centre Rech.
    Pau-SNPA, v. 5 suppl., p. 63-91.
Berggren, W. A. and Hollister, C. D., 1974. Paleogeography, paleobiogeo-
    graphy and history of the circulation in the Atlantic Ocean. In:
    W. W. Hay (edit.) Studies in paleo-oceanography. Soc. Econ. Paleont.
    Mineral. sp. publ. no. 20, p. 126-186.
Bertels, A., 1968a. Huantraiconella n. gen. (Ostracoda, Buntoniinae) del
    Tertiario inferior (Daniano) de Argentina. Ameghiana, v. 5, no. 8,
    p. 252-6.
_____, 1968b. Micropaleontología y estratigrafía del limite cretetácico-
    tertiario en Huantrai-co (Provincia de Neuquén). Ibid., v. 5, no. 8,
    p. 279-295.
_____, 1969a. Estratigrafía del limite Cretácico-Terciario en Patagonia
    septentrional. Rev. Asoc. Geol. Argent., v. 24, no. 1, p. 41-54.

Bertels, A., 1969b. "Rocaleberidinae", nueva subfamilia ("Ostracoda, Crustacea") del limite Cretácico-Tertiario de Patagonia septentrional. Ameghiniana, v. 6, no. 2, p. 146-171.

_____, 1969c. Micropaleontología y estratigrafía del limite Cretácico-Tertiario en Huantrai-Co (Provincia de Neuguén). Ameghiniana, v. 6, no. 4, p. 253-280.

_____, 1973. Ostracodes of the type locality of the Lower Tertiary (lower Danian) Rocanian Stage and Roca Formation of Argentina. Micropal., v. 19, no. 3, p. 308-340.

_____, 1975. Ostracode ecology during the upper Cretaceous and Cenozoic in Argentina. Bull. Amer. Paleont. v. 65, no. 282, p. 317-341.

Bold, W. A. van den, 1946. Contribution to the study of Ostracoda with special reference to the Tertiary and Cretaceous microfauna of the Caribbean Region. Diss. Utrecht Univ., DeBussy, Amsterdam, 167 pp.

_____, 1950a. Miocene Ostracoda from Venezuela. Jour. Paleont., v. 24, no. 1, p. 76-88.

_____, 1950b. A checklist of Cuban Ostracoda. Ibid. v. 24, no. 1, p. 107-109.

_____, 1957a. Ostracoda from the Paleocene of Trinidad. Micropal., v. 3, no. 1, p. 1-18.

_____, 1957b. Oligo-Miocene Ostracoda from southern Trinidad. Ibid. v. 3, no. 3, p. 231-254.

_____, 1958a. Distribution of freshwater ostracodes in Trinidad. Ibid. v. 4, no. 1, p. 71-74.

_____, 1958b. Ostracoda of the Brasso Formation of Trinidad. Ibid. v. 4, no. 4, p. 391-418.

_____, 1958c. *Ambocythere*, a new genus of Ostracoda. Ann. Mag. Nat. Hist., s. 12, v. 10, p. 801-813.

_____, 1960a. Eocene and Oligocene Ostracoda of Trinidad. Micropal., v. 6, no. 2, p. 145-196.

_____, 1960b. Determination of ostracodes, in: Pessagno, E. A., Stratigraphy and micropaleontology of the Cretaceous and Lower Tertiary of Puerto Rico. Micropal., v. 6, no. 1, p. 87-110, Ostracoda, p. 93; in: Mattson, P. H., Geology of the Mayaguez area, Puerto Rico. Bull. Geol. Soc. Am. v. 71, p. 310-362, Ostracoda, p. 347.

_____, 1961. Some new Ostracoda of the Caribbean Tertiary. Proc. Kon. Nederl. Akad. Wetensch. ser. B., v. 64, no. 5, p. 627-638.

_____, 1963a. The ostracode genus *Orionina* and its species. Jour. Paleont., v. 37, no. 1, p. 33-50.

_____, 1963b. Anomalous hinge structure in a new species of *Cytherelloidea*. Micropaleontology, v. 9, no. 1, p. 75-78.

_____, 1963c. Ostracods and the Tertiary stratigraphy of Guatemala. Bull. Amer. Assoc. Petr. Geol., vol. 47, no. 4, p. 696-698.

_____, Upper Miocene and Pliocene Ostracoda of Trinidad. Micropaleontology, v. 9, no. 4, p. 361-424.

_____, 1965a. *Pseudoceratina*, a new genus of Ostracoda from the Caribbean. Kon. Nederl. Akad. Wetensch., Proc., ser. B, v. 68, no. 3, 160-164.

_____, 1965b. Middle Tertiary Ostracoda from northwestern Puerto Rico. Micropaleontology, v. 11, no. 4, p. 381-414.

_____, 1966a. Ostracoda from the Pozón section, Falcón, Venezuela. Jour. Paleont., v. 40, no. 1, p. 177-185.

_____, 1966b. New species of the ostracode-genus *Ambocythere*. Ann. Mag. Nat. Hist., ser. 13, v. 8, no. 1, p. 1-18.

_____, 1966c. Ostracoda from Colon Harbour, Panama. Carib. Jour. Sci., v. 6, no. 1/2, p. 43-53.

_____, 1966d. Ostracode zones in Caribbean Miocene. Bull. Amer. Assoc, Petr. Geol. v. 50, no. 5, p. 1029-1031.

Bold, W. A. van den, 1966e. Miocene and Pliocene Ostracoda from Northeastern Venezuela. Verh. Kon. Nederl. Akad. Wetensch., ser. 1, v. 23, no. 3, p. 1-43.

_____, 1966f. Upper Miocene Ostracoda from the Tubará Formation (northern Colombia). Micropaleontology, v. 12, no. 3, p. 360-364.

_____, 1966g. Ostracoda from the Antigua Formation (Oligocene, Lesser Antilles). Jour. Paleont., v. 40, no. 5, p. 1233-1236.

_____, 1966h. Les ostracodes du Néogène du Gabon. Rev. Inst. franç. du Pétr. v. 21, no. 2, p. 155-176.

_____, 1966i. Répartition de certains ostracodes dans le Tertiaire des Caraïbes. 3rd sess. (Bern) Comm. Medit. Neogene Stratigraphy, p. 134-139.

_____, 1967a. Miocene Ostracoda from Costa Rica. Micropaleontology, v. 13, no. 1. p. 75-86.

_____, 1967b. Ostracoda of the Gatún Formation, Panama. Ibid. v. 13, no. 3, p. 306-318.

_____, 1968a. Distribution of Trachyleberidinae (Ostracoda) in the Neogene of the Caribbean. 4th sess. Comm. Medit. Neog. Strat., Proc. p. 55-66.

_____, 1968b. Ostracoda of the Yague Group (Neogene) of the northern Dominican Republic. Bull. Amer. Paleont., v. 54, no. 239, 106 pp.

_____, 1968c. Ostracodes du Néogène du Gabon et de l'Italie. Rev. Inst. franç. Pétr. v. 23, no. 10, p. 1327-1328.

_____, 1969a. Neogene Ostracodes from Southern Puerto Rico. Carib. Jour. Sci., v. 9, no. 3/4, p. 117-125.

_____, 1969b. *Messinella*, a new genus of Ostracoda in the Caribbean Cenozoic. Micropaleontology, v. 15, no. 4, p. 397-400.

_____, 1970a. The genus *Costa* (Ostracoda) in the Upper Cenozoic of the Caribbean Region. Ibid. v. 16, no. 1, p. 61-75.

_____, 1970b. Ostracoda of the Lower and Middle Miocene of St. Croix, St. Martin and Anguilla. Carib. Jour. Sci., v. 10, no. 1/2, p. 35-52.

_____, 1971a. Distribution of ostracodes in the Oligomiocene of the Northern Caribbean. Trans. 5th Carib. Geol. Conf., Bull. no. 5, p. 123-128.

_____, 1971b. Ostracoda of the Coastal Group of Formations of Jamaica. Trans. Gulf Coast Assoc. Geol. Soc. v. 21, p. 325-348.

_____, 1971c. Ostracode associations, salinity and depth of deposition in the Neogene of the Caribbean Region. Bull. Centre Rech. Pau-SNPA, 5 sup. p. 449-460.

_____, 1972a. Ostrácodos de; Post-Eoceno del Venezuela y regiones vecinas. Congr. Geol. Venezolano, v. 2, mem. 4, Sp. publ. 5, p. 999-1063.

_____, 1972b. Contribution of Ostracoda to the correlation of Neogene formations of the Caribbean Region. Mem-Trans. 6th Carib. Geol. Conf., p. 485-490.

_____, 1972c. Ostracoda of the La Boca Formation, Panama Canal zone. Micropaleontology, v. 18, no. 4, p. 410-442.

_____, 1973. Distribution of Ostracoda in the Oligocene and Lower and Middle Miocene of Cuba. Carib. Jour. Sci., v. 13, no. 3/4, p. 145-159.

_____, 1974a. Ostracode associations in the Caribbean Neogene. Verh. Naturf. Gesell. Basel, v. 84, no. 1, p. 214-221.

_____, 1974b. Ornate Bairdiidae in the Caribbean. Geosci. Man, v. 6, p.29-40.

_____, 1974c. Taxonomic status of *Cardobairdia* van den Bold, 1960, and *Abyssocypris* n. gen., two genera of deep-water ostracodes from the Caribbean Tertiary. Ibid., v. 6, p. 65-79.

_____, 1975a. Neogene biostratigraphy (Ostracoda) of southern Hispañola. Bull. Amer. Paleont., v. 66, no. 286, p. 549-625.

_____, 1975b. Ostracodes from the late Neogene of Cuba. Ibid., vol. 68, no. 289, p. 121-167.

516

Bold, W. A. van den, 1975c. Distribution of the *Radimella confragosa* Group
(Ostracoda, Hemicytherinae) in the late Neogene of the Caribbean.
Jour. Paleont., 49, no. 4, p. 692-701.

Brady, G. S., 1870. Description of Ostracoda, In: De Folin et Périer:
Les Fonds de la Mer, v. 1, p. 177-256.

_____, 1880. Report on the Ostracoda dredged by HMS Challenger during the
years 1873-1876. Report on the scientific results of the voyage of
HMS Challenger, Zool. v. 1, pt. 3, p. 1-184.

_____, 1887. Description of Ostracoda from the explorations of the "Tra-
vailleur" and the "Talisman". In: De Folin et Périer: Les Fonds de
la Mer, v. 4, p. 164-226.

_____, 1907. Crustacea, Pt. 5: Ostracoda. National Antarctic Expedition.
Nat. Hist., 1901-1904, v. 3, p. 1-9.

_____, 1911. Notes on marine Ostracoda from Madeira. Proc. Zool. Soc.,
London, 1911, p. 595-601.

Brönnimann, P. and Rigassi, D., 1963. Contribution to the geology and paleontology of the area of the City of La Habana, Cuba and its surroundings. Ecl. Geol. Helv. v. 56, no. 1, p. 193-480.

Dingle, R. V., 1976. Palaeogene ostracods from the continental shelf off
Natal, South Africa. Trans. Roy. Soc. S. Africa, vol. 42, pt. 1, p.
35-79.

Drooger, C. W., 1963. Microfauna and age of the Basses Plaines Formation
of French Guayana, I. and II. Ostracoda, p. 460-468; Kon. Nederl.
Akad. Wetensch., Proc. ser. B., v. 63, no. 4.

Drooger, C. W. and Kaasschieter, J. P. H., 1958. Foraminifera of the
Orinoco-Trinidad-Paria shelf. Rep. Orinoco Shelf Exp., v. 4. Verh.
Kon. Nederl. Akad. Wetensch., afd. Natuurk. ser. 1, v. 22. Ostracoda,
p. 88-92.

(Grékoff, N., 1958. Ostracodes du Bassin de Congo. III: Tertiaire. Ann.
Mus. Roy. Congo-Belge, v. 22, p. 5-36.)

Hartmann, Gerd, 1955. Neue marine Ostracoden der Familie Cypridae und der
Subfamilie Cytherideinae der Familie Cytheridae aus Brasilien. Zool.
Anz. v. 154, no. 5/6, p. 109-127.

_____, 1956. Weitere marinen Ostracoden aus Brasilien, In: Beiträge zur
neotropischen Fauna, p. 19-62 by E. Titchak and H. W. Koepeke, Gustav
Fischer Verlag, Jena.

_____, 1959a. Neue Ostracoden von Teneriffa. Zool. Anz. v. 162, no. 5/6,
p. 160-171.

_____, 1959b. Ostracodas de las agunas salobres de America latina y su
significación para la investigación palaeontológica. Act. y Trab.
primer Congr. Sudameric. zoologia (La Plata), v. 2, p. 169-174.

Hartmann, G. and Hartmann Schröder, G., 1975. Zoogeography and biology of
littoral Ostracoda from South Africa, Angola and Mozambique. Bull.
Amer. Paleont., v. 65, no. 282, p. 353-367.

(Jones, T. R., 1860. Fossil Entomostraca from Montserrate (Brazil), p. 266-
268 in: Alport, S., On the Discovery of some Fossil Remains near
Bahia in South America with Notes on the Fossils by Sir P. G. Egerton,
Prof. J. Morris and T. R. Jones. Quart. Jour. Geol. Soc. London, 1860,
pt. 1, p. 263-268.)

Keen, M. C., 1975. Some *Ruggieria*-like ostracods from the Tertiary and Recent of West Africa. Proc. 5th Afr. Coll. Micropal. Rev. Esp. Micropal., ser. 7, no. 3, p. 451-464.

Keij, A. J., 1954. Ostracoda: Identification and description of species
in: Van Andel and Postma: Recent sediments of the Gulf of Paria.
Reports Orinoco shelf Exp., vol. 1. Verh. Kon. Nederl. Akad. Wetensch.
afd. Natuurk. ser. 1, v. 20, no. 5, p. 218-231.

Keij, A. J., 1976. Note on *Havanardia* and *Triebelina* species (Ostracoda) Kon. Nederl. Akad. Wetensch. Proc. ser. B, vol. 70, no. 1, p. 36-44, pls. 1, 2, 4 textfigs.

Klie, W., 1939a. Ostracoden aus den marinen Salinen von Bonaire, Aruba und Curaçao. Capita Zool., v. 8, pt. 4, p. 1-19.

_____, 1939b. Brackwasserostracoden von Nordostbrasilien. Zool. Jahrb. Abt. Syst., v. 72, p. 359-372.

_____, 1940. Ostracoden von der Küste Deutsch-südwest Afrikas. Beitr. Fauna Eulitorals von d. SW Afrika. Kieler Meeresf., v. 3, no. 2, p. 404-448.

Kogbe, C. A., Mehes, K., Le-Calvez, Y., and Grekoff, N., 1975, Micro-bio-stratigraphy of lower Tertiary sediments from the south-eastern flanks of the Iullemmeden basin (northwest Nigeria). Proc. 5th Afr. Coll. Micropal., Rev. Esp. Micropal., ser. 7, no. 3, p. 523-538.

Krömmelbein, Karl, 1975. Remarks on marine Cretaceous Ostracodes of Gondwanic distribution. Proc. 5th Afr. Coll. Micropal., Rev. Esp. Micropal., ser. 7, no. 3, p. 539-551.

Macedo, Antonio C. Magãlhaez, 1965. As microfaunas do sambaqui de Sernam-betiba e do litoral de Magé, Est. Rio de Janeiro. Min. Min. e Energ. Dept. Nac. Prod. min., div. Geol. e Min., Notas preliminares e estudos no. 128, p. 2-63.

Maddocks, Rosalie F., 1969. Revision of Recent Bairdiidae (Ostracoda). U. S. Nat. Mus., Nat. Hist. Bull., v. 295, p. 1-126.

Malumian, Norberto, 1970. Bioestratigrafía del Terciario Marino del sub-suelo de la provincia de Buenos Aires (Argentina). Ameghiniana, v. 7, no. 2, p. 173-204.

Masoli, M., 1975. Ostracofaunes récentes du plateau continental de la côte d'Ivoire en tant qu'indicateurs écologiques. Proc. 5th Afr. Coll. Micropal., Rev. Esp. Micropal., ser. 7, no. 3, p. 623-633.

McKenzie, K. G., 1973. Cenozoic Ostracoda, in: Hallam A. (edit.), Atlas of Palaeobiogeography, Elsevier Publ., p. 477-487.

Müller, G. W., 1908. Die Ostracoden der deutschen Südpolar-Expedition 1901-1903, im Auftrag des Reichamtes des Inneren, herausgegeben von E. von Drygalski., vol. 10, Zool., no. 2, p. 51-181.

Neale, J. W., 1967. An ostracod fauna from Halley Bay, Coats Land, British Antarctic Territory. Brit. Antarct. Surv., Sci. Rep. no. 58, p. 1-50.

Noordermeer, Elly, J. and Wagner, C. W., 1969. Preliminary note on the ostracod fauna of the boring Alliance 28 in Surinam (Dutch Guiana). Geol. en Mijnb., v. 28, no. 2, p. 163-164.

Omatsola, M. Ebi., 1970a. Podocopid Ostracoda from the Lagos Lagoon, Ni-geria. Micropaleontology, v. 16, no. 4, p. 407-445.

_____, 1970b. On Structure and Morphologic Variation of Normal Pore System in Recent Cytherid Ostracoda (Crustacea), Acta Zool., v. 51, p. 115-124.

_____, 1970c. On the occurrence of Cytherellidae (Ostr. Crust.) in a brackish-water Environment. Bull. Geol. Inst. Univ. Uppsala, NS, v. 2, p. 91-96.

_____, 1970d. Notes on three new species of Ostracoda from the Niger Delta, Nigeria. Ibid., v. 2, p. 97-102.

_____, 1971. *Campylocythereis*, a new genus of the Campylocytherinae (Ostr., Crust.) and its muscle scar variation. Bull. Centre Rech. Pau-SNPA, v. 5, suppl., p. 101-123.

_____, 1972. Recent and Subrecent Trachyleberididae and Hemicytheridae (Ostr. Crust.) from the western Niger delta, Nigeria. Bull. geol. Inst. Uppsala, NS., vol. e, p. 37-120.

_____, 1972. Recent and Subrecent Trachyleberididae and Hemicytheridae

518

Pinto, Irajá Damiani and Ornellas, Lilia Pinto de, 1965. A new brackish-
water Ostracode *Cyprideis riograndensis* Pinto and Ornellas from southern Brazil and its ontogenetic carapace development. Esc. Geol. Univ.
do Rio Grande do Sul, publ. esp. no. 8, p. 7-23.
_____, 1970. A new brackishwater ostracode *Perissocytheridea krömmelbeini*
Pinto and Ornellas, sp. nov. from southern Brazil. Ibid. Publ. esp.
no. 2), p. 1-19.
Pinto, Irajá Damiani and Sanguinetti, Yvonne, T., 1962. A complete revision of the genera *Bisulcocypris* and *Theriosynoecum* (Ostracoda) with
the world geographical and stratigraphical distribution (including
Metacypris, Elpidium, Gomphocythere and *Cytheridella*). Ibid. publ.
esp. no. 4, 97 pp.
Pokorný, Vladimir, 1968. *Havanardia,* g. nov., a new genus of the Bairdii-
dae (Ostracoda, Crust.). Věst. ústr. úst. geol. v. 43, no. 1, p. 61-
63.
Puri, H. S., 1967. Ecologic distribution of Recent Ostracoda. Proc. Symp.
Crustacea, pt. 1, p. 457-495.
_____, 1971. Distribution of ostracodes in the oceans. The Micropalaeon-
tology of Oceans, Cambr. Univ. Press, p. 163-169.
Ramírez, F. C., 1967. Ostrácodos de Lagunas de la Provincia de Buenos
Aires. Rev. Mus. de la Plata, N.S., secc. Zool., v. 10, p. 5-54.
Ramírez, F. C. and Moguilevsky, A., 1971. Ostrácodos planktónicos hallados
en aguas oceánicas frente a la Provincia de Buenos Aires. Res. XLI
comissao oceanografica Costa Sul. Physis, v. 30, no. 81, p. 637-666.
Reyment, R. A., 1959. Die Ostracodengattung *Paijenborchella* im Unterozän
Nigeriens. Acta Univ. Stockholm., Stockholm Contr. in Geol., v. 3,
no. 7, p. 139-143.
_____, 1960. Notes on the Cretaceous-Tertiary boundary in Nigeria. Rep.
21st Sess. Int. Geol. Congr. Norden, pt. 5, sect. 3, proc.: The Cretaceous-Tertiary boundary, p. 131-135.
_____, 1963. Studies on Nigerian Upper Cretaceous and Lower Tertiary Os-
tracoda Pt. 2: Danian, Paleocene and Eocene Ostracoda. Acta Univ.
Stockholm. Stockholm Contr. in Geol., v. 10, p. 1-286.
_____, 1964. Biostratigraphie et micropaléontologie du Paléogène de la
Nigeria occidentale. Mem. B. R. G. M., v. 28, no. 2, p. 839-847.
_____, 1965. Quantitative morphologic variation and classification of some
Nigerian Paleocene Cytherellidae. Micropaleontology, v. 11, no. 4,
p. 457-465.
_____, 1966. Studies on Nigerian Upper Cretaceous and Lower Tertiary Os-
tracoda, Pt. 3: Stratigraphical, Paleoecological and biometrical con-
clusions. Acta Univ. Stockholm., Stockholm Contr. Geol., v. 14, 151 p.
Reyment, R. A. and Elofson, O., 1959. Zur Kenntnis der Ostracodengattung
Buntonia. Ibid., v. 3, no. 9, p. 157-164.
Reyment, R. A. and Reyment, E., 1959. *Bairdia ilaroensis* s. nov. aus dem
Paleozän Nigeriens und die Gültigkeit der Gattung *Bairdoppilata* (Ostr.
Crust.). Ibid., v. 3, no. 2, p. 59-68.
Reyment, R. A. and Van Valen, L., 1969. *Buntonia olokunudui* sp. nov. (Os-
stracoda, Crustacea). Ibid. N.S. v. 1, no. 3, p. 83-94.
Rossi de García, Elsa, 1966a. Sobre la presencia del genero *Cytheridea* en
la depresión de el Sampal (Chubut). Rev. Geol. Argent. v. 2, no. 2,
p. 118.
_____, 1966b. Contribución al conocimiento de los ostrácodos de la Argen-
tina I: Formación Entre Ríos, de Victoria, provincia de Entre Ríos.
Ibid. v. 21, no. 3, p. 194-208.
_____, 1967a. Un nuevo género de ostrácodos de la familia Trachyleberidi-
dae. Ibid. v. 22, no. 1, p. 95-98.

_____, 1967b. Contribución al conocimiento de los ostrácodos de la Argentina II: Ostrácodos del cordón litoral Loma de Tajamar. Ibid. v. 22, no. 3, p. 203-208.

_____, 1969. Algunos ostrácodos del Enterriense de Paraná, provincia de Entre Ríos. Ibid. v. 24, no. 3, p. 267-280.

_____, 1971. Ostracodes du Miocène de la République Argentina. Act. 4e Coll. Afr. Micropal., p. 391-431.

_____, 1972. Cuvillierina, nuevo genero de ostrácodos. Rev. Esp. Micropal., v. 4, no. 1, p. 23-26.

Scott, Th., 1899. Report on the marine and freshwater crustacea from Franz Joseph Land, collected by Mr. W. S. Bruce of the Jackson-Harmsworth Expedition. Linn. Soc. Journ. Zool. London, v. 27, p. 60-126.

_____, 1912. The Entomostraca of the Scottish National Antarctic Expedition. Trans. Roy. Soc. Edinburgh, v. 48, pt. 3, p. 521-600.

Teeter, J. W., 1975. Distribution of Holocene Ostracoda from Belize. In: Kenneth F. Wantland and Walter C. Pusey III: Belize Shelf-Carbonate Sediments, Clastic sediments and Ecology. Amer. Assoc. Petr. Geol, Studies in Geology, No. 2, p. 400-499.

Whatley, R. C. and Cholich, Teresa del Carmen, 1974, A new Quaternary ostracod genus from Argentina. Palaeontology, v. 17, no. 3, p. 669-684.

Whatley, R. C. and Moguilevsky, A., 1975. The family Leptocytheridae in Argentine waters. Bull. Amer. Paleont. v. 65, no. 282, p. 501-521.

CENOZOIC RADIOLARIANS OF THE ATLANTIC BASIN AND MARGINS

Richard E. Casey and Kenneth J. McMillen
Rice University, Houston, Texas

ABSTRACT

Living radiolarians in the Atlantic are indicative of specific water masses. Radiolarian skeletons occur only in the upper few centimeters of late Neogene sediments over most of the mid and lower latitudes of the Atlantic. This is in contrast to common occurrences of radiolarians in the sediments of these latitudes during much of the Paleogene and early Neogene. Paleogene to early Neogene radiolarian biostratigraphies are cosmopolitan due to the pan-tropical oceanic connection. Isolation of the Atlantic starting in Mid-Miocene resulted in changes in siliceous sedimentation and the development of provincial radiolarian faunas. The development of a new cosmopolitan radiolarian zonation based on tropical submergent forms may aid in interoceanic warm water correlations.

RÉSUMÉ

Les radiolaires actuels de l'Atlantique caractérisent des masses d'eau déterminées. Les squelettes de radiolaires se trouvent seulement dans les quelques centimètres supérieurs des sédiments du Néogène supérieur dans la plupart des latitudes moyennes et inférieures de l'Atlantique. Ceci forme un contraste avec leur abondance relative dans les sédiments de ces latitudes durant la plus grande partie du Paléogène et du Néogène Inférieur. Les biostratigraphies des radiolaires entre le Paléogène et le Néogène Inférieur sont cosmopolités à cause de la communication océanique pan-tropicale. A partir du milieu du Miocène l'isolement de l'Atlantique cause des changements dans le sédimentation de la silice et provoque le développement de faunes provinciales de radiolaires. Le développement, pour les radiolaires, une biostratigraphie nouvelle et cosmopolitaire basée sur les formes tropicales submergées aidera peut-être à déterminer des corrélations interocéaniques entre les régions tropicales.

RADIOLARIAN PRESERVATION IN THE ATLANTIC THROUGHOUT THE CENOZOIC AND ITS IMPLICATIONS

The Paleogene and early Neogene Atlantic appears to be favorable to radiolarian preservation as are the other major ocean basins of that time. A major change in radiolarian preservation occurred during the mid-Miocene (or at about _Dorcadospyris alata_ Zone time). From the mid-Miocene to the present radiolarians have been only rarely and sporadically preserved in Atlantic basin proper, and its marginal basins, from about 40 degrees north, although poleward of each of these latitudes radiolarians are commonly being preserved today as they have been throughout much of the Cenozoic. This pattern is unlike the Pacific pattern that appears to have remained relatively constant throughout the Cenozoic with radiolarian oozes forming underneath the equatorial divergences, radiolarian-diatom oozes forming poleward of the polar convergences and relatively high radiolarian contents (sometimes becoming radiolarian oozes) underneath the eastern and western boundary current systems. A major change seen in Pacific radiolarian sedimentation appears to be the increase in radiolarian (and siliceous forms in general) sediments beginning in the mid-Miocene (at the same time the radiolarians were dropping out of the picture in the Atlantic) and continuing but decreasing somewhat in magnitude to the present. Kobayashi (1944) has noticed an alternation of radiolarian rocks in that when they exhibit a great development in the eastern Asiatic region, time equivalent rocks in the Australian region from Paleozoic to Tertiary are radiolarian poor. Therefore the explanation of the decline of radiolarian sedimentation in the Atlantic and the intensification of radiolarian sedimentation in the Pacific (probably at the expense of radiolarian sedimentation in the Atlantic) during the Neogene may well have had earlier counterparts.

Although radiolarians are essentially absent from Neogene mid and low latitude Atlantic sediments they are quite common and in many cases the dominant biogenic element in many Paleo-

Figure 1 TECTONIC AND OCEANIC EVENTS IN THE ATLANTIC AND THEIR POSSIBLE EFFECTS ON RADIOLARIANS AND SILICEOUS SEDIMENTATION

EVENT	POSSIBLE EFFECT
1. Development of diatoms and rapid increase in diversity and "abundance" during the Cenozoic	1. Radiolarians adapt to development of diatoms and competitive pressure for silica by using less silica in test.
2. Separation of Australia and Antarctica about Late Paleogene (about 35 m.y.a.) (Kennett et al., 1972)	2. Development of a circum-Antarctic current resulting in thermal isolation of Antarctica and increased glaciation in Antarctica. Paleogene "Antarctic Bottom Water" formed and flows into Atlantic through newly-formed pass in the Rio Grande-Walvis Ridge changing the mid and low latitude Atlantic from a carbonate sink or lagoonal-type basin (that it had been during most of the Paleogene) to a silica sink or estuarine-type basin (with both silica and carbonate being deposited) similar to the present day Pacific. This persisted until early Miocene.
3. Closure of eastern Tethys (Africa-Arabia joined to S.W. Asia) in early Miocene about 18 to 20 m.y.a. and closure of western tethys (Europe-Africa join at Gibraltar) in mid-Miocene about 12-14 million years ago (Berggren and Van Couvering, 1974) and elevation of Panamanian block to about 1500 meters (Bandy and Casey, 1973)	3. Atlantic isolated from equatorial Indian Ocean and considerably restricted from the equatorial Pacific, resulting in the development of Neogene water mass regimes and radiolarian faunas endemic to these water masses (such as those referred to in living radiolarians in this paper). Periods of glaciation due to development of meridional circulation and periods of production of Mediterranean intermediate water resulting in saline waters reaching the Antarctic and lowering the freezing point depression and resulting in "interglacials" and periods of no generation of Mediterranean Intermediate Water resulting in Mediterranean evaporites and "freshening" of Antarctic waters and resulting "glacials". This period is the period of change from an Atlantic silica sink (estuarine type basin) to a carbonate basin (lagoonal type) which has persisted to the present. (Ryan, et al, 1974)
4. Elevation of the Panamanian Block to effective sill, stopping communication between the equatorial Atlantic and Pacific completely at about the Miocene-Pliocene boundary (4 to 5 m.y.a.)	4. Development of provincial radiolarian faunas in the low latitude oceans with "leaking" of low latitude Pacific and Indian Ocean faunas into the Atlantic around the Cape of Good Hope and the retention of a relict radiolarian fauna in the equatorial Atlantic and its marginal seas (Gulf of Mexico and Caribbean). The almost complete closure at Panama intensifies the Gulf Stream, carrying warm water to high latitudes, allow island hopping of vertebrate faunas between North and South America, and results in increased precipitation and the development of northern hemisphere glaciation at about 3 m.y.a. Increased glaciation in the Pleistocene may result from the more complete restriction through Panama at that time.

gene and earliest Neogene sediments of the same area. Our research on this subject is still in its early stages, however
with the previous discussion as background the following para-
graph presents a possible explanation for the Cenozoic stratigraphic and geographic distribution of radiolarians in the
Atlantic.

Radiolarian sediments appear to have been a common constituent of the world ocean sediments of the Paleogene and early
Miocene including the Atlantic. In the mid-Miocene (at about
Dorcalospyris alata Zone time) the sediment deposited in the
mid and low latitude Atlantic became essentially devoid of
radiolarians except for sporadic occurrences and some "near
shore" (Calvert Formation) and marginal basin (European
stratotypes) occurrences. The main reasons for this change appear to be isolation of the Atlantic Basin, paleoclimatic
changes, the change of the Atlantic Basin from an "estuarine
type" basin to a "lagoonal type" basin (in Berger's terms,
Berger, 1970) and the rapid development of diatoms and resultant "competition" of diatoms and radiolarians for available
silica   (Harper and Knoll, 1975).

The proposed events responsible for or correlated and related to this change in much of the Atlantic from a silica sink
basin and a proposed history for the Atlantic related to radiolarian history are listed on Figure 1   Conditions for optimum
radiolarian preservation and optimum radiolarian dissolution
in sediments are suggested along with present day and past examples of   Figure 2

CENOZOIC RADIOLARIAN BIOSTRATIGRAPHY OF THE ATLANTIC

Riedel and Sanfilippo have developed a "warm water" radiolarian biostratigraphy for the Cenozoic using material mainly
from Pacific Deep Sea Drilling Project cores (Riedel and Sanfilippo, in press). This zonation appears to work quite well
in "warm water" Paleogene and early Neogene Atlantic sediments.

Figure 2   CONDITIONS FOR OPTIMUM RADIOLARIAN PRESERVATION AND OPTIMUM RADIOLARIAN DISSOLUTION

OPTIMUM PRESERVATION

Low Latitudes
1. High Productivity (Upwelling areas, under boundary currents and equatorial divergences)
2. Low Dissolution Potential of waters (Estuarine type waters)
3. Rapid radiolarian sedimentation (fecal pellet sedimentation)
4. Relatively low benthonic activity per unit time
5. Low dissolution potential in sediments (high silica content in sediments and interstitial waters plus low pH)
6. High rate of terrigenous sedimentation

High Latitude
1. Medium productivity (away from polar divergences)
2. Rapid radiolarian sedimentation (fecal pellet sedimentation)
3. Relatively low benthonic activity per unit time
4. Low dissolution potential in sediments (high silica content in sediments and interstitial water plus low pH)
5. High rate of terrigenous sedimentation

OPTIMUM DISSOLUTION

Low Latitudes
1. Low Productivity (within mid-latitude gyres)
2. High dissolution potential of waters (lagoonal type waters)
3. Low radiolarian sedimentation (individual test sedimentation)
4. Relatively high benthonic activity per unit time
5. High dissolution potential in sediments (low silica content in sediments and interstitial waters high pH)
6. Low rate of terrigenous sedimentation

High Latitudes
1. High productivity (development of centric diatoms to exclusion of shallow radiolarians (at polar divergences) or low productivity)
2. Low radiolarian sedimentation (individual test sedimentation)
3. Relatively high benthonic activity per unit time
4. High dissolution potential in sediments (low silica content in sediments and interstitial waters plus high pH)
5. Low rate of terrigenous sedimentation

526

Post early Neogene (or more specifically post about <u>Dorcadospy-ris alata</u> Zone) sediments containing significant radiolarians for biostratigraphic purposes are usually lacking. The few oc-currences of "radiolarian rich sediments", including those from the late Neogene stratotype localities (Riedel and Sanfilippo, in press) are difficult to fit into Riedel and Sanfilippo's zonation. The reason for this difficulty is most likely due to the "isolation" of the Atlantic and Pacific-Indian Ocean faunas as mentioned previously in this chapter. As mentioned earlier we suggest that the first occurrence datum planes of Riedel and Sanfilippo's Zonation (post Mid-Miocene) may be reliable mark-ers although they may lag in time somewhat behind their appear-ance in the Pacific-Indian Ocean. Nigrini (1971) developed a Quaternary radiolarian zonation for the equatorial Pacific. Some members she used as Quaternary first occurrence datums such as <u>Buccinosphaera invaginata</u> Haeckel and <u>Amphirhopalum ypsilon</u> Haeckel occur in the living plankton of the low lati-tude Atlantic and their first occurrences might also be used in Atlantic. However the last occurrence datums appear to be unreliable due to the "one way" leakage mentioned earlier. The finding of living <u>Spongaster pentas</u> (Riedel and Sanfilippo) and related forms in plankton tows and surface sediments of the low latitude Atlantic and its marginal seas attests to this unre-liability of last occurrence datums in the Atlantic.

Hays (1965), Hays and Opdyke (1967), Bandy et al (1971) have developed radiolarian zonations for the Antarctic which should be usable for the Antarctic and southern Atlantic por-tions of the Atlantic. Theyer and Hammond (1974) have related the radiolarian stratigraphy of Riedel and Sanfilippo (1971) with magnetostratigraphy for the Neogene in the equatorial Pacific. Casey (in press b) and Johnson and Knoll (1975) have related Nigrini's zonation to the paleomagnetic scale in cores from the equatorial Pacific. These radiolarian zonations, sig-nificant datum planes and their relations to magnetostratig-raphy are illustrated on Figures 9 and 10. The general outcome of this correlation suggests that the Paleogene and early Neogene of the "warm water" Atlantic can be reliably zoned using Riedel and Sanfilippo's Zonation. The warm water mid and late Neogene of the Atlantic exhibits provincialism and the high latitude South Atlantic can be reliably zoned within the late Neogene.

Plankton tows and Holocene sediment samples collected in the tropical Atlantic, Gulf of Mexico and Caribbean have yielded radiolarians previously believed to have been extinct, and in fact used in Cenozoic "warm-water" radiolarian biostratigraphic zonations. Radiolarians collected in plankton tows and stained with Rose Bengal include Spongaster pentas, Spongaster berminghami, and "circular" and "elliptical" spongodiscids.

The evolution of Spongaster pentas from Spongaster berminghami occurred about 4.5 million years ago in the tropical Pacific (Theyer and Hammond, 1974) and is used to define the base of the Spongaster pentas Zone (Riedel and Sanfilippo, in press). Spongaster berminghami apparently became extinct shortly after the evolution of Spongaster pentas, and Spongaster pentas became extinct about 3.6 million years ago in the Pacific (Casey, in press b). Specimens of "circular" and "elliptical" spongodiscids believed to be the ancestors of Spongaster berminghami have also been collected in the plankton.

These species represent a relict radiolarian fauna in the tropical Atlantic, Gulf of Mexico and Caribbean. Their presence suggests some interesting consequences of both biostratigraphic and paleooceanographic significance. Of biostratigraphic significance is the conclusion that the geologic and geographic ranges of some of the species used in Riedel and Sanfilippo's zonation are provincial (they have suggested this themselves) (Sanfilippo and Riedel, 1974). This provinciality is a real problem because the late Neogene part of their zonation was mainly developed using tropical Pacific cores, and the findings here suggest that the radiolarian biostratigraphy (and perhaps other microfossil biostratigraphies) in the Atlantic, and the stratotype localities of the late Neogene in Europe should be quite different from the "warm-water" Pacific zonation of Riedel and Sanfilippo (Riedel and Sanfilippo, in press). Correlation attempts of the Pacific and European stratotype radiolarian assemblages have met with limited success, probably due in part to the problem of provinciality herein mentioned.

This provinciality for the late Neogene has not been obvious before due to the fact that the sediments of the low-latitude Atlantic and its marginal seas are essentially void of radiolarians post mid-Miocene. The upper few centimeters of Holocene sediments in the Gulf of Mexico and Caribbean do contain radiolarians and among them Spongaster pentas, Spongaster berminghami, and "circular" and "elliptical" spongodiscids.

The paleooceanographic significance is perhaps even more important. The Atlantic and Pacific appear to exhibit more or less "cosmopolitan", "warm-water" radiolarian biostratigraphies up to at least mid-Miocene. (See Figure 3). Sometime post mid-Miocene there appears to have been a divergence in radiolarian faunas and a development of greater provincialism. The reasons for this divergence are apparently related to geographic and climatic isolation and resultant allopatric speciation and differential geologic ranges of these isolated populations.

The authors of this paper believe that the geographic isolation of the tropical Pacific and Atlantic was due to the events listed on Figure 1 and culminating in the uplift of the Panamanian Block during the Miocene (Bandy and Casey, 1973) to "effective sill" at about 4.5 million years ago. Isolation is placed at about 4.5 million years ago for prior to this time the Spongaster faunas of the Gulf of Mexico and Caribbean resemble those of the Pacific but diverge shortly thereafter. At 4.5 million years ago, or at about the Miocene-Pliocene

Figure 3 RIEDEL AND SANFILIPPO'S ZONATION RELATED TO MAGNETOSTRATIGRAPHY

	RIEDEL AND SANFILIPPO (IN PRESS)		THEYER AND HAMMOND (1974)
EPOCHS	ZONES	RADIOLARIAN EVENTS	ESTIMATED AGE (M.Y.) AND POLARITY POSITION
QUAT.	LAMPROCYRTIS HAYSI	L. NEOHETEROPOROS TO L. HAYSI	1.6-1.7 top of Olduvai
PLIOCENE	PTEROCANIUM PRISMATIUM	L.O. STICHOCORYS PEREGRINA	2.5-2.6 latest Gauss
	SPONGASTER PENTAS	S. BERMINGHAMI TO S. PENTAS	4.5-4.6 Gilbert "c2" event
MIOCENE LATE	STICHOCORYS PEREGRINA	S. DELMONTENSIS TO S. PEREGRINA	6.2-6.3 Epoch 6, above event "a"
	OMMATARTUS PENULTIMUS	L.O. OMMATARTUS HUGHESI	8.6-8.7 base of Epoch 8
	OMMATARTUS ANTEPENULTIMUS	C. PETTERSSONI TO O. HUGHESI	10.7-10.8 top of Epoch 11
MIOCENE MID	CANNARTUS PETTERSSONI	F.O. C. PETTERSSONI	11.3-11.4 base of Epoch 11
	DORCADOSPYRIS ALATA	D. DENTATA TO D. ALATA	15.5-15.6 late Epoch 16
	CALOCYCLETTA COSTATA	F. O. CALOCYCLETTA COSTATA	about 18 early Epoch 17
MIOCENE EARLY	STICHOCORYS WOLFFII	F.O. STICHOCORYS WOLFFII	
	STICHOCORYS DELMONTENSIS	L.O. THEOCYRTIS ANNOSA	
	CYRTOCAPSELLA TETRAPERA	F.O. C. TETRAPERA	
	LYCHNOCANOMA ELONGATA	F.O. L. ELONGATA	
OLIGOCENE	DORCADOSPYRIS ATEUCHUS	TRISTYLOSPYRIS TRICEROS TO D. ATEUCHUS	
	THEOCYRTIS TUBEROSA	LITHOCYCLIA ARISTOTELIS GROUP TO L. ANGUSTA	
EOCENE LATE	THYSOCYRTIS BROMIA	F.O. CARPOCANISTRUM AZYX	
	PODOCYRTIS GOETHEANA	F.O. PODOCYRTIS GOETHEANA GOETHEANA	
	PODOCYRTIS CHALARA	PODOCYRTIS MITRA TO P. CHALARA	
EOCENE MID	PODOCYRTIS MITRA	PODOCYRTIS SINUOSA TO P. MITRA	
	PODOCYRTIS AMPLA	PODOCYRTIS PHYXIS TO P. AMPLA	
	THYRSOCYRTIS TRIANCANTHA	F.O. EUSYRINGIUM LANGENA LAGENA	
EOCENE EARLY	THEOCAMPE MONGOLFIERI	F.O. THEOCAMPE MONGOLFIERI	
	THEOCOTYLE CRYPTOCEPHALA CRYPTOCEPHALA	THEOCOTYLE CRYPTOCEPHALA NIGRINIAE TO T. CRYPTOCEPHALA CRYPTOCEPTHALA	
	PHORMOCYRTIS STRIATA STRIATA	F.O. THEOCORYS ANACLASTA	
PALEOCENE	BURYELLA CLINATA	BURYELLA TETRADICA TO B. CLINATA	
	BEKOMA BIDARFENSIS	F.O. BEKOMA BIDARFENSIS	

boundary, the sill depth of the Panamanian Block would have been about 500 meters (Bandy and Casey, 1973). Therefore, the isolation may well be twofold; restricted circulation due to the emergence of the Panamanian Block, and cooling that resulted in the initiation and development of Neogene glaciations and water mass regimes (Casey, 1973).

Our current research suggests that water mass regimes and radiolarian faunas similar to today's were established by mid-Miocene and that Atlantic and Pacific warm-water faunas have been "isolated" from one another since about the base of the Spongaster pentas Zone, or about 4.5 million years ago, or about the Miocene-Pliocene boundary. This isolation is a leaky and "one way" isolation. The Atlantic still (since 4.5 million years ago) receives newly evolved warm water forms from the Pacific-Indian Ocean faunas via the filter bridge of the Cape of Good Hope (waters with temperature as high as 25°C come round the Cape presently and at warmer intervals in the past warmer faunas should have migrated into the Atlantic). However warm water radiolarian faunas evolving in the Atlantic, post closure of Panama, cannot migrate into the Pacific around the Tierra del Fuego.

Therefore it appears the "warm water" zonation of Riedel and Sanfilippo will work in part in the Atlantic, especially first occurrences may lag a bit in time having to await establishment in the Atlantic, most likely being established during world-wide warm intervals.

The R-mode cluster of live radiolarians from the South Texas outer continental shelf (Figure 2) separates the relict radiolarians from the others (they are not associated with any season and only associate at a low similarity level with anything). Spongaster pentas attaches at a low (and probably insignificant) level with the winter group which is interesting for it is within the winter group that Spongaster cruciferus associates. However Spongaster cruciferus associates at a

"high level" with a few others and again this high level is due
to few occurrences so this may be thrown out with more sampl-
ing. Spongaster? pentas, and the "circular" and "elliptical"
spongasters all cluster out together between the spring upwelling (SU) and summer (S) radiolarian assemblages.

We believe that this "throwing out" of the radiolarian seasonal cluster groups means that either the relict radiolarians
can get along with any group (which would be a way to survive)
or that they have an unspecialized niche (can eat a variety of
nanophytoplankton or are detritus feeders or contain symbiotic
zooxanthellae) and have been able to survive as the rest of
the populations have evolved "around them". Plankton tows
taken in the low latitude Atlantic in April of 1976 revealed
that the spongasters (Spongaster tetras, S. pentas, and the
"circular" and "elliptical" spongasters) appear red in fresh
live unstained material. We suggest that this red colour is
due to the presence of symbiotic algae and suggests a very unspecialized and perhaps a very conservative niche.

As has been noted many radiolarian datum planes and zonations are either provincial, such as Riedel and Sanfilippo's
"warm water" Cenozoic, and Nigrini's equatorial Quaternary
zonations, or are time transgressive, such as has been shown
by Petrushevskaya (1972). This is especially true for the late
Neogene of the Atlantic. Our studies suggest that at least
some radiolarians exhibit fairly consistent datum planes of a
more or less cosmopolitan nature and has been suggested by
Casey (Casey, in press b). These were apparently shallow living polar and/or temperate forms that were "endemic" to diving
water masses and exhibit tropical submergence. Living radiolarians exhibiting this type of distribution have been described by Casey (1971b and in press, a). Living radiolarians
exhibiting this type of distribution include Spongopyle
osculosa Dreyer, and Siphocampe erucosa Haeckel which appear
to be endemic to the diving Central Water Masses of the Pacific,

Cyrtopera laguncula Haeckel and *Cornutella profunda* Ehrenberg
which appear to be endemic to the diving Intermediate Water
Masses of the Pacific and *Peripyramis circumtexta* Haeckel
which appears to occur in both Central and Intermediate Water
Masses (Casey, 1971 and in press, a). Fossil radiolarians believed to have exhibited this type of distribution are *Lamprocyclas heteroporos* Haus probably exhibiting an endemic pattern
to the diving Central Water Masses. ?*Prunopyle titan* Campbell
and Clark and ?*Lychnocanium grande* Campbell and Clark are believed to be diving Intermediate Water Mass forms and *Eucyrtidium calvertense* s.s. Martin appears to have occurred in both
Central and Intermediate Water Masses due to its relatively
common occurrence in V-24-59 and its occurrence as far south
as *P. titan* in Weaver's cores (1973). These fossil forms may
well prove to exhibit "cosmopolitan" datum planes and a biostratigraphic zonation based on radiolarians of this type may
be the best cosmopolitan biostratigraphic zonation possible.
These fossil species are considered to exhibit this tropical
submergent pattern of distribution because their occurrence
in fossil sediments mirrors the occurrences of the living
tropical submergent forms in recent sediments (being dominant
in high latitutdes but rare in tropical sediments). Their
consistency as useful datum planes is suggested by comparing
their extinctions and evolutions in high (core E-14-8) and
low (core V-24-59) latitudes with magnetostratigraphy (as
shown on Figure 4). *Lamprocyclas heteroporos* and perhaps
Eucrytidium calvertense s.s. last occur at the Olduvai or
Gilsa event, and *Lychnocanium grande* and *Prunopyle titan*
consistently last occur within the upper Gauss; whereas
Lamprocyclas heteroporos apparently first occurs in a primitive form in the Gilbert. Using this same line of reasoning
it is possible to suggest that *Theocyrtis redonodoensis*
Campbell and Clark exhibited a bipolar pattern of distribution
instead of a tropical submergent pattern for it appears at

high and mid latitudes (the Antarctic and southern California)
during the late Neogene but not at low latitudes (such as core
V-24-59).

These ideas are still in an embryonic stage of development
and more work will have to be done to confirm or deny their
validity. However we suggest that the future use of this
"cosmopolitan radiolarian zonation" may be of great help in
correlating provincial radiolaria and other microfossil zonations.

A few selected examples of radiolarians used in the various
zonations mentioned are illustrated on plate 2 along with comments in the legend as to what zonations they apply.

ACKNOWLEDGEMENTS

Research on Recent radiolarian ecology and distribution in
the water column and sediments has been supported mainly by
the Oceanographic Section, National Science Foundation, NSF
Grant DES 74-21805 and in part by the Bureau of Land Manage-
ment Contract #08550-CTS-17. Research on the Cenozoic history
of the Atlantic and the paleooceanography and biostratigraphy
and evolution of radiolarians has been supported by a grant
from the Petroleum Research Fund of the American Chemical
Society, PRF #8657-AC 2 to which the authors are most grateful.
The authors also acknowledge the editorial and typing assistance of B. Hawkins, typing assistance of J. Cisneros, the research assistance of M. Bauer and J. Gevirtz, and abstract
translation by J.-C. De Bremaecker, all of the Geology Department, Rice University. Lamont-Doherty Geological Observatory
and Florida State University kindly provided core material.
Curating facilities at Lamont are supported by NSF grant GA
35454 and ONR contract N00012-67-A-0108-0004, and an NSF-
Office of Polar Programs grant GV-27549 to Florida State
University.

534

REFERENCES

Bandy, O. L., and Casey, R. E., 1973. Reflector horizons and paleobathymetric history, eastern Panama. Geol. Soc. Amer. Bull., v. 84, p. 3081-3086.

Bandy, O. L., Casey, R. E., and Wright, R. C., 1971. Late Neogene planktonic zonation, magnetic reversals, and radiometric dates, Antarctic to the tropics. In Reid, J. L., (Ed.), Antarctic Oceanography 1, Antarctic Research Series, v. 15: Washington (American Geophysical Union), p.1-26.

Berger, W. H., 1968, Radiolarian skeletons: solution at depth; Science, v. 159, p. 1237-1238.

_____, 1970, Biogenous deep-sea sediments: fraction - by deep-sea circulation. Geol. Soc. Amer. Bull., v. 81, no. 5, p. 1385-1402.

Berggren, W. A., and Van Couvering, J. A., 1974. The Late Neogene, Amsterdam (Elsevier Scientific Publishing Co.), p. 1-216.

Casey, R. E., 1971a. Distribution of polycystine radiolarians in the oceans in relation to physical and chemical conditions. In The Micropaleontology of Oceans. B. M. Funnell and W. R. Riedel (Eds.). Cambridge (Cambridge Univ. Press). 151.

_____, 1971b. Radiolarians as indicators of past and present water masses. In The Micropaleontology of Oceans. B. M. Funnell and W. R. Riedel (Eds.). Cambridge (Cambridge Univ. Press). 331.

_____, 1973, Radiolarian evidence for the initiation and development of neogene glaciations and the neogene water mass regimes: (discussion paper in) Geol. Soc. Amer. Meet., Dallas, Texas, Geol. Soc. Am. Ann. Meeting, Dallas, 1973, Abstracts, p. 570-571.

_____, in press. a. The ecology and distribution of recent Radiolaria. In Ramsay, A.T.S. (Ed.) Oceanic micropaleontology.

_____, in press. b. Late Neogene radiolarian biostratigraphy related to magnetostratigraphy, Polar to Tropics: O. L. Bandy Memorial Volume.

Cifelli, R., and Sachs, K. N., 1966. Abundance relationships of planktonic Foraminifera and Radiolaria. Deep-Sea Res., v. 13, p. 751-753.

Edmund, J. M. and Anderson,G. C., 1970. On the structure of North Atlantic Deep Water; Deep-Sea Res., v. 18, p. 127-138.

Fanning, K. A., and Schink,D. R., 1969. Interaction of marine sediments with dissolved silica; Limnol. and Oceanogr., v. 14, p. 59-68.

Goll, R. M. and Björklund, K. R., 1971. Radiolaria in surface sediments of the North Atlantic Ocean. Micropaleont. 17, (4), 434-457.

Goll, R. M. and Bjørklund, K. R., 1974. Radiolaria in surface
    sediments of the South Atlantic; Micropaleontology, v. 20,
    no. 1, p. 38-75.
Grant, A. B., 1968. Atlas of oceanographic sections, Davis
    Strait-Labrador Basin-Denmark Strait-Newfoundland Basin:
    1965-1967; Atlantic Ocean Lab, Bedford Inst. Rept., AOL
    v. 68, no. 5, 80 p.
Harper, H. E., Jr. and Knoll, A. H., 1975. Silica, diatoms,
    and Cenozoic radiolarian evolution. Geology, vol. 3, no.
    4, 175-177.
Hays, J. D., 1965. Radiolaria and late Tertiary and Quater-
    nary history of Antarctic Seas. Biology of Antarctic Seas
    II. Am. Geophys. Union, Antarctic Research Ser. 5, 125.
Hays, J. D. and Opdyke, N. D., 1967, Antarctic Radiolaria,
    Magnetic Reversals, and Climatic Change. Science 158,
    (3804), 1001.
Johnson, T. C., 1974. The dissolution of siliceous microfos-
    sils in surface sediments of the eastern tropical Pacific.
    Deep-Sea Res., v. 21, p. 851-864.
Johnson, D. A., and Knoll, A. H., 1975. Absolute ages of
    Quaternary radiolarian datum levels in the equatorial
    Pacific. Quaternary Research 5, p. 99-110.
Kennett, J. P., Burns, R. E., Andrews, J.E., Churkin, H. Jr.,
    Davis, T. A., Dumitrica, P., Edwards, A. R., Galehouse,
    J.S., Packham, G. H., and van der Lingen, G. J., 1972,
    Australian-Antarctic Continental Drift, Palaeocirculation
    Changes and Oligocene Deep Sea Erosion. Nature, vol. 239,
    p. 51-55.
Kobayashi, T., 1944. Reciprocal development of radiolarian
    rocks as between Asiatic and Australian sides. Proc.
    Imp. Acad. Tokyo, vol. 21, no. 4, p. 234-238.
McMillen, K. J., 1975. Quaternary lebensspuren and their re-
    lationship to depositional environments in the Caribbean
    Sea, the Gulf of Mexico, and the eastern and central
    North Pacific, M. A. thesis, Rice University, Houston.
_____, 1976. Ecology, distribution and preservation
    of Polycystine Radiolaria in the Gulf of Mexico and Carib-
    bean Sea, Ph. D. thesis, Rice University, Houston.
Nigrini, C. A., 1971, Radiolarian zones in the Quaternary of
    the Equatorial Pacific Ocean. In Funnell, B. M. and Reidel
    W. R. (Eds.). The Micropaleontology of Oceans: Cambridge
    (Cambridge Univ. Press), p. 443-461.
Petrushevskaya, M. G., 1971. Spumellarian and nassellarian
    Radiolaria in the plankton and bottom sediments of the
    Central Pacific. In The Micropalaeontology of Oceans
    (B.M. Funnell and W.R. Riedel - Eds.) Cambridge University
    Cambridge, 309-317.
_____, 1972. Biostratigraphy of deep-water
    Quaternary deposits from radiolarian analysis. Oceanology,
    v. 12, no. 1, 57-70.
Riedel, W.R. and Sanfilippo, A., 1971. Radiolaria. In

Winterer, E. L., Riedel, W. R., et al. Initial Reports of the Deep Sea Drilling Project, Volume 7; Washington (U.S. Government Printing Office), p. 1529.

Riedel, W. R. and Sanfilippo, A., in press, Cenozoic Radiolaria. In Ramsay, A.T.S. (Ed.) Oceanic micropalaeontology.

Ryan, W.B.F., Cita, M.B., Dreyfus Rawson, M., Burckle, L. H., and Saito, T. 1974. A paleomagnetic assignment of Neogene stage boundaries and the development of isochronous datum planes between the Mediterranean, the Pacific and Indian Oceans in order to investigate the response of the World Ocean to the Mediterranean "salinity crisis". Riv. Ital. Paleont., vol. 80, pp. 631-688.

Sanfilippo, A., Burckle, L.H., Martini, E., and Riedel, W.R., 1973, Radiolarians, diatoms, silicoflagellates and calcareous nannofossils in the Mediterranean Neogene: Micropaleontology, v. 19, no. 2, p. 209-234.

Theyer, F., and Hammond, S.R., 1974, Paleomagnetic polarity sequence and radiolarian zones, Brunhes to polarity epoch 20: Earth Planet. Sci. Lett., v. 22, p. 307-319.

Weaver, F.M., 1973, Pliocene paleoclimatic and paleoglacial history of East Antarctica recorded in deep sea piston cores: Sedimentology Research Laboratory, Department of Geology, Florida State Univ., Contrib. 36.

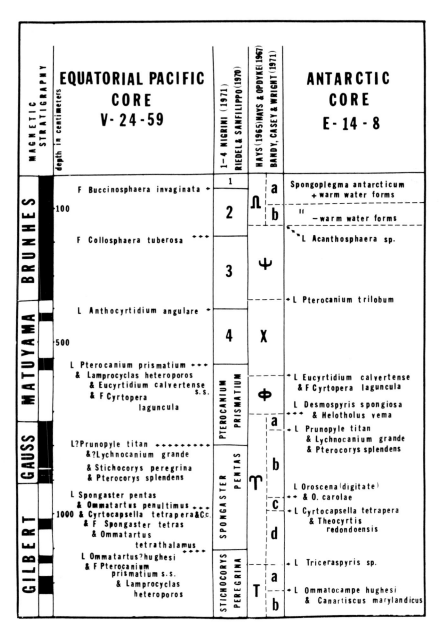

Fig. 4. Radiolarian biostratigraphy.

PLATE 1

Radiolarians indicative of specific water masses from the Caribbean Sea, Gulf of Mexico, and Atlantic Ocean. The black bar
equals about 100 microns. The letter p indicates specimens
from plankton tows, the letter s indicates individuals from
sediment samples.
Surface Water - Figures 1 to 5
    Figure 1 Pterocorys zancleus (Muller) s; Figure 2 Pteroca-
    nium praetextum praetextum (Ehrenberg) s; Figure 3 Spong-
    aster tetras tetras Ehrenberg s; Figure 4 Choenicosphaera
    murrayana Haeckel p with protoplasm; Figure 5 Euchitonia
    elegans (Ehrenberg)s

Surface Water, relict fauna - Figures 6 to 7
    Figure 6 "Circular" spongodiscid s; Figure 7 Spongaster
    pentas Riedel and Sanfilippo s

Subtropical Underwater - Figures 8 to 12
    Figure 8 Clathrocanium diadema s; Figure 9 Lithelius minor
    Jorgenson p; Figure 10 Spongotrochus glacialis Popofsky s;
    Figure 11 Amphirhopalum ypsilon Haeckel s; Figure 12 Sty-
    lochlamidium asteriscus Haeckel p

Antarctic Intermediate Water - Figures 13 to 17
    Figure 13 Actinomma sp. s; Figure 14 Cubotholus c.f. rhom-
    bicus Haeckel s; Figure 15 Carpocanium evacuatum Haeckel
    s; Figure 16 Tholonid p; Figure 17 Dictyophimus crisiae
    Ehrenberg s

North Atlantic Deep Water - Figures 18 to 20
    Figure 18 Spongopyle osculosa Dreyer s; Figure 19 Ptero-
    corys bicornis Popofsky s; Figure 20 Clathrocorys sp. p

PLATE 2

Radiolarians used as datums in Riedel and Sanfilippos "warm
water" Cenozoic Pacific Zonation (Figures 1, 2, 3, 4, 5 and 8),
Nigrini's Quaternary Equatorial Pacific Zonation (Figures 6
and 7) and Casey's Developing Tropical Submergent Cosmopolitan
Zonation (Figures 9 through 15).

    Figure 1 Pterocanium prismatium Riedel, Core V-24-59, 1111
    cm. First occurrence datum for base of Spongaster pentas
    Zone of Riedel and Sanfilippo (in press)
    Figure 2 Ommatartus? hughesi (Campbell and Clark), Core
    V-24-59, 1191 cm.
    Figure 3 Ommatartus hughesi (Campbell and Clark), Core E-
    14-8, 1750-1752 cm. Last occurrence used in Riedel and
    Sanfilippo (in press) as datum for base of Ommatartus pen-
    ultimus Zone.
    Figure 4 Ommatartus penultimus (Riedel), Core V-24-59,
    1241 cm. Evolves within the lower portion of the Ommatar-
    tus penultimus Zone of Riedel and Sanfilippo (in press)
    Figure 5 Cyrtocapsella tetrapera Haeckel, Core E-14-8,
    1750-1752 cm. First occurrence defines the bottom of the
    new Cyrtocapsella tetrapera Zone of Riedel and Sanfilippo

(in press)

Figure 6 Amphirhopalum ypsilon Haeckel, Core V-24-59, ? cm.
This form of the species is indicative of the Amphirhopalum
ypsilon Assemblage Zone (Zone 3) of Nigrini (1971)

Figure 7 Collosphaera tuberosa Haeckel, Core V-24-59, ? cm.
Defines the bottom of Collosphaera tuberosa Concurrent
Range Zone (Zone 2) of Nigrini (1971)

Figure 8 Stichocorys peregrina (Riedel), Core V-24-59, 1008
cm. Evolutionary bottom (more than 50% of S. peregrina
when compared to ancestor S. delmontense) defines the base
of the Stichocorys peregrina Zone (Riedel and Sanfilippo,
in press), last occurrence defines bottom of Pterocanium
prismatium Zone (Riedel and Sanfilippo, in press). Appears
to be useful over wide latitudinal area.

Figure 9 Eucyrtidium calvertense Martin, Core #-14-8, 1750-
1752 cm. Last occurrence at Pliocene-Pleistocene boundary
in Antarctic (Hays, 1965) (Bandy et al. 1971). A probably
tropical submergent "cosmopolitan" species (Casey, in press
b)

Figure 10 Theocyrtis redondoensis Campbell and Clark, Core
E-14-8, 1728-1730 cm. Appears to exhibit a bipolar rather
than tropical submergent distributional pattern (Casey, in
press b)

Figure 11 ?Prunopyle titan Campbell and Clark, Core E-14-8,
809 cm. Found in the tropics, considered a tropical submergent form, its last occurrence is indicative of upper
Gauss magnetic time over large geographical area (Casey, in
press b)

Figure 12 Prunopyle titan Campbell and Clark, Experimental
Mohole, EM-8-15, 242-245 cm. The higher latitude "subspecies" of the low latitude tropical submergent "subspecies"
illustrated in Figure 11 of this plate.

Figure 13 Oroscena with digitate spines, Core E-14-8, 1750-
1752 cm. May be tropical submergent "cosmopolitan" species whose last occurrence is indicative of the upper Gilbert. May be good for "red clay" biostratigraphy.

Figure 14 Lychnocanium grande Campbell and Clark, Core E-
14-8, 1750-1752 cm. Considered tropical submergent form
commonly last occurring with Prunopyle titan in the upper
Gauss.

Figure 15 Lamprocylas heteroporos Hays, Core V-24-59, 809
cm. Considered tropical submergent whose range is apparently the range of the Pliocene.

PLATE 1

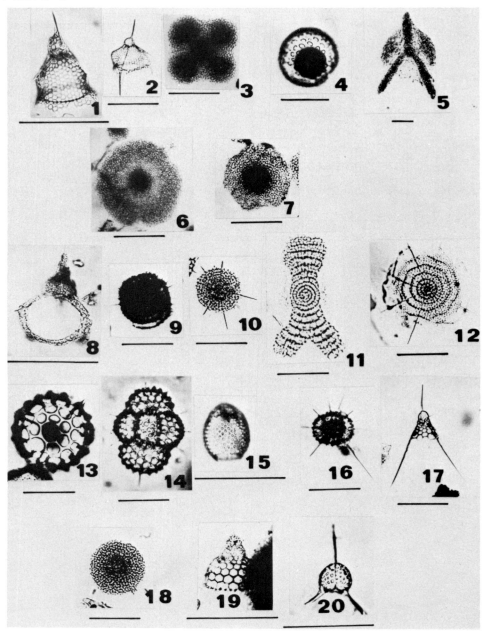

PLATE 2

Discussion

Dr. D. Habib: When you say the cold-water forms dive, are they diving alive or are the tests being transported?

Casey: Yes, they're diving alive. These organisms will take the protoplasmic stain in preserved plankton tows from these deeper waters.   Now, of course these water masses have taken hundreds of years to get from one place to another, so it's their great great great grandfather that was in shallow water.

Dr. I. G. Sohn: What animals eat radiolarians?

Casey: Well they're not tremendously abundant in the water column, the radiolarians, so I would think that any kind of a filter feeder would be important in eating them.   You do find them in fecal pellets of copepods and that sort of thing.   It appears that radiolarians are important intermediaries in the open ocean food web between nannoplankton and other microplankton and mesoplankton.

Sohn: The reason for the question is your statement that when you put radiolarians in acid and take them out they dissolve. Several years ago I collaborated with Kornicker in a study where we fed live ostracodes to fish and some of the valves developed holes.   Could digestive acid remove a protective coating thus causing the radiolarian to dissolve?

Casey: This might be; we've considered this.   One of the things that's interesting about many of the copepods, one of the main feeders on radiolarians, is that if they get a chance they'll act as gluttons, say when they're grazing on diatoms, and the fecal pellets will actually have the protoplasm still in the diatoms, even the chlorophyll, so they'll just go right through. So although many of them may be dissolved in this manner, or be etched and ready to dissolve, probably many will come through.   We think, therefore, that in general being eaten by something would be an aid, because you've got a sand-sized particle now, and its going to fall through the water in a couple days.   Based on our settling rate studies on the radiolarians we know that in average oceanic depths the average radiolarian falling individually through the water column takes about a year to hit bottom.   So it's the couple days compared to a year that would help out.

Dr. W. A. Berggren: Do you believe fecal pellet sedimentation is important in forming radiolarian oozes?

Casey: Yes, in general we think that the radiolarians are better preserved under places where they are eaten and then reach the bottom in fecal pellets.   In other words, fecal pellet sedimentation is one of the main ways we think we can get radiolarian occurrences in the sediments.   In areas where the radiolarians just die individually and fall through the water column those are the places where they will usually just dissolve away.   But the whole question of radiolarian sedimentation is one weak spot in radiolarian research.   We may see very fragile collosphaerids in the bottom sediments and even in sediments where other radiolarians have dissolved out.   In the Gulf of Mexico and Caribbean in general radiolarians occur in the upper 10 cm and then they're just gone as though there's perhaps a dissolution horizon rising through the sediment.   It may be that 10 cm is the depth of active burrowing by benthic organisms.

Berggren: I imagine the disappearance of your species Spongaster pentas is no doubt associated with the elevation of
the Isthmus of Panama, but when you say you're not sure about
how to date it I think you can date it pretty accurately.
You can do it with Foraminifera.

Casey: About 3.5 million years ago.

Berggren: Yes, but the reason you can do it is there are certain taxa which appear in both oceans which evolved about
4 million years ago. There's another taxon which evolved
around 3.5 million years ago and it's presently in the Atlantic but never occurs in the Pacific, so it had to happen
somewhere between that time. The genus Pulleniatina which
has little left and right coiling wiggles in the Atlantic
and Pacific through the Pliocene disappears in the Atlantic
at exactly 3.5 million years ago. It's no longer present
but it continues wiggling left and right in the Pacific
right up to the present day. It reappears in the Atlantic
at 2.3 million years and we don't know why that happens
but it probably came around the south part of the Agulhas
Current again somehow or other. But I think your Spongaster
pentas Zone the upper limit of which is, what, 4 million
years from the magnetics you have on a scale there? ...

Casey: Died out before the end of the Zone.

Berggren: Yes, it is probably related in some way to the
changing geometry of the Isthmus of Panama but that event
can be dated pretty accurately.

Casey: I think so, but I was concerned about suggesting that
at 3.5 million years ago an animal could walk across the
Isthmus. The elevating Isthmus has to be a lumpy sort of
thing, and there may have been an effective sill that stop
ed the connection of radiolarian populations; perhaps stop-
ed it for one group earlier than it did for another (a
shallow-living group vs. a deep-living group).

Dr. S. Gartner: I'll address this point both to you Dick and
to Bill Berggren, and let me say first of all I take except-
tion to your view with respect to the Isthmus of Panama.
About six or seven years ago we did a study of Blake Plateau sediments and established that the sedimentary regime
prevailing there at the present time, which is largely controlled by the Gulf Stream, was established something in
excess of 5 million years ago. This would indicate that the
Gulf Stream would have had its present configuration. The
Gulf Stream would be dependent on the Isthmus of Panama, and
if the latter did not come into existence until 3.5 million
years ago, necessarily the Gulf Stream would have been different at that time. But the Gulf Stream did establish its
present configuration at least over the Blake Plateau in excess of 5 million years ago. Now the evidence that you have
presented abd that Bill presented is suggestive that it
might have happened say 3.5 million years ago at presumably
about the time that Antarctic glaciation occurred. It seems
to me that you're not separating these two factors. It
could be that what you are looking at is the influence of
Antarctic glaciation all by itself being felt 3.5 million
years ago and that the Isthmus did in fact come into existence som time before that. Now perhaps I misunderstood
either or both of you. Do you have any positive evidence
that suggests that the Isthmus had to be in existence or

could have not been in existence until 3.5 million years ago?
Casey: In a paper that Bandy and I published in 1973, mainly
on benthonic forams, the elevation of it is shown. What I'm
thinking is that all of these can be correlated with the elevation of the Isthmus during the Miocene, say about the time
that the main elevation took place. I think that various
things happened because of the elevation; the shutting off
of the faunas as it reached effective sill depth, effective
sills for faunal isolation. It might have also reached some
effective sill depth earlier to divert and to intensify the
Gulf Stream because the latter is a very big and deep sort
of thing. So it may be that it was wide enough and deep
enough so that a major body of water was diverted north, before the shallow water radiolarian isolations took place. I
don't think they all have to happen at the same time. The
sort of sewuence I would like to see is that it comes up, it
starts to divert the Gulf Stream and then this intensifies,
with more elevation, the warm water pushed north causing
cooling in Ewing-Donn sense, and radiolarian paleotemperature curves that I've done suggest that it does get cold at
say 4 million years ago or thereabouts. But then you get
into Pleistocene glaciations and they're even colder; and
that might be due to even a more complete isolation and
deflection of water by the complete emergence of Panama.
Land faunas could move between North and South America by
hopping prior to this perhaps. It doesn't have to be a
complete and total emergence at an instant. I think it
came up and then I think the points you bring up are important ones. Then you've got to start wondering about eustatic
rises and falls of sea level, etc.

ATLANTIC CENOZOIC SILICOFLAGELLATES, POTENTIAL FOR
BIOSTRATIGRAPHIC AND PALEOECOLOGIC STUDIES

Richard E. Casey, Rice University, Houston, Texas

ABSTRACT

Silicoflagellates have been used only occasionally in the
Atlantic region mainly due to poor preservation in much of
this region. However, silicoflagellates do possess potential
for biostratigraphic and paleoecologic studies in the Atlantic.

RÉSUMÉ

Les silicoflagellés n'ont été utilisés que rarement dans
la région atlantique, principalement à cause de leur médiocre
état de conservation dans cette région. Néanmoins les silicoflagellés possèdent un potentiel à l'utilisation pour des
études biostratigraphiques et paléoécologiques dans l'Atlantique.

INTRODUCTION

Silicoflagellates evolved in the Cretaceous and have been
represented by about 12 genera and 50 species belonging to
two families. About 6 to 12 species and their varieties (perhaps amounting to 60 entities) belonging to about 3 to 6 genera (mainly the genera Dictyocha, Distephanus and Mesocena)
and one family live in the present-day seas. Silicoflagellates have been used in biostratigraphic and paleotemperature
studies mainly in the Pacific and Antarctic. Silicoflagellate studies in the Atlantic and its margins have been few
mainly because of the poor preservation of these fossils in
most of the Atlantic throughout most of the Cenozoic. Silicoflagellates are usually only common in rocks and sediments
containing large amounts of biogenic silica (Lipps, 1970) and
therefore are subject to some of the same preservation problems as are the radiolarians in the Atlantic (see Casey and

McMillen in this volume). The relatively low diversity of
silicoflagellates has and will limit their usefulness in biostratigraphic and paleoecologic studies; however, the following sections deal with their potential for such use (especially
for use in the Atlantic), and with a short review of important
silicoflagellate literature.

SILICOFLAGELLATE BIOSTRATIGRAPHY AND THEIR POTENTIAL USE IN THE ATLANTIC

Hanna (1928) suggested that silicoflagellates were excellent biostratigraphic indices. However, only within the last
25 years (and especially since the start of the Deep Sea Drilling Project) has this group been seriously considered for biostratigraphic correlation. Mandra (1960) recognized silicoflagellate complexes characteristic of Late Eocene, Miocene and
Pliocene ages using statistical analysis. Papp (1959) noted
that silicoflagellates can be easily used to distinguish Paleogene from Neogene deposits.

The presence of the silicoflagellate genus Lyramula appears
to be indicative of upper Cretaceous and does not penetrate
into the Cenozoic. The genera Corbisema, Naviculopsis and
Paradictyocha appear to be indicative of the Paleogene, whereas the genera Cannopilus and Rocella appear to be restricted
to the Miocene along with 5 other Miocene genera (Loeblich et
al, 1968) with the late Neogene being dominated by the genera
Dictyocha, Mesocena and Distephanus.

Martini (1971) set up the first silicoflagellate zonation
derived from continuous sequences. These were from DSDP sites
and Swedish Deep-Sea core 76 from the equatorial Pacific.
Bukry (1975) in alluding to the silicoflagellate zonation
"philosophies" states that "silicoflagellate zonation, unlike
the one employed for coccoliths that relies exclusively on
occurrence, requires quantitative data and more flexible

definition." Bukry continues stating "consistency and abundance of species are generally more significant in silicoflagellate zonation than are absolute first and last occurrences
-- the apparent skeletal plasticity of silicoflagellate species
reflecting high ecological responsiveness, also contributes to
a need for lowered reliance on first or last occurrences, in
favor of full-assemblage analysis and cosmopolitan-trend
determination -- part of this necessary caution results from
the present lack of multiple successions preventing determination of the regional versus cosmopolitan character of assemblages."

Of special interest in Bukry (1975) is his comparison of
tropical, cosmopolitan and nontropical silicoflagellate zonations. From this work and a rapid survey of other silicoflagellate biostratigraphic works a cosmopolitan zonation appears to function well for the Paleogene to mid-Miocene. Post
mid-Miocene the silicoflagellate floras and biostratigraphy
diverge into tropical and nontropical components. It is suggested that this divergence is caused by the same factors involved in the similar divergence in radiolarian faunas and
biostratigraphy at the same time (refer to paper by Casey
and McMillen in this volume for detailed discussion and to
their figure on tectonic and oceanic events and their effects
on Cenozoic siliceous sedimentation and radiolarians in the
Atlantic). It is therefore suggested that the silicoflagellate zonation of Bukry (1975) may be used in most of the
Atlantic for the Paleogene and early Neogene but that the
silicoflagellate floras of the Atlantic in the mid and late
Neogene exhibit provincialism (as do the floras referred to
during this time in Bukry (1975), and the Atlantic radiolarian faunas discussed in the chapter by Casey and McMillen in
this volume). Martini (Sanfilippo et al, 1973) attempted to
zone portions of the Mediterranean Neogene using his Neogene
silicoflagellate zonation established in the equatorial Pacific

(Martini, 1971). Martini was only able to place his samples of
Mediterranean in late Neogene tentatively but his samples from
mid-Miocene (about the Dorcadospyris alata Zone of Riedel and
Sanfilippo, in press) and older were zoned with apparent ease.
Here then it might be possible to suggest that the development of
a detailed warm water Atlantic Neogene silicoflagellate zonation is in order if enough material in continuous enough sections from mid and low latitudes can be found.

SILICOFLAGELLATE PALEOECOLOGIC STUDIES AND THEIR
POTENTIAL USE IN THE ATLANTIC

Colom (1952) attempted to reconstruct paleooceanographic
condition in Aquitanian-Burdigalian deposits in Spain. More
recently attempts have been made at using silicoflagellates
to determine paleotemperatures. Mandra and Mandra (1969) described a technique using the ratios of Dictyocha (warm) to
Distephanus (cold) to determine paleotemperatures in the Tertiary of the Antarctic. Weaver and Ciesielski (1974) suggest
coolings of 8 to 10°C in the Antarctic Pliocene. Earlier work
on this interval from cores in the Southern Ocean using ratios
of radiolarians suggested a drop of about 15°C (Bandy et al.,
1971). Apparently silicoflagellates can be used for paleotemperature analysis in warm waters also as has been done by
Martini (1971) for the Neogene of the equatorial Pacific.

Silicoflagellates have not been related to water mass distributions as have many of the other shelled nannoplanktonic
and microplanktonic groups, but a review of the distributions
of recent and fossil species published by Glezer (1970) suggests that these surface living forms exhibit both cosmopolitan
and provincial distributional patterns. For example Distephanus speculum (Ehrenberg) appears to occur in all oceans but
becomes rare nearshore where salinities approach 10 ppt. In
contrast Distephanus fibula Ehrenberg appears to be able to

exist at these lower salinities (Glezer, 1970). Therefore
silicoflagellates may be useful paleooceanographic indices
not only for paleotemperature but paleowater-mass, paleosalinity and other studies as well.

IMPORTANT SILICOFLAGELLATE LITERATURE WITH NOTES ON CONTENTS

Monographic works on silicoflagellates include: Glezer's
(1970) dealing with biology, systematics, distribution and
ecology and biostratigraphic and geologic occurrences, a must
for silicoflagellate work; Loeblich's et al. (1968) which is an
annotated index of fossil and recent silicoflagellates and
ebridians with descriptions and illustrations another must;
and Ling's (1972) work on upper Cretaceous and Cenozoic silicoflagellates and ebridians.

Some important recent publications concerned with silicoflagellate zonates include: Mandra (1951, 1954 and 1960) on
California material; Stradner (1961) on Oligocene silicoflagellates from Austria; Bachmann, Papp and Stradner (1963)
on Tertiary silicoflagellates of the Vienna basin; Bachmann
and Ichikawa (1962) on the silicoflagellates of the Neogene
of Japan; Zhuze (1949, 1951, 1955) on Upper Cretaceous and
Paleogene of the eastern slope of the Urals and the West
Siberian plain; Glezer (1970); and especially the works published in the Deep Sea Drilling Project too numerous to mention here except for the paper by Dumitrica (1972) on the
Mediterranean.

Some important recent publications concerned with silicoflagellate ecology, paleoecology and paleooceanography include:
Mandra and Mandra (1969), Jendrzejewski and Zarillo (1971),
Martini (1971), Ciesielski and Weaver (1973) and Weaver and
Ciesielski (1973 and 1974) on silicoflagellate paleotemperatures. Glezer (1970) gives considerable detail on the distribution and ecology of living species and refers to many

valuable references on these aspects.

Bukry is the most prolific silicoflagellate biostratigraph-
er widely published in the Deep Sea Drilling Project volumes.
In his DSDP Leg 34 paper (Bukry, 1976) he gives a brief but
excellent discussion of silicoflagellate stratigraphy, paleoecology and terminology of silicoflagellate morphology plus
some important references not given here.

ACKNOWLEDGEMENTS

Research on this paper has been supported by the Oceanographic
Section, National Science Foundation, NSF Grant DES 74-21805
and by a grant from the Petroleum Research Fund of the American
Chemical Society, PRF #8657-AC 2 to which the author is most
grateful. The authors also acknowledge the editorial and typing assistance of B. Hawkins, typing assistance of J. Cisneros,
and abstract translation by J. C. DeBremaecker, all of the
Geology Department, Rice University.

REFERENCES

Bachmann, A. and Ichikawa, W., 1962. The silicoflagellides in
    the Wakura Beds, Nanao City, Prefecture Ishikawa, Japan:
    Kanazawa Univ. Sci. Rept., v 8, p. 161.
Bachmann, A., Papp, A., and Stradner, H., 1963. Mikropaläontolo-
    gische Studien im "Badener Tegel" von Frattingsdorf N.O. -
    Mitt. geol. Ges. Wien, v. 56, no. 1, p. 117-210.
Bandy, O. L., Casey, R. E., and Wright, R. C., 1971. Late Neo-
    gene planktonic zonation, magnetic reversals, and radio-
    metric dates, Antarctic to the tropics. In Reid, J. L.,
    (Ed.), Antarctic Oceanography 1, Antarctic Research Series,
    v. 15: Washington (American Geophysical Union), p. 1-26.
Bukry, D., 1975. Silicoflagellate and coccolith stratigraphy,
    Deep Sea Drilling Project Leg 29. In Kennett, J. P.,
    Houtz, R. E., et al., Initial Reports of the Deep Sea Dril-
    ling Project, Volume 29: Washington (U.S. Government Prin-
    ting Office), p. 845-872.
    _____, 1976. Silicoflagellate and coccolith stratigraphy,
    southeastern Pacific Ocean, Deep Sea Drilling Project Leg
    34. In Yeat, R. S., Hart, S. R., et al., Initial Reports
    of the Deep Sea Drilling Project, Volume 34: Washington
    (U.S. Government Printing Office), p. 715-735.

Ciesielski, P. F. and Weaver, F. M., 1973. Southern Ocean
     Pliocene paleotemperatures based on silicoflagellates from
     deep-sea cores: Antarctic J. U. S., v. 8, no. 5, p. 295-
     297.
Colom, G., 1952. Aquitan - burdigalien Diatom Deposits of the
     North Betic Strait. J. Paleont., v. 26, no. 6, p. 867-885.
Dumitrica, P., 1972. Miocene and Quaternary silicoflagellates
     in sediments from the Mediditerranean Sea. In Ryan, W. B.
     F., Hsu, K. J. et al., Initial Reports of the Deep Sea
     Drilling Project, Volume 13: Washington (U. S. Government
     Printing Office), p. 902.
Glezer, Z. I., 1966. Silicoflagellatophyceae. In Gollerbakh,
     M. M. (Ed.), Cryptogamic plants of the U.S.S.R.: Akad.
     Nauk SSSR, V. A. Komarova Bot. Inst. (Translated from Rus-
     sian by Israel Program for Scientific Translations Ltd.,
     Jerusalem, 1970), v. 7, p. 1-363.
Hanna, G. D. 1928, Silicoflagellatae from Cretaceous of Cali-
     fornia, J. Paleont., vol. 1, no. 4, p. 259-264.
Jendrzejewski, J. P. and Zarillo, G. A., 1971. Late Pleisto-
     cene paleotemperatures: Silicoflagellates and foramini-
     feral frequency changes in a Subantarctic deep-sea core:
     Antarctic J. U.S., v. 6, p. 178-179.
Ling, H. Y., 1972. Upper Cretaceous and Cenozoic silicoflagel-
     lates and ebridians: Am. Paleontol. Bull. v. 62, p. 135-
     229.
Lipps, J. H., 1970. Ecology and evolution of silicoflagel-
     lates. Proc. North Amer. Paleo. Conv. Sept. 1969. PARTG.
     p. 965-993.
Loeblich, A. R., III, Loeblich, L. A., Tappan, H., and Loeblich,
     A. R., Jr., 1968. Annotated index of fossil and Recent
     silicoflagellates and ebridians with descriptions and il-
     lustrations of validly proposed taxa: Geol. Soc. Am.,
     Mem. 106, 319 p.
Mandra, Y. T., 1951. Preliminary stratigraphic report on some
     California Eocene Silicoflagellates. Bull. Geol. Soc. Am.,
     Part 2, v. 62, no. 12, p. 1523.
_____, 1954. Silicoflagellata, a new tool for the
     Geologist. Bull. Geol. Soc. Am., Part 2, v. 65, no. 12,
     p. 1396.
_____, 1960. Fossil silicoflagellates from California,
     U.S.A. rept. of Twenty-First Session Intern. Geol. Con-
     gress, Part 6, Proc. Sect. 6, Pre-Quatern. Micropaleontol.,
     p. 77-89.
Mandra, Y. T. and Mandra, H., 1969. Silicoflagellates: A new
     tool for the study of Antarctic Tertiary climates: Ant-
     arctic J. U.S., v. 4, p. 172-174.
Martini, E., 1971. Neogene silicoflagellates from the equa-
     torial Pacific. In Winterer, E. L., Riedel, W. R., et al.,
     Initial Reports of the Deep Sea Drilling Project Volume
     VII: Washington (U.S. Government Printing Office), p.
     1695-1708.

552

Papp, A., 1959. Handbuch der stratigraphischen Geologie. vol.
    3 no. 1, p. 328.
Riedel, W. R. and Sanfilippo, A., in press. Cenozoic Radio-
    laria. In Ramsay, A.T.S. (Ed.), Oceanic micropalaeonto-
    logy.
Sanfilippo, A., Burckle, L. H., Martini, E., and Riedel, W. R.,
    1973. Radiolarians, diatoms, silicoflagellates and calcar-
    eous nannofossils in the Mediterranean Neogene: Micropal-
    eontology, v. 19, p. 209.
Stradner, H., 1961. Uber fossile Silicoflagelliden und die
    Möglich-keit ihrer Verwendung in der Erdölstratigraphie:
    Erdöl und Kohle, v. 14, no. 2., p. 87-92.
Weaver, F. M., and Ciesielski, P. F., 1973. Pliocene paleo-
    climatic history recorded in antarctic deep sea cores.
    Geological Society of America, Abstracts with Programs:
    p. 856-857.
_____, 1974. Pliocene paleo-
    temperatures and regional correlations, southern ocean.
    Ant. Jour. of the United States, v. 9, no. 5, p. 251-253.
Zhuze, A. P., 1949. New Diatoms and Silicoflagellates from
    Upper Cretaceous Argillaceous Sands of the Basin of Bol'
    shoi Aktai River (Eastern Slope of the Northern Urals). --
    Botanicheskie Materialy Otdela Sporovykh Rastenii Botani-
    cheskogo Instituta AN SSSR, 6(1/6): 65-78, Moskva-Lenin-
    grad. (in Russian)
_____, 1951. Diatoms and Silicoflagellates of Upper
    Cretaceous of the Northern Urals. -- Botanicheskie Materi-
    aly Otdela Sporovykh Rastenii Botanicheskogo Instituta, AN
    SSSR, Vol. 7: 42-65, Moskva - Leningrad. (in Russian)
_____, 1955. Silicoflagellates of the Paleogene. --
    Botanicheskie Materialy Otdela Sporovykh Rastenii Bota
    nicheskogo Instituta, AN SSSR, Vol. 10: 77-81. Moskva -
    Leningrad. (in Russian)

North American Microtektites, Radiolarian Extinctions and the
Age of the Eocene-Oligocene Boundary

B. P. Glass and Michael J. Zwart, Geology Department, University of Delaware, Newark, DE 19711

Abstract

North American microtektites have been found in one Gulf of Mexico and two Caribbean cores. The microtektite layer occurs in upper Eocene sediment and coincides with the apparent extinction of five radiolarian species. Based on fission-track dating of the microtektites and K-Ar and fission-track dating of North American tektites, the North American microtektites have an absolute age of ~34 m.y., indicating an age of less than 35 m.y. for the Eocene-Oligocene boundary.

Résumé

Les microtechtites de l'Amérique du Nord ont été trouvées dans un coeur du Golfe du Mexique et dans deux du Caraïbe. La couche microtechtite se trouve dans les dépôts de l'éocène supérieure et coïncide avec l'extinction de cinq espèces radiolariennes. Basés sur fission-track dating, des microtechtites, et les K-Ar et fission-track dating des techtites de l'Amérique du Nord, les microtechtites de l'Amérique du Nord ont un âge absolue de ~ 34 m.a., ce qui indique un âge de moins de 35 m.a. pour la limite éocène-oligocène.

Introduction

Tektites are small, generally 2-4 cm diameter, jet black to translucent green glass bodies found at several localities, referred to as strewnfields, on the earth's surface. Tektites are similar to obsidian, but can be distinguished from terrestrial igneous glasses by their chemistry (tektites have higher MgO and less Na_2O and H_2O than terrestrial igneous glasses with similar SiO_2 contents) and general lack of crystalline inclusions. There are four generally accepted tektite strewnfields: Australasian (Australia, Indochina and Northern

Philippines), Ivory Coast of Africa, Czechoslovakian and North
American (Texas and Georgia). The tektites from these strewnfields have ages of approximately 0.7, 1.1, 14.7 and 35 m.y.,
respectively (O'Keefe, 1976).

Microscopic tektites (<1 mm diameter), called microtektites, have been found in deep-sea sediments adjacent to three
of the known tektite strewnfields: Australasian, Ivory Coast
and North American (Fig. 1). Identification of the microtektites and association with a given strewnfield is based on
geographic location, age (stratigraphic and absolute), appearance, petrography, physical properties and chemical composition (Glass, 1969a; Glass, 1972; Gentner et al., 1970;
Glass et al., 1973).

Although there is not unanimous agreement concerning the
origin of tektites, and therefore microtektites, most investigators believe that they are the result of a meteorite impact
on the earth's surface (Barnes and Barnes, 1973). The microtektite layers are apparently, therefore, the result of in-
stantaneous events (geologically speaking) and thus form
chronostratigraphic layers in deep-sea sediments. The microtektites are generally concentrated in a layer that is 20 to
40 cm thick. The thickness of the layer indicates the degree
of reworking that has occurred since their deposition on the
ocean floor (Glass, 1969b). Interestingly, the Australasian
and Ivory Coast microtektite layers coincide with the Brunhes-
Matuyama reversal boundary and Jaramillo magnetic event, respectively (Glass and Heezen, 1967; Glass, 1975). Whether
this is merely a coincidence or whether there is a causal relationship is not known.

Microtektites associated with the North American tektite
strewnfield have been found in a Lamont-Doherty Geological
Observatory piston core (RC9-58) taken in the Caribbean Sea,
and two Deep Sea Drilling Project (DSDP) cores, site 94 in
the Gulf of Mexico, and site 149, in the Caribbean. The North
American microtektite layer is apparently not coincident with
a reversal of the earth's magnetic field (Zwart and Glass,
1975); however, the layer does appear to coincide with the
extinction of five Radiolaria (Maurrasse and Glass, in press;
Zwart, unpublished Master's Thesis). It was, in fact, this

correlation that aided us in finding the microtektite layer
in the two DSDP cores.

The purpose of this paper is to discuss the stratigraphic
and absolute age of the North American microtektite layer and
to discuss the implications of this data concerning the age
of the Eocene-Oligocene boundary.

Core Locations and Descriptions

Piston core RC9-58 is from the Venezuelan Basin
(14o 33.4' N. Lat., 70° 48.6'W. Long.). It is approximately
490 cm long and is composed of siliceous and calcareous lut-
ite. It contains, in addition to Radiolaria and calcareous
nannofossils, negligible amounts of benthic calcareous and
arenaceous Foraminifera, badly corroded planktonic Foraminifera and fish teeth. Radiolaria are well-preserved and predominant in the coarse fraction greater than 38 μm. Inter-
mittent, conspicuous ash falls, blurred by burrow mottling,
occur at ∼160, 310 and 470 cm. Except for a slight darkening
effect at the ash levels, the sediment is homogeneous pale
orange (10YR8/2) throughout.

Site 94 of the DSDP is located on the continental slope
of the Yucatan platform (24o 31.64' N. Lat., 88o 28.16' W.
Long.). Core 15, barrels 3 and 4, contains the microtektite
layer. This section of the core is described as a strongly
burrowed soft Foraminifera-rich nannofossil chalk of greenish
white (5G9/1) color, containing volcanic ash (Worzel et al.,
1973).

Site 149 is located in the central Venezuelan Basin
(15° 06.25' N. Lat., 69° 21.85 ' W. Long.) not far from where
core RC9-58 was taken. Core 31 contains the microtektite
layer and is described as a crumbly semi-indurated richly
calcareous radiolarian ooze of grayish orange (10YR7/4) to
dark yellowish orange (10YR6/6) color (Edgar et al., 1973).
Sparse volcanic glass and plagioclase were found throughout,
along with disseminated pumice pebbles up to 1.5 cm in diameter. A few microtektites were found in the washed residue
of the core catcher (Donnelly and Chao, 1973).

The Microtektite Layer

The microtektites in core RC9-58 occur in a single well-

defined layer with over 90% of the recovered microtektites
(~6000 with diameters greater than 125 μm) coming from a 30-
40 cm thick zone at a depth of ~250 cm in the core (Fig. 2).
Many of the arenaceous Foraminifera recovered from this core
contained microtektites in their tests (Baker and Glass,
1974).

At site 149 a few microtektites were found at the bottom
of core 30, but most were recovered from core 31, with the
peak abundance occurring at the top of the core. Unfortunately the DSDP initial report shows a gap of ~7 meters between the two cores; the actual length of missing sediment
is not known.

Microtektites occur in core 15, sections 3 and 4, of
site 94. The peak abundance is at ~416.18 m subbottom depth
in section 3, but a smaller peak is at 417.64 m subbottom
depth in section 4 (Fig. 3). The double peak in microtektite
abundance cannot be explained. If it documents two separate
events, both would be expected to occur in the other cores,
where only one peak is observed. It is suspected that the
double peak is an artifact of the drilling method used.

Stratigraphic Age of the Microtektite Layer

Except for some mixing of Eocene and Recent sediments at
the very top, core RC9-58 is entirely Eocene in age (L. Burckle, personal communication). The microtektite-rich layer
was identified as middle Upper Eocene in age based on Radiolaria (R. Goll, personal communication). Florentin Maurrasse,
of the Florida International University, worked out a detailed
radiolarian stratigraphy for this core. Maurrasse also assigned a latest Eocene age to the microtektite layer based on
Radiolaria (Maurrasse and Glass, in press). Correlation of
the microtektite data with the radiolarian stratigraphy shows
that the peak abundance of the microtektites occurs at a
level in the core where at least four species of Radiolaria,
Thyrsocyrtis bromia Ehrenberg, 1873, T. tetracantha (Ehrenberg) Riedel and Sanfilippo, 1970, T. triacantha (Ehrenberg)
Riedel and Sanfilippo, 1970, and Calocyclas turris Ehrenberg,
1873 become suddenly extinct or erratic (Fig. 2) (Maurrasse
and Glass, in press). The microtektite layer in the cores

from DSDP sites 94 and 149 also appears to coincide with the
extinction or sharp decrease in abundance of these four species and a fifth species, Thyrsocyrtis finalis Ehrenberg, 1873
(Zwart, unpublished Master's Thesis) (Figs. 3 & 4).

Core 15 of site 94, containing the microtektite layer(s),
was assigned a late Eocene age in the Initial Reports of the
Deep Sea Drilling Project, Vol. X (Worzel et al., 1973).
Core 14 is also assigned a late Eocene age. Core 13, which
ends ∼ 29 m above the major microtektite layer, is assigned an
early Oligocene age. Although based on widely spaced samples,
in many cases, the section of core containing the microtektite
layer is assigned to the Isthmolithus recurvus or Sphenolith-
us pseudoradians calcareous nannofossil Zone (Hay, 1973), the
Discoaster barbadiensis Zone (Bukry, 1973), the Thysocyrtis
bromia radiolarian Zone (Foreman, 1973; Sanfilippo and Riedel,
1973), the Globigerapsis semiinvoluta or Globorotalia cerroazulensis foraminiferal Zone (McNeely, 1973),and the P15
foraminiferal Zone (Worzel et al., 1973).

The microtektite layer at the top of core 31, DSDP site
149, also occurs in sediment identified as late Eocene in the
Initial Reports of the Deep Sea Drilling Project, Vol. XV
(Edgar et al., 1973). Core 30, directly above core 31, was
assigned a middle Oligocene age (Edgar et al., 1973). Core
31 is assigned to the Discoaster saipanensis (?) Zone (Hay
and Beaudry, 1973), and the Thyrsocyrtis bromia and/or
Podocyrtis goetheana radiolarian Zone (Riedel and Sanfilippo,
1973).

Thus, based on studies by numerous investigators invol-
ving Radiolaria, Foraminifera and nannofossils, the microtektite layer in all three cores occurs in sediments of late
Eocene age. The biostratigraphic age of the North American
microtektite layer is, therefore, well established.

Absolute Age of the North American Microtektites

The microtektites from core RC9-58 have been dated using
the fission-track method by D. Storzer and G. A. Wagner of
the Max-Planck-Institut in Heidelberg, Germany. They obtained an age, after correcting for annealing effects, of
34.6 ± 4.2 m.y. (Glass et al., 1973) The large error is

attributed to bad counting statistics due to the small surface
area available for counting. However, the microtektites in
core RC9-58 and the two DSDP cores are part of the North American tektite strewnfield as indicated by their appearance,
geographical location, stratigraphic and absolute age, petrography, physical properties and major element chemistry
(Glass et al., 1973; Zwart, unpublished Master's Thesis).
Further, the North American tektites have been dated by both
fission-track and K-Ar methods.

Ten K-Ar ages for North American tektites obtained by
Zähringer (1963) and Gentner et al. (1969) have an average of
34.2 m.y. with a standard deviation of 0.48. Seven published
fission-track ages (Fleischer and Price, 1964; Fleischer et
al., 1965; Storzer and Wagner, 1971; Garlick et al., 1971;
Storzer et al., 1973) have an average of 34.6 m.y. with a
standard deviation of 0.48. Considering all of the above
data we would assign an age of 34.4 ± 0.5 m.y. for the age of
the North American tektites and microtektites. Thus, the
microtektite layer provides a well-dated chronostratigraphic
layer.

Age of the Eocene-Oligocene Boundary

Evernden and Evernden (1970) indicated an age of about
36-37 m.y. for the Ducheunian-Chardomian boundary which is
equivalent to the Eocene-Oligocene boundary. Berggren (1972)
assigned a similar age of 37.5 m.y. for the Eocene-Oligocene
boundary, which relies in part, at least, on a K-Ar glauconite date of 37.5 ± 3 m.y. (Odin et al., 1969) for the sands
of Neerrepen (Belgian Basin), which is latest Eocene or
earliest Oligocene in age. The North American microtektite
layer, however, occurs in upper Eocene sediments based on
radiolarian, foraminiferal and nannofossil data for three
cores and has an absolute age of 34.4 ± 0.5 m.y. based on
fission-track and K-Ar ages of the North American tektites
and microtektites. In conclusion, therefore, the microtektite data indicate that the Eocene-Oligocene boundary must be
less than 35 m.y. which is 2-3 million years younger than the
age assigned to this boundary by Berggren (1973), but is within the limits of accuracy of the K-Ar age of the sands of

Neerrepen given by Odin et al. (1969).

Acknowledgements

J. A. Glass helped prepare the manuscript. F. Maurrasse helped with radiolarian identification. Research was supported by NSF grant OCE72-01439 A04.

References

Baker, R.N., and Glass, B.P., 1974, Microtektites as test components of Caribbean arenaceous Foraminifera, Micropaleontology, vol. 20, pp. 231-235.

Barnes, V.E., and Barnes, M.A., editors, 1973, Tektites. In: R.W. Fairbridge (Series Editor), Benchmark Papers in Geology, Dowden, Hutchinson and Ross, Stroudsburg, Pa., 445 pp.

Berggren, W.A., 1972, A Cenozoic time scale - some implications for regional geology and paleobiogeography, Lethaia, vol. 5, pp. 195-215.

Bukry, D., 1973, Coccolith Stratigraphy. In: Worzel, J.L., Bryant, W., et al., 1973, Initial Reports of the Deep Sea Drilling Project, vol. 10, Washington (U.S. Government Printing Office), pp. 385-406.

Donnelly, T.W. and Chao, E.C.T., 1973, Microtektites of late Eocene Age from the eastern Caribbean Sea. In: Edgar, N.T., Saunders, J.B., et al., 1973, Initial Reports of the Deep Sea Drilling Project, vol. 15, Washington (U.S. Government Printing Office), pp. 1031-1037.

Edgar, N.T., Saunders, J.B., et al., 1973, Initial Reports of the Deep Sea Drilling Project, vol. 15, Washington (U.S. Government Printing Office), 1137 pp.

Evernden, J.F. and Evernden, R.K.S., 1970, The Cenozoic time scale, Geological Society of America, Special Paper 124, 71 pp.

Fleischer, R.L., and Price, P.B., 1964, Fission-track evidence for the simultaneous origin of tektites and other natural glasses. Geochimica et Cosmochimica Acta, vol. 28, pp. 755-760.

Fleischer, R.L., Price, P.B., and Walker, R.M., 1965, On the simultaneous origin of tektites and other natural glasses. Geochimica et Cosmochimica Acta, vol. 29, pp. 161-166.

Foreman, H.P., 1973, Radiolaria of Leg 10 with systematics and ranges for the families Amphipyndacidae, Artostrobiidae, and Theoperidae. In: Worzel, J.L., Bryant, W., et al., 1973, Initial Reports of the Deep Sea Drilling Project, vol. 10, Washington (U.S. Government Printing Office), pp. 407-473.

Garlick, G.D., Naeser, C.W., and O'Neil, J.R., 1971, A Cuban Tektite. Geochimica et Cosmochimica Acta, vol. 35, pp. 731-734.

Gentner, W., Storzer, D., and Wagner, G.A., 1969, New fission-track ages of tektites and related glasses. Geochimica et Cosmochimica Acta, vol. 33, pp. 1075-1081.

Glass, B.P., 1969a, Chemical composition of Ivory Coast microtektites, Geochimica et Cosmochimica Acta, vol. 33, pp. 1135-1147.

Glass, B.P., 1969b, Reworking of deep-sea sediments as indicated by the vertical dispersion of the Australasian and Ivory Coast microtektite horizons, Earth and Planetary Science Letters, vol. 6, pp. 409-415.

Glass, B.P., 1972, Australasian Microtektites in deep-sea sediments. In: Hayes, D.E., (editor), Antarctic Oceanology II: The Australasian-New Zealand Sector, Antarctic Research Series, vol. 19, Washington (American Geophysical Union), pp. 335-348.

Glass, B.P., 1975, Geomagnetic reversals and tektites. In: Fisher, R.M., Fuller, M., Schmidt, V.A. and Wasilewski, P.J. (Editors). Proceedings of the Takesi Nagata Conference, June 3-4, 1974, Greenbelt, Maryland (Goddard Space Flight Center), pp. 225-229.

Glass, B.P., and Heezen, B.C., 1967, Tektites and geomagnetic reversals. Scientific American, vol. 217, pp. 32-38.

Glass, B.P., Baker, R.N., Storzer, D. and Wagner, G.A., 1973, North American microtektites from the Caribbean Sea and their fission track age. Earth and Planetary Science Letters, vol. 19, pp. 184-192.

Hay, W.W., 1973, Preliminary dating by fossil calcareous nannoplankton, Deep Sea Drilling Project: Leg 10. In: Worzel, J.L., Bryant, W., et al., 1973, Initial Reports of the Deep Sea Drilling Project, vol. 10, Washington (U.S. Government Printing Office), pp. 375-383.

Hay, W.W., and Beaudry, F.M., 1973, Calcareous nannofossils - Leg 15, Deep Sea Drilling Project. In: Edgar, N.T., Saunders, J.B., et al., 1973, Initial Reports of the Deep Sea Drilling Project, vol. 15, Washington (U.S. Government Printing Office), pp. 625-683.

Maurrasse, F. and Glass, B.P., In press, Radiolarian stratigraphy and North American microtektites in Caribbean core RC9-58: Implications concerning late Eocene radiolarian chronology and the age of the Eocene-Oligocene boundary. Proceedings of the 7th Caribbean Geological Conference, July, 1974, Guadeloupe.

McNeely, B.W., 1973, Biostratigraphy of the Mesozoic and Paleogene pelagic sediments of the Campeche Embankment area. In: Worzel, J.L., Bryant, W., et al., 1973, Initial Reports of the Deep Sea Drilling Project, vol. 10, Washington (U.S. Government Printing Office), pp. 679-695.

Odin, C.S., Gulinck, M., Bodelle, J., and Lay, C., 1969, Géochronologie de niveaux glauconieux tertiares du basin de Belgique (méthode potassium-argon). C.R. Somm Séances Soc. Géol. France, vol. 6, p. 198.

O'Keefe, J.A., 1976, Tektites and their origin, New York (Elsevier), 254 pp.

Riedel, W.R., and Sanfilippo, A., 1973, Cenozoic Radiolaria from the Caribbean, Deep Sea Drilling Project, Leg 15. In: Edgar, N.T., Saunders, J.B., et al., 1973, Initial Reports of the Deep Sea Drilling Project, vol. 15, Washington (U.S. Government Printing Office), pp. 705-751.

Sanfilippo, A., and Riedel, W.R., 1973, Cenozoic Radiolaria (Exclusive of Theoperids, Artostrobids and Amphipyndacids) from the Gulf of Mexico, Deep Sea Drilling Project Leg 10. In: Worzel, J.L., Bryant, W., et al., 1973, Initial Reports of the Deep Sea Drilling Project, vol. 10, Washington (U.S. Government Printing Office), pp. 475-611.

Storzer, D., and Wagner, G.A., 1971, Fission track ages of North American tektites. Earth and Planetary Science Letters, vol. 10, pp. 435-444.

Storzer, D., Wagner, G.A., and King, E.A., 1973, Fission track ages and stratigraphic occurrence of Georgia tektites. Journal of Geophysical Research, vol. 78, pp. 4915-4919.

Worzel, J.L., Bryant, W., et al., 1973, Initial Reports of the Deep Sea Drilling Project, vol. 10, Washington (U.S. Government Printing Office), 748 pp.

Zwart, M.J., 1976, North American microtektites from Deep Sea Drilling Project Cores (Unpublished Master's Thesis, University of Delaware, Newark, DE).

Zwart, M.J. and Glass, B.P., 1975, North American microtektites from DSDP cores: associated radiolarian extinctions and paleomagnetic stratigraphy. American Geophysical Union, Transactions, vol. 56, p. 385 (abstract).

Zähringer, J., 1963, K-Ar measurements of tektites. In: Radioactive Dating. Proceedings Symposium, Athens, Nov. 19-23, 1962, International Atomic Energy Agency, Vienna, pp. 289-305.

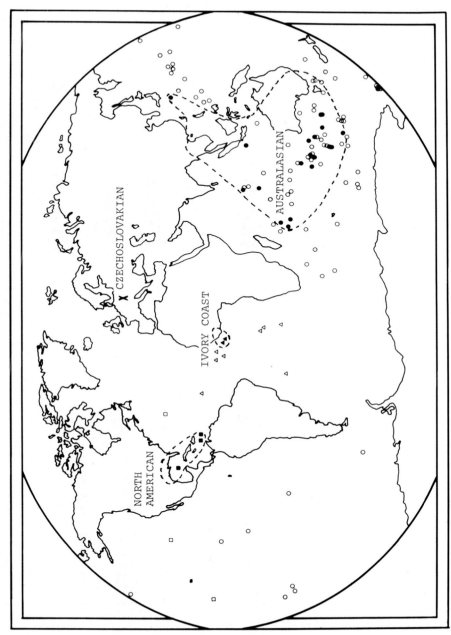

Fig. 1. Tektite strewnfields. Deep-sea cores searched for Australasian, Ivory Coast and North American microtektites indicated by circles, triangles and squares, respectively. Cores found to contain microtektites indicated by solid figures.

RC9-58

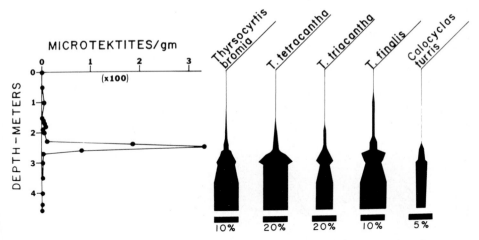

Fig. 2. Vertical distribution of microtektites and abundance of five Radiolaria species in core RC9-58.

DSDP SITE 94

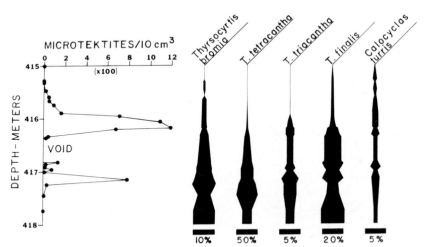

Fig. 3. Vertical distribution of microtektites and abundance of five Radiolaria species in sections 3 and 4 of core 15, DSDP site 94. Depth is from sediment-water interface.

565

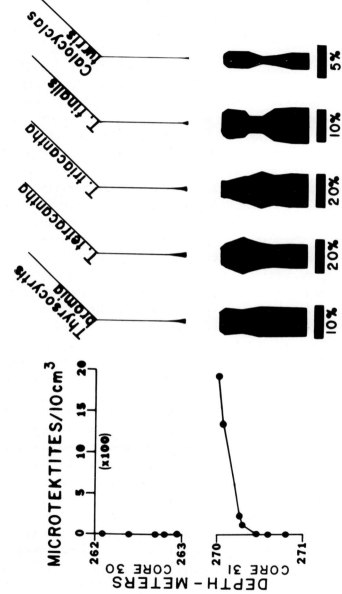

Fig. 4. Vertical distribution of microtektites and abundance of five Radiolaria species in section 2, core 30 and section 1, core 31, DSDP site 149. Depth is from sediment-water interface.

Discussion

Dr. B. K. Holdsworth: There seems some doubt whether the
Thyrsocyrtis spp. extinction level in deep-sea cores really
is the Eocene/Oligocene boundary. It would conventionally be
taken at the base of the T. tuberosa Zone - which is a higher
level.

Dr. B. P. Glass: I did not mean to imply that the radiolarian
extinctions occurred at the Eocene/Oligocene boundary. In
the DSDP reports, the extinctions do occur below the Eocene/
Oligocene boundary. This means that the microtektites give
an upper limit to the age of the Eocene/Oligocene boundary.

Dr. F. Michael: How are the extinction of the Radiolaria and
the microtektite zone related to paleomagnetic reversals?

Glass: The Australian and Ivory Coast microtektites appear to
be associated with the Brunhes/Matuyama boundary and Jaramillo
event, respectively. The North American microtektites do not
appear to be associated with a reversal. The sections of
cores containing the microtektite layer appear to be entirely normally magnetized. However, it's very difficult to get
good paleomagnetic stratigraphy from these enclosed, or semi-
enclosed basins, like the Caribbean and the Gulf of Mexico.
They generally have a normal overprint on top of the magnetic
stratigraphy that's very difficult to remove. So we don't
know the answer to your question. I don't know the relationship between North American microtektites and the paleomagnetic stratigraphy at this time.

Dr. L. Shishkevish: Were microtektites found in moon samples?

Glass: That' a hard one to answer. Glass beads were found in
lunar samples with approximately the same shape and size distribution as the microtektites. Chemically they are distinct
from microtektites and tektite material. Microtektites and
tektites generally have silica contents greater than 60%,
whereas most lunar material has silica contents less than 60%.
So chemically they are distinct. They may be the same in
that they may both have been formed by the same process; that
is meteorite impact. So if you want to call glass beads produced by meteorite impact microtektites, then yes, microtektites have been found on the moon. If you want a stricter definition to chemistry, then, no, they're not found on the moon.

Dr. M. C. Keen: First of all I'd like to point out that we do
know where the Eocene-Oligocene boundary is, of course. We
may not know where it is in DSDP cores, but we certainly know
where it is in relationship to the stratotype. Just before I
came here I saw some age dates that had been, I think fairly
new ones, by Prof. Currie. Unfortunately I can't remember
them exactly, but I have a feeling that they're more in line
with what you want. But, I just can't remember; I seem to
remember seeing 34 m.y., but it's just from memory.

Glass: If you could get a hold of those and send them to me
I'd be very happy.

No more questions?

Dr. G. Williams: Can we assume that the meteorites didn't hit
the Earth before the Eocene. Or is it a problem of...

Glass: Before? In other words you'd like to put them in cold
storage for a while before you deposit them on the earth is
that it?

Williams: No, I was wondering why...

Glass: Okay, in other words, you mean no tektites or microtektites would be found older than 34 million years. I don't
know if that's a problem of just not finding them or whether
it's a problem of them undergoing solution or being destroyed
over that time period. As you know, or may not know
volcanic glasses are fairly unstable and as you get back
pretty far in the time scale they seem to be lacking. I
think this is mostly a weathering problem, being destroyed.
Whether or not they actually existed at one time, or we just
haven't found them, or they've been dissolved by solution,
or what, I don't know. But certainly none have been found
older than about 34 million years. I don't know the answer.

Williams: Could you tell me then how you concentrate them
please and how you get them out of the sediments? Is it like
foram preparation?

Glass: Yes, it's very simple. You take a sediment, disaggre-
gate it and wet sieve it and look at the greater than 125
micron size fraction. Since we're dealing with pelagic sediments for the most part they are they only thing in there
larger than 125 microns, unless you have forams, or something,

in which case, I hate to say it here in front of micropaleontologists, but I use a little acid and get rid of them.
The only things left are the microtektites.

Index to Authors

Index to Genera and Species

588

592